Current Topics in Membranes, Volume 47

Amiloride-Sensitive Sodium Channels

Physiology and Functional Diversity

Current Topics in Membranes, Volume 47

Series Editors

Douglas M. Fambrough
Department of Biology
The Johns Hopkins University
Baltimore, Maryland

Dale J. Benos
Department of Physiology and Biophysics
University of Alabama
Birmingham, Alabama

Current Topics in Membranes, Volume 47

Amiloride-Sensitive Sodium Channels

Physiology and Functional Diversity

Edited by

Dale J. Benos
Department of Physiology and Biophysics
University of Alabama
Birmingham, Alabama

ACADEMIC PRESS
San Diego London Boston New York Sydney Tokyo Toronto

Front cover photograph (paperback edition only): In situ hybridization of normal human bronchus hybridized to probes for αENaC. See Chapter 12 Figure 4 on color insert for details.

This book is printed on acid-free paper. ∞

Academic Press
a division of Harcourt Brace & Company
525 B Street, Suite 1900, San Diego, California 92101-4495, USA
http://www.apnet.com

Academic Press
24-28 Oval Road, London NW1 7DX, UK
http://www.hbuk.co.uk/ap/

International Standard Book Number: 0-12-153347-6 (case)
International Standard Book Number: 0-12-089030-5 (pb)

PRINTED IN THE UNITED STATES OF AMERICA
99 00 01 02 03 04 EB 9 8 7 6 5 4 3 2 1

To my wife, Kim, and our daughters, Kaitlin and Emilee,
whose love and strength are my inspiration.

Contents

Contributors

Numbers in parentheses indicate the pages on which the authors' contributions begin.

Dale J. Benos (3, 339), Department of Physiology and Biophysics, University of Alabama at Birmingham, Birmingham, Alabama 35294

Bakhrom K. Berdiev (3, 37, 351), Department of Physiology and Biophysics, University of Alabama at Birmingham, Birmingham Alabama 35294

M. Blot-Chabaud (51), INSERM U478, X. Bichat, BP 416, 75870 Paris Cedex 18, France

J. P. Bonvalet (51), INSERM U478, X. Bichat, BP 416, 75870 Paris Cedex 18, France

M. Brouard (51), INSERM U478, X. Bichat, BP 416, 75870 Paris Cedex 18, France

James K. Bubien (155), Department of Physiology and Biophysics, University of Alabama at Birmingham, Birmingham, Alabama 35924

Horacio Cantiello (177), Renal Unit, Massachusetts General Hospital East, Charlestown, Massachusetts 02129

Chun Cheng (25), Howard Hughes Medical Institute and Departments of Internal Medicine and Physiology and Biophysics, University of Iowa College of Medicine, Iowa City, Iowa 52242

S. Djelidi (51), INSERM U478, X. Bichat, BP 416, 75870 Paris Cedex 18, France

Monica Driscoll (297), Department of Molecular Biology and Biochemistry, The State University of New Jersey–Rutgers, Piscataway, New Jersey 08855

Michael D. DuVall (219), Department of Anesthesiology, University of Alabama at Birmingham, Birmingham, Alabama 35233

R. S. Edinger (95), Department of Medicine, University of Pittsburgh, Pittsburgh, Pennsylvania 15213

B. Escoubet (51), INSERM U426, X. Bichat, BP 416, 75870 Paris Cedex 18, France

Nicolette Farman (51), INSERM U478, X. Bichat, BP 416, 75870 Paris Cedex 18, France

Catherine M. Fuller (3), Department of Physiology and Biophysics, University of Alabama at Birmingham, Birmingham, Alabama 35294

Timothy A. Gilbertson (315), Pennington Biomedical Research Center, Louisiana State University, Baton Rouge, Louisiana 70808

Sandra Guggino (279), Johns Hopkins University Medical School, Baltimore, Maryland 21205

Toru Ishikawa (65), The Hospital for Sick Children and Biochemistry Department, University of Toronto, Toronto, Ontario, Canada M5G 1X8

Iskander I. Ismailov (3, 37, 351), Department of Physiology and Biophysics, University of Alabama at Birmingham, Birmingham, Alabama 35294

John P. Johnson (95), Department of Medicine, University of Pittsburgh, Pittsburgh, Pennsylvania 15213

Sue C. Kinnamon (315), Department of Anatomy and Neurobiology, Colorado State University, Fort Collins, Colorado 80523; and Rocky Mountain Taste and Smell Center, University of Colorado Health Sciences Center, Denver, Colorado 80262

Thomas R. Kleyman (37), Departments of Medicine and Physiology, University of Pennsylvania, Philadelphia, Pennsylvania 19104

Farhad Kosari (37), Departments of Medicine and Physiology, University of Pennsylvania, Philadelphia, Pennsylvania 19104

Ahmed Lazrak (219), Department of Anesthesiology, University of Alabama at Birmingham, Birmingham, Alabama 35233

Jinqing Li (37), Departments of Medicine and Physiology, University of Pennsylvania; and VA Medical Center, Philadelphia, Pennsylvania 19104

Li Li (109), Department of Physiology and Biophysics, University of Alabama at Birmingham, Birmingham, Alabama 35294

Bernd Lindemann (315), Department of Physiology, Saar University, D-66421 Homburg, Germany

Itzhak Mano (297), Department of Molecular Biology and Biochemistry, The State University of New Jersey–Rutgers, Piscataway, New Jersey 08855

Yoshinori Marunaka (255), The Hospital for Sick Children, University of Toronto, Toronto, Ontario, Canada M5G 1X8

Sadis Matalon (219), Department of Anesthesiology, University of Alabama at Birmingham, Birmingham, Alabama 35233

Ryan G. Morris (109), Departments of Anesthesiology, Physiology and Biophysics, and Pediatrics, University of Alabama at Birmingham, Birmingham, Alabama 35294

Naomi Niisato (255), The Hospital for Sick Children, University of Toronto, Toronto, Ontario, Canada M5G 1X8

Hugh O'Brodovich (239), Department of Pediatrics, University of Toronto, The Hospital for Sick Children, Toronto, Ontario, Canada M5G 1X8

Olli Pitkänen (239), Hospital for Children and Adolescents, University of Helsinki, Helsinki, Finland

Pamela Plant (65), The Hospital for Sick Children and Biochemistry Department, University of Toronto, Toronto, Ontario, Canada M5G 1X8

Bijan Rafii (239), Program in Lung Biology Research, The Hospital for Sick Children, University of Toronto, Toronto, Ontario, Canada M5G 1X8

Daniela Rotin (65), The Hospital for Sick Children and Biochemistry Department, University of Toronto, Toronto, Ontario, Canada M5G 1X8

Sukhvinder Sahota (297), Department of Molecular Biology and Biochemistry, The State University of New Jersey–Rutgers, Piscataway, New Jersey 08855

Sarah Sariban-Sohraby (87), Laboratoire de Physiologie, Université Libre de Bruxelles, 1070 Brussels, Belgium

James A. Schafer (109), Department of Physiology and Biophysics, University of Alabama at Birmingham, Birmingham, Alabama 35294

Laurent Schild (65), Institut de Pharmacologie, Université de Lausanne, CH-1015 Lausanne, Switzerland

Shaohu Sheng (37), Departments of Medicine and Physiology, University of Pennsylvania, Philadelphia, Pennsylvania 19104

V. Gh. Shlyonsky (3), Department of Physiology and Biophysics, University of Alabama at Birmingham, Birmingham, Alabama 35294

Peter R. Smith (133), Department of Physiology and Biophysics, University of Alabama at Birmingham, Birmingham, Alabama 35294

Peter M. Snyder (25), Department of Internal Medicine, University of Iowa College of Medicine, Iowa City, Iowa 52242

Olivier Staub (65), The Hospital for Sick Children and Biochemistry Department, University of Toronto, Toronto, Ontario, Canada M5G 1X8

Duo Sun (109), Department of Physiology and Biophysics, University of Alabama at Birmingham, Birmingham, Alabama 35294

Colleen R. Talbot (197), Department of Biology, California State University, San Bernardino, California 92407

A. Keith Tanswell (239), MRC Group in Lung Development, The Hospital for Sick Children, Toronto, Ontario, Canada M5G 1X8

Heather A. Thieringer (297), Department of Molecular Biology and Biochemistry, The State University of New Jersey–Rutgers, Piscataway, New Jersey 08855

J. M. Wang (95), Department of Medicine, University of Pittsburgh, Pittsburgh, Pennsylvania 15213

Michael J. Welsh (25), Howard Hughes Medical Institute and Departments of Internal Medicine and Physiology and Biophysics, University of Iowa College of Medicine, Iowa City, Iowa 52242

Teresa W. Wilborn (109), Department of Physiology and Biophysics, University of Alabama at Birmingham, Birmingham, Alabama 35294

Preface

Since the publication of the first subunit of the cloned epithelial Na^+ channel (ENaC) in 1993, many laboratories throughout the world have focused attention on this important ion channel. Progress has accelerated, and a plethora of novel discoveries concerning the channel's structure–function relationships and regulatory influences have been forthcoming. For example, Fig. 1 highlights the sheer number of new members of the ENaC/degenerin superfamily that have been identified in four years. Moreover, the direct involvement of ENaC in important physiological processes and human disease has been demonstrated.

This volume is a compilation of current work in the general area of amiloride-sensitive Na^+ channels. Researchers from many of the leading laboratories working in this area have contributed chapters. The volume is organized into five main sections: ENaC structure–function relationships, regulation of amiloride-sensitive Na^+ channels, sodium channels in the lung, sensory and mechanical transduction, and involvement of Na^+ channels in disease. These sections deal with the following questions:

1. What is the molecular basis for the functional and biochemical diversity of amiloride-sensitive Na^+ channels?
2. What are the biophysical properties of the ENaCs?
3. What is the stoichiometry of a functional ENaC?

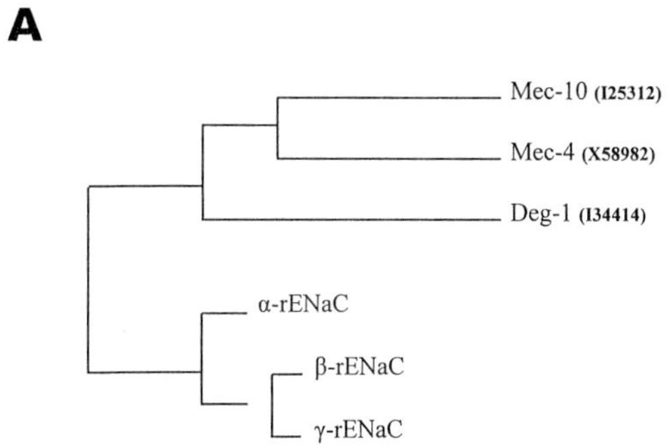

FIGURE 1

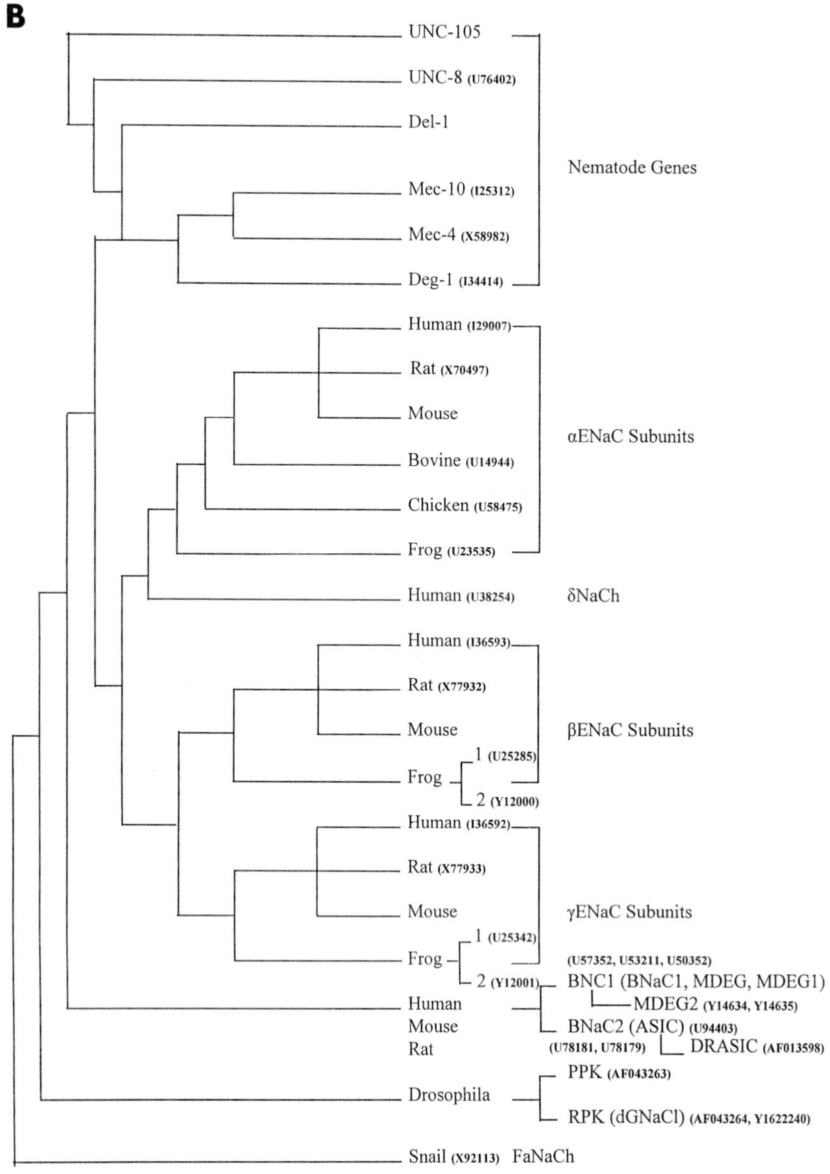

FIGURE 1—*Continued*

4. What are the mechanisms underlying physiological regulation of amiloride-sensitive Na^+ channels?

5. Do increases in intracellular cAMP and Liddle's disease mutations increase single-channel P_o, increase surface density of channels, or both?

6. What are the biosynthetic and intracellular processing pathways for the ENaCs?

7. What is(are) the role(s) of the cytoskeleton in amiloride-sensitive Na^+ channel function and regulation?

8. Is there a direct physical interaction between ENaC and other ion channels?

9. What role do amiloride-sensitive Na^+ channels play in sensory transduction?

10. Are ENaCs mechanosensitive?

11. What is the involvement of amiloride-sensitive Na^+ channels in pathological conditions?

Although these questions are being experimentally addressed, they are far from being answered. Moreover, with increasing knowledge and experimental activity, new questions arise that demand attention. For example,

1. What does an ENaC pore look like?
2. What gates ENaC?
3. What constitutes an amiloride binding site?
4. What are the signaling pathways controlling ENaC expression and function?
5. What are the molecular sites of interaction between ENaC and other cellular components?
6. What other diseases involve ENaC?

I thank all of the authors for their outstanding contributions. I especially acknowledge my administrative associate, Mrs. Cathleen Guy, for her tireless and cheerful effort in ensuring the completion of this volume, and Dr. Emelyn Eldredge, Acquisition Editor at Academic Press, for her constant encouragement. Above all, I acknowledge all of the very fine investigators who toil purposefully and selflessly to uncover new information about this class of ion channel in order to make each of our lives better.

Dale J. Benos

Previous Volumes in Series

Current Topics in Membranes and Transport

Volume 23 Genes and Membranes: Transport Proteins and Receptors* (1985)
Edited by Edward A. Adelberg and Carolyn W. Slayman

Volume 24 Membrane Protein Biosynthesis and Turnover (1985)
Edited by Philip A. Knauf and John S. Cook

Volume 25 Regulation of Calcium Transport across Muscle Membranes (1985)
Edited by Adil E. Shamoo

Volume 26 Na^+–H^+ Exchange, Intracellular pH, and Cell Function* (1986)
Edited by Peter S. Aronson and Walter F. Boron

Volume 27 The Role of Membranes in Cell Growth and Differentiation (1986)
Edited by Lazaro J. Mandel and Dale J. Benos

Volume 28 Potassium Transport: Physiology and Pathophysiology* (1987)
Edited by Gerhard Giebisch

Volume 29 Membrane Structure and Function (1987)
Edited by Richard D. Klausner, Christoph Kempf, and Jos van Renswoude

Volume 30 Cell Volume Control: Fundamental and Comparative Aspects in Animal Cells (1987)
Edited by R. Gilles, Arnost Kleinzeller, and L. Bolis

Volume 31 Molecular Neurobiology: Endocrine Approaches (1987)
Edited by Jerome F. Strauss, III, and Donald W. Pfaff

Volume 32 Membrane Fusion in Fertilization, Cellular Transport, and Viral Infection (1988)

* *Part of the series from the Yale Department of Cellular and Molecular Physiology*

Edited by Nejat Düzgünes and Felix Bronner

Volume 33 Molecular Biology of Ionic Channels* (1988)
Edited by William S. Agnew, Toni Claudio, and Frederick J. Sigworth

Volume 34 Cellular and Molecular Biology of Sodium Transport* (1989)
Edited by Stanley G. Schultz

Volume 35 Mechanisms of Leukocyte Activation (1990)
Edited by Sergio Grinstein and Ori D. Rotstein

Volume 36 Protein–Membrane Interactions* (1990)
Edited by Toni Claudio

Volume 37 Channels and Noise in Epithelial Tissues (1990)
Edited by Sandy I. Helman and Willy Van Driessche

Current Topics in Membranes

Volume 38 Ordering the Membrane–Cytoskeleton Trilayer* (1991)
Edited by Mark S. Mooseker and Jon S. Morrow

Volume 39 Developmental Biology of Membrane Transport Systems (1991)
Edited by Dale J. Benos

Volume 40 Cell Lipids (1994)
Edited by Dick Hoekstra

Volume 41 Cell Biology and Membrane Transport Processes* (1994)
Edited by Michael Caplan

Volume 42 Chloride Channels (1994)
Edited by William B. Guggino

Volume 43 Membrane Protein–Cytoskeleton Interactions (1996)
Edited by W. James Nelson

Volume 44 Lipid Polymorphism and Membrane Properties (1997)
Edited by Richard Epand

Volume 45 The Eye's Aqueous Humor: From Secretion to Glaucoma (1998)
Edited by Mortimer M. Civan

Volume 46 Potassium Ion Channels: Molecular Structure, Function, and Diseases
Edited by Yoshihisa Kurachi, Lily Yeh Jan, and Michel Lazdunski

PART I

Structure–Function Relations of Amiloride-Sensitive Sodium Channels

CHAPTER 1

Mapping Structure/Function Relations in αbENaC

C. M. Fuller, I. I. Ismailov, B. K. Berdiev, V. Gh. Shlyonsky, and D. J. Benos
Department of Physiology and Biophysics, University of Alabama at Birmingham, Birmingham, Alabama 35294

I. INTRODUCTION

Epithelial ion channels permit the passage of ions and ultimately water across the epithelial barrier. The polarized nature of epithelial cells allows them to play a major role in the regulation of whole-body salt and water homeostasis by separating compartments and maintaining ion gradients. The recent cloning of several members of the amiloride-sensitive epithelial Na^+ channel family, ENaC, represents a major advance in our knowledge about how salt reabsorbing and secreting epithelia such as those in the kidney, lung, and colon accomplish this task at the molecular level. The ENaC channel family comprises a diverse array of polypeptides ranging from distant cousins in the *mec* proteins of *C. elegans* (Driscoll and Chalfie, 1991) to the "prototypical" ENaC channels cloned from rat colon (Canessa

1063-5823/99 $30.00

et al., 1993, 1994b). Members of the family have also been identified in mammalian and gastropod neural tissue (Garcia-Anoveros *et al.*, 1997; Lingueglia *et al.*, 1995; Price *et al.*, 1996; Waldmann *et al.*, 1995b). The focus of the present chapter is to describe structure/function relations in one ENaC isoform, αbENaC, cloned from the bovine renal papilla. While this isoform shares many features with the other members of the family, there are some distinct differences. These will be discussed in light of what is known about sequences in other ENaCs that regulate basic channel properties such as gating, amiloride binding and ion selectivity.

II. CLONING AND EXPRESSION OF αbENaC

A. Background

Prior to the cloning of the first ENaC subunit (αrENaC) by Rossier and colleagues (Canessa *et al.*, 1993, 1994b), our laboratory had previously identified from the bovine renal papilla a large complex protein of M_r 700,000 that on reducing SDS–PAGE migrated as 6 subunits ranging in size from 45 to 330 kDa (Benos *et al.*, 1986, 1987). A protein complex of similar size was also purified from A6 cells, an amphibian cell line derived from the toad kidney (Benos *et al.*, 1986; Sariban-Sohraby and Benos, 1986; Kleyman *et al.*, 1991). Antibodies raised against the heterooligomeric complex labeled the apical membranes of principal cells in the rat cortical collecting duct and the surface of A6 epithelial cells, i.e., precisely those membranes where a Na^+ channel would be predicted to be localized (Brown *et al.*, 1989; Sorscher *et al.*, 1988; Tousson *et al.*, 1989). However, one discrepancy between bilayer and patch-clamp analysis was that the bovine renal protein complex formed an amiloride-sensitive 39-pS Na^+ channel when reconstituted into planar lipid bilayers, a channel of substantially larger single channel conductance than was observed by patch-clamp analysis of native A6 epithelia (~5–9 pS) (Hamilton and Eaton, 1985; Marunaka and Eaton, 1991; Oh *et al.*, 1993). Thus, the molecular nature of the epithelial Na^+ channel was uncertain. However, it is important to note that even αβγrENaC incorporated into lipid bilayers displays a 39-pS major conductance state (Benos *et al.*, 1997).

Cloning of the rat ENaC protein from rat colon by Canessa *et al.* (1993, 1994b) demonstrated that the primary translated product (αrENaC) was a polypeptide of approximately 70 kDa that migrated at approximately 90 kDa when fully glycosylated. Identification of this subunit was quickly followed by the cloning of the related β and γ subunits. Coexpression of all three subunits in *Xenopus* oocytes promptly demonstrated that a greatly augmented current was generated compared to the current recorded from oocytes expressing αrENaC alone. The expressed current was sensitive to amiloride in the micromolar range, consistent with what had been observed for recordings from A6 epithelial cells. The predicted topology of the channel (later confirmed by mapping experiments) was that each subunit would span the membrane twice with each transmembrane domain being separated by a large (>400 amino acids) extracellular loop. Both the N and the C termini were predicted to be intracellular (Canessa *et al.*, 1994a; Snyder *et al.*, 1994).

Using the full-length αrENaC cDNA as a probe, we screened a bovine renal papillary cDNA expression library and isolated a 3112-bp fragment containing a 1.95-kb open reading frame encoding a polypeptide of 650 amino acids (Fuller *et al.*, 1995). The protein had an M_r of approximately 80,000 (in the absence of glycosylation) following *in vitro* translation and separation on reducing SDS–PAGE. This polypeptide, which we termed αbENaC (for *b*ovine *e*pithelial *Na⁺* channel), had a high homology to αrENaC and the human homolog (hENaC (80 and 84% at the nucleotide level, respectively). However, this isoform differed significantly at the N terminal, missing the first 47 amino acids found in αrENaC and having 9 nonconservative amino acid substitutions clustered at the C terminal (Fig. 1, see color insert). Hydrophobicity plotting suggested a topological arrangement for αbENaC identical to that proposed for both αrENaC and αhENaC (Canessa *et al.*, 1994a; Snyder *et al.*, 1994).

When heterologously expressed in *Xenopus* oocytes, αbENaC was associated with approximately 30 μA of amiloride-sensitive current. However, the magnitude of the current was greatly enhanced by coexpression of the β and γ subunits of rENaC. These observations indicated that not only were β and γENaC important for maximum phenotypic expression, as had been found for the other homologs, but also that αbENaC was capable of forming a fully functional chimeric Na^+ channel with subunits isolated from different and distantly related species (Fig. 2).

B. Phosphorylation

The primary translated amino acid sequence of αbENaC predicts 8 glycosylation sites, 12 consensus sequences for phosphorylation by protein kinase C (PKC), but no sites for protein kinase A (PKA) phosphorylation. This latter finding is similar to that for motif analysis of αrENaC, which shows two sites for PKA-dependent phosphorylation located in the extracellular loop and which are therefore presumably nonfunctional. Consistent with the motif analysis, a cAMP-elevating cocktail of forskolin + IBMX had no effect on the magnitude of amiloride-sensitive current when αbENaC

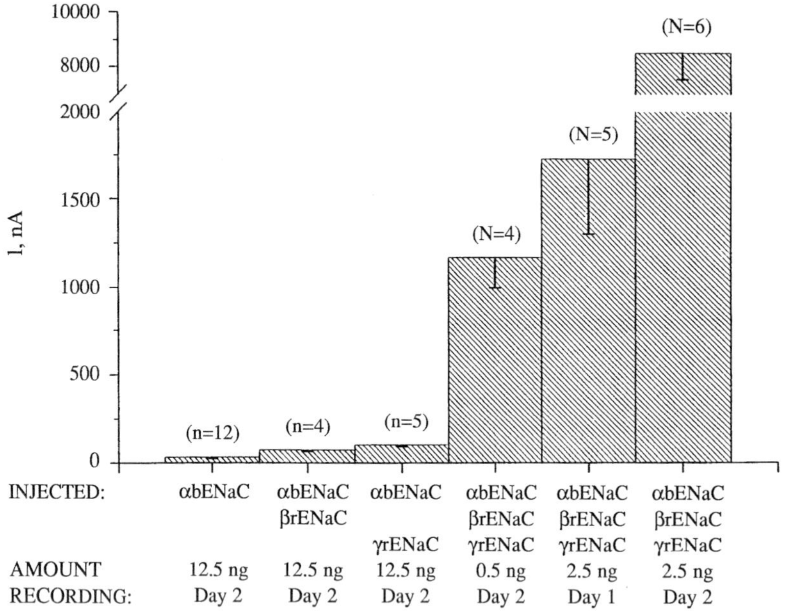

FIGURE 2 Amiloride-sensitive currents in oocytes injected with a combination of αbENaC and β and γrENaC. A greater amiloride-sensitive current was observed in oocytes expressing all three subunits than in those expressing αbENaC alone. Oocytes injected with 2.5 ng of each cRNA exhibited a fivefold increase in current following 48 h of expression. Results are ±SEM. Reprinted with permission from Fuller *et al.* 1995.

was expressed in oocytes. Intracellular PKA consensus sequences have been identified only in the γ subunits of the rat and *Xenopus* isoforms; the human ENaC has no predicted PKA-sensitive sites (Garty and Palmer, 1997). However, the ability of vasopressin/adenylate cyclase to activate amiloride-sensitive Na^+ transport in rat renal and A6 epithelia is well known (Garty and Palmer, 1997). Similarly, cAMP-mediated agonists could activate amiloride-sensitive Na^+ channels in fibroblasts and MDCK cells transfected with αβγrENaC observed under physiological conditions (Stutts *et al.,* 1995, 1997). Furthermore, when incorporated into planar lipid bilayers, the 700-kDa Na^+ channel complex purified from the bovine renal papilla also showed an increase in P_o in response to addition of PKA to the bilayer chamber (Ismailov *et al.,* 1994, 1996b; Oh *et al.,* 1993). *In vivo* phosphorylation experiments showed that the 315-kDa polypeptide subunit of this complex isolated from A6 cells was a PKA-sensitive substrate (Sariban-Sohraby *et al.,* 1988). However, it has recently been proposed that both β and γrENaC stably expressed in the canine kidney cell line MDCK are themselves basally phosphorylated and that this phosphorylation is increased by exposing the cells to forskolin (Shimkets *et al.*, 1998). Phosphopeptide mapping revealed that residues in the C terminal of βrENaC, which by consensus were substrates for phosphorylation by caesin kinase II and protein kinase C, were phosphorylated by insulin and aldosterone, although unfortunately this analysis was not carried out following forskolin stimulation of the MDCK monolayer. Whether these sequences in βrENaC were also targets for phosphorylation by PKA was not determined. However, despite these findings, others have reported no consistent effect of PKA on channel P_o (Chalfant *et al.*, 1996; Frings *et al.*, 1988; Lester *et al.*, 1988). Similarly, αβγrENaC incorporated into planar lipid bilayers or coexpressed in *Xenopus* oocytes was not activated by cAMP or PKA (Awayda *et al.*, 1996). These somewhat disparate observations suggest that phosphorylation-dependent activation of this channel may also require other associated proteins in addition to whatever direct effects phosphorylation may have on the channel. One recently proposed candidate for this role is cytosolic actin (Berdiev *et al.*, 1996; Prat *et al.*, 1993). A recent report by Warnock and collaborators has shown that while cAMP-mediated activation of αβγhENaC expressed in *Xenopus* oocytes was highly variable, coexpression of the human channel with γ-actin consistently resulted in activation of an amiloride-sensitive current when oocytes were exposed to cpt-cAMP (Tamba *et al.*, 1998).

In contrast to the proposed effects of PKA, protein kinase C (in the presence of diacylglycerol and Ca^{2+}) or the phorbol ester PMA has been shown to cause a direct reduction in the open probability of αrENaC and αβγrENaC incorporated into planar lipid bilayers or expressed in *Xenopus* oocytes (Awayda *et al.*, 1996). Similar results have been obtained for the protein complex purified from the renal papilla when incorporated into planar bilayers (Oh *et al.*, 1993). However, direct phosphorylation of the αrENaC subunit expressed in MDCK cells by PMA has not been successfully demonstrated, although both the β and γrENaC subunits could be phosphorylated by phorbol ester when expressed in MDCK cells (Shimkets *et al.*, 1998). In αbENaC, four of the proposed consensus sites for PKC-dependent phosphorylation are predicted to be intracellular. However, the ability of PKC to inhibit Na^+ conductance is controversial, as both a lack of effect and inhibition of Na^+ transport have been reported in different tissues (Rouch *et al.*, 1993; Silver *et al.*, 1993; Yanase and Handler, 1986). The ability of cytochalasin B to restore 80% of the PMA-inhibited αβγrENaC-associated current in oocytes suggests that the actin cytoskeleton may also play a role in mediating the kinase C dependent inhibition of current, possibly by facilitating translocation of the kinase to the membrane (Awayda *et al.*, 1996). In contrast, in oocytes expressing αbENaC alone, PMA was found to increase the amiloride-sensitive current as determined by dual-electrode voltage-clamp experiments (Fuller *et al.*, 1995). This observation may reflect a specific difference in how αbENaC is regulated when expressed by itself; however, similar experiments on the chimeric bENaC/rENaC channel needed to address this question have not yet been performed.

III. THE C TERMINUS OF αbENaC: A KINETIC SWITCH?

Whereas the effect of cellular messengers on the activity of αbENaC have largely been carried out on expressing oocytes studied by dual-electrode voltage clamp, a more detailed analysis of the intrinsic properties of the channel require recording at the single-channel level. Our laboratory has used a technique first developed for the study of the large conductance K^+ channels from *Drosophila* (Perez *et al.*, 1994). In essence, channel cRNA is expressed in *Xenopus* oocytes (usually 50–100 oocytes per group), which are allowed to express for 24–48 h. At that time the oocytes are homogenized and membrane vesicles prepared by centrifugation over a discontinuous sucrose gradient. The resultant vesicles are collected and washed and can then be fused to the lipid bilayer for single-channel recording. Using this technique, we have been able to study the complex behaviors of rENaC and bENaC under a variety of experimental conditions.

Despite the high homology between the α subunits of bENaC and rENaC, single-channel recordings revealed that the two proteins had very different gating behavior (Fuller *et al.*, 1996). As shown in Fig. 3, single-channel

FIGURE 3 Single channel records of wild-type and C-terminal truncated αbENaC and αrENaC. The cRNAs for each construct (25 ng/oocyte) were injected and allowed to express in *Xenopus* oocytes for 48 h. Membrane vesicles were then prepared from the oocytes and fused to planar lipid bilayers for single-channel recording. (A) Wild-type αbENaC exhibited predominantly single-step transitions of 39 pS, whereas R567X αbENaC had an almost constitutively open 13-pS conductance level, on top of which were frequent 26-pS transitions to a 39-pS conductance main state. Wild-type αbENaC also showed a pronounced burst pattern of gating that was lost in the truncation mutant. In contrast, both wild-type and R613X αrENaC showed an identical gating pattern, characterized as in truncated αbENaC by an almost continuously open 13-pS state on top of which were frequent 26-pS transitions. (B) Effect of actin on αbENaC. Addition of 0.6 μM G-actin to αbENaC incorporated into the bilayer was associated with a reduction in single-channel conductance from 39 pS to approximately 18–20 pS, together with a concomitant increase in single-channel P_o. There was no change in single-channel kinetics under these conditions. The reducing agent DTT further resolved the channel into 3 independently gated subconductance states, each with a conductance of approximately 6 pS. Each panel is representative of at least 6 separate experiments. The holding potential was +100 mV. The dashed lines represent the zero current level. Panel A reprinted with permission from Fuller *et al.* (1996).

A

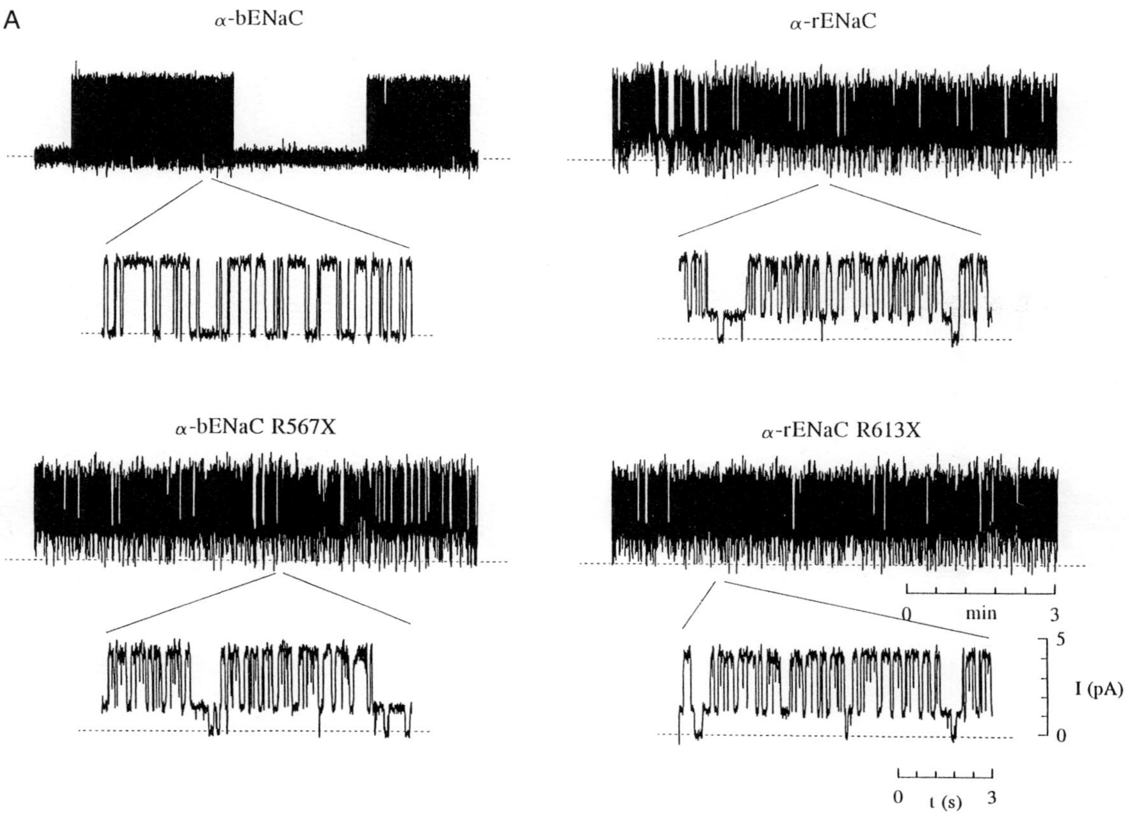

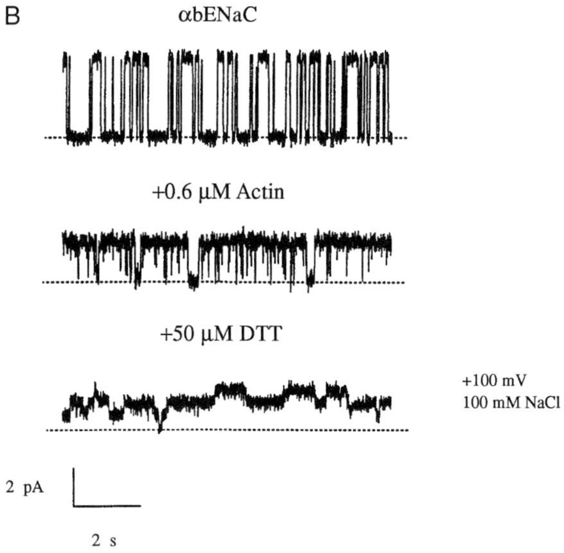

FIGURE 3 (*Continued*)

records of αbENaC revealed a single step transition of 39 pS. The channel also exhibited a pronounced bursting behavior, with intense periods of activity followed by periods of quiescence that could last for several minutes. In contrast, αrENaC exhibited two steps: an almost constitutively open 13-pS transition on which was superimposed 26-pS openings. This pattern did not represent two separate channels, as the steps were not independently gated (the steps did not follow a binomial distribution), but rather the 13 and 26-pS transitions represented subconductance states that gated cooperatively. Furthermore, αbENaC also interacted with actin. As shown in Fig. 3B, addition of G-actin to the channel incorporated into the lipid bilayer reduced the single-channel conductance by approximately half, with no change in gating pattern, similar to previous studies of αrENaC (Berdiev *et al.,* 1996).

Given the high homology of bENaC and rENaC, where might the difference in gating between the two isoforms be located? On consideration of the primary sequence, the only regions with a pronounced dissimilarity seem to reside at the N and C termini. The bovine sequence is missing nearly 50 amino acids from the N terminal that are present in αrENaC. At the C terminal, the cDNA predicts 9 nonconservative substitutions in bENaC (compared to rENaC) and a 3-residue gap. Initial sequencing of αbENaC had suggested that the divergence at the C terminal between αbENaC and αrENaC was much higher than later realized due to a severe C compression that had not been recognized and which resulted in a double frameshift. We therefore focused on the C terminal as the most likely site for the channel gate to be located and designed a construct that would result in a truncation of αbENaC at residue 567, by insertion of a premature stop codon. This mutation (R567X) was upstream of the compression and in a region bearing high homology to αrENaC. The corresponding mutation in αrENaC was R613X. In the revised sequence of αbENaC, the site of these mutations is followed by a string of 37 largely conserved amino acids (see Fig. 1).

However, when incorporated into the lipid bilayer, we found that R567X αbENaC now gated identically to wild-type αrENaC; not only had the long periods of closure disappeared, but the single channel conductance was now manifested as an almost constitutively open 13-pS state, on top of which was observed a 26-pS transition. The equivalent truncation in αrENaC (R613X) had no effect on the gating of this αENaC isoform (Fig. 3A). Thus it was apparent that the kinetic switch governing the unique gating pattern of αbENaC was located in the C terminal, although as the channel continued to open and close, this suggested that the role of the C terminal was a more subtle modification of the gating pattern.

We have previously proposed that αrENaC channels consist of 3 conductive barrels (Ismailov *et al.,* 1996c). When wild-type αrENaC was exposed to the disulfide reducing agent DTT, the channel was converted into a form consisting of 3 conductive states each of about 13 pS that now followed a binomial distribution, i.e., were gated independently. In contrast, exposure to high salt which would be predicted to disrupt hydrophobic interactions, changed the gating pattern such that the 13- and 26-pS states gated independently. These findings were interpreted on the basis of a "triple-barrel" model such that the smallest functional unit for an αrENaC channel would consist of 2 barrels linked by disulfide bonds and 1 barrel linked by hydrophobic interaction; disruption of the disulfides would free the 26-pS double barrel into 2 × 13-pS barrels and, at the same time, prevent the interaction of the third barrel. On the other hand, high salt would merely prevent the third single 13-pS barrel from interacting with the double barrel, resulting in the appearance of 2 independent channels, 1 with a conductance of 26 pS and 1 with a conductance of 13 pS (Benos *et al.,* 1997). However, despite the convenient conceptual model, these findings should not be interpreted as evidence for the interaction of 3 αENaC polypeptide chains as the basis for the stoichiometric assembly of the functional channel. A pure αENaC channel has been suggested to consist of 4 α subunits (this volume) while the stoichiometry of αβγrENaC may range from 2α, 1β, and 1γ, to 3α, 3β, and 3γ (Firsov *et al.,* 1998; Snyder *et al.,* 1998). The exact assembly of this channel at the membrane is therefore still to be determined. Interestingly however, when the native renal channel incorporated into the bilayer was reduced by DTT, i.e., the same experimental maneuver that revealed the individual subunit composition of the native channel by gel electrophoresis, the conductance of the native channel was reduced from about 40 to 13 pS, identical to that of $\alpha\beta\gamma$ rENaC studied under the same conditions.

A triple-barrel type behavior also underlies the activity of wt-αbENaC and αbENaC R567X (Fuller *et al.*, 1996). When treated with DTT, both constructs resolved into 3 × 13-pS unit conductances that obeyed the binomial distribution for independent ion channels (Fig. 4). Similarly, both were resolved into single 13- and 26-pS independent conductances when exposed to high salt. Cross-linking R567X αbENaC with DTNB converted the gating pattern back to that of wild-type αbENaC, i.e., a single, 39-pS step transition, as opposed to the cooperative gating of 13- and 26-pS states observed in the nonreduced mutant. Identical experiments with both wild-type and truncated αrENaC had the same results. As shown in Fig. 3B, DTT also resolved the αbENaC channel into its constituent subconductance states in the presence of actin, each individual state having a conductance of approximately 6 pS.

The role of the C terminal in αbENaC is therefore to modify the cooperative gating of all 3 barrels so that they open synchronously rather than sequentially. Other properties intrinsic to the channel such as ion selectivity and amiloride affinity were unaffected by truncation, the P_{Na}/P_K ratio being approximately 10 : 1 for each isoform and mutant construct. The $K_{i\ amils}$ values for wild-type and R567X αbENaC were 110 ± 32 and 113 ± 32 nM, respectively, while those for wt- and R613X αrENaC were 169 ± 46 and 176 ± 49 nM, respectively.

Where might the kinetic switch be located? As previously indicated, the C terminal of αbENaC contains a cluster of nonconserved amino acids that

FIGURE 4 Effect of sulfhydryl-active agents and high salt on the kinetic behavior of R567X αbENaC. Exposure of truncated αbENaC to the reducing agent DTT changed channel gating from two main conductance levels that appeared to gate cooperatively, to three independently gated 13-pS conductance states. The addition of 1.5 M NaCl also changed the gating pattern of R567X αbENaC such that the 13- and 26-pS subconductance states gated independently. The disulfide cross-linking agent DTNB restored the wild-type gating pattern to R567X αbENaC. Each panel is representative of at least 5 separate experiments, the holding potential was +100 mV, and the dashed lines represent the zero current level. Reprinted with permission from *Fuller, C. M., et al. (1996).*

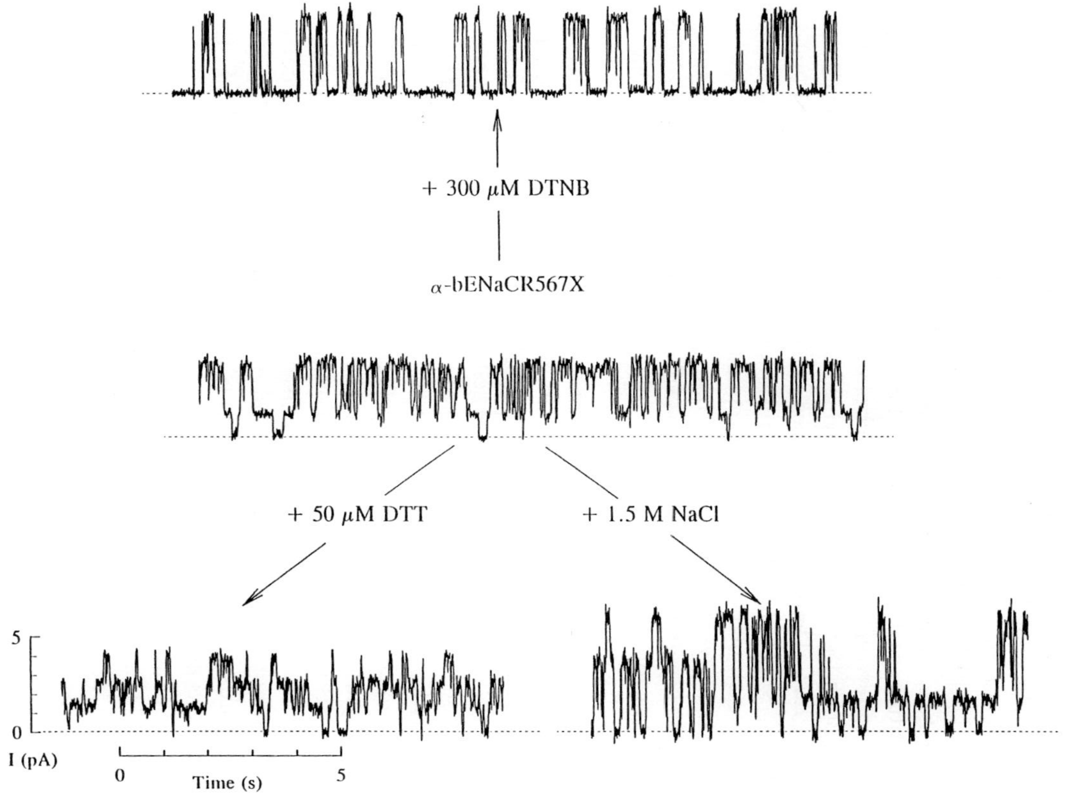
+ 300 μM DTNB
α-bENaCR567X
+ 50 μM DTT
+ 1.5 M NaCl
5
0
I (pA)
0
Time (s)
5

fall within the last 20 residues of the polypeptide, as well as a prominent gap in the alignment. In particular, a histidine and two prolines have been substituted for cysteine and leucines present in αrENaC (Fig. 5, see color insert). Proline in particular is a strong breaker of β sheets; thus it is conceivable that this residue in the C terminal may disrupt the tertiary structure of the protein necessary for gating. The human isoform of αrENaC, αhENaC has a much closer homology in the extreme C terminus to αbENaC than it does to αrENaC. However, preliminary experiments reconstituting αhENaC in planar lipid bilayers have suggested that the gating pattern of this subunit is much closer to αrENaC than it is to αbENaC, i.e., an open 13-pS state on top of which is seen a 26-pS transition (Bradford *et al.,* 1997). Therefore differences in sequence between αbENaC and αhENaC, which include αbENaC residues H645 and P647 (C664 and L666 in αhENaC), may be key sites of regulation.

Interestingly, the C terminals of both β and γENaCs have been shown to influence gating, although in a less subtle manner than that for αbENaC. Truncation of the C-terminal tails of either of these subunits has been associated with Liddle's syndrome, a rare autosomal dominant form of inherited hypertension that results in end stage renal disease (Hansson *et al.,* 1995a,b; Shimkets *et al.,* 1994). Premature truncation of β or γENaC subunits results in heterotrimeric αβγENaC channels that exhibit both an increased P_o and are retained at the cell surface for longer, as the retrieval recognition sequence (PPPXPY) is lost in the truncation (Schild *et al.,* 1995, 1996; Snyder *et al.,* 1995; Tamura *et al.,* 1996; Staub *et al.,* 1996). A similar increase in channel P_o and whole-cell current has been identified both in lymphocytes and in purified lymphocyte Na^+ channels isolated from Liddle's patients (Bubien *et al.,* 1996; Ismailov *et al.,* 1996a). The net result is increased renal Na^+ reabsorption; the hypertensive disease can be corrected by renal transplant. The etiology and functional consequences of Liddle's disease are reviewed more fully elsewhere in this volume. The PPPXPY sequence is also found in the α subunits, including αbENaC. However, to date no spontaneous mutation in the C terminal of this subunit has been identified in a patient. Furthermore, as described above loss of this sequence in R613X αrENaC did not result in an α channel of noticeably different characteristics from the wild-type channel. Incorporation of R567X αrENaC into planar lipid bilayers was associated with an increase in single channel P_o, as shown, although this increase was more reflective of the reduction in the inter-burst interval. Whether the retention of αbENaC is increased in the C terminal truncation mutant, R567X, is also unknown at present. However, synthetic peptides corresponding to the last 10–30 amino acids of β or γrENaC act synergistically to block αβγrENaC channels that have had their β and/or γ C-terminal tails truncated, raising the possibility that a "ball and chain" mechanism, reminiscent of that described to block Shaker K^+ channels of *Drosophila,* may act as an intrinsic inactivation particle of the channel; other peptides containing either different residues or the same residues but in a different order were without effect on channel gating (Bubien *et al.,* 1996; Ismailov *et al.,* 1996a; Zagotta *et al.,* 1990).

Some control of channel gating probably also resides in the N-terminal region. A single point mutation in βrENaC (G37S) was detected in a patient diagnosed with pseudo-hypoaldosteronism (PHA), a condition associated with severe salt wasting and therefore a mirror image of Liddle's disease (Chang *et al.,* 1996). An homologous mutation (G95S) in αrENaC results in a dramatic reduction in channel open time (Grunder *et al.,* 1997). Consistent with this observation, the deletion of the entire N terminal (Δ1-109T → M), dramatically altered single αβγ rENaC channel τ_{open} and τ_{closed} values (Stanton *et al.,* 1997). However, the ability of the channel to open and close, even if these transitions were infrequent, was not totally abolished. Taken together, these findings suggest that both the N and C termini of αENaC contribute to channel gating. Gating in αbENaC is obviously hierarchical, as the truncated/mutated channels still open and close, albeit with a different kinetic than is seen in the wild-type protein. Detecting the primary gate will be difficult unless a mutation that results in constitutive opening is found; loss of function mutations that result in a permanently closed channel will be nearly impossible to identify using techniques such as dual electrode voltage clamp of expressing oocytes. In accordance with this, the Δ1-109T → M mutation is not associated with increased current when studied in expressing oocytes (Chalfant *et al.,* 1997). Identification of the secondary gate will give us insights into how the primary amino acid sequence of a protein can translate into the subtle structural motifs that impact gating in the intact channel.

IV. A REGION IN THE EXTRACELLULAR LOOP INFLUENCES GATING, ION SELECTIVITY, AND AMILORIDE AFFINITY

Although the C-terminal tails of αbENaC and αrENaC are different, much of the primary sequence of the two proteins is either identical or highly conserved. To gain further insight into structure/function relationships in the ENaC family as a whole, we have also examined a region in the extracellular loop prior to M2 that is highly conserved between the mammalian and amphibian α subunit isoforms cloned to date (Fuller *et al.,* 1997). The sequence between amino acid residues 495–516 in αbENaC contains a repeating lysine residue at intervals of 5–6 amino acids (underlined) (IKNKRDGVA**K**LNIFFKELNY**K**S). A similar repeating motif of arginines is found in the Shaker K^+ channel of *Drosophila,* where it forms part of the voltage sensor and plays a role in channel gating (Perozo *et al.,* 1994). We therefore used site-directed mutagenesis to reverse the fixed positive charge on lysines 504 and 515 (shown in bold above) to fixed negative charges by converting these residues to glutamic acid (E). Expression and single-channel analysis of both K504E and K515E constructs were associated with changes in channel gating, ion selectivity, and amiloride affinity. In general terms, K504E was associated with the most dramatic shifts in channel characteristics; whereas K515E channels showed a reduction in the interburst interval (and a consequent increase in P_o), in K504E the burst pattern was eradicated, the channel continually transitioning between open and closed states (Fig. 6). However, in contrast to the C-terminal truncation, the size of the individual transitions was not affected by these mutations, remaining as single 39-pS steps.

Wild-type αbENaC exhibits an ion selectivity ratio (P_{Na}/P_K) of approximately 10:1, very similar to that reported for αrENaC studied under the same conditions (Fuller *et al.,* 1996). In contrast, both K504E and K515E αbENaC had lower ion selectivities for Na^+ over K^+ of 2:1 and 3:1, respectively, as determined under biionic conditions, although neither mutation affected the single-channel conductance (Fig. 7). These results suggest that this pre-M2 region may contribute to the selectivity filter of the channel. Each mutation also exhibited a much lower affinity for amiloride; the K_i for amiloride for wt-αbENaC was 0.11 μM, whereas that for the two mutants was 0.54 and 0.95 μM for K515E and K504E, respectively, a substantial rightward shift in the dose–response curve (Fig. 8).

This latter observation raises several interesting questions. The most simple interpretation is that the amino acid sequence in this region encompasses the amiloride binding site. However, it seems highly likely that a region in the extracellular loop between residues 278 and 283 (post M1) may constitute the principal binding site as mutations in this region dramatically

FIGURE 6 Single-channel records of wild-type (WT) and mutated (K515E and K504E) αbENaC. The cRNAs for each construct were injected into *Xenopus* oocytes, and membrane vesicles were then fused to planar lipid bilayers as described. In K515E and K504E, the interval between bursts of activity that were prominent in wt αbENaC was reduced, such that in K504E αbENaC, the channel was nearly continuously active. However, neither mutant exhibited any change in single channel conductance as compared to the wild-type channel. Each record is representative of at least 8 separate experiments. The holding potential was +100 mV. Dashed lines represent zero current. Reprinted with permission from Fuller *et al.* (1997).

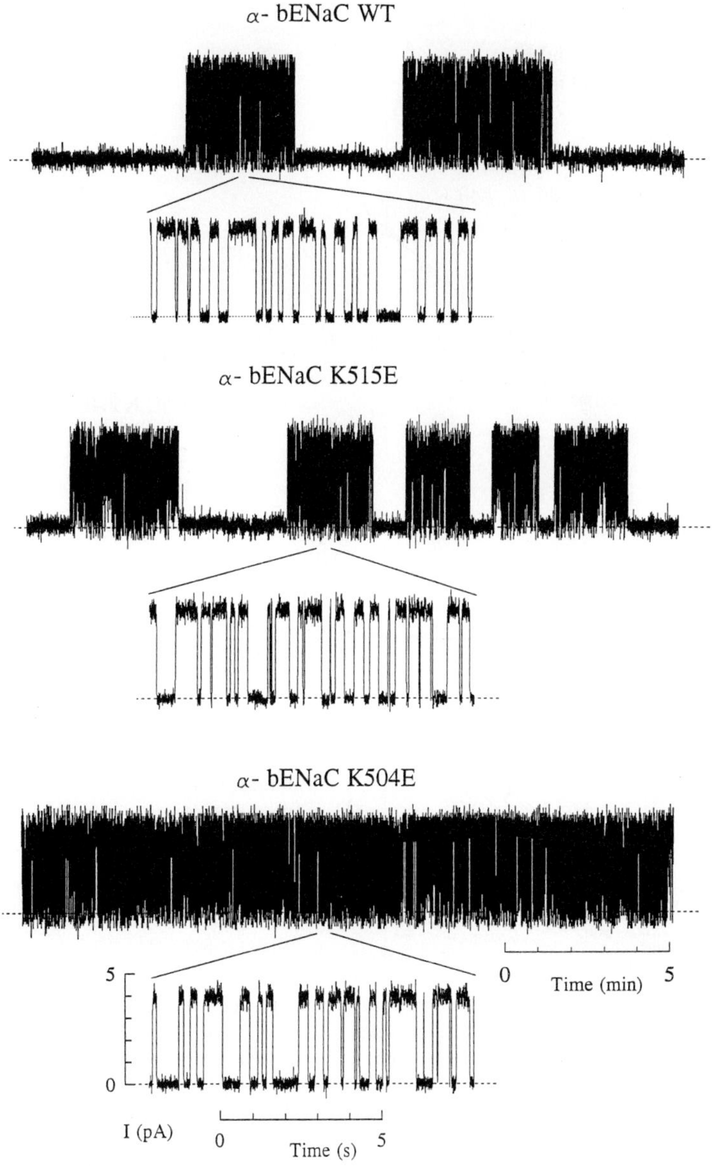
α- bENaC WT
α- bENaC K515E
α- bENaC K504E
5
0
I (pA)
0
Time (s)
5
0
Time (min)
5

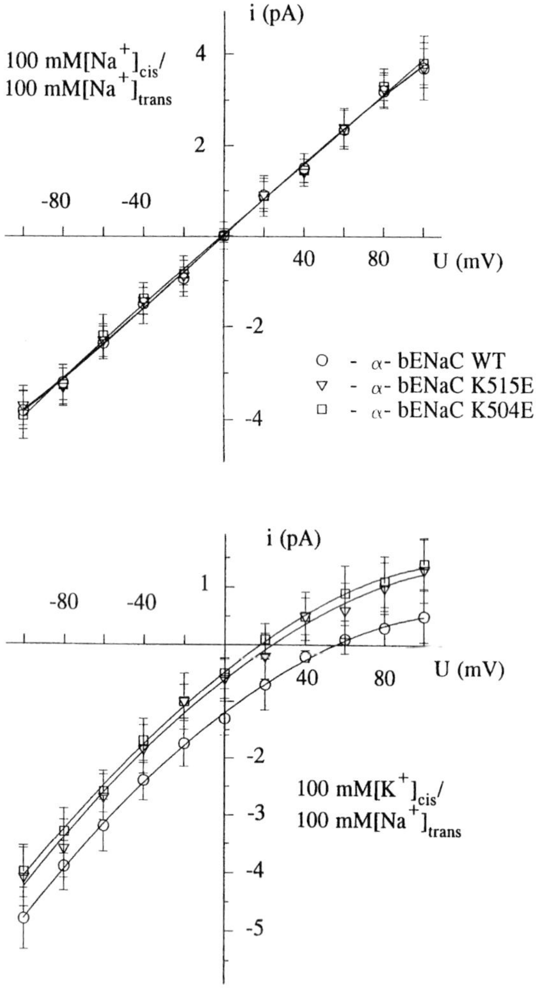

FIGURE 7 Current/voltage relationship and ion selectivity for wild-type, K504E, and K515E αbENaC. The current/voltage relations for wt αbENaC and K504E and K515E αbENaC are shown on the top panel. Neither mutation affected single-channel conductance recorded under symmetrical conditions. The ion selectivity of wild-type and mutant ENaC subunits was determined under bi-ionic conditions, with 100 mM Na^+ in the *trans* compartment and 100 mM K^+ in the *cis* chamber, and is shown in the bottom panel. Values are mean ± SD for 4 to 7 experiments at each voltage, as calculated from events histogram analysis. The calculated permeability ratio (P_{Na}/P_K) was approximately 10:1 for wild-type αbENaC and fell to 2:1 and 3:1 for K504E, and K515E, αbENaC, respectively. Under symmetrical conditions (100 mM Na^+), the current voltage relationship of all three αbENaC subunits was linear. Reprinted with permission from Fuller *et al.* (1997).

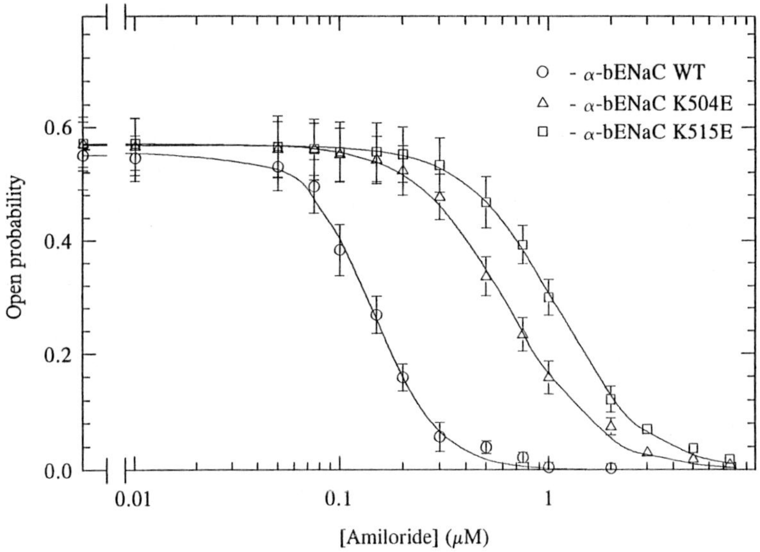

FIGURE 8 Amiloride dose–response curves for wild-type, K504E, and K515E αbENaC. Although all three channels were sensitive to amiloride, both mutants had much attenuated responses to the diuretic. These data suggest that more than one binding site in αbENaC may influence the interaction of the drug with the channel. Reprinted with permission from Fuller *et al.* (1997).

affected the affinity of the channel for the drug, while excision of the site abolished the amiloride-induced inhibition of the channel. Furthermore, an alternately spliced form of αrENaC (αrENaCa) that is missing the last 199 residues of αrENaC (including the lysine rich region) binds ^{3}[H] phenamil equally as well as the wild-type protein (Li *et al.,* 1995). Our finding that a region downstream of that identified by Kleyman and collaborators (Ismailov *et al.,* 1997) affects amiloride binding suggests that the interaction of the extracellular loop with the drug is much more complex than simple interaction at a single site. This conclusion has also been confirmed by others who observed that mutations in the β and γrENaC subunits can also influence the affinity of amiloride for the channel (see below).

Closed and open time amplitude histogram analysis of the wild-type and mutated channels in the presence of amiloride could be fitted with a double exponential, suggesting that the channel could have two open states (open and open-bound). However, because the open-bound state (accounting for the first exponent) was only observed at very low amiloride concentrations

(<0.2 μM for wt and <1 μM for the mutants) it is likely that at higher concentrations of amiloride, a single amiloride-bound state equivalent to an open-channel block is dominant. A closed $\rightarrow$ open $\rightarrow$ blocked model has been previously described for the amiloride-sensitive block of native Na^+ channels in A6 cells and in the rat cortical collecting duct (Hamilton and Eaton, 1985; Palmer and Frindt, 1986a,b). The ability of low concentrations of amiloride to bind to the channel without blocking it suggests the presence of additional sites for amiloride binding. Interestingly, the ratio of the k_{on} and k_{off} rate constants calculated as the inverse of the shorter τ_{open} (first exponential) gave values for the K_d for amiloride very close to the K_i values determined from the dose–response curve (0.13, 0.55, and 1.06 μM for wt, K515E, and K504E, respectively). This finding is consistent with the hypothesis that more than a single site in the channel may interact with amiloride. A reduction in mean τ_o and an increase in mean τ_c is in agreement with previously described effects of amiloride on single Na^+ channels (Hamilton and Eaton, 1985; Palmer and Frindt, 1986a,b). The study by Kleyman and collaborators (Ismailov *et al.*, 1997) reported a Hill coefficient of 2.4 for amiloride inhibition of both αrENaC and $\alpha\beta\gamma$rENaC, also consistent with multiple binding sites for the drug. As the stoichiometry of the channel has not been conclusively demonstrated, this result may reflect either multiple sites in the α subunit, or a single site on each α, β, and γ subunit. Replacement of M1 or M2 in αrENaC with the corresponding membrane spanning domains of Mec-4, a homolog of ENaC from *C. elegans,* dramatically reduced the amiloride K_i, suggesting that these regions may also contribute to inhibitor binding (Waldmann *et al.*, 1995a). Mutations in the pre-M2 region of β and γ rENaC subunits (at residues G525C (β) and G537C (γ)) significantly reduced the amiloride binding affinity for the heterotrimeric channel. The corresponding mutation in αrENaC (S583C) increased amiloride affinity (Schild *et al.*, 1997). Clearly, the interaction of amiloride with ENaC is highly complex, probably involving much of the extracellular loop and the transmembrane domains for the maintenance of high-affinity binding.

V. SUMMARY

Studies of αbENaC in the planar lipid bilayer have demonstrated that despite high overall homology with other members of the family, this particular isoform has some unique characteristics, in addition to features shared with related isoforms. The utility of the planar lipid bilayer technique resides in the opportunity it provides to study channel activity in biophysical terms in the absence of a complicating and sometimes confounding cell system. Although the subtleties of channel gating, amiloride affinity, and ion selectivity may be hidden when investigating macroscopic cell function, the study of these parameters allows us to begin to focus on the individual amino acid residues that regulate channel and ultimately cellular behavior.

Acknowledgment

The authors were supported by NIH Grant DK 37206 from the National Institute of Diabetes, Digestive, and Kidney Diseases.

References

Awayda, M. S., Ismailov, I. I., Berdiev, B. K., Fuller, C. M., and Benos, D. J. (1996). Protein kinase regulation of a cloned epithelial Na^+ channel. *J. Gen. Physiol.* **108,** 49–65.

Benos, D. J., Saccomani, G., Brenner, B. M., and Sariban-Sohraby, S. (1986). Purification and characterization of the amiloride-sensitive sodium channel from A6 cultured cells and bovine renal papilla. *Proc. Natl. Acad. Sci. U.S.A.* **83,** 8525–8529.

Benos, D. J., Saccomani, G., and Sariban-Sohraby, S. (1987). The epithelial sodium channel. Subunit number and location of the amiloride binding site. *J. Biol. Chem.* **262,** 10613–10618.

Benos, D. J., Fuller, C. M., Shlyonsky, V. G., Berdiev, B. K., and Ismailov, I. I. (1997). Amiloride-sensitive Na^+ channels: Insights and outlooks. *News Physiol. Sci.* **12,** 55–61.

Berdiev, B. K., Prat, A. G., Cantiello, H. F., Ausiello, D. A., Fuller, C. M., Jovov, B., Benos, D. J., and Ismailov, I. I. (1996). Regulation of epithelial sodium channels by short actin filaments. *J. Biol. Chem.* **271,** 17704–17710.

Bradford, A. L., Berdiev, B. K., Ismailov, I. I., Fuller, C. M., and Benos, D. J. (1997). Functional effects of mutations in the membrane spanning regions of the α-subunit of the amiloride-sensitive Na^+ channel (ENaC). *Physiologist* **40,** A–7 (abstr. 7.6).

Brown, D., Sorscher, E. J., Ausiello, D. A., and Benos, D. J. (1989). Immunocytochemical localization of Na^+ channels in rat kidney medulla. *Am. J. Physiol.* **256,** F366–F369.

Bubien, J. K., Ismailov, I. I., Berdiev, B. K., Cornwell, T., Lifton, R. P., Fuller, C. M., Achard, J. M., Benos, D. J., and Warnock, D. G. (1996). Liddle's disease: Abnormal regulation of amiloride-sensitive Na^+ channels by β-subunit mutation. *Am. J. Physiol.* **270,** C208–C213.

Canessa, C. M., Horisberger, J.-D., and Rossier, B. C. (1993). Epithelial sodium channel related to proteins involved in neurodegeneration. *Nature* (*London*) **361,** 467–470.

Canessa, C. M., Merrilat, A.-M., and Rossier, B. C. (1994a). Membrane topology of the epithelial sodium channel. *Am. J. Physiol.* **267,** C1682–C1690.

Canessa, C. M., Schild, L., Buell, G., Thorens, B., Gautschi, I., Horisberger, J.-D., and Rossier, B. C. (1994b). Amiloride sensitive epithelial Na^+ channel is made of three homologous subunits. *Nature* (*London*) **367,** 463–467.

Chalfant, M. L., Peterson-Yantorno, K., O'Brien, T. G., and Civan, M. M. (1996). Regulation of epithelial Na^+ channels from M-1 cortical collecting duct cells. *Am. J. Physiol.* **271,** F861–F870.

Chalfant, M. L., Karlson, K. H., McCoy, D. E., Halpin, P. A., Loffing, D. N., and Stanton, B. A. (1997). Possible role of the N-terminus of α-rENaC in sodium channel function. *Physiologist* **40,** A–8 (abstr. 7.11).

Chang, S. S., Grunder, S., Rosler, A., Mathew, P. M., Hanukoglu, I., Schild, L., Lu, Y., Shimkets, R. A., Nelson-Williams, C., Rossier, B. C., and Lifton, R. P. (1996). Mutations in subunits of the epithelial sodium channel cause salt wasting with hyperkalaemic acidosis, pseudohypoaldosteronism type I. *Nat. Genet.* **12,** 248–253.

Driscoll, M., and Chalfie, M. (1991). The mec-4 gene is a member of a family of *Caenorhabditis elegans* genes that can mutate to induce neuronal degeneration. *Nature* (*London*) **349,** 588–593.

Firsov, D., Gautschi, I., Merillat, A. M., Rossier, B. C., and Schild, L. (1998). The heterotetrameric architecture of the epithelial sodium channel (ENaC). *EMBO J.* **17,** 344–352.

Frings, S. M., Purves, R. D., and MacKnight, A. D. C. (1988). Single channel recordings from the apical membrane of the toad urinary bladder epithelial cell. *J. Membr. Biol.* **106,** 157–172.

Fuller, C. M., Awayda, M. S., Arrate, M. P., Bradford, A. L., Morris, R. G., Canessa, C. M., Rossier, B. C., and Benos, D. J. (1995). Cloning of a bovine renal epithelial Na^+ channel subunit. *Am. J. Physiol.* **269,** C641–C654.

Fuller, C. M., Ismailov, I. I., Berdiev, B. K., Shlyonsky, V. G., and Benos, D. J. (1996). Kinetic interconversion of rat and bovine homologs of the α subunit of an amiloride-sensitive Na^+ channel by C-terminal truncation of the bovine subunit. *J. Biol. Chem.* **271,** 26602–26608.

Fuller, C. M., Berdiev, B. K., Shlyonsky, V. G., Ismailov, I. I., and Benos, D. J. (1997). Point mutations in αbENaC regulate channel gating, ion selectivity and sensitivity to amiloride. *Biophys. J.* **72,** 1622–1632.

Garcia-Anoveros, J., Derfler, B., Neville-Golden, J., Hyman, B. T., and Corey, D. P. (1997). BNaC1 and BNaC2 constitute a new family of human neuronal sodium channels related to degenerins and epithelial sodium channels. *Proc. Natl. Acad. Sci. U.S.A.* **94,** 1459–1464.

Garty, H., and Palmer, L. (1997). Epithelial sodium channels: Function, structure and regulation. *Physiol. Rev.* **77,** 359–396.

Grunder, S., Firsov, D., Chang, S. S., Jaeger, N. F., Gautschi, I., Schild, L., Lifton, R. P., and Rossier, B. C. (1997). A mutation causing pseudohypoaldosteronism type I identifies a conserved glycine that is involved in the gating of the epithelial sodium channel. *EMBO J.* **16,** 899–907.

Hamilton, K. L., and Eaton, D. C. (1985). Single-channel recordings from amiloride-sensitive epithelial sodium channel. *Am. J. Physiol.* **249,** C200–C207.

Hansson, J. H., Nelson-Williams, C., Suzuki, H., Schild, L., Shimkets, R., Lu, Y., Canessa, C., Iwaski, T., Rossier, B., and Lifton, R. P. (1995a). Hypertension caused by a truncated epithelial sodium channel gamma subunit: Genetic heterogeneity of Liddle syndrome. *Nat. Genet.* **11,** 76–82.

Hansson, J. H., Schild, L., Lu, Y., Wilson, T. A., Gautschi, I., Shimkets, R., Nelson-Williams, C., Rossier, B. C., and Lifton, R. P. (1995b). A de novo missense mutation of the beta subunit of the epithelial sodium channel causes hypertension and Liddle syndrome, identifying a proline-rich segment critical for regulation of channel activity. *Proc. Natl. Acad. Sci. U.S.A.* **92,** 11495–11499.

Ismailov, I. I., McDuffie, J. H., and Benos, D. J. (1994). Protein kinase A phosphorylation and G protein regulation of purified renal Na^+ channels in planar bilayer membranes. *J. Biol. Chem.* **269,** 10235–10241.

Ismailov, I. I., Berdiev, B. K., Fuller, C. M., Bradford, A. L., Lifton, R. P., Warnock, D. G., Bubien, J. K., and Benos, D. J. (1996a). Peptide block of constitutively activated Na^+ channels in Liddle's disease. *Am. J. Physiol.* **270,** C214–C223.

Ismailov, I. I., Berdiev, B. K., Bradford, A. L., Awayda, M. S., Fuller, C. M., and Benos, D. J. (1996b). Associated proteins and renal epithelial Na^+ channel function. *J. Membr. Biol.* **149,** 123–132.

Ismailov, I. I., Awayda, M. S., Berdiev, B. K., Bubien, J. K., Lucas, J. E., Fuller, C. M., and Benos, D. J. (1996c). Triple barrel organization of ENaC, a cloned epithelial Na^+ channel subunit. *Am. J. Physiol.* **271,** 807–816.

Ismailov, I. I., Kieber-Emmons, T., Lin, C., Berdiev, B. K., Shlyonsky, V. G., Patton, H. K., Fuller, C. M., Worrell, R., Zuckerman, J. B., Sun, W., Eaton, D. C., Benos, D. J., and Kleyman, T. R. (1997). Identification of an amiloride binding domain within the α-subunit of the epithelial Na^+ channel. *J. Biol. Chem.* **272,** 21075–21083.

Kleyman, T. R., Kraehenbuhl, J. P., and Ernst, S. A. (1991). Characterization and cellular localisation of the epithelial Na^+ channel. Studies using an anti-Na^+ channel antibody raised by an anti-idiotypic route. *J. Biol. Chem.* **266,** 3907–3915.

Lester, D. S., Asher, C., and Garty, H. (1988). Characterization of cAMP-induced activation of epithelial Na^+ channels. *Am. J. Physiol.* **254,** C802–C808.

Li. X.-J., Xu, R.-H., Guggino, W. B., and Snyder, S. H. (1995). Alternatively spliced forms of the α subunit of the epithelial sodium channel: Distinct sites for amiloride binding and channel pore. *Mol. Pharmacol.* **47,** 1133–1140.

Lingueglia, E., Champigny, G., Lazdunski, M., and Barbry, P. (1995). Cloning of the amiloride-sensitive FMRFamide peptide-gated sodium channel. *Nature (London)* **378,** 730–733.

Marunaka, Y., and Eaton, D. C. (1991). Effects of vasopressin and cAMP on single amiloride-blockade Na channels. *Am. J. Physiol.* **260,** C1071–C1084.

Oh, Y. S., Smith, P. R., Bradford, A. L., Keeton, D., and Benos, D. J. (1993). Regulation by phosphorylation of purified epithelial Na channels in planar lipid bilayers. *Am. J. Physiol.* **265,** C85–C91.

Palmer, L. G., and Frindt, G. (1986a). Amiloride-sensitive Na^+ channels from the apical membrane of the rat cortical collecting tubule. *Proc. Natl. Acad. Sci. U.S.A.* **83,** 2767–2770.

Palmer, L. G., and Frindt, G. (1986b). Epithelial sodium channels: Characterization by using the patch-clamp technique. *Fed. Proc., Fed. Am. Soc. Exp. Biol.* **45,** 2708–2712.

Perez, G., Lagrutta, A., Adelman, J. P., and Toro, L. (1994). Reconstitution of expressed K_{Ca} channels from *Xenopus* oocytes to lipid bilayers. *Biophys. J.* **66,** 1022–1027.

Perozo, E., Santacruz-Toloza, L., Stefani, E., Bezanilla, F., and Papazian, D. M. (1994). S4 mutations alter gating currents of Shaker K channels. *Biophys. J.* **66,** 345–354.

Prat, A. G., Bertorello, A. M., Ausiello, D. A., and Cantiello, H. F. (1993). Activation of epithelial Na^+ channels by protein kinase A requires actin filaments. *Am. J. Physiol.* **265,** C224–C233.

Price, M. P., Snyder, P. M., and Welsh, M. J. (1996). Cloning and expression of a novel human brain Na^+ channel. *J. Biol. Chem.* **271,** 7879–7882.

Rouch, A. J., Chen, L., Kudo, L. H., Bell, P. D., Fowler, B. C., Corbitt, B. D., and Schafer, J. A. (1993). Intracellular Ca^{2+} and PKC activation do not inhibit Na^+ and water transport in the rat CCD. *Am. J. Physiol.* **265,** F569–F577.

Sariban-Sohraby, S., and Benos, D. J. (1986). Detergent solubilization, functional reconstitution, and partial purification of an epithelial amiloride-binding protein. *Biochemistry* **25,** 4639–4640.

Sariban-Sohraby, S., Sorscher, E. J., Brenner, B. M., and Benos, D. J. (1988). Phosphorylation of a single subunit of the epithelial Na^+ channel protein following vasopressin treatment of A6 cells. *J. Biol. Chem.* **263,** 13875–13879.

Schild, L., Canessa, C. M., Shimkets, R. A., Gautschi, I., Lifton, R. P., and Rossier, B. C. (1995). A mutation in the epithelial sodium channel causing Liddle disease increases channel activity in the *Xenopus laevis* oocyte expression system. *Proc. Natl. Acad. Sci. U.S.A.* **92,** 5699–5703.

Schild, L., Lu, Y., Gautschi, I., Schneeberger, E., Lifton, R. P., and Rossier, B. C. (1996). Identification of a PY motif in the epithelial Na channel subunits as a target sequence for mutations causing channel activation found in Liddle syndrome. *EMBO J.* **15,** 2381–2387.

Schild, L., Schneeberger, E., Gautschi, I., and Firsov, D. (1997). Identification of amino acid residues in the α, β, and γ subunits of the epithelial sodium channel (ENaC) involved in amiloride block and ion permeation. *J. Gen. Physiol.* **109,** 15–26.

Shimkets, R. A., Warnock, D. G., Bositis, C. M., Williams, C. N., Hansson, J. H., Schamelan, M., Gill, J. R. J., Ulick, S., Milora, R. V., Findling, J. W., Canessa, C. M., Rossier, B. C., and Lifton, R. P. (1994). Liddle's syndrome: Heritable human hypertension caused by mutations in the β subunit of the epithelial sodium channel. *Cell (Cambridge, Mass.)* **79,** 407–414.

Shimkets, R. A., Lifton, R., and Canessa, C. M. (1998). In vitro phosphorylation of the epithelial sodium channel. *Proc. Natl. Acad. Sci. U.S.A.* **95,** 3301–3305.

Silver, R. B., Frindt, G., Windhager, E. E., and Palmer, L. G. (1993). Feedback regulation of Na channels in rat CCT. I. Effects of inhibition of the Na pump. *Am. J. Physiol.* **264,** F557–F564.

Snyder, P. M., McDonald, F. J., Stokes, J. B., and Welsh, M. J. (1994). Membrane topology of the amiloride-sensitive epithelial sodium channel. *J. Biol. Chem.* **269,** 24379–24383.

Snyder, P. M., Price, M. P., McDonald, F. J., Adams, C. M., Volk, K. A., Zeiher, B. G., Stokes, J. B., and Welsh, M. J. (1995). Mechanism by which Liddle's syndrome mutations increase activity of a human epithelial Na^+ channel. *Cell (Cambridge, Mass.)* **83,** 969–978.

Snyder, P. M., Cheng, C., Prince, L. S., Rogers, J. C., and Welsh, M. J. (1998). Electrophysiological and biochemical evidence that DEG/ENaC cation channels are composed of nine subunits. *J. Biol. Chem.* **273,** 681–684.

Sorscher, E. J., Accavitti, M. A., Keeton, D., Steadman, E., Frizzell, R. A., and Benos, D. J. (1988). Antibodies against purified epithelial sodium channel protein from bovine renal papilla. *Am. J. Physiol.* **255,** C835–C843.

Stanton, B. A., Chalfant, M. L., Ismailov, I. I., Karlson, K., Loffing, D., Halpin, P., Berdiev, B. K., Fuller, C. M., Shlyonsky, V. G., and Benos, D. J. (1997). Regulation of ENaC by the amino terminus of α-rENaC. *Physiologist* **40,** A–4 (abstr. 4.5).

Staub, O., Dho, S., Henry, P. C., Correa, J., Ishikawa, T., McGlade, J., and Rotin, D. (1996). WW domains of Nedd 4 bind to the proline-rich PY motifs in the epithelial Na^+ channel deleted in Liddle's syndrome. *EMBO J.* **15,** 2371–2380.

Stutts, M. J., Canessa, C. M., Olsen, J. C., Hamrick, M., Cohn, J. A., Rossier, B. C., and Boucher, R. C. (1995). CFTR as a cAMP-dependent regulator of sodium channels. *Science* **269,** 847–850.

Stutts, M. J., Rossier, B. C., and Boucher, R. C. (1997). CFTR inverts PKA-mediated regulation of epithelial sodium channel single channel kinetics. *J. Biol. Chem.* **272,** 14037–14040.

Tamba, K., Tucker, J. K., Ma, H., Saxena, S., Unlap, T., Quick, M., Oh, Y., and Warnock, D. G. (1998). Human ENaC expressed in Xenopus oocytes is activated by cyclic AMP. *FASEB J.* **12,** A372 (abstr. 2161).

Tamura, H., Schild, L., Enomoto, N., Matsui, N., Marumo, F., Rossier, B. C., and Sasaki, S. (1996). Liddle disease caused by missense mutation of β subunit of the epithelial Na^+ channel gene. *J. Clin. Invest.* **97,** 1780–1784.

Tousson, A., Alley, C. D., Sorscher, E. J., Brinkley, B. R., and Benos, D. J. (1989). Immunochemical localization of amiloride-sensitive sodium channels in sodium transporting epithelia. *J. Cell Sci.* **93,** 349–362.

Waldmann, R., Champigny, G., and Lazdunski, M. (1995a). Functional degenerin-containing chimeras identify residues essential for amiloride-sensitive Na^+ channel function. *J. Biol. Chem.* **270,** 11735–11737.

Waldmann, R., Champigny, G., Bassilana, F., Voilley, N., and Lazdunski, M. (1995b). Molecular cloning and functional expression of a novel amiloride-sensitive Na^+ channel. *J. Biol. Chem.* **270,** 27411–27414.

Yanase, M., and Handler, J. S. (1986). Activators of protein kinase C inhibit sodium transport in A6 epithelia. *Am. J. Physiol.* **250,** C517–C522.

Zagotta, W. N., Hoshi, T., and Aldrich, R. W. (1990). Restoration of inactivation in mutants of Shaker potassium channels by a peptide derived from ShB. *Science* **250,** 568–571.

CHAPTER 2

Membrane Topology, Subunit Composition, and Stoichiometry of the Epithelial Na^+ Channel

Peter M. Snyder,* Chun Cheng,† and Michael J. Welsh†
*Department of Internal Medicine, University of Iowa College of Medicine, Iowa City, Iowa 52242
†Departments of Internal Medicine and Physiology and Biophysics, Howard Hughes Medical Institute, University of Iowa College of Medicine, Iowa City, Iowa 52242

I. INTRODUCTION

The epithelial Na^+ channel, ENaC, plays an important role in Na^+ homeostasis (Benos *et al.,* 1995; Garty and Palmer, 1997). ENaC is expressed at the apical membrane of epithelia in the kidney cortical collecting duct, the lung, and the distal colon. Na^+ enters the cell through ENaC, moving down its electrochemical gradient, and Na^+ is then pumped out of the cell at the basolateral membrane by the Na^+, K^+-ATPase, resulting in net Na^+ absorption. Three ENaC subunits (α, β, and γ) have been cloned from rat (Canessa *et al.,* 1993, 1994b), human (McDonald *et al.,* 1994, 1995), and *Xenopus* (Puoti *et al.,* 1995), and are part of a rapidly growing "Deg/ENaC" family of ion channels. In *C. elegans,* MEC-4, MEC-10, and DEG-1, play

1063-5823/99 $30.00

a role in mechanotransduction, and some gain-of-function mutations cause neurodegeneration (Huang and Chalfie, 1994; Tavernarakis and Driscoll, 1997). In *H. aspersa,* the FMRFamide-gated channel (FaNaCh) functions as a neurotransmitter receptor (Lingueglia *et al.,* 1995). Three family members have recently been identified in mammalian nervous system: BNC1 (Price *et al.,* 1996), BNaC2 (ASIC) (Garcia-Anoveros *et al.,* 1997; Waldmann *et al.,* 1997a), and DRASIC (Waldmann *et al.,* 1997b). Although their functions are not certain, BNaC2 and DRASIC may play a role in sensing acidic pH.

Altered function or regulation of the human epithelial Na^+ channel (hENaC) disrupts Na^+ homeostasis and results in disease. Mutations that delete or disrupt a conserved motif (PPPXYXXL) in the C-terminus of β or γhENaC cause Liddle's syndrome, a rare inherited form of hypertension (Shimkets *et al.,* 1994; Hansson *et al.,* 1995; Schild *et al.,* 1995; Snyder *et al.,* 1995). These mutations increase renal Na^+ absorption, at least in part by increasing the number of Na^+ channels at the apical membrane (Snyder *et al.,* 1995; Firsov *et al.,* 1996). Conversely, loss-of-function mutations in α, β, or γhENaC cause pseudohypoaldosteronism type I, a neonatal Na^+ wasting disorder (Chang *et al.,* 1996; Strautnieks *et al.,* 1996). In cystic fibrosis, loss of cystic fibrosis transmembrane conductance regulator function increases amiloride-sensitive current (Boucher *et al.,* 1986), which may contribute to the altered NaCl composition in the airway that contributes to the pathophysiology of cystic fibrosis. To learn how mutations produce altered channel function and disease, we investigated the structure of ENaC. In this chapter, we review the membrane topology, subunit composition, and stoichiometry of ENaC.

II. TOPOLOGY OF ENaC

To begin to define the structure of ENaC, we determined the membrane topology of the rat α subunit (αrENaC) (Snyder *et al.,* 1994). *In vitro* translation of αrENaC produced a 73-kDa protein, and an additional 93-kDa protein was produced when αrENaC was translated in the presence of microsomal membranes, corresponding to the core-glycosylated form of the protein (Fig. 1). We took advantage of glycosylation to begin to determine the topology of αrENaC, since glycosylation of a site localizes it to the extracellular side of the plasma membrane. αrENaC contains eight potential sites for N-linked glycosylation (NXS/T, Fig. 1B). Mutation of six of these sites ($Asn_{190-538}$) abolished glycosylation of αrENaC (Fig. 1A). This suggested that two additional sites near the N-terminus (Asn_{89} and Asn_{90}) were not glycosylated. By individually mutating $Asn_{190-538}$, we found

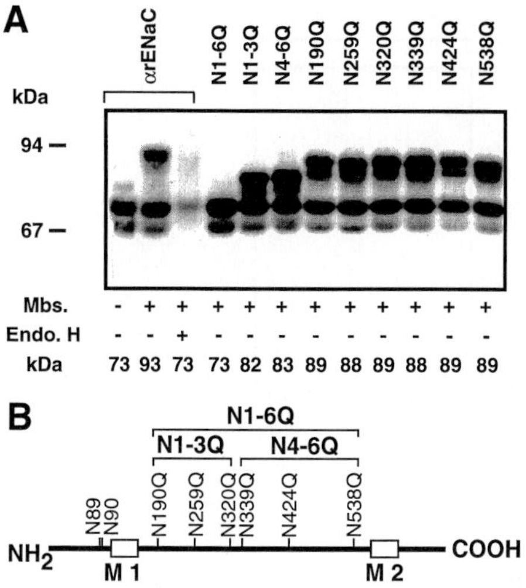

FIGURE 1 Glycosylation of αrENaC. (A) Wild-type or mutant αrENaC was transcribed and translated *in vitro* in the presence or absence of microsomal membranes and incubated with or without endoglycosidase H, as indicated. Proteins were separated by SDS-PAGE and ^{35}S-labeled proteins were detected by fluorography. The molecular mass (kDa) of the largest major band in each lane is indicated below each lane. (B) Schematic of αrENaC indicating the eight potential sites for N-linked glycosylation, and the identity of the mutations tested in (A). Adapted from Snyder *et al.* (1994).

that each of the six was glycosylated (Fig. 1A), suggesting that they all lie within an extracellular domain.

To localize the C-terminus with respect to the plasma membrane, we placed a Flag epitope at the C-terminus ($F_{C\text{-term}}$). Following *in vitro* translation and incorporation of the tagged αrENaC into microsomal membranes, we determined the sensitivity of the epitope to proteinase K. The C-terminal epitope was not protected from protease digestion (Fig. 2A), indicating that it was oriented on the outside of the microsome, and hence, is located on the intracellular side of the plasma membrane. In contrast, when a Flag epitope replaced Asn_{190} (extracellular based on its glycosylation, F_{EC}), the epitope was protected from digestion, confirming that it was extracellular (Fig. 2A). Similar results were obtained when αrENaC was expressed in *Xenopus* oocytes, suggesting that the topology *in vitro* reflects its topology in cells. We also localized the N-terminus as intracellular based on sensitivity of an N-terminal epitope ($F_{N\text{-term}}$) to proteinase K digestion (Fig. 2B). This is consistent with the absence of a signal sequence in αrENaC. Thus, our results (Snyder *et al.,* 1994), and the results of others (Renard *et al.,* 1994; Canessa *et al.,* 1994a), indicate that αrENaC contains two transmembrane

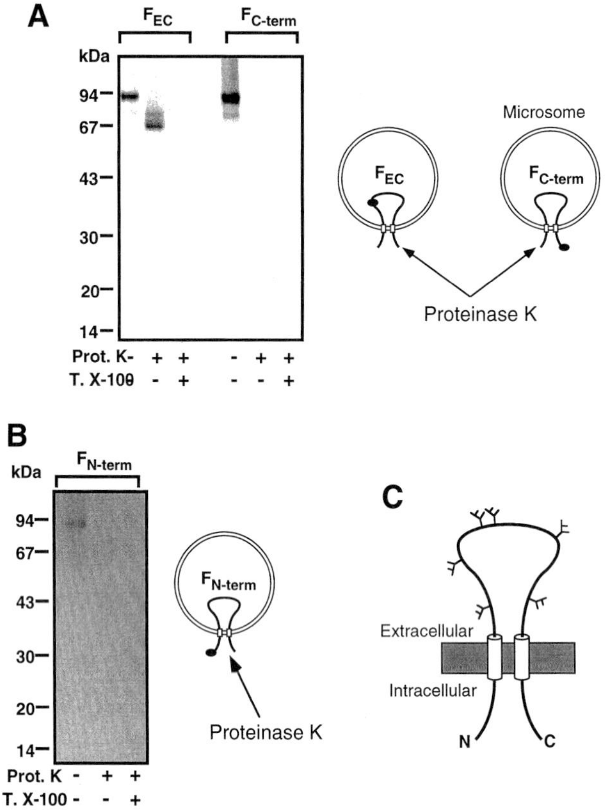

FIGURE 2 Localization of the C- and N-Termini by protease sensitivity. αrENaC containing a Flag epitope (A) at the C-terminus ($F_{C\text{-term}}$), in the extracellular domain replacing Asn_{190} (F_{EC}), or (B) at the N-terminus ($F_{N\text{-term}}$) was translated *in vitro* in the presence of microsomal membranes. Some of the samples were treated with proteinase K (200 μg/ml) and/or Triton X-100 (1%), as indicated. Tagged αrENaC was immunoprecipitated with anti-Flag M2 monoclonal antibody, separated by SDS-PAGE, and detected by fluorography. Diagrams illustrate the orientation of the Flag epitopes (filled oval) with respect to the microsomal membrane. (C) Topology of αrENaC. Adapted from Snyder *et al.* (1994).

segments, a large glycosylated extracellular domain and the N-terminus and C-terminus lie within the cytoplasm of the cell (Fig. 2C). Because of their high degree of sequence similarity, it is likely that β and γENaC share this topology. A similar topology has recently been found for the degenerin MEC-4 (Lai *et al.*, 1996).

Defining its topology has helped us to begin to learn about functional domains within ENaC. The C-terminus is cytoplasmic, where it functions in the regulation of ENaC, at least in part by controlling the number of channels at the cell surface (Snyder *et al.*, 1995). Although the mechanism of this regulation is unclear, it may involve the interaction of the C-terminus with one or more cytoplasmic proteins. The N-terminus is also cytoplasmic, where it plays a role in subunit–subunit interactions (Adams *et al.*, 1997). M2 forms a transmembrane segment, and residues within M2 line the channel pore (Waldmann *et al.*, 1995; Schild *et al.*, 1997; P. M. Snyder and M. J. Welsh, unpublished). Finally, the extracellular domain contains a number of conserved cysteines, bearing a superficial resemblance to receptor proteins. Although an extracellular ligand has not been identified for ENaC, the related channel FaNaCh is gated by the neurotransmitter FMRFamide, possibly by interacting with the extracellular domain.

III. SUBUNIT COMPOSITION OF hENaC

Three ENaC subunits have been identified in epithelia of rat and human: α, β, and γENaC. We found that expression of the human α subunit alone in *Xenopus* oocytes produced amiloride-sensitive Na^+ current (Fig. 3) (McDonald *et al.*, 1994). However, this current was very small. In contrast, expression of β and γhENaC, either alone or in combination, did not produce Na^+ current. Coexpression of γ with α increased current 4-fold, and coexpression of all three subunits increased current 20-fold compared to αhENaC alone (McDonald *et al.*, 1995). These results are in agreement

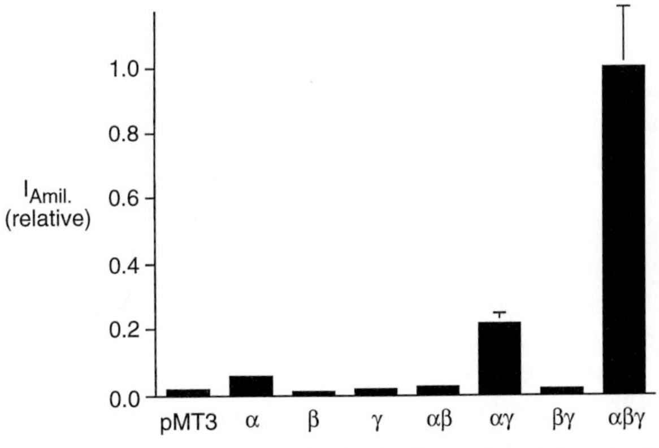

FIGURE 3 Expression of hENaC subunits in *Xenopus* oocytes. Amiloride-sensitive whole-cell Na^+ current in oocytes (−60 mV) one day after injection of either vector (pMT3) or hENaC subunits, as indicated. Current is expressed relative to cells expressing $\alpha\beta\gamma$hENaC. Adapted from McDonald *et al.* (1995).

with expression of rat ENaC (Canessa *et al.,* 1994b) and suggest that all three subunits contribute to the functional Na^+ channel complex.

IV. STOICHIOMETRY OF hENaC

A. Electrophysiological Analysis

The finding that coexpression of α, β, and γENaC is required for maximal Na^+ current suggests that ENaC is a heteromultimeric channel. Genetic evidence from *C. elegans* suggests that the degenerins also function as heteromultimers (Tavernarakis and Driscoll, 1997). To determine the stoichiometry of hENaC, we first asked how many γ subunits contribute to the channel complex (Snyder *et al.,* 1998). We and others (Schild *et al.,* 1997) previously found that a residue within the second transmembrane segment of γhENaC (Gly_{536}) lines the channel pore. Wild-type hENaC is insensitive to the cysteine-reactive compound MTSET (Fig. 4A). However, when Gly_{536} was replaced by cysteine (γ_{G536C}), MTSET irreversibly inhibited hENaC (87%, Fig. 4A). We took advantage of this to determine the number of γ subunits in the channel complex. When we coexpressed equal amounts of wild-type γ and γ_{G536C} cDNA (with wild-type α and β), the majority of current (77%) was inhibited by MTSET (Fig. 4B). When we increased the proportion of wild-type cDNA relative to mutant γ (0.8 : 0.2), MTSET inhibited 48% of the Na^+ current. Thus, MTSET inhibited hENaC out of proportion to the fraction of γ subunits that contained the G536C mutation. This suggests that hENaC contains more than one γ subunit and that a single mutant γ within a channel complex is sufficient to confer sensitivity to MTSET (i.e., the mutant phenotype is dominant).

If we assume that the wild-type and mutant γ subunits express equally and associate randomly into the channel complex, the composition of the channel will follow a binomial distribution. This assumption seems valid since we found that the wild-type and mutant subunits produced equal whole-cell Na^+ currents (not shown). Therefore, we can calculate the number of γ subunits (n) using the equation $Inh = f^n_{\mathrm{wt}} Inh_{\mathrm{wt}} + (1 - f^n_{\mathrm{wt}}) Inh_{\mathrm{mut}}$, where Inh_{wt} is the fraction of current inhibited in channels containing only wild-type subunits, Inh_{mut} is the fraction of current inhibited in channels containing n mutant subunits, and f is the fraction of subunits expressed that are wild-type or mutant, as indicated. The data suggest that hENaC contains three γ subunits. Figure 4C illustrates the possible combinations of wild-type and mutant γ subunits if the channel contains three γ subunits and we coexpress equal amounts of each (0.5 : 0.5). If channels containing at least one mutant γ are sensitive to MTSET, then the fraction of current

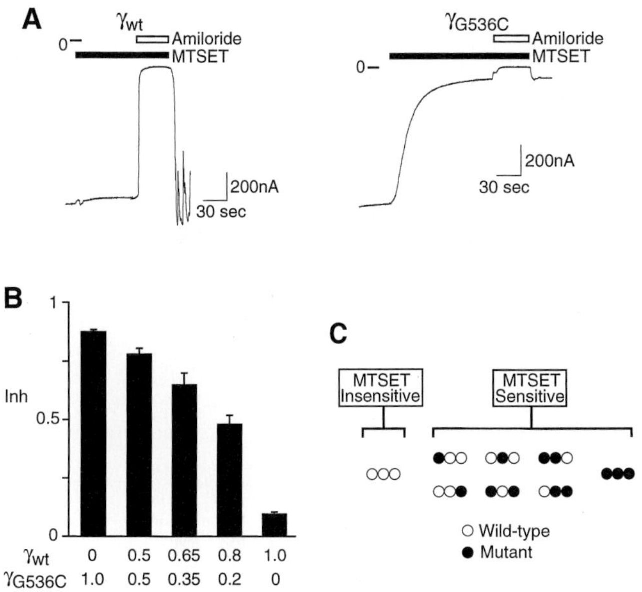

FIGURE 4 Stoichiometry of γhENaC using γ_{G536C}. Wild-type α andβhENaC were coexpressed in *Xenopus* oocytes with γ_{wt}, γ_{G536C}, or a mixture, as indicated. (A) Representative time course of current. Amiloride (3 mM) and MTSET (3 mM) added as indicated (*bars*). (B) Fraction of amiloride-sensitive current inhibited by MTSET (Inh) following expression of γ_{wt}, γ_{G536C}, or combinations of both in the indicated ratios. (C) Possible combinations of wild-type and mutant γ subunits resulting from coexpression of equal concentrations (0.5 : 0.5) of each. Channels predicted to be sensitive and insensitive to MTSET are indicated. Adapted from Snyder *et al.* (1998).

inhibited by MTSET should be 7/8 that of the completely mutant channel. This is in close agreement with our results.

To confirm the number of γ subunits, we used a second mutation which had the opposite effect. In contrast to MTSET, a smaller cysteine-reactive compound (MTSEA) inhibits wild-type hENaC 54% by modifying one or more cysteine within the channel pore (P. M. Snyder, and M. J. Welsh, unpublished). Mutation of Ser_{529} to cysteine in γ abolished inhibition by MTSEA (Fig. 5A). This residue is equivalent to the "Deg" residue in *C. elegans* degenerins, the site of neurodegeneration causing mutations. Modification of this cysteine by MTSEA may prevent channel inhibition by preventing further MTSEA molecules from entering the channel pore. When we coexpressed combinations of wild-type γ and γ_{S529C}, MTSEA inhibited Na^+ current less than would be expected from the proportion of wild-type subunits (Fig. 5A). This suggested that a single mutant γ subunit

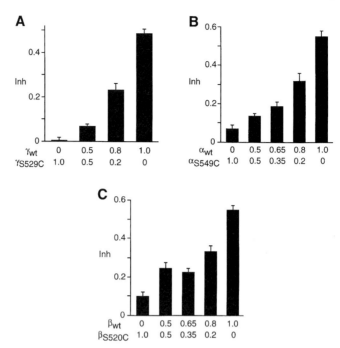

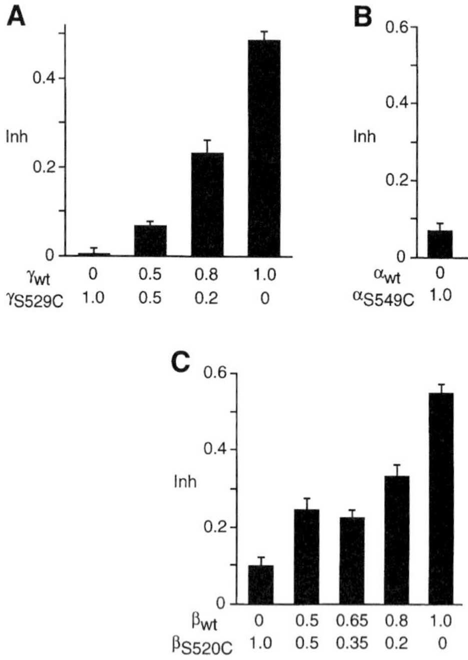

FIGURE 5 Stoichiometry of hENaC using the "Deg" mutation. Wild-type and mutant hENaC subunits were coexpressed (along with the other two wild-type subunits) in the indicated ratios, and the fraction of Na^+ current inhibited by MTSEA (1 mM) is plotted. Adapted from Snyder *et al.* (1998).

within a channel complex was sufficient to abolish sensitivity to MTSEA. Further, wild-type and mutant γ subunits (when coexpressed with α and βhENaC) produced equal Na^+ currents, suggesting that the channel composition fits a binomial distribution. Using the same equation shown above, the data for γ_{S529C} suggested that hENaC contains three γ subunits. Thus, two independent assays produced equivalent results.

Mutations of the equivalent residues in the α (α_{S549C}) and β (β_{S520C}) subunits also disrupted channel inhibition by MTSEA (Figs. 5B and 5C). We therefore used an identical strategy to determine the number of α and β subunits in hENaC. Our results are consistent with three α and three β subunits in the channel complex, suggesting that hENaC contains nine total subunits. MTSEA inhibits hENaC by modifying one or more pore-lining cysteine in the γ subunit. Because mutations in the α and β subunits prevented modification of γ by MTSEA, this suggests that all nine subunits contribute to a common pore.

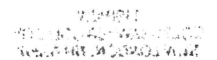

B. Sucrose Gradient Analysis of hENaC

We used a biochemical assay to determine whether the mass of hENaC was consistent with a complex containing nine subunits (Snyder *et al.,* 1998). Because αhENaC can form a functional channel by itself, we determined the mass of *in vitro* expressed α by sucrose gradient sedimentation. The unglycosylated form of the protein sedimented in relatively light fractions, similar to a standard of 240 kDa (Fig. 6). In contrast, the more mature glycosylated protein was found in heavier fractions, similar to a 950 kDa standard. This was consistent with a complex of at least nine α subunits.

To determine the mass of hENaC containing all three subunits, we coexpressed α, β, and γhENaC in COS7 cells and pulse-labeled cells with [^{35}S]methionine. Following sucrose gradient sedimentation, we detected α subunits by immunoprecipitation. Immediately after the pulse, most of the α was found in fractions 4–5, similar to a 240-kDa standard (Fig. 7). With increasing time after the pulse, an increasing amount of α was found in fractions 8 and 9, consistent with complexes containing nine hENaC subunits. We were also able to detect β subunits that coimmunoprecipitated with α. After a 7 h chase, a large fraction of β was also found in fractions 8 and 9. This was an important finding, since we could only detect β subunits that were part of a heteromeric complex with α. Thus, the mass of the heteromeric hENaC complex is consistent with a channel containing at least nine subunits.

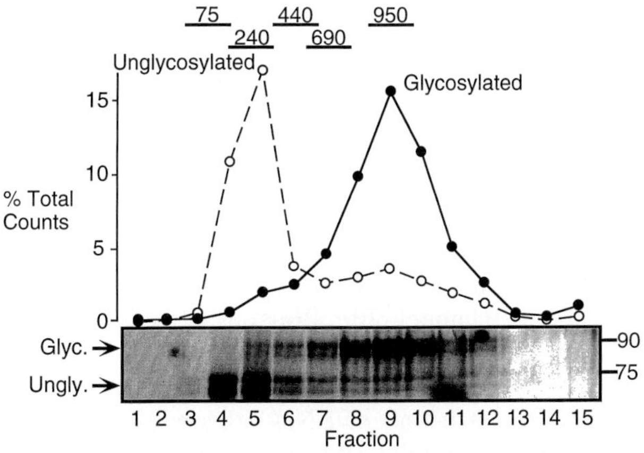

FIGURE 6 Sucrose gradient sedimentation of αhENaC expressed *in vitro*. Autoradiogram and quantitation of αhENaC following sedimentation on 10% (fraction 1) to 45% (fraction 15) sucrose gradient and SDS-PAGE. Glycosylated and unglycosylated αhENaC and sedimentation of standard protein markers (kDa) are indicated. Adapted from Snyder *et al.* (1998).

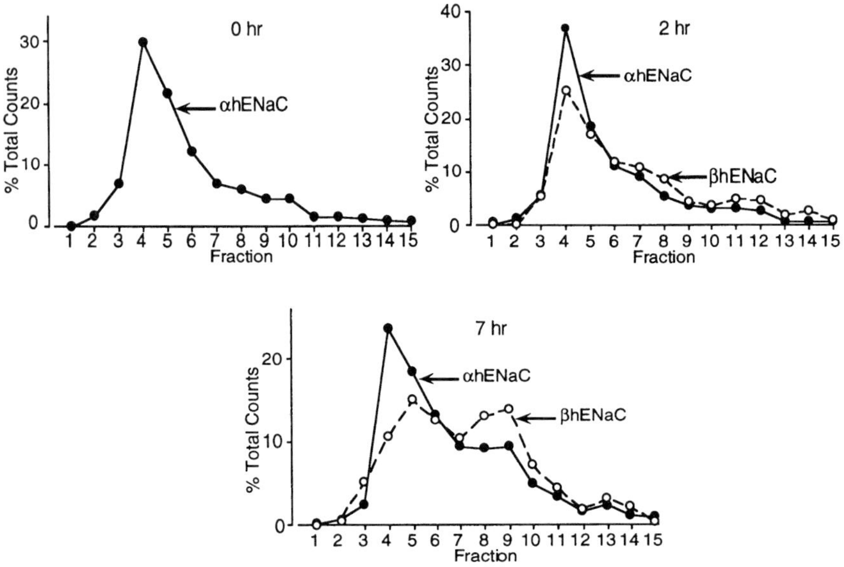

FIGURE 7 Sucrose gradient sedimentation of α and β expressed with γhENaC in COS7 cells. Following sedimentation on 10–45% sucrose gradient, αhENaC was immunoprecipitated, separated by SDS-PAGE, and the glycosylated form in each fraction was quantitated by phosphorimaging. βhENaC that coprecipitated with αhENaC was also quantitated (at 0 h, insufficient β was present for quantitation). Reproduced from Snyder *et al.* (1998).

V. CONCLUSION

The results reviewed above suggest that ENaC is a heteromultimeric channel complex, composed of three α, three β, and three γ subunits. This is a novel stoichiometry for an ion channel, suggesting a unique structure. How nine subunits assemble to form a Na^+ selective pore will be an important question for future work. Interestingly, Guy and Durell independently proposed a model for ENaC containing three α, three β, and three γ subunits, based on theoretical considerations (Guy and Durell, 1995). In their model, residues adjacent to M1 and M2 in each subunit form $\alpha\beta$ barrels which line the channel pore. Previous work from our lab and from others has shown that all three subunits contribute to formation of the pore, and that residues within the predicted $\alpha\beta$ barrel line the pore. As we further define the structure of ENaC, this will help us to understand the molecular basis for a number of its properties: high selectivity of ENaC for Na^+ over K^+, slow gating kinetics, and channel block by amiloride. These studies will also help us to understand how mutations cause hypertension and other diseases of Na^+ homeostasis.

Acknowledgments

We thank our laboratory colleagues for support and helpful discussions, and Diane Olson, Dan Bucher, and Ellen Tarr for technical assistance. We thank the University of Iowa DNA Core for assistance. P.M.S. is supported by a Fellowship from the Roy J. Carver Charitable Trust and by the NHLBI and NIDDK, National Institutes of Health. M.J.W. is supported by the Howard Hughes Medical Institute.

References

Adams, C. M., Snyder, P. M., and Welsh, M. J. (1997). Interactions between subunits of the human epithelial sodium channel. *J. Biol. Chem.* **272,** 27295–27300.

Benos, D. J., Awayda, M. S., Ismailov, I. I., and Johnson, J. P. (1995). Structure and function of amiloride-sensitive Na^+ channels. *J. Membr. Biol.* **143,** 1–18.

Boucher, R. C., Stutts, M. J., Knowles, M. R., Cantley, L., and Gatzy, J. T. (1986). Na^+ transport in cystic fibrosis respiratory epithelia. Abnormal basal rate and response to adenylate cyclase activation. *J. Clin. Invest.* **78,** 1245–1252.

Canessa, C. M., Horisberger, J.-D., and Rossier, B. C. (1993). Epithelial sodium channel related to proteins involved in neurodegeneration. *Nature* (*London*) **361,** 467–470.

Canessa, C. M., Merillat, A. M., and Rossier, B. C. (1994a). Membrane topology of the epithelial sodium channel in intact cells. *Am. J. Physiol.* **267,** C1682–C1690.

Canessa, C. M., Schild, L., Buell, G., Thorens, B., Gautschi, I., Horisberger, J.-D., and Rossier, B. C. (1994b). Amiloride-sensitive epithelial Na^+ channel is made of three homologous subunits. *Nature* (*London*) **367,** 463–467.

Chang, S. S., Grunder, S., Hanukoglu, A., Rösler, A., Mathew, P. M., Hanukoglu, I., Schild, L., Lu, Y., Shimkets, R. A., Nelson-Willams, C., Rossier, B. C., and Lifton, R. P. (1996). Mutations in subunits of the epithelial sodium channel cause salt wasting with hyperkalaemic acidosis, pseudohypoaldosteronism type 1. *Nat. Genet.* **12,** 248–253.

Firsov, D., Schild, L., Gautschi, I., Merillat, A. M., Schneeberger, E., and Rossier, B. C. (1996). Cell surface expression of the epithilial Na^+ channel and a mutant causing Liddle syndrome: A quantative approach. *Proc. Natl. Acad. Sci. U.S.A.* **93,** 15370–15375.

Garcia-Anoveros, J., Derfler, B., Neville-Golden, J., Hyman, B. T., and Corey, D. P. (1997). BNaC1 and BNaC2 constitute a new family of human neuronal sodium channels related to degenerins and epithelial sodium channels. *Proc. Natl. Acad. Sci. U.S.A.* **94,** 1459–1464.

Garty, H., and Palmer, L. G. (1997). Epithelial sodium channels: Function, structure, and regulation. *Physiol. Rev.* **77,** 359–396.

Guy, H. R., and Durell, S. R. (1995). Structural model of ion selective regions of epithelial sodium channels. *Biophys. J.* **68,** A243.

Hansson, J. H., Nelson-Williams, C., Suzuki, H., Schild, L., Shimkets, R., Lu, Y., Canessa, C., Iwasaki, T., Rossier, B., and Lifton, R. P. (1995). Hypertension caused by a truncated epithelial sodium channel gamma subunit: Genetic heterogeneity of Liddle syndrome. *Nat. Genet.* **11,** 76–82.

Huang, M., and Chalfie, M. (1994). Gene interactions affecting mechanosensory transduction in *Caenorhabditis elegans. Nature* (*London*) **367,** 467–470.

Lai, C.-C., Hong, K., Kinnell, M., Chalfie, M., and Driscoll, M. (1996). Sequence and transmembrane topology of MEC-4, an ion channel subunit required for mechanotransduction in *Caenorhabditis elegans. J. Cell Biol.* **133,** 1071–1081.

Lingueglia, E., Champigny, G., Lazdunski, M., and Barbry, P. (1995). Cloning of the amiloride-sensitive FMRFamide peptide-gated sodium channel. *Nature* (*London*) **378,** 730–733.

McDonald, F. J., Snyder, P. M., McCray, P. B. Jr., and Welsh, M. J. (1994). Cloning, expression, and tissue distribution of a human amiloride-sensitive Na^+ channel. *Am. J. Physiol.* **266,** L728–L734.

McDonald, F. J., Price, M. P., Snyder, P. M., and Welsh, M. J. (1995). Cloning and expression of the beta- and gamma-subunits of the human epithelial sodium channel. *Am. J. Physiol.* **268,** C1157–C1163.

Price, M. P., Snyder, P. M., and Welsh, M. J. (1996). Cloning and expression of a novel human brain Na^+ channel. *J. Biol. Chem.* **271,** 7879–7882.

Puoti, A., May, A., Canessa, C. M., Horisberger, J.-D., Schild, L., and Rossier, B. C. (1995). The highly selective low-conductance epithelial Na^+ channel of *Xenopus laevis* A6 kidney cells. *Am. J. Physiol.* **269,** C188–C197.

Renard, S., Lingueglia, E., Voilley, N., Lazdunski, M., and Barbry, P. (1994). Biochemical analysis of the membrane topology of the amiloride-sensitive Na^+ channel. *J. Biol. Chem.* **269,** 12981–12986.

Schild, L., Canessa, C. M., Shimkets, R. A., Gautschi, I., Lifton, R. P., and Rossier, B. C. (1995). A mutation in the epithelial sodium channel causing Liddle disease increases channel activity in the *Xenopus laevis* oocyte expression system. *Proc. Natl. Acad. Sci. U.S.A.* **92,** 5699–5703.

Schild, L., Schneeberger, E., Gautschi, I., and Firsov, D. (1997). Identification of amino acid residues in the α, β, and γ subunits of the epithelial sodium channel (ENaC) involved in amiloride block and ion permeation. *J. Gen. Physiol.* **109,** 15–26.

Shimkets, R. A., Warnock, D. G., Bositis, C. M., Nelson-Williams, C., Hansson, J. H., Schambelan, M., Gill, J. R., Jr., Ulick, S., Milora, R. V., Findling, J. W., Canessa, C. M., Rossier, B. C., and Lifton, R. P. (1994). Liddle's syndrome: Heritable human hypertension caused by mutations in the beta subunit of the epithelial sodium channel. *Cell* (*Cambridge, Mass.*) **79,** 407–414.

Snyder, P. M., McDonald, F. J., Stokes, J. B., and Welsh, M. J. (1994). Membrane topology of the amiloride-sensitive epithelial sodium channel. *J. Biol. Chem.* **269,** 24379–24383.

Snyder, P. M., Price, M. P., McDonald, F. J., Adams, C. M., Volk, K. A., Zeiher, B. G., Stokes, J. B., and Welsh, M. J. (1995). Mechanism by which Liddle's syndrome mutations increase activity of a human epithelial Na^+ channel. *Cell* (*Cambridge, Mass.*) **83,** 969–978.

Snyder, P. M., Cheng, C., Prince, L. S., Rogers, J. C., and Welsh, M. J. (1998). Electrophysiological and biochemical evidence that DEG/ENaC cation channels are composed of nine subunits. *J. Biol. Chem.* **273,** 681–684.

Strautnieks, S. S., Thompson, R. J., Gardiner, R. M., and Chung, E. (1996). A novel splice-site mutation in the γ subunit of the epithelial sodium channel gene in three pseudohypoaldosteronism type 1 families. *Nat. Genet.* **13,** 248–250.

Tavernarakis, N., and Driscoll, M. (1997). Molecular modeling of mechanotransduction in the nematode *Caenorhabditis elegans. Annu. Rev. Physiol.* **59,** 659–689.

Waldmann, R., Champigny, G., and Lazdunski, M. (1995). Functional degenerin-containing chimeras identify residues essential for amiloride-sensitive Na^+ channel function. *J. Biol. Chem.* **270,** 11735–11737.

Waldmann, R., Champigny, G., Bassilana, F., Heurteaux, C., and Lazdunski, M. (1997a). A proton-gated cation channel involved in acid-sensing. *Nature* (*London*) **386,** 173–177.

Waldmann, R., Bassilana, F., de Weille, J., Champigny, G., Heurteaux, C., and Lazdunski, M. (1997b). Molecular cloning of a non-inactivating proton-gated Na^+ channel specific for sensory neurons. *J. Biol. Chem.* **272,** 20975–20978.

CHAPTER 3

Subunit Stoichiometry of Heterooligomeric and Homooligomeric Epithelial Sodium Channels

Farhad Kosari,* Bakhram K. Berdiev,† Jinqing Li,* Shaohu Sheng,* Iskander Ismailov,† and Thomas R. Kleyman*

*Departments of Medicine and Physiology, University of Pennsylvania, and VA Medical Center, Philadelphia, Pennsylvania, 19104,
†Department of Physiology and Biophysics, University of Alabama at Birmingham, Birmingham, Alabama 35294

I. INTRODUCTION

Epithelial Na^+ channels (ENaC's) have a key role in the regulation of urinary Na^+ reabsorption, extracellular fluid volume homeostasis, and blood pressure, and are a major site of action of volume regulatory hormones (Benos *et al.,* 1995; Garty and Palmer, 1997; Verrey, 1995). ENaCs are composed of three homologous subunits, termed α, β, and γ (Canessa *et al.,* 1994). One of the fundamental questions pertaining to the structure of the Na^+ channel is the determination of its subunit stoichiometry. This question has recently attracted the attention of several laboratories, although conflicting results have been reported (Firsov *et al.,* 1998; Kosari *et al.,* 1998; Snyder *et al.,* 1998). To determine the subunit stoichiometry of ENaC, we used a biophysical assay utilizing mutant subunits that display

1063-5823/99 $30.00

significant differences in sensitivity to channel blockers from the wild-type channel (Kosari *et al.*, 1998). Our results suggest that ENaCs are a tetrameric channel, with an $\alpha_2\beta\gamma$ stoichiometry. The α-subunit itself is sufficient to form functional channels (Canessa *et al.*, 1993; Fuller *et al.*, 1995; Lingueglia *et al.*, 1993). Our results suggest that similar to heterooligomeric $\alpha\beta\gamma$ Na^+ channels, homooligomeric α-subunit Na^+ channels are also tetrameric in structure.

A biophysical approach, initially described by MacKinnon in 1991, was used to determine the subunit stoichiometry of $\alpha\beta\gamma$ Na^+ channels. As an example of this approach, consider the situation in which an ion channel is formed by a single polypeptide "X." The exact number of X subunits that form an individual ion channel, i.e., subunit stoichiometry of the channel, is to be determined. We assume that the channel has been cloned, can be functionally expressed in *Xenopus* oocytes, and is inhibited by the compound "Y." We also assume that a mutant channel has been identified that is insensitive to Y, and that the functional characteristics of the wild-type and mutant channel are otherwise similar. To determine the stoichiometry of this channel, equal amounts of wild-type and mutant X cRNAs are mixed and are injected into *Xenopus* oocytes. Channels composed of drug-sensitive and/or drug-insensitive subunits will be expressed at the plasma membrane, and the composition of these channels will depend on the stoichiometry. Two key assumptions regarding the assembly and blocking behavior of channels are (1) the assembly of wild-type and the mutant subunits is a random event, and (2) the presence of a single drug-sensitive (i.e., wild-type) subunit in a channel complex is sufficient to render the channel drug sensitive. At a sufficient inhibitor concentration to block wild-type channels, the remaining current is carried by fully mutant channels. If the channel has a subunit stoichiometry of one, there will be only two species of channel on the oocyte surface: wild type and mutant. In this instance, inhibitor-insensitive channels will carry 50% of the total current. The composition of channels increases in complexity as the stoichiometry increases. If two subunits participate in channel formation, three distinct species of channels will be expressed. One group of channels will have two wildtype subunits, a second group will have two mutant subunits, and finally there will be a third group of channels that contain one wild-type and one mutant subunit (i.e., a hybrid channel). Inhibitor-insensitive channels will carry only 25% of the total current. If the stoichiometry of the channel is three, there will be four species of channels on the oocyte surface, and the inhibitor-insensitive channels will transport 13% of the total current. As the subunit stoichiometry increases, the inhibitor-insensitive component of whole-cell Na^+ current decreases in a predictable manner. We have adopted this scheme to determine the subunit stoichiometry of ENaC.

II. STOICHIOMETRY OF HETEROOLIGOMERIC $\alpha\beta\gamma$ Na^+ CHANNELS

Schild and co-workers (1997) recently characterized selected mutations within conserved residues preceding the second transmembrane domains of ENaC subunits. Several of these mutants exhibit differences in sensitivities to channel blockers when compared with wild-type ENaCs. We generated corresponding mutations within subunits of mouse ENaC, including αS583C, βG525C, and γG542C, in order to determine ENaC subunit stoichiometry. Table I illustrates the functional properties of wild-type Na^+ channels, as well as Na^+ channels containing one of the mutant subunits. α,βG525C,γmENaC, and α,β,γG542CmENaC were inhibited by amiloride with IC's$_{50}$ approximately 500-fold greater than the IC_{50} of wild-type mENaC (Table I). In contrast, αS583C,β,γmENaC was blocked by amiloride with an IC_{50} that was sufficiently close to the IC_{50} of wild-type mENaC to preclude the use of amiloride in the determination of α-subunit stoichiometry (Table I). Other than differences in amiloride sensitivity, all three mutant channels displayed functional characteristics that are hallmarks of ENaC: long opening and closing times of the order of seconds when examined under a cell attached patch-clamp mode and a selectivity of Na^+ over K^+ by more than 20-fold. In addition, single-channel conductances of the three mutant channels were fairly similar to that of the wild-type ENaC (Table I). These observations suggest that the assembly of Na^+ channel subunits in oocytes co-injected with both a wild type and the corresponding mutant ENaC cRNAs occurs randomly as dictated by the binomial distribution, as has been confirmed by Firsov and co-workers (1998).

III. α-SUBUNIT STOICHIOMETRY

Substitution of the serine residue with cysteine in αS583C,β,γENaC rendered the mutant channel susceptible to a rapid, irreversible inhibi-

TABLE I

Properties of Wild-Type and Mutant mENaCs

mENaC species	K_i amiloride	Hill coefficient	Single channel conductance (pS)
Wild type	92.8 nM (n = 7)	0.99	4.7
αS583C,β,γ	600 nM (n = 2)	n.d.[a]	4.7
α,βG525C,γ	73 μM (n = 8–10)	0.75	2.8
α,β,γG542C	43μM (n = 6–7)	0.84	3.2

[a] n.d., not determined.

Note Reprinted with permission from Kosari *et al.,* 1998, The Journal of Biological Chemistry.

tion by low concentrations (0.5 mM) of the sulfhydryl reactive reagent 2-(aminoethyl)methanethiosulphonate (MTSEA). In contrast, wild-type ENaCs were largely insensitive to MTSEA (Fig. 1). This differential sensitivity to MTSEA provided an experimental approach to determine α-subunit stoichiometry.

Whole-cell current responses to MTSEA of oocytes injected with 1:1, 2:1, and 4:1 ratios of α,β,γmENaC: αS583C,β,γmENaC cRNAs are illustrated in Fig. 2, as are the predicted responses for an α-subunit stoichiometry of 1, 2, or 3. The results were consistent with the predicted response for an α-subunit stoichiometry of 2. In generating the predicted responses for α-subunit stoichiometry, it was assumed that the hybrid channels (i.e., channels containing both wild-type and mutant α-subunits) were blocked by MTSEA. However, if MTSEA induces a partial block of the hybrid channels, the measured response could be consistent with an α-subunit

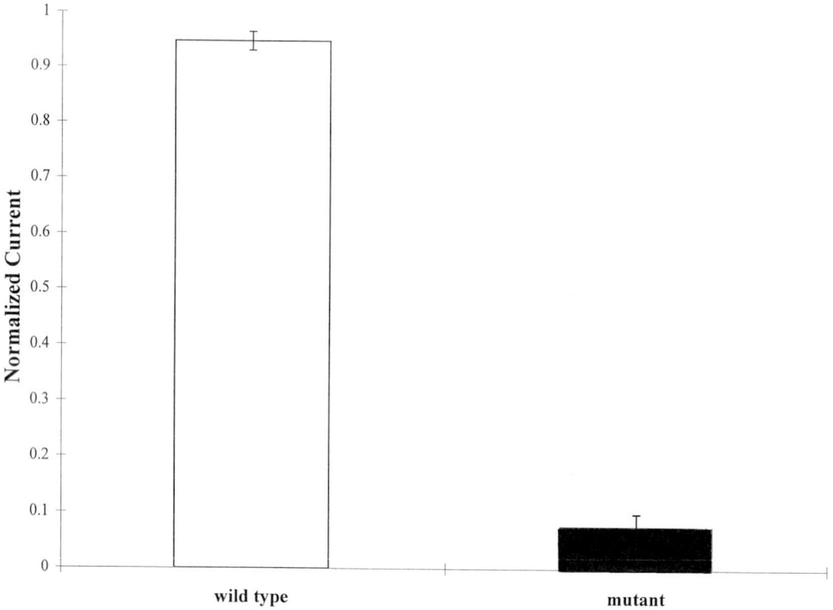

FIGURE 1 Response of α,β,γmENaC and αS583C,β,γmENaC to MTSEA. Oocytes were injected with wild-type α,β,γmENaC or with the mutant αS583C,β,γmENaC cRNAs. Normalized amiloride-sensitive currents are shown. Wild-type channels (open bar) were insensitive to 0.5 mM MTSEA, whereas αS583C,β,γmENaC mutant (closed bar) was largely blocked by the drug. Reprinted with permission from Kosari *et al.*, 1998, The Journal of Biological Chemistry.

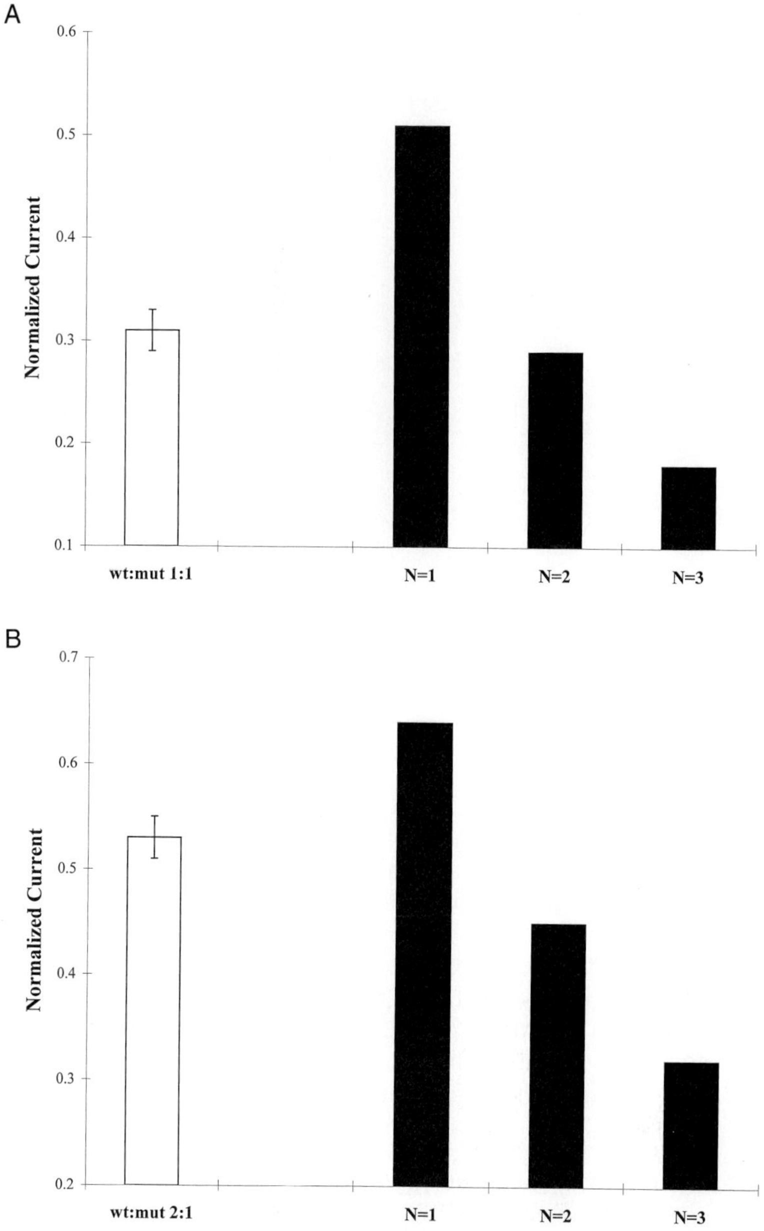

FIGURE 2 Stoichiometry of $\alpha\beta\gamma$ Na^{+} channels: α-subunit stoichiometry. Oocytes were co-injected with a 1:1 (A), 2:1 (B), or 4:1 (C) ratio of wild-type α,β,γmENaC and mutant αS583C,β,γmENaC cRNAs (open bar). Normalized amiloride-sensitive currents in response to MTSEA are shown. The right sides of the panels illustrate the predicted current response to MTSEA, assuming an α-subunit stoichiometry (N) of 1, 2, or 3. The responses to MTSEA are most consistent with an α-subunit stoichiometry of 2. Figures 2A and 2C are reprinted with permission from Kosari *et al.,* 1998, The Journal of Biological Chemistry.

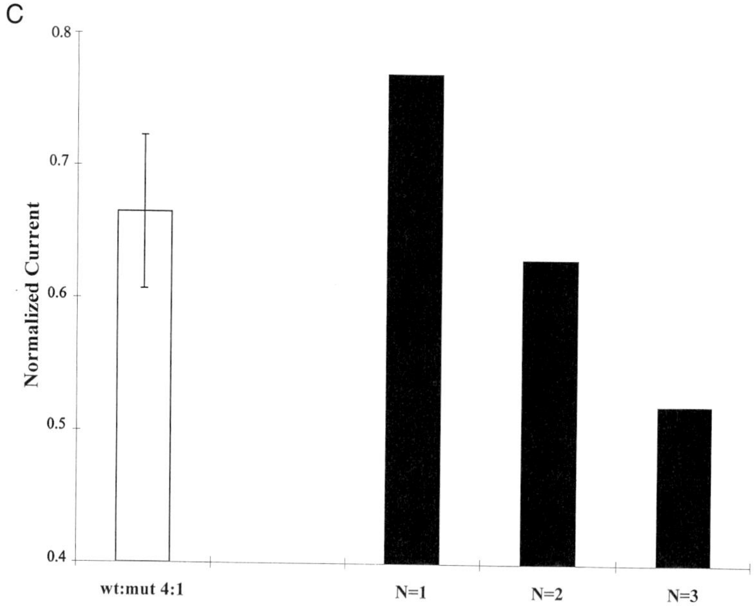

FIGURE 2 *(Continued)*

stoichiometry of greater than 2. We used cell-attached patch clamp to examine whether MTSEA-induced block of channels composed of a wild-type and mutant α-subunits was complete or partial (Fig. 3). Wild-type α,β,γ Na^+ channels detected in the absence of MTSEA were also observed following MTSEA infusion into the patch pipette. In contrast, αS583C,β,γ Na^+ channels detected in the absence of MTSEA were completely blocked following MTSEA infusion. Of the 28 Na^+ channels observed in oocytes injected with a 1:1 mixture of α,β,γmENaC and αS583C,β,γmENaC cRNAs, no prolonged subconductance states, suggestive of a partial Na^+ channel block, were observed following MTSEA infusion.

IV. β-SUBUNIT AND γ-SUBUNIT STOICHIOMETRY

The differential sensitivities of wild-type and mutant mENaCs to amiloride were used to determine β-subunit and γ-subunit stoichiometry. The whole-cell current responses to amiloride of oocytes injected with a 1:1 or a 1:4 mixture of wild-type α,β,γmENaC and mutant α,βG525C,γmENaC cRNA's are illustrated in Fig. 4A. The whole-cell current responses to

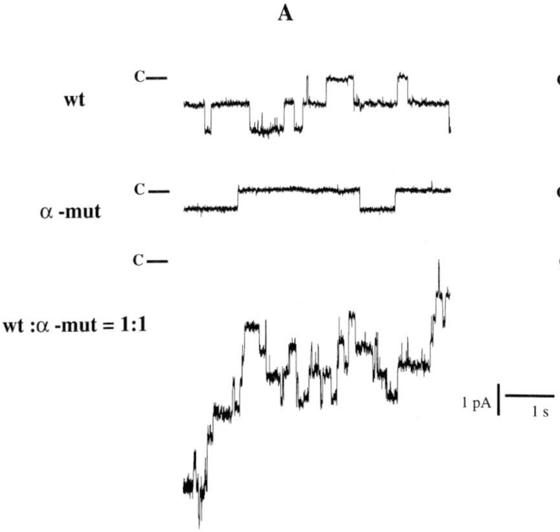

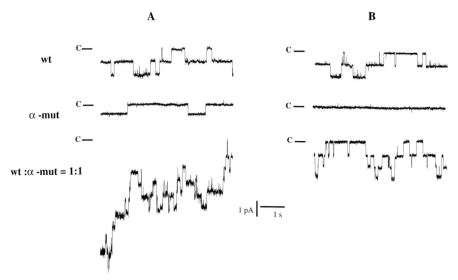

FIGURE 3 Single-channel analyses of MTSEA-induced block of mutant (i.e., αS583C, β, γ) mENaCs. Single-channel currents were analyzed by cell attached patch clamp in oocytes injected with either (1) wild type α,β,γmENaC cRNA (wt), (2) αS583C,β,γmENaC cRNA (α-mut), (3) or oocytes co-injected with a 1:1 ratio of α,β,γmENaC and αS583C,β,γmENaC cRNAs (wt: α-mut = 1:1). Closed state is as indicated (C). Representative tracings are illustrated prior to MTSEA (A), and following infusion of 5 mM MTSEA into the patch pipette (B). No prolonged subconductance state, suggestive of a partial block of Na^+ channels, was observed in oocytes injected with a 1:1 ratio of α,β,γmENaC and αS583C,β,γmENaC cRNAs. Reprinted with permission from Kosari *et al.*, 1998, The Journal of Biological Chemistry.

amiloride of oocytes injected with a 1:1 or a 1:4 mixture of wild-type α,β,γmENaC and mutant α,β,γG542CmENaC cRNA's are illustrated in Fig. 4B. The predicted responses to amiloride for β-subunit and γ-subunit stoichiometries (N) of 1 and 2 are also depicted in Fig. 4. In generating the predicted amiloride dose–response curves for $N = 2$, differences in single-channel conductances between the mutant and the wild-type channels were taken into account. The values for the K_i for amiloride, Hill coefficient, and single-channel conductance for the hybrid channels (i.e., channels that have both a wild-type and a mutant β- or γ-subunit) were obtained by minimizing the χ^2 error of the predicted response for stoichiometry of 2 to the experimental data. Experimental data were most consistent with a β-subunit and a γ-subunit stoichiometry of 1. In summary, analyses of inhibition of wild-type and mutant Na^+ channels suggest that ENaCs have a quaternary structure and are composed of two α-subunits, one β-subunit, and one γ-subunit.

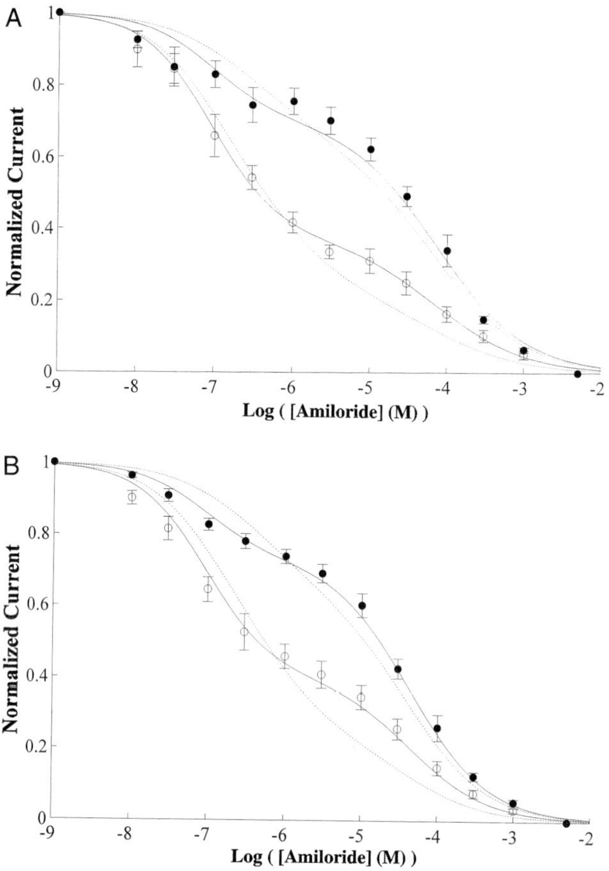

FIGURE 4 Stoichiometry of $\alpha\beta\gamma$ Na^+ channels: β-subunit and γ-subunit stoichiometry. (A) The amiloride dose–response curve derived from oocytes co-injected with a 1 : 1 (open circles) or a 1 : 4 mixture (closed circles) of wild-type (i.e., α,β,γ) and mutant (i.e., α,βG525C,γ) mENaC cRNA's (Fig. 4A). (B) The amiloride dose response derived from oocytes co-injected with a 1 : 1 (open circles) or a 1 : 4 mixture (closed circles) of wild-type (i.e., α,β,γ) and mutant (i.e., α,β,γG542C) mENaC cRNAs. Normalized amiloride-sensitive currents are shown. Solid lines illustrate the predicted amiloride dose response assuming a subunit stoichiometry of 1. Dashed lines illustrate the predicted amiloride dose response assuming a subunit stoichiometry of 2. Reprinted with permission from Kosari *et al.*, 1998, The Journal of Biological Chemistry.

V. STOICHIOMETRY OF HOMOOLIGOMERIC α-SUBUNIT Na^+ CHANNELS

The α-subunit alone is sufficient to induce expression of Na^+ currents in *Xenopus* oocytes that are inhibited by amiloride with an apparent inhibitory constant (K_i) nearly identical to that observed with expression of α,β,γENaC (Canessa *et al.*, 1993, 1994). However, current levels observed with expression of αENaC in oocytes are approximately 100-fold lower than current levels observed in oocytes co-expressing α,β,γENaC. To circumvent the problem of low macroscopic Na^+ currents in oocytes expressing αENaC (Canessa *et al.*, 1994), planar lipid bilayers were used to examine the properties of wild-type and mutant homooligomeric α-subunit Na^+ channels (Fuller *et al.*, 1995; Ismailov *et al.*, 1996). *In vitro* translated α-subunits formed Na^+ selective channels in planar lipid bilayers which were blocked by amiloride with a K_i of 0.2 μM (Fig. 5) (Ismailov *et al.*, 1997).

We previously identified a 6-amino-acid residue tract within the extracellular domain of the α-subunit of ENaC that participates in amiloride binding (Ismailov *et al.*, 1997; Kieber-Emmons *et al.*, 1995; Lin *et al.*, 1994). The identification of this 6-residue tract was based on defining the interaction of an anti-amiloride antibody with amiloride and consequent sequence homology between the anti-amiloride antibody and a residue tract within the extracellular domain of the αENaC. Deletion of this tract (αrENaC Δ278–283) resulted in a loss of sensitivity of homooligomeric α-subunit Na^+ channels to submicromolar concentrations of amiloride (K_i = 27 μM, Fig. 5), although the single-channel condutance and cation selectivity of the wild-type and mutant αrENaC were identical. This amiloride-binding site within the extracellular may interact with the substituted pyrazine ring moiety of amiloride and stabilize the binding of amiloride to the channel (Ismailov *et al.*, 1997; Kieber-Emmons *et al.*, 1995; Lin *et al.*, 1994).

A biophysical approach was used to determine the stoichiometry of α-subunit Na^+ channels. A mixture of equal amounts of *in vitro* translated wild-type αrENaC and αrENaC Δ278–283 was used to reconstitute Na^+ channels in planar lipid bilayers. These reconstituted channels had similar single-channel conductances, ion selectivities, and kinetics, but differed in their sensitivities to amiloride (Ismailov *et al.*, 1997). Five populations of channels with distinct sensitivities to amiloride were identified. Typical current records and responses to amiloride representative of each of these populations are illustrated in Fig. 5B. Amiloride dose–response curves were fit to the first-order Michaelis–Menten equation and are shown in Fig. 5C. The apparent inhibitory constants (K_i) determined for each of the 68 channels that were analyzed are shown in Fig. 5D. The distributions of K_i values are consistent with a homooligomeric α-subunit Na^+ channel stoichiometry of 4 (Table II).

In summary, analyses of inhibition of wild-type and mutant mENaCs suggest that $\alpha\beta\gamma$ Na^+ channels are composed of two α-subunits, one β-subunit, and one γ-subunit; and that α-subunit Na^+ channels are composed of four α-subunits. These channels have a quaternary structure, similar to

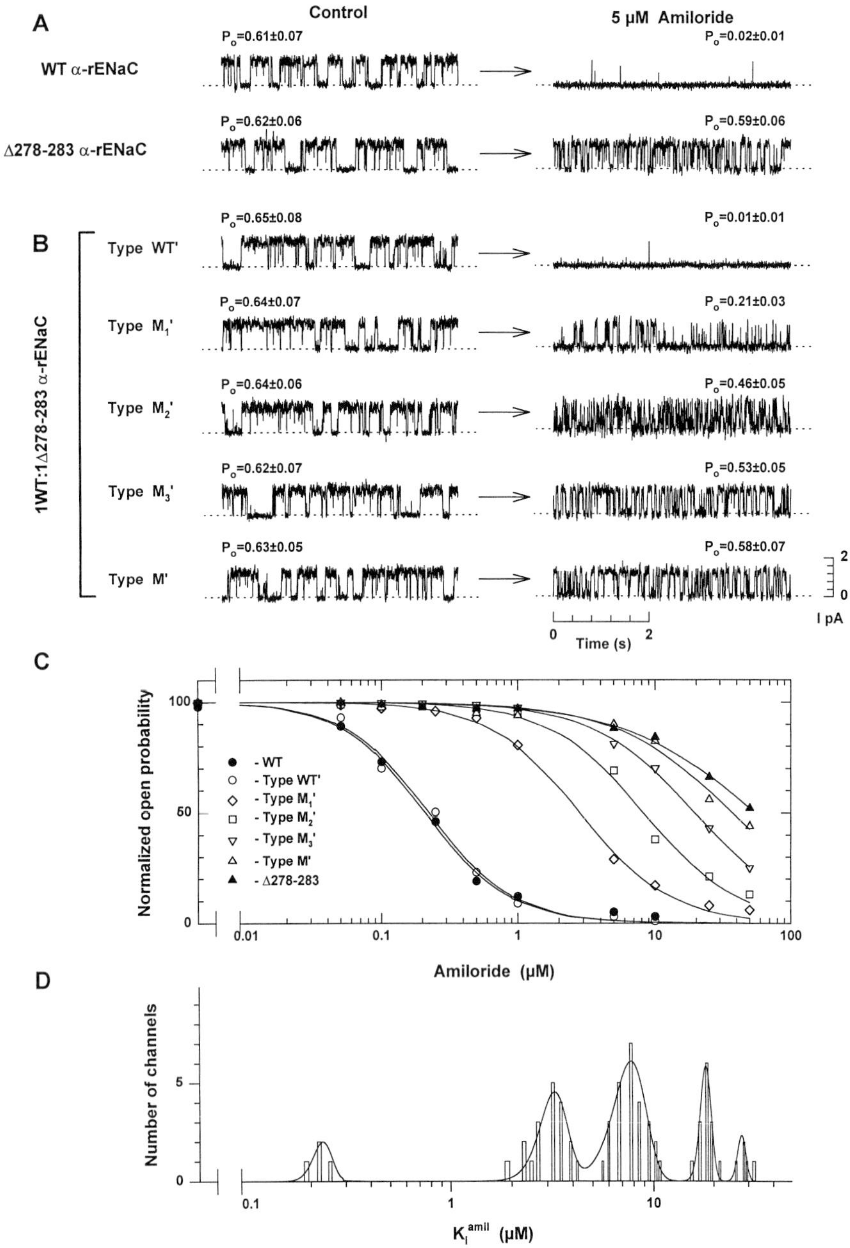
A
Control
5 μM Amiloride
WT α-rENaC
Po=0.61±0.07
Po=0.02±0.01
Δ278-283 α-rENaC
Po=0.62±0.06
Po=0.59±0.06
B
1WT:1Δ278-283 α-rENaC
Type WT'
Po=0.65±0.08
Po=0.01±0.01
Type M1'
Po=0.64±0.07
Po=0.21±0.03
Type M2'
Po=0.64±0.06
Po=0.46±0.05
Type M3'
Po=0.62±0.07
Po=0.53±0.05
Type M'
Po=0.63±0.05
Po=0.58±0.07
2
0
I pA
0
2
Time (s)
C
Normalized open probability
100
50
0
- WT
- Type WT'
- Type M1'
- Type M2'
- Type M3'
- Type M'
- Δ278-283
0.01
0.1
1
10
100
Amiloride (μM)
D
Number of channels
5
0
0.1
1
10
Ki amil (μM)

TABLE II
Distribution of Channels Formed from a 1:1 Mixture of *In Vitro* Translated Wild-Type αrENaC and αrENac Δ278–283

Channel type (n = 68)	Observed (%)	Expected (%) (stoichiometry of 4)
Type I (wild-type αrENaC)	6	6.25
Type II	28	25
Type III	38	37.5
Type IV	22	25
Type V (αrENaC Δ278–283)	6	6.25

that reported for Kv, Kir, and voltage-gated Na^+ and Ca^{2+} channels (Na^+ and Ca^{2+} channels are composed of a large polypeptide with 4-fold internal symmetry) (Catterall, 1993; Liman *et al.,* 1992; MacKinnon, 1991; Yang *et al.,* 1995). Our results are in agreement with work recently published by Firsov and co-workers (1998) and Coscoy and co-workers (1998) demonstrating that ENaC and the related FMRFamide-activated Na^+ channel are tetramers.

Acknowledgments

This work was supported by grants from the Department of Veterans Affairs, the Cystic Fibrosis Foundation, and the National Institutes of Health (DK51391, HL07027).

FIGURE 5 Alpha subunit Na^+ channel stoichiometry. Wild-type αrENaC and αrENaC Δ278–283 were individually *in vitro* translated and reconstituted into proteoliposomes. (A) Typical records of channels following reconstitution and responses to 5 μM amiloride. Alternatively, individually *in vitro* translated wild-type αrENaC and αrENaC Δ278–283 were mixed at a 1:1 ratio and then reconstituted into proteoliposomes in the presence of 50 μM DTT. (B) Typical records of channels noted following Na^+ channel reconstitution. Five distinct responses of these channels to 5 μM of amiloride were observed: channels identical to wild-type αrENaC (type WT'), channels identical to αrENaC Δ278–283 (type M'), and channel types M1 through M3 that have an intermediate sensitivity to amiloride. (C) Amiloride dose–response curves were obtained for each population of channels. Solid lines in the amiloride dose–response graph represent best fits of the experimental data points to the first order Michaelis–Menten equation. (D) The distribution of channel phenotypes as a bar graph. The solid line is a fifth-order Gaussian fit of the histogram. Reprinted with permission from Berdiev *et al.,* 1998, The Biophysical Journal.

References

Benos, D. J., Awayda, M. S., Ismailov, I. I., and Johnson, J. P. (1995). Structure and function of amiloride-sensitive Na^+ channels. *J. Membr. Biol.* **143,** 1–18.

Berdiev, B. K., Karlson, K. H., Loffina, D., Halpin, P., Stanton, B. A., Kleyman, T. R., and Ismailov, I. I. (1998). Subunit stoichiometry of a core conductive element in a cloned epithelial amiloride-sensitive Na^+ channel. *Biophys. J.* **75,** 2292–2301.

Canessa, C. M., Horisberger, J.-D., and Rossier, B. C. (1993). Epithelial sodium channel related to proteins involved in neurodegeneration. *Nature (London)* **361,** 467–470.

Canessa, C. M., Schild, L., Buell, G., Thorens, B., Gautschl, I., Horisberger, J.-D., Rossier, B. C. (1994). Amiloride-sensitive epithelial Na^+ channel is made of three homologous subunits. *Nature (London)* **36,** 463–467.

Catterall, W. A. (1993). Structure and function of voltage-gated ion channels. *Trends Neurosci.* **16,** 500–506.

Coscoy, S., Lingueglia, E., Lazdunski, M., and Barbry, P. (1998). The Phe-Met-Arg-Phe-amide-activated sodium channel is a tetramer. *J. Biol. Chem.* **273,** 8317–8322.

Firsov, D., Gautschi, I., Merillat, A. M., Rossier, B. C., and Schild, L. (1998). The heterotetrameric architecture of the epithelial sodium channel (ENaC). *EMBO J.* **17,** 344–352.

Fuller, C. M., Awayda, M. S., Arrate, M. P., Bradford, A. L., Morris, R. G., Canessa, C. M., Rossier, B. C., and Benos, D. J. (1995). Cloning of a bovine renal epithelial Na^+ channel subunit. *Am. J. Physiol.* **269,** C641–C654.

Garty, H., and Palmer, L. G. (1997). Epithelial sodium channels: Structure, function, and regulation. *Physiol. Rev.* **77,** 359–396.

Ismailov, I. I., Awayda, M. S., Berdiev, B. K., Bubien, J. K., Lucas, J. E., Fuller, C. M., and Benos, D. J. (1996). Triple-barrel organization of ENaC, a cloned epithelial Na^+ channel. *J. Biol. Chem.* **271,** 807–816.

Ismailov, I. I., Kieber-Emmons, T., Lin, C., Berdiev, B. K., Schlonsky, V. G., Patton, H. K., Fuller, C. M., Worrell, R., Zuckerman, J. B., Sun, W., Eaton, D. C., Benos, D. J., and Kleyman, T. R. (1997). Identification of an amiloride binding domain within the alpha subunit of the epithelial Na^+ channel. *J. Biol. Chem.* **272,** 21075–21083.

Kieber-Emmons, T., Lin, C., Prammer, K., Villalobos, A., Kosari, F., and Kleyman, T. R. (1995). Defining topological similarities among ion transport proteins with anti-amiloride antibodies. *Kidney Int.* **48,** 956–964.

Kosari, F., Sheng, S., Li, J., Mak, D. D., Foskett, J. K., and Kleyman, T. R. (1998). Subunit stoichiometry of the epithelial sodium channel. *J. Biol. Chem.* **273,** 13469–13474.

Liman, E. R., Tytgat, J., and Hess, P. (1992). Subunit stoichiometry of a mammalian K^+ channel determined by construction of multimeric cDNAs. *Neuron* **9,** 861–871.

Lin, C., Kieber-Emmons, T., Villalobos, A. P., Foster, M. H. Wahlgren, C., and Kleyman, T. R. (1994). Topology of an amiloride binding protein. *J. Biol. Chem.* **269,** 2805–2813.

Lingueglia, E., Voilley, N., Waldmann, R., Lazdunski, M., and Barbry, P. (1993). Expression cloning of an epithelial amiloride-sensitive Na^+ channel. *FEBS Lett.* **318,** 95–99.

MacKinnon, R. (1991). Determination of the subunit stoichiometry of a voltage activated potassium channel. *Nature (London)* **350,** 232–235.

Schild, L., Schneeberger, E., Gautschi, I., and Firsov, D. (1997). Identification of amino acid residues in the α, β, and γ subunits of the epithelial sodium channel (ENaC) involved in amiloride block and ion permeation. *J. Gen. Physiol.* **109,** 15–26.

Snyder, P. M., Cheng, C., Prince, L. S., Rogers, J. C., and Welsh, J. M. (1998). Electrophysiological and biochemical evidence that DEG/ENaC catio channels are composed of nine subunits. *J. Biol. Chem.* **273,** 681–684.

Verrey, F. (1995). Transcriptional control of sodium transport in tight epithelia by adrenal steroids. *J. Membr. Biol.* **143,** 93–110.

Yang, J., Jan, Y. N., and Jan, L. Y. (1995). Determination of the subunit stoichiometry of an inwardly rectifying potassium channel. *Neuron* **15,** 1441–1447.

PART II

Regulation of Sodium Channels

CHAPTER 4

Cell-Specific Expression of ENaC and Its Regulation by Aldosterone and Vasopressin in Kidney and Colon

N. Farman,* S. Djelidi,* M. Brouard,* B. Escoubet,† M. Blot-Chabaud,* and J. P. Bonvalet*
INSERM U478* and U426,† IFR "Cellules Epithéliales," Faculté de Médecine X, Bichat, BP 416, 75870 Paris Cedex 18, France

I. INTRODUCTION

The amiloride-sensitive sodium channel is the rate-limiting step for sodium reabsorption in epithelia with high electrical resistance (Garty and Palmer, 1997; Voilley *et al.,* 1997). Functional studies have established that its expression is in the apical membrane of some epithelial cells, allowing sodium entry into the cell and subsequent sodium extrusion at the basolateral side of the cell, via the Na^+,K^+-ATPase. With the molecular characterization (Canessa *et al.,* 1993, 1994; Lingueglia *et al.,* 1993; McDonald *et al.,* 1995; Voilley *et al.,* 1994, 1995) of three subunits (α, β and γ) of ENaC (for epithelial sodium channel), tools became available to examine in detail its expression and regulation at the mRNA and at the protein level in various tissues and cell types.

1063-5823/99 $30.00

II. CELL-SPECIFIC EXPRESSION

Initial studies (Duc *et al.*, 1994) were focused on epithelial cells in which functional channels are known to exert a physiological function. They include mainly the terminal portions of the nephron, the distal colon, and the ducts of sweat and salivary glands, which were examined at the cellular level by *in situ* hybridization and immunohistochemistry (Duc *et al.*, 1994). Figure 1 illustrates the expression of the mRNA encoding for the β subunit of ENaC in a section of rat kidney cortex of rat fed a standard diet. It appears clearly that ENaC expression is restricted to the renal distal tubule and collecting duct. High signal was also present in surface cells of distal colon epithelium. In contrast, no ENaC mRNA was detected in the renal proximal tubule or loop of Henle or in the crypt cells of the distal colon. The pattern of expression of α and γ subunits was similar to that of the β subunit in a series of epithelia with high resistance, which exhibit hormone-regulated sodium transport. Similar cell specificity was found when ENaC expression was examined at the protein level, using antibodies against fusion proteins between part of ENaC subunits and glutathione *S*-transferase. Moreover, immunolocalization made it possible to show that each of the subunits is indeed expressed at the apical membrane of the collecting duct

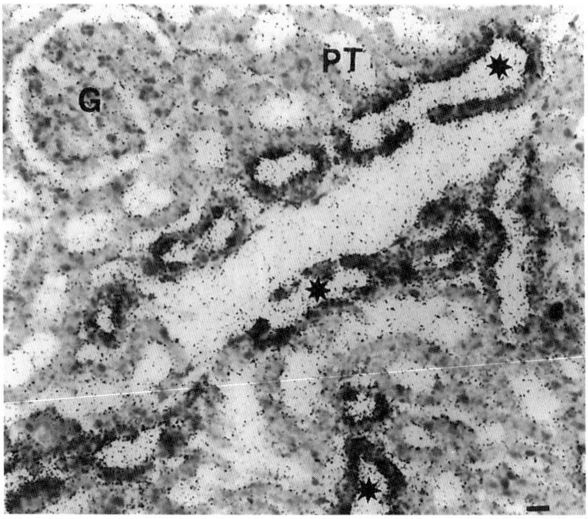

FIGURE 1 Renal expression of the β subunit of ENaC, at the mRNA level. *In situ* hybridization was performed with a β-subunit specific ^{35}S-cRNA probe on a rat kidney cortex section. The signal (silver grains) is over distal tubules and cortical collecting ducts (*); glomeruli (G) and proximal tubules (PT) exhibit background labeling. Bar, 10 μm.

principal cells. Table I summarizes the expression pattern of rat ENaC subunits (Duc *et al.*, 1994) in kidney, along the digestive tract and in sweat and salivary glands. ENaC expression appears to be clearly restricted to specific cell types, which reabsorb sodium; in all these cell types, we have found all three subunits, at comparable levels, although perhaps with some prevalence of α-subunit expression. It should also be noted that α, β and γ mRNAs were found in rats fed a normal diet, i.e., in the absence of stimulation of aldosterone secretion. This is at variance with results obtained by others (Renard *et al.*, 1995), who claimed that a low-sodium diet was necessary to detect the transcripts encoding for at least some subunits; this discrepancy may be due to the different sodium contents of so-called "normal" diets. On the whole, there is a general agreement on the expression of ENaC in the cell types listed above. It is also true that a very small number of transcripts may be undetectable by *in situ* hybridization, while PCR allows detection of them, for example, in the inner medullary collecting duct (Ono *et al.*, 1997; Volk *et al.*, 1995).

TABLE I

Cell-Specific Expression of ENaC (mRNA and Protein) in Epithelial Cells (Kidney, Digestive Tract, and Excretory Glands)

Coordinate expression of α, β, and γ ENaC (mRNA and/or protein) in	
Kidney	Distal convoluted tubule
	Cortical collecting duct
	Outer medullary collecting duct
Digestive tract	Surface cells of distal colon
Salivary glands	Mucous tubules
	Striated ducts
Sweat glands	Excretory ducts
No expression of α, β, and γ ENaC in (i.e., absent or very low levels)	
Kidney	Glomerulus
	Proximal tubule (convoluted, straight)
	Thin limbs of Henle's loop
	Thick ascending limb (medullary and cortical)
	Inner medullary collecting duct
Digestive tract	Jejunum
	Ileon
	Proximal colon
	Crypts of distal colon
Salivary glands	Acini
Sweat glands	Secretory coil
Liver	Hepatocytes

We have also observed ENaC expression in a series of epithelial cells of the respiratory tract (Farman *et al.*, 1997). The mRNAs encoding for ENaC subunits were found in rat nasal and tracheal surface epithelium, in nasal and tracheal gland ducts and acini, in bronchiolar epithelium, and in a minority of alveolar cells, presumably type II pneumocytes. This pattern of expression is discussed in more detail by C. Talbot in this volume, as is a comparison of ENaC expression along the mouse and human respiratory tract. In addition to this specific cellular localization, we emphasize an apparent difference in stochiometry of expression of the subunits in some respiratory cells compared to tight epithelia of the distal nephron and colon. Indeed, while all three subunit mRNAs appear to be coexpressed at comparable levels in these latter tissues, this was not the case in tracheal and nasal epithelia or in alveolar cells: in these cells very low (or undetectable) levels of β-subunit transcripts were found (using the same cRNA probes as in kidney and colon). We have checked that this was not due to expression of a truncated β transcript, as reported in human respiratory tissues (Voilley *et al.*, 1995) : the absence of the β signal was confirmed with a 5′-specific probe. The cell types listed in Table II appear to express much less (or no) β transcripts than α or γ. Whether these findings reflect actual changes in the stochiometry of expression of the sodium channel is not known at present. However, this expression pattern may be important for its function (McNicholas and Canessa, 1997). For example, cells with less (or no) β subunit may have altered stability of the channel in the membrane; alternatively, interaction of an incomplete sodium channel with other ion channels or transporters (such as CFTR) may need to be reevaluated (Stutts *et al.*, 1995).

TABLE II

ENaC mRNA Expression along the Rat Respiratory Tract

Cell types that coexpress all three subunits (at apparently equivalent levels)
Bronchiolar epithelium
Nasal gland ducts
Cell types with low (or undetectable) β transcripts
Nasal epithelium (ciliated or nonciliated)
Nasal gland acini
Tracheal epithelium
Tracheal gland acini and ducts
Alveolar cells (type II pneumocytes)

We were surprised to observe expression of ENaC subunit mRNAs in nontransporting epithelial cells of the skin, i.e., in keratinocytes (Roudier-Pujol *et al.,* 1996). Indeed, in keratinocytes of the sole skin, as well as in those of the hair follicle, there was a clear signal with both α and γ probes, while the β signal was very low (as in respiratory tissues). In addition, the transcripts encoding for each subunit were expressed at higher levels in large suprabasal differentiated epidermal keratinocytes than in small basal undifferentiated ones. This finding was unexpected, because mammalian epidermal keratinocytes do not transport sodium. Sodium channel expression in keratinocytes may be linked to other functions; in particular, because these cells are not polarized, one can expect that sodium channel activity could result in cell swelling, i.e., could modify cell volume. It should be noted that ENaC belongs to a large gene family, including the *Caenorhabditis elegans* genes *mec-4, deg-1* and *mec-10,* which encode for products which are involved in sensory touch transduction (Driscoll and Chalfie, 1991). Mutations in these genes result in touch-insensitive phenotypes, with neuronal swelling preceding cell death. Whether sodium channel expression is related to keratinocyte differentiation and/or to the cell size increase observed in keratinocytes that undergo terminal differentiation is not known. Human epidermal keratinocytes in primary culture appear to be an interesting model for such purpose. This approach is now used in the laboratory, and preliminary results show an actual expression of ENaC in these cells.

The esophagal mucosa, another tissue with stratified multilayer epithelium with analogies to keratinocytes, also expresses sodium channel transcripts, without any known function.

III. REGULATION OF ENaC EXPRESSION BY ALDOSTERONE AND VASOPRESSIN

Final adjustements of sodium reabsorption in the kidney (Breyer and Ando, 1994; Schafer, 1994; Verrey, 1995) are tightly regulated by both corticosteroid hormones (aldosterone and glucocorticoids) and arginine-vasopressin (AVP). Corticosteroid hormones act in their target cells via binding to their receptors (mineralocorticoid receptor, MR, or glucocorticoid receptor, GR); the MR or GR acts as ligand-dependent transcription factors to regulate gene expression (Fig. 2). This leads to delayed (1–2 h) regulation of sodium transport (Verrey, 1995). The molecular determinants of transepithelial sodium reabsorption are (i) ENaC , located in the apical membrane, controlling the limiting step for transport from the tubular lumen, and (ii) the Na^+,K^+-ATPase, in the basolateral membrane, which

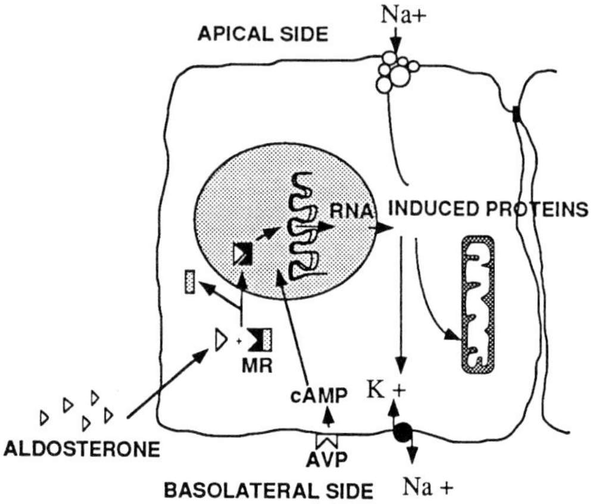

FIGURE 2 Schematic representation of a principal cell of the collecting duct. Transcriptional regulation pathways by aldosterone and arginine-vasopressin (AVP) in a principal cell of the collecting duct. MR, mineralocorticoid receptor.

extrudes sodium from the cell to extracellular space. While it is clear that the activity of these transporters depends on aldosterone, it remains to be established whether they are primary induced genes or whether their biosynthesis (and their activity) depend on unidentified primary induced proteins. In contrast with this long-term regulation of sodium transport, AVP stimulates rapidly sodium (and water) reabsorption, through binding to membrane receptors, production of cAMP, and a cascade of events leading to activation of prexisting sodium channels (Breyer and Ando, 1994) (Fig. 2).

We have evaluated the *in vivo* role of aldosterone in the control of ENaC subunit mRNAs in two target tissues, the kidney and the colon. We have also examined whether AVP could influence the sodium channel subunits transcription rate, in addition to short-time regulation of ENaC activity, in a renal cortical collecting duct cell line.

A. Regulation of ENaC Subunit Transcripts by Aldosterone

Levels of endogenous aldosterone may be manipulated in rats by adrenalectomy and aldosterone infusion via minipumps to restore a physiological concentration of hormone. We have measured the level of mRNA encoding for each subunit, in the kidney and the distal colon, by an RNase protection assay (Escoubet *et al.,* 1997). Adrenalectomy (adx) results in dissociated effects on ENaC subunits, which vary with the duration of hormonal deprivation, as shown in Fig. 3. After 7 days of adx, α transcripts are moderately downregulated in the kidney ; this downregulation appears to be "spontaneously" corrected, at least partially, 12 days after adx, probably due to compensatory endogenous regulations. In contrast, no major alterations of β or γ subunit mRNAs were apparent in the kidney of adx rats. In the distal colon, γ transcripts levels are transiently downregulated to very low levels after 7 days adx. The level of β transcripts is also decreased after adx, whereas α levels are unaltered. Thus, the absence of corticosteroid hormones affects α ENaC subunit transcripts in the kidney, and β and γ in the distal colon. Infusion of low physiological doses of aldosterone reversed the observed changes in both tissues (Fig. 4). It should be noted that a major effect of aldosterone is apparent only after a long delay of corticosteroid withdrawal (12 days adx) in the colon, leading to a 3- to 4-fold increase in β and γ transcripts. In contrast in the kidney, α transcripts

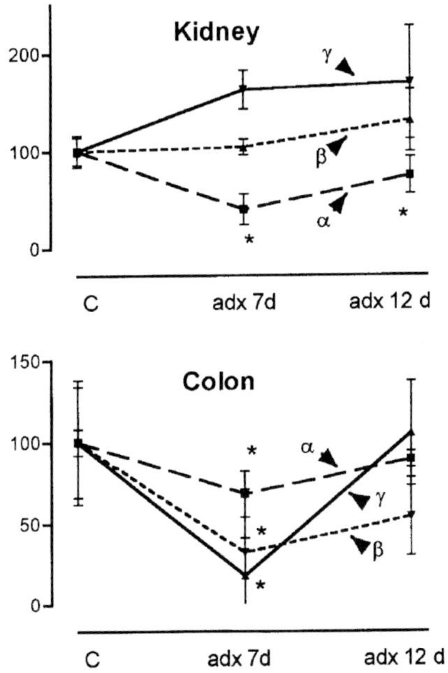

FIGURE 3 Effects of adrenalectomy (adx) on ENaC subunit transcripts in rat kidney and colon. ENaC subunit mRNAs were measured by RNase protection assay in the kidney and colon from rats that were adrenalectomized (adx) for 7 days (7d) or 12 days (12d). Values are normalized as percentages of those observed in control rats. In the kidney, only α-subunit transcripts were significantly (*) downregulated by adx. In the colon, all three subunit transcripts were transiently downregulated by 7 days adx; this effect was much more marked for β and γ subunits than for α.

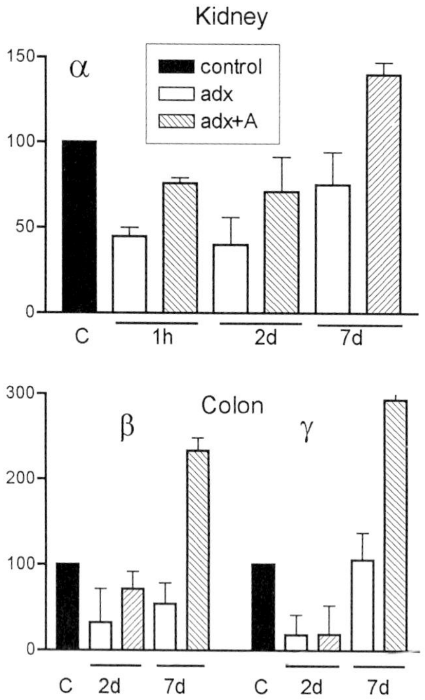

FIGURE 4 Effect of aldosterone infusion on the level of ENaC subunit transcripts in rat kidney and colon. ENaC subunit mRNAs are expressed as percentages of control rats (black bar). Each series of aldosterone infused rats is compared to their respective adrenalectomized (adx) rats. In the kidney, aldosterone upregulated α transcripts as early as 1 h after aldosterone administration, while β and γ transcripts were unaltered. In the colon, an aldosterone upregulation of β and γ transcripts was apparent only after long-term treatment (7 days); α transcripts were not modified significantly.

are increased by a factor of 2 as early as 1 h after aldosterone administration. These results are in general agreement with other reports from the literature (Asher *et al.*, 1996; Volk *et al.*, 1995; Renard *et al.*, 1995, Escoubet *et al.*, 1997), despite differences in the protocols used to modify aldosterone status of the animals, or the use of cultured versus native cells. At variance, Denault *et al.* (1996) found an aldosterone-dependent upregulation of γ transcripts in rabbit cultured cortical collecting duct cells. No information is available yet on regulation of channel expression at the protein level, in part because of the lack of an adequate cellular model of mammalian collecting duct cells sensitive to aldosterone. In the amphibian A6 cell line, the kinetics of aldosterone-induced changes in ENaC expression at the mRNA and at the protein level have been established (May *et al.*, 1997).

In A6 cells grown on filters, aldosterone elicited a delayed increase in all three subunit ENaC mRNAs. Interestingly aldosterone increased the rate of synthesis of the α subunit as early as 1 h after hormonal exposure, just at the time when sodium transport begins to rise. From these results and others, it was thus postulated that synthesis of the α subunit of ENaC could be rate-limiting in the action of aldosterone (May *et al.,* 1997).

B. *Regulation of ENaC Expression by Arginine-Vasopressin*

Arginine-vasopressin is well known as a rapid (within a few minutes) regulator of sodium and water reabsorption in the kidney (Breyer and Ando, 1994). The hormone acts in principal cells of the distal nephron, i.e., in the same cells as those sensitive to aldosterone (Farman *et al.,* 1991). This effect of AVP is mediated via binding to its V2 receptors, activating immediately the cAMP cascade. A series of experiments showed that AVP effects are amplified in aldosterone-treated animals: this cooperation was evidenced by the AVP-induced short-term increase in sodium or water reabsorption (Hawk *et al.,* 1996; Reif *et al.,* 1986; Schafer, 1994); such cooperation is also apparent to constitute and activate a latent pool of Na^+,K^+-ATPase pumps (Coutry *et al.,* 1995). However, little information is available concerning putative long-term effects of AVP, which could eventually interfere positively with nuclear events associated to binding of MR to its target genes. It is difficult to evaluate the respective roles of aldosterone and AVP *in vivo*: in some cases (such as sodium depletion or dehydration) plasma levels of both hormones are high; in other situations (as adrenalectomy), aldosterone plasma levels are suppressed, while AVP increases.

To gain insight into the long-term effects of AVP in the absence of corticosteroid hormones, we have used a collecting duct cell line that maintains several properties of the native collecting duct, but lacks aldosterone responsiveness (Blot-Chabaud *et al.,* 1996). The rat RCCD1 cell line forms domes, is highly polarized, and has a high electrical transepithelial resistance (2000–4000 ohms/cm^2); transepithelial sodium transport (from the apical to the basolateral side) can be stimulated by AVP, but not by aldosterone (due to the lack of significant expression of mineralocorticoid receptor). Initial experiments showed that transepithelial sodium transport (measured by short-circuit current Isc on cells grown on filters) was transiently increased by AVP (few minutes) and then returned to baseline. However, after 5–6 h in the presence of AVP, a secondary and sustained increased in Isc was apparent (Djelidi *et al.,* 1997). This late effect of AVP is indeed, at least partly, due to an increased sodium transport, since it was sensitive to amiloride. The late effect of AVP was preceded by an increase in mRNA steady-state levels of β and γ subunits of ENaC, while α transcripts were unchanged. After a 1-h metabolic labeling of $RCCD_1$ cells with ^{35}S-methionine in the presence or absence of AVP, immunoprecipitation experiments showed that both β and γ subunit *de novo* translation rates were increased 1–3 h after AVP, with no change in α-subunit neosynthesis. In terms of the kinetics of events (Fig 5), it appears that the earliest effect of AVP is the *de novo* synthesis of the γ subunit (1 h after addition of the hormone) and then *de novo* synthesis of the β subunit (3 h after AVP),

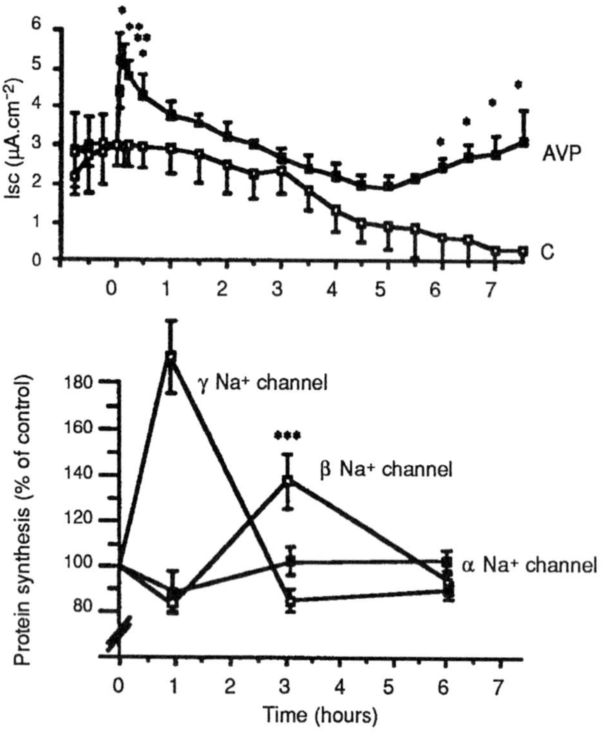

FIGURE 5 Effects of arginine-vasopressin (AVP) on the renal collecting duct cell line RCCD1. Upper panel, short-circuit current (Isc) monitored across RCCD1 cells grown on filters. Addition of AVP at zero time results in a rapid and transient rise in Isc, followed by a delayed and sustained stimulation of Isc after 5–6 h of AVP treatment. Lower panel, time course of changes in *de novo* protein synthesis of ENaC subunits after AVP treatment. At each time studied (1, 3, and 6 h), cells have been metabolically labeled with ^{35}S-methionine for 1 h. AVP treatment resulted in an early (1 h) stimulation of the synthesis of the γ subunit, and β-subunit neosynthesis was increased at 3 h AVP. In contrast α-subunit neosynthesis was not modified.

TABLE III

Respective Roles of Aldosterone and Arginine Vasopressin in Regulating ENaC Transcription Rate in Kidney Cells

	α subunit	β subunit	γ subunit
Aldosterone	↗	—	—
AVP	—	↗	↗

while the α-subunit protein synthesis is unaltered at any time. Thus, AVP augments the synthesis of some (β and γ) subunits of ENaC at the mRNA and at the protein level within 1–3 h after addition of AVP (Djelidi *et al.,* 1997). It precedes the AVP-induced delayed stimulation of sodium transport. Such long-term effects of AVP may well be physiologically relevant in the control of renal sodium reabsorption, in particular because they are sustained for several hours.

This phenomenon may be integrated in the view according to which the cAMP-dependent signaling cascade can regulate gene expression (Foulkes and Sassonecorsi, 1996; Vallejo, 1994). Such a mechanism has been shown to depend on cAMP-dependent DNA binding factors, such as CREB, CREM, and ATF1; these proteins bind to cAMP-response elements (CRE) which are found in the promoter region of several genes. Interestingly, a CRE sequence has been evidenced in the ENaC γ-subunit promoter (Thomas *et al.,* 1996) (no information is available yet on the promoter region of α and β ENaC).

On the whole, we suggest that transcriptional control of ENaC depends on both aldosterone and arginine-vasopressin. As summarized in Table III, the two hormones act in synergy to regulate distinct subunits of ENaC; the presence of these two hormones ensures a coordinated expression of all three subunits. This overlapping network of gene expression may be of importance to ensure regulated sodium channel expression in a large range of physiological situations. Precisely which intermediate steps are involved and whether other transcription factors participate in the fine tuning of ENaC activity to adapt sodium excretion to sodium homeostasis remain to be established.

References

Asher, C., Wald, H., Rossier, B. C. and Garty, H. (1996). Aldosterone-induced increase in the abundance of Na^+ channel subunits. *Am. J. Physiol.* **40,** C605–C611.

Blot-Chabaud, M., Laplace, M., Cluzeaud, F., Capurro, C., Cassingena, R., Vandewalle, A., Farman, N., and Bonvalet, J. P. (1996). Characteristics of a rat cortical collecting duct cell line that maintains high transepithelial resistance. *Kidney Int.* **50,** 367–376.

Breyer, M. D., and Ando, Y. (1994). Hormonal signaling and regulation of salt and water transport in the collecting duct. *Annu. Rev. Physiol.* **56,** 711–739.

Canessa, C. M., Horisberger, J. D., and Rossier, B. C. (1993). Epithelial sodium channel related to proteins involved in neurodegeneration. *Nature* (*London*) **361,** 467–470.

Canessa, C. M., Schild, L., Buell, G., Thorens, B., Gautschi, I., Horisberger, J.-D., and Rossier, B. C. (1994). Amiloride-sensitive epithelial Na^+ channel is made of three homologous subunits. *Nature* (*London*) **367,** 463–467.

Coutry, N., Farman, N., Bonvalet, J. P., and Blot-Chabaud, M. (1995). Synergistic action of vasopressin and aldosterone on basolateral Na^+-K^+-ATPase in the cortical collecting duct. *J. Membr. Biol.* **145,** 99–106.

Denault, D. L., Fejestoth, G., and Naray-Fejestoth, A. (1996). Aldosterone regulation of sodium channel gamma-subunit mRNA in cortical collecting duct cells. *Am. J. Physiol.* **40,** C423–C428.

Djelidi, S., Fay, M., Cluzeaud, F., Escoubet, B., Eugene, E., Capurro, C., Bonvalet, J. P., Farman, N., and Blot-Chabaud, M. (1997). Transcriptional regulation of sodium transport by vasopressin in renal cells. *J. Biol. Chem.* **272,** 32919–32924.

Driscoll, M., and Chalfie, M. (1991). The mec-4 gene is a member of a family of *Caenorhabditis elegans* genes that can mutate to induce neuronal degeneration. *Nature* (*London*) **349,** 588–593.

Duc, C., Farman, N., Canessa, C. M., Bonvalet, J. P., and Rossier, B. C. (1994). Cell-specific expression of epithelial sodium channel alpha, beta, and gamma subunits in aldosterone-responsive epithelia from the rat: Localization by in situ hybridization and immunocytochemistry. *J. Cell. Biol.* **127,** 1907–1921.

Escoubet, B., Coureau, C., Bonvalet, J. P., and Farman, N. (1997). Noncoordinate regulation of epithelial Na channel and Na pump subunit mRNAs in kidney and colon by aldosterone. *Am. J. Physiol.* **272,** C1482–C1491.

Farman, N., Oblin, M. E., Lombes, M., Delahaye, F., Westphal, H. M., Bonvalet, J. P., and Gasc, J. M. (1991). Immunolocalization of gluco and mineralocorticoid receptors in the rabbit kidney. *Am. J. Physiol.* **260,** C226–C233.

Farman, N., Talbot, C. R., Boucher, R., Fay, M., Canessa, C., Rossier, B., and Bonvalet, J. P. (1997). Non coordinated expression of alpha , beta-, and gamma-subunit mRNAs of epithelial Na+ channel along rat respiratory tract. *Am. J. Physiol.* **41,** C131–C141.

Foulkes, N. S., and Sassonecorsi, P. (1996). Transcription factors coupled to the cAMP-signalling pathway. *Biochim. Biophys. Acta* **1288,** F101–F121.

Garty, H., and Palmer, L. G (1997). Epithelial sodium channels: Function, structure, and regulation. *Physiol. Rev.* **77,** 359–396.

Hawk, C. T., Li, L., and Schafer, J. A. (1996). AVP and aldosterone at physiological concentrations have synergistic effects on Na transport in rat CCD. *Kidney Int.* **50,** S35–S41.

Lingueglia, E., Voilley, N., Waldmann, R., Lazdunski, M., and Barbry, P. (1993). Expression cloning of an epithelial amiloride-sensitive Na^+ channel. A new channel type with homologies to *Caenorhabditis elegans* degenerins. *FEBS Lett.* **318,** 95–99.

May, A., Puoti, A., Gaeggeler, H. P., Horisberger, J.-D., and Rossier, B. C. (1997). Early effect of aldosterone on the rate of synthesis of the epithelial sodium channel alpha subunit in A6 renal cells. *J. Am. Soc. Nephrol.* **8,** 1813–1822.

McDonald, F. J., Price, M. P., Snyder, P. M., and Welsh, M. J. (1995). Cloning and expression of the beta- and gamma-subunits of the human epithelial sodium channel. *Am. J. Physiol.* **37,** C1157–C1163.

McNicholas, C. M., and Canessa, C. M. (1997). Diversity of channels generated by different combinations of epithelial sodium channel subunits. *J. Gen. Physiol.* **109,** 681–692.

Ono, S., Kusano, E., Muto, S., Ando, A., and Asano, Y. (1997). A low-Na+ diet enhances expression of mRNA for epithelial Na+ channel in rat renal inner medulla. *Pfluegers Arch.* **434,** 756–763.

Reif, M. C., Troutman, S. L., and Schafer, J. A. (1986). Sodium transport by rat cortical collecting tubule. Effects of vasopressin and desoxycorticosterone. *J. Clin. Invest.* **77,** 1291–1298.

Renard, S., Voilley, N., Bassilana, F., Lazdunski, M., and Barbry, P. (1995). Localization and regulation by steroids of the alpha, beta and gamma subunits of the amiloride sensitive Na^+ channel in colon, lung and kidney. *Pfluegers Arch.* **430,** 299–307.

Roudier-Pujol, C., Rochat, A., Escoubet, B., Eugene, E., Barrandon, Y., Bonvalet, J. P., and Farman, N. (1996). Differential expression of epithelial sodium channel subunit mRNAs in rat skin. *J. Cell Sci.* **109,** 379–385.

Schafer, J. A. (1994). Salt and water homeostasis—is it just a matter of good bookkeeping? *J. Am. Soc. Nephrol.* **4,** 1933–1950.

Stutts, M. J., Canessa, C. M., Olsen, J. C., Hamrick, M., Cohn, J. A., Rossier, B. C., and Boucher, R. C. (1995). CFTR as a cAMP-dependent regulator of sodium channels. *Science* **269,** 847–850.

Thomas, C. P., Doggett, N. A., Fisher, R., and Stokes, J. B. (1996). Genomic organization and the 5′ flanking region of the gamma subunit of the human amiloride-sensitive epithelial sodium channel. *J. Biol. Chem.* **271,** 26062–26066.

Vallejo, M. (1994). Transcriptional control of gene expression by cAMP-response element binding proteins. *J. Neuroendocrinol.* **6,** 587–596.

Verrey, F. (1995). Transcriptional control of sodium transport in tight epithelia by adrenal steroids. *J. Membr. Biol.* **144,** 93–110.

Voilley, N., Lingueglia, E., Champigny, G., Mattei, M. G., Waldmann, R., Lazdunski, M., and Barbry, P. (1994). The lung amiloride-sensitive Na^+ channel—biophysical properties, pharmacology, ontogenesis, and molecular cloning. *Proc. Natl. Acad. Sci. U.S.A.* **91,** 247–251.

Voilley, N., Bassilana, F., Mignon, C., Merscher, S., Mattei, M., G., Carle, G., F., Lazdunski, M., and Barbry, P. (1995). Cloning chromosal localization and physical linkage of the beta and gama subunits (SCNN 1B and SCNN 1G) of the human epithelial amiloride-sensitive sodium channel. *Genomics* **28,** 560–565.

Voilley, N., Galibert, A., Bassilana, F., Renard, S., Lingueglia, E., Coscoy, S., Champigny, G., Hofman, P., Lazdunski, M., and Barbry, P. (1997). The amiloride-sensitive Na^+ channel—from primary structure to function. *Comp. Biochem. Physiol.* **118,** 193–200.

Volk, K. A., Sigmund, R. D., Snyder, P. M., McDonald, F. J., Welsh, M. J., and Stokes, J. B. (1995). rENaC is the predominant Na^+ channel in the apical membrane of the rat renal inner medullary collecting duct. *J. Clin. Invest.* **96,** 2748–2757.

CHAPTER 5

Regulation of ENaC by Interacting Proteins and by Ubiquitination

Olivier Staub,*† Pamela Plant,* Toru Ishikawa,* Laurent Schild,† and Daniela Rotin*

*The Hospital for Sick Children, Toronto, Ontario, Canada, M5G 1X8, and Department of Biochemistry, University of Toronto, Toronto, Canada.

†Insitut de Pharmacologie et de Toxicologie, Université de Lausanne, CH-1015 Lausanne, Switzerland

I. INTRODUCTION

The apically located amiloride-sensitive epithelial Na^+ channel plays a critical role in Na^+ and fluid absorption in several epithelia, including those of the lung, distal colon, and distal nephron (reviewed in Garty and Palmer, 1997). A major breakthrough in our understanding of the function of the epithelial Na^+ channel came several years ago, when Canessa, Rossier, and colleagues, as well as other groups, isolated the rat epithelial Na^+ channel (rENaC) by expression cloning in *Xenopus* oocytes (Canessa *et al.,* 1993, 1994a; Linguеglia *et al.,* 1993, 1994). Subsequently, ENaC homologues were isolated from several other species, including human (hENaC), *Xenopus*

1063-5823/99 $30.00

(xENaC), bovine (bENaC), and chicken (McDonald *et al.,* 1994, 1995; Voilley *et al.,* 1994; Puoti *et al.,* 1995; Fuller *et al.,* 1995; Goldstein *et al.,* 1997; Killik and Richardson, 1997), and ENaC-related genes, the degenerins, isolated from *C. elegans* (Chalfie and Wolinsky, 1990; Driscoll and Chalfie, 1991; Hong and Driscoll, 1994; Liu *et al.,* 1996).

ENaC is composed of 3 similar subunits, α, β, and γ (Fig. 1), which when assembled together yield an active channel with single-channel characteristics of a highly Na^+ selective, low-conductance (~5 pS), and high amiloride sensitivity (K_i~ 10–100 nM) (Canessa *et al.,* 1994a; Ishikawa *et al.,* 1998). Each ENaC subunit is composed of two transmembrane (TM) domains, a large ectodomain, and short intracellular N and C termini (Canessa *et al.,* 1994b; Renard *et al.,* 1994; Snyder *et al.,* 1994) (Fig. 1). We have previously identified proline-rich sequences within the C termini of each ENaC subunit and have focused our studies on identifying and characterizing proteins that bind to these sequences, with the hope of understanding how they may modulate channel function.

II. PROLINE-RICH REGIONS IN ENaC AND THEIR INTERACTING PROTEINS

Close examination of the C termini of $\alpha\beta\gamma$ENaC reveals two proline-rich regions within each subunit, which we have termed P1 and P2 (Fig. 1).

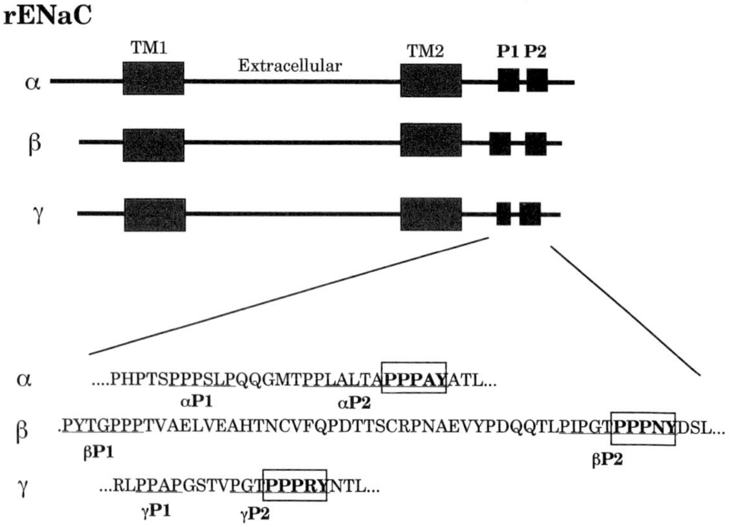

FIGURE 1 Schematic representation of $\alpha\beta\gamma$rENaC highlighting the proline-rich regions (P1 and P2) in each subunits. Boxed are the PY motifs. TM1 and TM2 are the 2 transmembrane domains.

Some of these conform to putative SH3 binding sequences (PxxP). In addition, the P2 regions include a highly conserved sequence now known as the PY motif (xPPxY, Fig. 1, boxed). Because proline-rich sequences are often implicated in protein:protein interactions, we have embarked on investigations of putative proteins that interact with these sequences in ENaC.

A. Interactions of αrENaC with α-Spectrin

Earlier studies have suggested an association between the biochemically purified epithelial Na^+ channel and the cytoskeleton, including α-spectrin (Smith *et al.,* 1991; Cantiello *et al.,* 1991), and more recent studies with ENaC have demonstrated the involvement of cytoskeletal components (actin, gelsolin) in regulating channel activity (Ismailov *et al.,* 1997). Because α-spectrin contains a SH3 domain, and because ENaC subunits contain proline-rich regions which conform to putative SH3 binding sequences, we initially tested whether α-spectrin–SH3 domain binds the proline-rich C terminus of αrENaC. Our work indeed demonstrated such an association *in vitro,* which took place between the α-spectrin–SH3 domain and the P2 region of αrENaC (Rotin *et al.,* 1994). Accordingly, deleting the αrENaC-P2 sequence led to abrogation of association with full-length α-spectrin endogenously expressed in primary cultured fetal distal lung epithelial (FDLE) cells (Figs. 2A,B). An *in vitro* association between SH3 domains of several proteins and the P2 region of αhENaC was demonstrated as well by Welsh and colleagues (McDonald and Welsh, 1995). In addition, we were able to coimmunoprecipitate α-spectrin and αrENaC from living cells. Moreover, microinjection of the proline-rich C terminus of αrENaC into FDLE cells led to an exclusive apical membrane localization of the protein, with distribution which parallels that of α-spectrin in these cells (Rotin *et al.,* 1994). Future studies are necessary to determine whether the C terminus of αrENaC contributes to the stabilization of the protein at the apical membrane.

B. Isolation of Nedd4 as a Protein Interacting with ENaC

Despite possessing putative SH3 binding sequences within their P2 regions (Fig. 1), neither βrENaC nor γrENaC demonstrated any significant binding to α-spectrin (Fig. 2C) nor to isolated SH3 domains from several proteins (Staub *et al.,* 1996). We had therefore focused our attention on trying to identify putative proteins that interact with the proline-rich regions

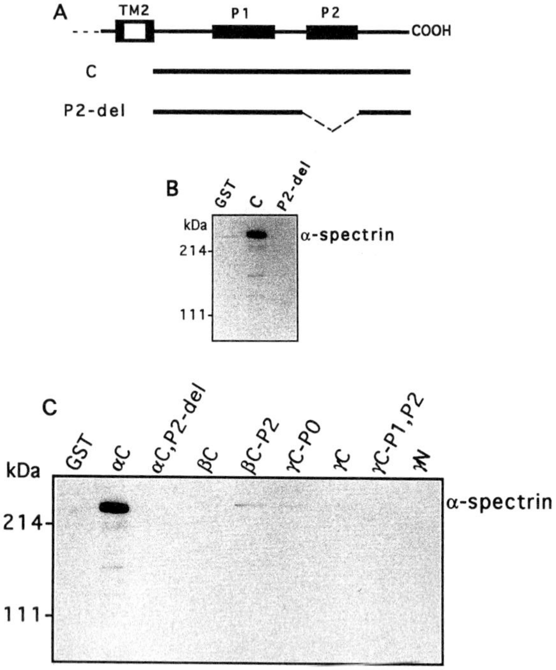

FIGURE 2 Coprecipitation of α-spectrin from primary cultured alveolar epithelial cell lysate with GST fusion proteins encompassing the C terminus of αrENaC (αC). Note reduced or lack of binding to α-spectrin by a P2-deleted αrENaC C terminus (αC,P2-del), βrENaC C terminus (βC), βrENaC-P2 region (βC-P2), γrENaC C terminus (γC), a C terminal region in γrENaC upstream of P1 (γC-P0), the 2 proline-rich regions of γrENaC (γC-P1P2), and the N terminus of γrENaC (γN). In this GST pull down experiment, GST fusion proteins were incubated with the above lysate and following thorough washes, proteins were separated on SDS–PAGE and immunoblotted with monoclonal anti α-spectrin antibodies.

of βrENaC or γrENaC. This search became particularly pertinent as it was then discovered by Lifton and colleagues, using genetic linkage analysis, that Liddle syndrome is caused by an effective deletion of the P2 region of βhENaC (Shimkets *et al.*, 1994) or γhENaC (Hansson *et al.*, 1995a). Liddle syndrome is a rare form of hereditary hypertension characterized by hyperactivity of the epithelial Na^+ channel (Liddle *et al.*, 1963; Botero-Valez *et al.*, 1994). Accordingly, expression of truncated βrENaC or γrENaC (which lack the P2 regions) in *Xenopus* oocytes led to an elevated channel activity (Schild *et al.*, 1995) which was caused by an increase in both the number of channels at the cell surface and the open probability of the channel (Firsov *et al.*, 1996). Because deletion of the P2 region in βrENaC or γrENaC caused an increase in channel activity, we embarked on a search for a putative suppressor protein(s) that interacts with these regions. Using a yeast 2 hybrid screen of rat lung library with the P2 region of βrENaC (βP2) as a bait (identical between hENaC and rENaC at the key amino acids; see below), we identified Nedd4 as a protein interacting with βP2 (Staub *et al.*, 1996).

III. Nedd4: ITS DOMAINS AND MODE OF INTERACTION WITH ENaC

Nedd4 (neuronal precursor cells expressed developmentally downregulated 4; Kumar *et al.*, 1992) is a ubiquitin protein ligase composed of an N-terminal C2 domain, 3 (or 4) WW domains, an E2 binding region, and a C-terminal ubiquitin protein ligase Hect domain (Fig. 3A). The function of the Nedd4–C2 domain is discussed in Section V. As the smaller of the Nedd4 2 clones identified in the yeast 2 hybrid screen contained only the region encompassing the 3 WW domain (and part of the E2 binding region), we hypothesized that the WW domains of Nedd4 are responsible for the association with ENaC.

A. *Nedd4–WW Domain Interaction with the PY Motifs of ENaC*

WW domains are recently described protein:protein interaction modules ~40 amino acids in length (Bork and Sudol, 1994; André and Springael,

A rNedd4

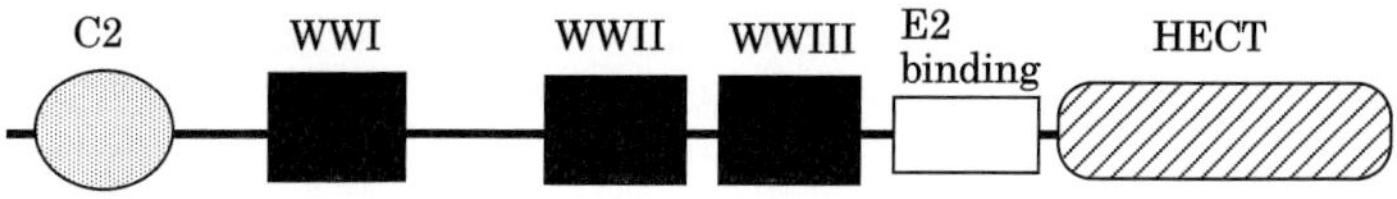

B PY MOTIFS IN ENaC

	α	β	γ
rENaC	PPPAY	PPPNY	PPPRY
hENaC	PPPAY	PPPNY	PPPKY
xENaC	PPPAY	PPPNY	PPPKY

FIGURE 3 (A) Schematic representation of rat Nedd4 depicting its C2 domain, 3 WW domains, an E2 binding region and the ubiquitin protein ligase Hect domain. (B) A list of the conserved PY motifs in the P2 regions of rat (r), human (h), and *Xenopus* (x) rENaC.

1994; Hofmann and Bucher, 1995; Staub and Rotin, 1996) which bind short proline-rich sequences conforming either to the sequence xPPxY (the PY motif) in the case of Nedd4, YAP, and dystrophin WW domains (Chen and Sudol, 1995; Staub *et al.,* 1996; Jung *et al.,* 1995) or to PPLP in the case of formin binding proteins or FE65 WW domains (Bedford *et al.,* 1997; Ermekova *et al.,* 1997). As shown in Fig. 3B, a highly conserved PY motif is found in all 3 of the ENaC subunits. Using a series of techniques, including (i) quantitative yeast 2 hybrid binding (β-gal) assays (Fig. 4A), (ii) *in vitro* binding using fusion proteins encoding GST–Nedd4–WW domains and His–ENaC–PY motifs, (iii) *in vitro* peptide competition experiments using a synthetic peptide encompassing the ENaC–PY motifs, and (iv) coprecipitation and coimmunoprecipitation from mammalian cells (Figs. 4B,4C), we were able to demonstrate that indeed the WW domains of Nedd4 mediate binding to the PY motifs of ENaC (Staub *et al.,* 1996), with stronger binding to the PY motifs of βrENaC or γrENaC (Fig. 4A). More recent quantitative experiments using internal fluorescence have identified the highest affinity of interactions (K_d <20 μM) between Nedd4–WWIII and the PY motif of βrENaC (V. Kanelis, J. Forman-Kay, and D. Rotin, unpublished). In accordance with our binding studies listed above, we have recently demonstrated that at least in some Na^+-absorptive tissues, such as the distal nephron and lung epithelia, Nedd4 is localized to the same regions and cells previously shown to express ENaC (Staub *et al.,* 1997a). For example, in the cortical collecting tubules and outer medullary collecting ducts, Nedd4 is strongly expressed in the principal but not in the intercalating cells, similar to the expression of ENaC (Duc *et al.,* 1994). Similarly, Nedd4 is expressed in the epithelia lining the airway, distal epithelium, and submucosal glands and ducts, with distribution which resembles that of ENaC (Matsushita *et al.,* 1996; Farman *et al.,* 1997). This suggests that the ENaC–Nedd4 interactions may have biological importance in selected Na^+-absorbing tissues.

In addition to the above mentioned cases of Liddle syndrome in which the PY motif of βENaC or γENaC was effectively deleted (Shimkets *et al.,* 1994; Hansson *et al.,* 1995a), individuals with Liddle syndrome were recently identified who possess specific point mutations in the PY motif of βENaC: either the C-terminal proline (xP$\underline{\text{P}}$xY) or the tyrosine (xPPx$\underline{\text{Y}}$) is mutated (Hansson *et al.,* 1995b; Tamura *et al.,* 1996). In agreement, mutations of the same amino acids to alanine residues led both to elevation of ENaC activity when expressed in *Xenopus* oocytes (Snyder *et al.,* 1995; Schild *et al.,* 1996) and to abrogation of binding to the Nedd4–WW domains (Fig. 4D) (Staub *et al.,* 1996). Thus, although not yet proven conclusively, it is possible that binding of Nedd4 to ENaC via its WW domains may allow the Nedd4–Hect domain to ubiquitinate the channel, thus accelerating its degradation. We can speculate that this ubiquitination may be at least partially impaired in Liddle syndrome, thus allowing increased accumulation of channels at the cell surface, leading to increased channel activity associated with the disease. However, a recent report by Canessa and coworkers (Shimkets *et al.,* 1997) has identified the PY motifs of βENaC or γENaC as internalization signals which are responsible for clathrin-mediated endocytosis of the channel. They therefore suggested that the defective internalization signal in Liddle syndrome may lead to an accumulation of channels at the cell surface, as previously suggested (Snyder *et al.,* 1995). As discussed below, the two hypotheses may be complementary, as numerous recent reports have demonstrated a link between ubiquitination and endocytosis. Neither defective ubiquitination nor impaired endocytosis can explain, however, the increase in channel opening associated with the Liddle defect.

IV. REGULATION OF ENaC STABILITY AND FUNCTION BY UBIQUITINATION

Our demonstration of association between ENaC and a ubiquitin protein ligase raises the possibility that ENaC is regulated by ubiquitination. Ubiquitination serves to tag proteins for rapid degradation (reviewed in Ciechanover, 1994; Jentsch and Schlenker, 1995: Hilt and Wolf, 1996). Several enzymes are involved in this process, including a ubiquitin-activating enzyme (E1), a ubiquitin-conjugating enzyme (E2), and a ubiquitin protein ligase (E3) such as Nedd4. The E3 enzyme is usually responsible for the attachment of ubiquitin, or multiubiquitin chains, on lysine residues at the target protein. In most studies described to date, ubiquitinated proteins such as cell cycle proteins, transcription factors, and ER-associated proteins are degraded by the 26S proteasome, a cytosolic threonine protease complex (reviewed in Ciechanover, 1994; Jentsch and Schlenker, 1995: Hilt and Wolf, 1996). In recent years, however, it has become apparent that some transmembrane proteins also become ubiquitinated and that ubiquitination is involved in their subsequent endocytosis and degradation by the endosome/lysosome (Hicke and Riezman, 1996; Hochstrasser, 1996). Examples include several tyrosine kinase receptors, the T cell and growth hormone receptors, and several yeast receptors and transporters. Interestingly, the yeast amino acid permeases Gap1 and Fur4 (Hein *et al.,* 1995; Galan *et al.,* 1996) have been shown to become ubiquitinated, a necessary step for their subsequent endocytosis and degradation in the vacuoles (the yeast lysosomes); this ubiquitination and subsequent degradation is dependent on the presence of intact Rsp5, the yeast homologue of Nedd4 (Galan *et al.,* 1996). The interdependence of ubiquitination and endocytosis in

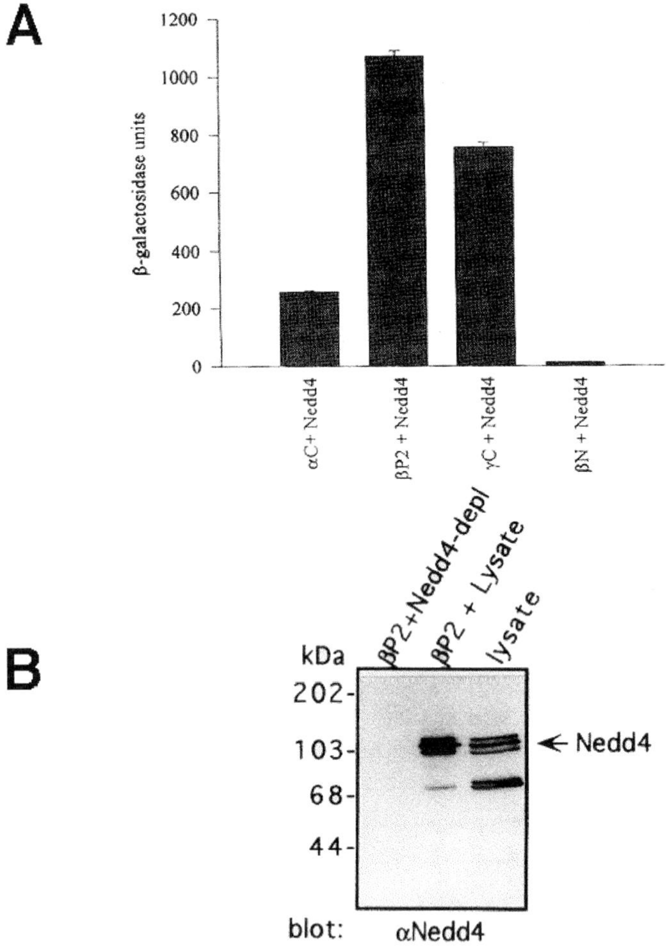

FIGURE 4 Binding of Nedd4 to WT ENaC but not to Liddle βENaC. (A) Yeast 2 hybrid binding assays using quantitative β-gal analyses depicting association of Nedd4–WW domains (WWI–WWIII) with the PY-motif containing C termini of αrENaC (αC), γrENaC (γC), and the P2 region of βrENaC (βP2). The N terminus of βrENaC (βN) was used as a negative control. (B) Coprecipitation of Nedd4 endogenously expressed in MDCK cells with βP2. The His-tagged βP2 region, immobilized on Ni agarose beads, was incubated with MDCK cell lysate, and following extensive washes, proteins on beads were separated on SDS–PAGE and immunoblotted with anti-Nedd4 antibodies (anti-WWII, Staub *et al.,* 1996), to test for coprecipitated Nedd4. The right lane represents Nedd4 expressed in MDCK cells. Coprecipitation of Nedd4 (middle lane) was abolished in cell lysate depleted of Nedd4 (Nedd4-depl., left lane).

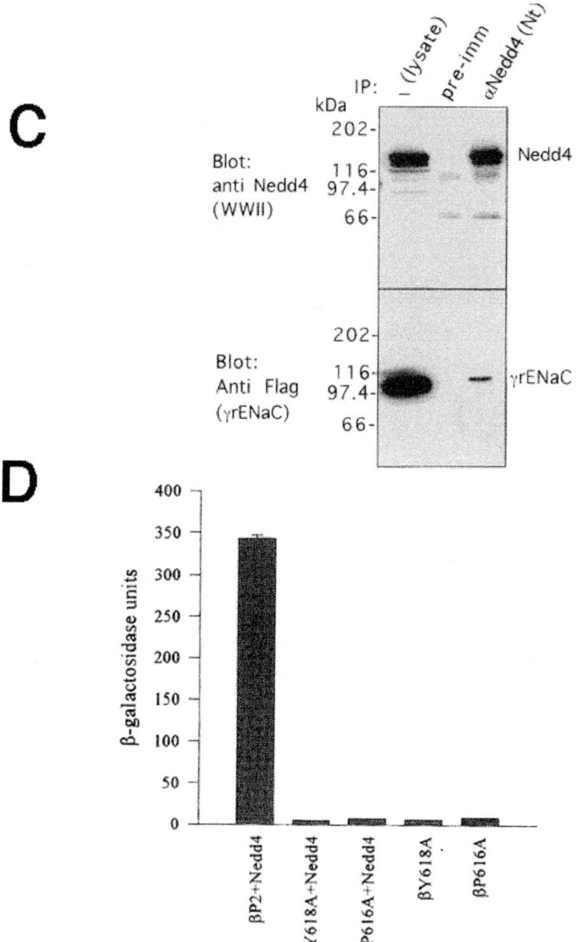

FIGURE 4 (*Continued*) (C) Coimmunoprecipitation of Nedd4 and γrENaC. Endogenously expressed Nedd4 was immunoprecipitated from MDCK cell lysate with antibodies directed against a region N terminal to the first WW domain (anti-Nt). Coimmunoprecipitated Flag-tagged γrENaC was detected with anti-Flag antibodies. (D) Loss of Nedd4 binding to mutants bearing point mutations in the PY motif of βrENaC, at the same sites identified in Liddle patients. Yeast 2 hybrid binding assays were performed to test binding of Nedd4 (WWI-WWIII) to βP2 (control) or to βP2 bearing mutations to Ala of either the third proline (βP616A) or the tyrosine (βY618A). Panels A and D were reproduced from Staub *et al.* (1997b), *EMBO J.* **16,** 6325–6336, by permission of Oxford University Press.

mammalian cells has been also demonstrated recently for the growth hormone receptor (Strous *et al.*, 1996; Govers *et al.*, 1997). How ubiquitination signals endocytosis / lysosomal degradation of transmembrane proteins is currently unknown (Hochstrasser, 1996).

One of the hallmarks of proteins regulated by ubiquitination is their rapid turnover. Accordingly, we have demonstrated that the half-life of the total cellular pool of rENaC heterologously expressed in MDCK or NIH-3T3 cells was very short, about 1 h (Staub *et al.*, 1997b), which is similar to the half-life of xENaC endogenously expressed in A6 cells (May *et al.*, 1997). Moreover, we have recently demonstrated *in vivo* ubiquitination of α and γ, but not β, rENaC when the 3 subunits are coexpressed in mammalian cells (Fig. 5) (Staub *et al.*, 1997b).

To study the possible consequences of ENaC ubiquitination on channel function, we have generated a series of rENaC mutants in which conserved N-terminal Lys residues in the α, β, and γ, subunits, putative targets for ubiquitination, were mutated individually or in clusters to arginine residues (Fig. 6A). This was done to eliminate putative ubiquitin attachment sites while retaining the positive charge of the replaced amino acids. Each Lys $\rightarrow$ Arg mutant was then introduced into *Xenopus* oocytes together with the remaining two wild-type (WT) chains, and amiloride-sensitive Na^+ current analyzed by voltage clamping. As seen in Fig. 6B, mutating any of the conserved lysines in αrENaC or βrENaC alone had no effect on channel activity. In contrast, mutating a cluster of Lys residues in γrENaC (γK6,8,10, 12,13R = γK6-13R) led to a 2- to 3-fold increase in channel activity (Fig. 6B), which was later confirmed in MDCK cells stably expressing this γK6-13R mutant along with WT $\alpha\beta$rENaC (T. Ishikawa, O. Staub, Y. Marunaka, and D. Rotin, unpublished). Moreover, even though none of the Lys mutations in αrENaC had any effect on their own, the αK47,50R mutant greatly potentiated the effect of the γK6-13R mutant when coexpressed in *Xenopus* oocytes (Fig. 6B) (Staub *et al.*, 1997b). To determine whether the increase in ENaC activity associated with the Lys $\rightarrow$ Arg mutations was indeed caused by increased channel numbers at the plasma membrane, a FLAG tag was introduced at the ectodomain of each WT or mutated (αK47,50R or γK6-13R) ENaC chains to allow quantitation of channel numbers at the cell surface by surface labeling with iodinated anti-FLAG antibodies, as previously described (Firsov *et al.*, 1996). Using this approach, which allows direct comparison of channel activity and numbers at the plasma membrane in each oocyte, we were able to conclusively determine that all the increase in channel activity associated with the above Lys $\rightarrow$ Arg mutations originated from an increase in channel numbers at the cell surface (Fig. 6C), with no effect on open probability or single channel conductance (Staub *et al.*, 1997b). To verify that this increase in channel numbers is related to

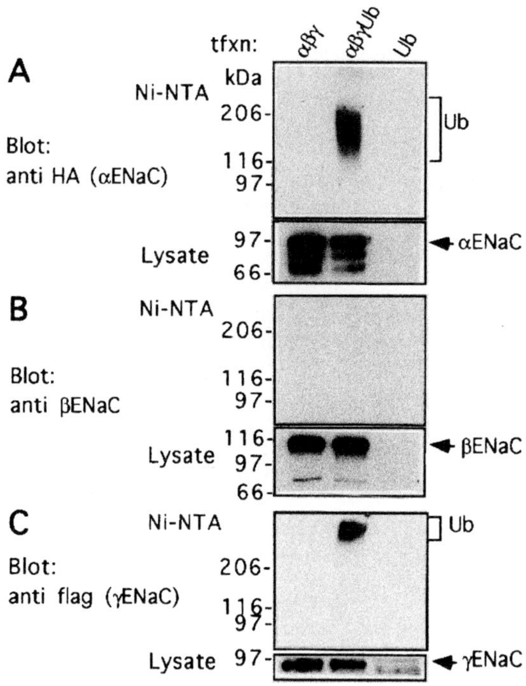

FIGURE 5 Ubiquitination of α and γ ENaC. HA-tagged αrENaC, βrENaC, and Flag-tagged γrENaC were cotransfected into Hek-293 cells alone ($\alpha\beta\gamma$) or together with a construct encoding 8 His-Ub (Ub). Cells were then lysed and lysate incubated with Ni agarose beads (Ni-NTA) to precipitate all histidinated (hence ubiquitinated) proteins. After extensive washes of the beads, proteins were separated on SDS–PAGE and immunoblotted with antibodies against HA (αrENaC), βrENaC, or Flag (γrENaC). Ubiquitinated species appear as high molecular weight species (square brackets). Lower panels depict expression of the transfected ENaC subunits. Reproduced from Staub *et al.* (1997b), *EMBO J.* **16,** 6325–6336, by permission of Oxford University Press.

loss of ubiquitination, we have compared ubiquitination of WT and Lys → Arg mutants of rENaC. An example of such an experiment is depicted in Fig. 6D, which shows a clear loss of ubiquitination of the γK6-13R mutant relative to the WT γrENaC. Thus, we have identified a cluster of Lys residues implicated in ubiquitination of ENaC and showed that the same mutations in ENaC which cause loss of ubiquitination (γK6-13R) lead to an increase in the number of channels at the cell surface and hence to an elevation of channel activity.

To determine whether the increase in ENaC numbers at the plasma membrane is due to increased arrival of newly synthesized channels, increased retention of channels at the cells surface, or both, we treated the

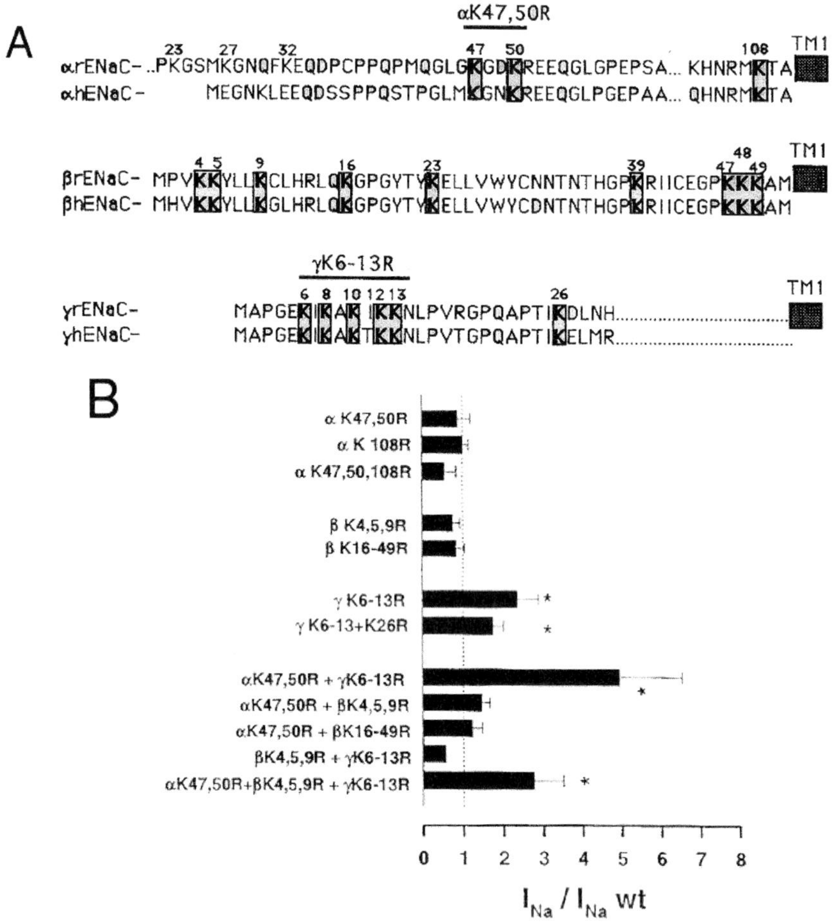

FIGURE 6 Sites and role of ENaC ubiquitination. (A) Conserved (boxed) Lys residues at the N termini of $\alpha\beta\gamma$ENaC which were mutated to Arg residues either individually or in clusters. (B). Amiloride-sensitive Na^{+} current (INa) of Lys → Arg mutants of rENaC analyzed in *Xenopus* oocytes. In all cases, the mutant subunit(s) was expressed together with WT remaining subunits. Note the increased channel activity with the γK6-13R mutant, which was potentiated by the αK47,50R mutant. (C) Correlation between INa and channel number at the cell surface in the same oocytes expressing WT, the γK6-13R mutant, or the γK6-13R plus αK47,50R double mutant. (D) Loss of ubiquitination of the γK6-13R mutant ($\alpha\beta\gamma_R$Ub). The ubiquitination experiment was performed as described in Fig. 5 above. Panels A–C were reproduced from Staub *et al.* 1997b, *EMBO J.* **16,** 6325–6336, by permission of Oxford University Press.

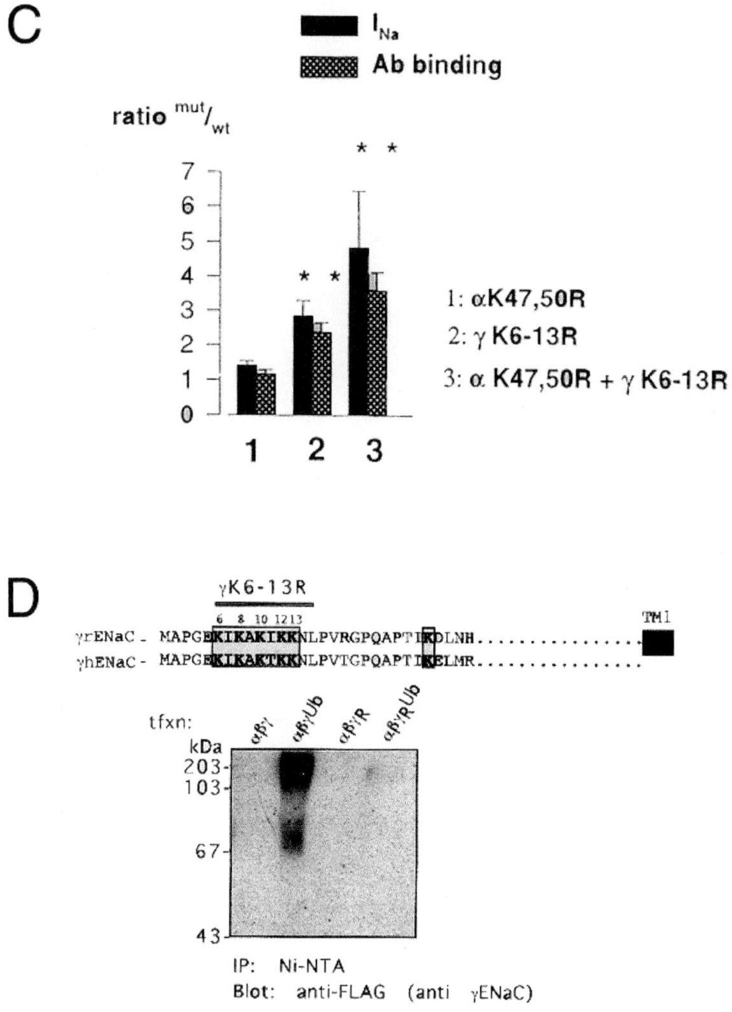

FIGURE 6 (*Continued*)

oocytes with Brefeldin A (BFA) to block ER to Golgi transport, and compared amiloride-sensitive channel activity of the WT and the Lys → Arg double mutant (αK47,50R plus γK6-13R). The outcome of these experiments revealed that while most of the WT channel disappeared from the cell surface by 8 h after the addition of BFA, about half of the Lys → Arg mutant appeared to be stuck at the plasma membrane (Fig. 7). Based on these results, we have proposed that ENaC ubiquitination regu-

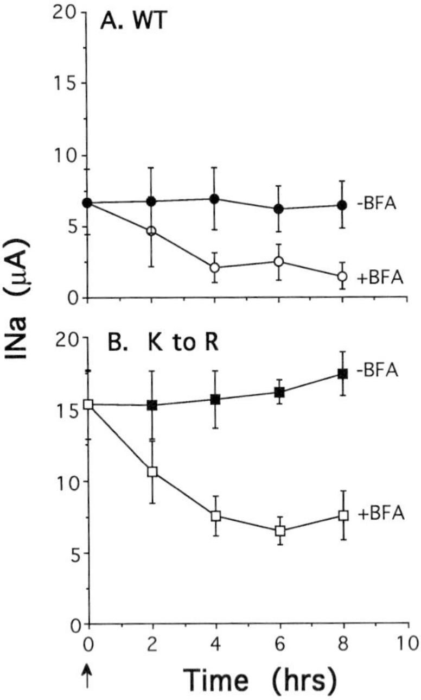

FIGURE 7 Effect of Brefeldin A on function of the ubiquitination-defective ENaC. *Xenopus* oocytes were injected with either WT (A) or the γK6-13R plus αK47,50R double mutant (K to R) (B) and INa measured at the indicated times following the addition of Brefeldin A (μg/ml BFA, Arrow). Reproduced from Staub *et al.* (1997b), *EMBO J.* **16,** 6325–6336, by permission of Oxford University Press.

lates both the arrival of new channels to and the retention of these channels at the plasma membrane (Staub *et al.,* 1997b).

In a separate set of experiments, we investigated the route(s) by which ENaC is degraded, using either proteasome inhibitors (lactacystin, LLnL, MG132) or lysosomal inhibitors (Chloroquine, NH_4Cl). To our surprise, we found that degradation of the total cellular pool of $\alpha\beta\gamma$ENaC stably expressed in MDCK or NIH-3T3 cells was sensitive to both proteasome and lysosome inhibitors, but that a stably transfected αrENaC alone was degraded by the proteasome only (Staub *et al.,* 1997b). In view of our inability to detect significant Na^+ currents in MDCK cells expressing αrENaC alone (unlike $\alpha\beta\gamma$ENaC-expressing MDCK cells, Ishikawa *et al.,* 1998), we speculate that in the absence of $\beta\gamma$ENaC, most of the α chain is targeted for proteasomal degradation and never reaches the plasma membrane. The properly assembled $\alpha\beta\gamma$ chains are probably degraded in the lysosomes.

V. ROLE OF THE C2 DOMAIN OF Nedd4 IN Ca^{2+}-DEPENDENT MEMBRANE TARGETING

In addition to WW and Hect domains, Nedd4 also contains a C2 domain. C2 domains in other proteins, such as Ca^{2+}-responsive isoforms of PKC, have been demonstrated to bind membrane and phospholipids in a Ca^{2+}-dependent fashion (reviewed in Nalefski and Falke, 1996; Ponting and Parker, 1996). We have therefore investigated the role of the Nedd4–C2 domain. Our initial studies have demonstrated that Nedd4, endogenously expressed in MDCK cells, redistributes from the soluble to the particulate fraction following treatment of cells with Ca^{2+} plus ionomycin, which leads to elevation of intracellular Ca^{2+} levels. Accordingly, immunolocalization of endogenous Nedd4 in these cells revealed rapid (within 5 min) translocation of the protein to the plasma membrane following the same Ca^{2+} plus ionomycin treatment (Fig. 8), which persisted for 30–45 min. More interestingly, confocal analysis has demonstrated that Nedd4 was localized mainly to the apical and lateral membranes, but not to the basal membrane of these cells (Fig. 8, see color insert) (Plant *et al.,* 1997). This plasma membrane association was mediated by the C2 domain, because deletion of this domain led to abrogation of targeting to the plasma membrane (Fig 9). Moreover, the Nedd4–C2 domain alone, generated as a GST fusion protein, was able to associate with membranes and purified phospholipids *in vitro* in response to Ca^{2+} (Plant *et al.,* 1997). Despite the ability of the C2 domain on its own to bind phospholipids, we anticipate that other proteins may be involved in the polarized distribution of Nedd4 in MDCK cells, possibly by associating with the C2 domain.

We have recently performed an extensive electrophysiological characterization of $\alpha\beta\gamma$rENaC heterologously expressed in epithelial MDCK cells, using patch-clamp analysis (Ishikawa *et al.,* 1998). This work has demonstrated an inhibition of ENaC activity in response to elevation of intracellular Ca^{2+}, in agreement with numerous previous studies demonstrating Ca^{2+}-mediated inhibition of Na^+ channels in native epithelia (reviewed in Garty and Palmer, 1997). This inhibition was biphasic (Fig. 10), suggesting it consists of a rapid response ($<$5 min) possibly mediated by effects on channel gating, and a slow response ($>$5 min), which may be mediated by a decrease in channel numbers. Although speculative at this point, it may be possible that the latter slow inhibition

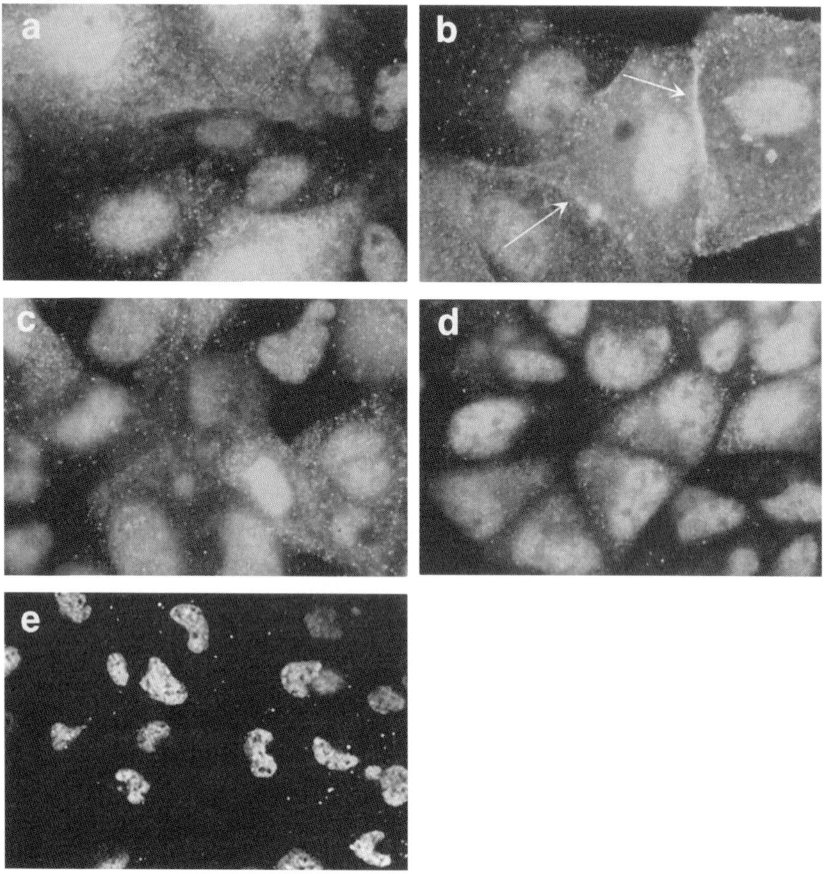

FIGURE 9 Loss of Ca^{2+}-dependent membrane localization of C2-deleted Nedd4. Stable MDCK cell lines expressing T7-tagged full length Nedd4 (a,b) or T7-tagged C2-deleted Nedd4 (c,d) were treated (b,d) or not (a,c) with Ca^{2+} (1 mM) plus ionomycin (1 μM), fixed as described in Fig. 8 and stained with monoclonal anti-T7 antibodies followed by secondary antibodies conjugated to Texas Red. Confocal sections taken 3 μm from the apical surface are shown. Panel e depicts staining of untransfected MDCK cells (revealing cross reactivity of the anti-T7 antibody with unidentified nuclear protein(s)). Reproduced with permission from Plant *et al. (1997)*.

could involve proteins such as Nedd4, which is targeted to the apical membrane in response to elevation of intracellular Ca^{2+} levels (Plant *et al.*, 1997). This would then allow Nedd4 to bind ENaC, ubiquitinate it, and enhance its degradation (Fig. 11).

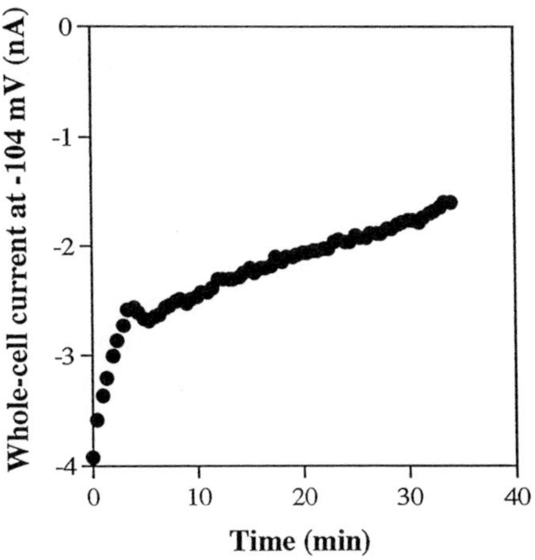

FIGURE 10 Inhibition of $\alpha\beta\gamma$rENaC expressed in MDCK cells by Ca^{2+}. Whole-cell current following cytoplasm perfusion with 1 μM free Ca^{2+}. Ramp command voltages were applied from −104 to +76 mV every 30 s. The time constants corresponding to the two slopes are 1.3 and 64.3 min. Reproduced with permission from Ishikawa *et al.* (1998).

VI. SUMMARY

In this review we have provided evidence that specific proteins, particularly Nedd4, interact with ENaC and may regulate its function. We suggest

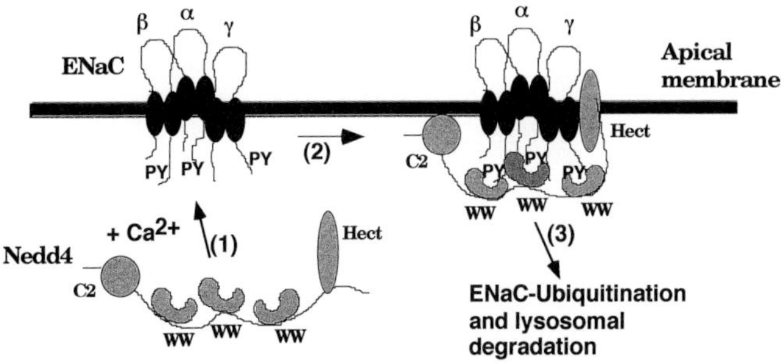

FIGURE 11 A possible model to explain Nedd4 effect on ENaC. Nedd4 is primarily a cytosolic protein. In response to elevated intracellular Ca^{2+}, it is translocated to the plasma membrane (mainly to the apical membrane in polarized epithelial cells) where ENaC is located. This allows the WW domains of Nedd4 to interact with the PY motifs of ENaC (with as yet an unknown stoichiometry), thus bringing the Nedd4–Hect domain in close proximity to the channel, in a position to ubiquitinate it. ENaC ubiquitination is then involved in the endocytosis/lysosomal degradation of the channel.

that in epithelial cells, Nedd4 is translocated to the apical membrane in response to elevated intracellular Ca^{2+}, to the same cellular compartment where ENaC is located. This then allows binding of the Nedd4–WW domains to the ENaC-PY motifs, thus bringing the ubiquitin protein ligase Hect domain in close proximity to the channel, in a position to ubiquitinate it, hence facilitating its degradation (Fig. 11). What needs to be demonstrated now is that the ubiquitination of ENaC that we have described is mediated, at least in part, by Nedd4. Moreover, it is necessary to demonstrate that Nedd4 indeed regulates ENaC function and to test whether both this putative regulation and ENaC ubiquitination are impaired in Liddle syndrome. Our own preliminary results, as well as those presented by Welsh and colleagues in this meeting, are clearly supportive of this notion, strengthening our view of the importance of Nedd4 to the regulation of ENaC. Whether Nedd4 also participates in regulating ENaC function by means other than ubiquitination (e.g., by its WW domain blocking the ENaC-PY motifs) awaits future investigation.

Acknowledgments

Work from the authors' laboratories was supported by the Medical Research Council of Canada, the Canadian Cystic Fibrosis Foundation, the International Human Frontier Science Program and the Swiss National Foundation for Scientific Research.

References

André, B., and Springael, J.-Y. (1994). WWP, a new amino acid motif present in single or multiple copies in various proteins including dystrophin and the SH3-binding Yes-associated protein YAP65. *Biochem. Biophys. Res. Commun.* **205,** 1201–1205.

Bedford, M. T., Chan, D. C., and Leder, P. (1997). FBP WW domains and the abl SH3 domain bind to a specific class of proline-rich ligands. *EMBO J.* **16,** 2376–2383.

Bork P., and Sudol, M. (1994). The WW domain: A signalling site in dystrophin. *Trends Biochem. Sci.* **19,** 531–533.

Botero-Velez, M., Curtis, J. J., and Warnock, D. G. (1994). Brief report: Liddles's syndrome revisited. *New Engl. J. Med.* **330,** 178–181.

Canessa, C. M., Horisberger, J.-D., and Rossier, B. C. (1993). Epithelial sodium channel related to proteins involved in neurodegeneration. *Nature* (*London*) **361,** 467–470.

Canessa, C. M., Schild, L., Buell, G., Thorens, B., Gautschi, I., Horisberger, J.-D., and Rossier, B. C. (1994a). Amiloride-sensitive epithelial Na^+ channel is made of three homologous subunits. *Nature* (*London*) **367,** 463–467.

Canessa, C. M., Mérillat, A.-M., and Rossier, B. C. (1994b). Membrane topology of the epithelial sodium channel in intact cells. *Am. J. Physiol.* **267,** C1682–C1690.

Cantiello, H. F., Stow, J. L., Prat, A. G., and Ausiello, D. A. (1991). Actin filaments control epithelial Na^+ channel activity. *Am. J. Physiol.* **261,** C882–C888.

Chalfie, M., and Wolinsky, E. (1990). The identification and suppression of inherited neurodegeneration in *Caenorhabditis elegans. Nature* (*London*) **345,** 410–416.

Chen, H. I., and Sudol, M. (1995). The WW domain of Yes-associated protein binds a novel proline-rich ligand that differs from the consensus established for SH3-binding modules. *Proc. Natl. Acad. Sci. U.S.A.* **92,** 7819–7823.

Ciechanover, A. (1994). The ubiquitin-proteasome proteolytic pathway. *Cell* (*Cambridge, Mass.*) **79,** 13–21.

Driscoll, M., and Chalfie, M. (1991). The mec-4 gene is a member of a family of *Caenorhabditis elegans.* Genes that can mutate to induce neuronal degeneration. *Nature* (*London*) **349,** 588–593.

Duc, D., Farman, N., Canessa, C. M., Bonvalet, J.-P., and Rossier, B. C. (1994). Cell-specific expression of epithelial sodium channel α,β and γ subunits in aldosterone-responsive epithelia from the rat: Localization by in situ hybridization and immunocytochemistry. *J.Cell Biol.* **127,** 1907–1921.

Ermekova, K. S., Zambrano, N., Linn, H., Minopoli, G., Gertler, F., Russo, T., and Sudol, M. (1997). The WW domain of neural protein FE65 interacts with proline-rich motifs in Mena, the mammalian homolog of *Drosophila* enabled. *J. Biol. Chem.* **272,** 32869–32877.

Farman, N., Talbot, C. R., Boucher, R., Canessa, C. M., Rossier, B. C., and Bonvalet, J. P. (1997). Non-coordinated expression of α, β, and γ subunit mRNAs of the epithelial Na^+ channel along the rat respiratory tract. *Am. J. Physiol.* **272,** C131–C141.

Firsov, D., Schild, L., Gautschi, I., Mérillat, A.-M., Schneeberger, E., and Rossier, B. C. (1996). Cell surface expression of the epithelial Na^+ channel and a mutant causing Liddle syndrome: A quantitativc approach. *Proc. Natl. Acad. Sci. U.S.A.* **93,** 15370–15375.

Fuller, C. M., Awayda, M. S., Arrate, M. P., Bradford, A. L., Morris, R. G., Canessa, C. M., Rossier, B. C., and Benos, D. J. (1995). Cloning of a bovine renal epithelial Na^+ channel subunit. *Am. J. Physiol.* **269,** C641–C654.

Galan, J. M., Moreau, V., André, B., Volland, C., and Haguenauer-Tsapis, R. (1996). Ubiquitination mediated by the Npi1p/Rsp5p ubiquitin-protein ligase is required for endocytosis of the yeast uracil permease. *J. Biol. Chem.* **271,** 10946–10952.

Garty, H., and Palmer, L. G. (1997). Epithelial sodium channels: Function, structure, and regulation. *Physiol. Rev.* **77,** 359–396.

Goldstein, O., Asher, C., and Garty, H. (1997). Cloning and induction by low NaCl intake of avian intestine Na channel subunits. *Am. J. Physiol.* **272,** C270–C277.

Govers, R., van Kerkhof, P., Schwartz, A. L., and Strous, G. J. (1997). Linkage of the ubiquitin-conjugating system and the endocytosis pathway in ligand-induced internalization of the growth hormone receptor. *EMBO J.* **16,** 4851–4858.

Hansson, J. H., Nelson-Williams, C., Suzuki, H., Schild, L., Shimkets, R., Lu, Y., Canessa, C. M., Iwasaki, T., Rossier, B. C., and Lifton, R. P. (1995a). Hypertension caused by a truncated epithelial sodium channel gamma subunit: Genetic heterogeneity of Liddle syndrome. *Nat. Genet.* **11,** 76–82.

Hansson, J. H., Schild, L., Lu, Y., Wilson, T. A., Gautschi, I., Shimkets, R. A., Nelson-Williams, C., Rossier, B. C., and Lifton, R. P. (1995b). A de novo missense mutation of the β subunit of the epithelial sodium channel causes hypertension and Liddle syndrome, identifying a proline-rich segment critical for regulation of channel activity. *Proc. Natl. Acad. Sci. U.S.A.* **25,** 11495–11499.

Hein, C., Springael, J. Y., Volland, C., Haguenauer-Tsapis, R., and André, B. (1995). NPI1, an essential yeast gene involved in induced degradation of Gap1 and Fur4 permeases, encodes the Rsp5 ubiquitin-protein ligase. *Mol. Microbiol.* **18,** 77–87.

Hicke, L., and Riezman, H. (1996). Ubiquitination of a yeast plasma membrane receptor signals its ligand-stimulated endocytosis. *Cell* (*Cambridge, Mass.*) **84,** 277–287.

Hilt, W., and Wolf, D. H. (1996). Proteasomes: destruction as a programme. *Trends Biochem. Sci.* **21,** 96–102.

Hochstrasser, M. (1996). Protein degradation or regulation: Ub the judge.*Cell* (*Cambridge, Mass.*) **84,** 813–815.

Hofmann, K., and Bucher, P. (1995). The rsp5-domain is shared by proteins of diverse functions. *FEBS Lett.* **358,** 153–157.

Hong, K., and Driscoll, M. (1994). A transmembrane domain of the putative channel subunit MEC-4 influences mechanotransduction and neurodegeneration in *C. elegans. Nature* (*London*) **367,** 470–473.

Ishikawa T., Marunaka, Y., and Rotin, D. (1998). Electrophysiological characterization of the rat epithelial Na^+ channel (rENaC) expressed in Madin-Darby canine kidney cells. Effects of Na^+ and Ca^{2+}. *J. Gen. Physiol.* **111,** 825–846.

Ismailov, I. I., Berdiev, B. K., Shlyonsky, V. G., Fuller, C. M., Prat, A. G., Jovov, B., Cantiello, H. F., Ausiello, D. A., and Benos, D. J. (1997). Role of actin in regulation of epithelial sodium channels by CFTR. *Am. J. Physiol.* **272,** C1077–C1086.

Jentsch, S., and Schlenker, S. (1995). Selective protein degradation: A journey's end within the proteasome. *Cell* (*Cambridge, Mass.*) **82,** 881–884.

Jung, D., Yang, B., Meyer, J., Chamberlain, J. S., and Campbell, K. P. (1995). Identification and characterization of the dystrophin anchoring site on β-dystroglycan. *J. Biol. Chem.* **270,** 27305–27310.

Killik, R., and Richardson, G. (1997). Isolation of chicken alpha ENaC splice variants from a cochlear cDNA library. *Biochim. Biophys. Acta Gene Struct. Expression* **1350,** 33–37.

Kumar, S., Tomooka, Y., and Noda, M. (1992). Identification of a set of genes with developmentally down-regulated expression in the mouse brain. *Biochem. Biophys. Res. Commun.* **185,** 1155–1161.

Liddle, G. W., Bledsoe, T., and Coppage, W. S., Jr. (1963). A familial renal disorder simulating primary aldosteronism but with negligible aldosterone secretion. *Trans. Assoc. Am. Physicians* **76,** 199–213.

Lingueglia, E., Voilley, N., Waldmann, R., Lazdunski, M., and Barbry, P. (1993). Expression cloning of an epithelial amiloride-sensitive Na^+ channel. A new channel type with homologies to *Caenorhabditis elegans* degenerins. *FEBS Lett.* **318,** 95–99.

Lingueglia, E., Renard, S., Waldmann, R., Voilley, N., Champigny, G., Plass, H., Lazdunski, M., and Barbry, P. (1994). Different homologous subunits of the amiloride-sensitive Na^+ channel are differently regulated by aldosterone. *J. Biol. Chem.* **269,** 13736–13739.

Liu, J., Schrank, B., and Waterson, R. H. (1996). Interaction between a putative mechanosensory membrane channel and a collagen. *Nature* (*London*) **273,** 361.

Matsushita, K., McCray, P. B., Jr., Sigmund, R. D., Welsh, M. J., and Stokes, J. B. (1996). Localization of epithelial sodium channel subunit mRNAs in adult rat lung by in situ hybridization. *Am. J. Physiol.* **271,** L332–L339.

May, A., Puoti, A., Gaeggeler, H. P., Horisberger, J. D., and Rossier, B. C. (1997). Early effect of aldosterone on the rate of synthesis of the epithelial Na channel α subunit in A6 renal cells. *J. Am. Soc. Nephrol.* **8,** 1813–1822.

McDonald, F. J., and Welsh, M. J. (1995). Binding of the proline-rich region of the epithelial Na^+ channel to SH3 domains and its association with specific cellular proteins. *Biochem J.* **312,** 491–497.

McDonald, F. J., Snyder, P. M., McCray, P. B., Jr., and Welsh, M. J. (1994). Cloning, expression, and distribution of a human amiloride-sensitive Na^+ channel. *Am. J. Physiol.* **266,** L728–L734.

McDonald, F. J., Price, M. P., Snyder, P. M., and Welsh, M. J. (1995). Cloning and expression of the β- and γ-subunits of the human epithelial sodium channel. *Am. J. Physiol.* **268,** C1157–C1163.

Nalefski, E. A., and Falke, J. J. (1996). The C2 domain calcium-binding motif: Structural and functional diversity. *Protein Science* **5,** 2375–2390.

Plant, P. J., Yeger, H., Staub, O., Howard, P., and Rotin, D. (1997). The C2 Domain of the ubiquitin protein ligase Nedd4 mediates Ca^{2+}-dependent plasma membrane localization. *J. Biol. Chem.* **272,** 32329–32336.

Ponting, C. P., and Parker, P. J. (1996). Extending the C2 domain family: C2s in PKCs delta, epsilon, eta, theta, phospholipases, GAPs, and perforin. *Protein Sci.* **5,** 162–166.

Puoti, A., May, A., Canessa, C. M., Horisberger, J.-D., Schild, L., and Rossier, B. C. (1995). The highly selective low-conductance epithelial Na^+ channel of *Xenopus laevis* A6 kidney cells. *Am. J. Physiol.* **38,** C188–C197.

Renard, S., Lingueglia, E., Voilley, N., Lazdunski, M., and Barbry, P. (1994). Biochemical analysis of the membrane topology of the amiloride-sensitive Na^+ channel. *J. Biol. Chem.* **269,** 12981–12986.

Rotin, D., Bar-Sagi, D., O'Brodovich, H., Merilainen, J., Lehto, P. V., Canessa, C. M., Rossier, B. C., and Downey, G. P. (1994). A SH3 binding region in the epithelial Na^+ channel (ENaC) mediates its localization at the apical membrane. *EMBO J.* **13,** 4440–4450.

Schild, L., Canessa, C. M., Shimkets, R. A., Warnock, D. G., Lifton, R. P., and Rossier, B. C. (1995). A mutation in the epithelial sodium channel causing Liddle's disease increases channel activity in the *Xenopus laevis* oocyte expression system. *Proc. Natl. Acad. Sci. U.S.A.* **92,** 5699–5703.

Schild, L., Lu, Y., Gautschi, I., Schneeberger, E., Lifton, R. P., and Rossier, B. C. (1996). Identification of a PY motif in the epithelial Na^+ channel subunits as a target sequence for mutations causing channel activation found in Liddle syndrome. *EMBO J.* **15,** 2381–2387.

Shimkets, R. A., Warnock, D. G., Bositis, C. M., Nelson-Williams, C., Hansson, J. H., Schambelan, M., Gill, J. R., Ulick, S., Milora, R. V., Findling, J. W., Canessa, C. M., Rossier, B. C., and Lifton, R. P. (1994). Liddles's syndrome: Heritable human hypertension caused by mutations in the β subunit of the epithelial sodium channel. *Cell (Cambridge, Mass.)* **79,** 407–414.

Shimkets, R. A., Lifton, R. P., and Canessa, C. M. (1997). The activity of the epithelial sodium channel is regulated by clathrin-mediated endocytosis. *J. Biol. Chem.* **272,** 25537–25541.

Smith, P. R., Saccomani, G., Joe, E. H., Angelides, K. J., and Benos, D. J. (1991). Amiloride-sensitive sodium channel is linked to the cytoskeleton in renal epithelial cells. *Proc. Natl. Acad. Sci. U.S.A.* **88,** 6971–6975.

Snyder, P. M., McDonald, F. J., Stokes, J. B., and Welsh, M. J. (1994). Membrane topology of the amiloride-sensitive epithelial sodium channel. *J. Biol. Chem.* **269,** 24379–24383.

Snyder, P. M., Price, M. P., McDonald, F. J., Adams, C. M., Volk, K. A., Zeiher, B. G., Stokes, J. B., and Welsh, M. J. (1995). Mechanism by which Liddle's syndrome mutations increase activity of a human epithelial Na^+ channel. *Cell (Cambridge, Mass.)* **83,** 969–978.

Staub, O., and Rotin, D (1996). WW domains. *Structure* **4,** 495–499.

Staub, O., Dho, S., Henry, P. C., Correa, J., Ishikawa, T., McGlade, J., and Rotin, D. (1996). WW domains of Nedd4 bind to the proline-rich PY motifs in the epithelial Na^+ channel deleted in Liddle's syndrome. *EMBO J.* **15,** 2371–2380.

Staub, O., Yeger, H., Plant, P., Kim, H., Ernst, S., and Rotin, D. (1997a). Immunolocalization of the ubiquitin-protein ligase Nedd4 in tissues expressing the epithelial Na^+ channel (ENaC). *Am. J. Physiol.* **272,** C1871–C1880.

Staub, O., Gautschi, I., Ishikawa, T., Breitschopf, K., Ciechanover, A., Schild, L., and Rotin, D. (1997b). Regulation of stability and function of the epithelial Na^+ channel (ENaC) by ubiquitination. *EMBO J.* **16,** 6325–6336.

Strous, G. J., Van Kerkhof, P., Govers, R., Ciechanover, A., and Schwartz, A. L. (1996). The ubiquitin conjugation system is required for ligand-induced endocytosis and degradation of the growth hormone receptor. *EMBO J.* **15,** 3806–3812.

Tamura, H., Schild, L., Enomoto, N., Matsui, N., Marumo, F., Rossier, B. C., and Sasaki, S. (1996). Liddle disease caused by a missense mutation of beta subunit of the epithelial sodium channel gene. *J. Clin. Invest.* **97,** 1780–1784.

Voilley, N., Lingueglia, E., Champigny, G., Mattéi, M.-G., Waldmann, R., Lazdunski, M., and Barbry, P. (1994). The lung amiloride-sensitive Na^+ channel: Biophysical properties, pharmacology, ontogenesis, and molecular cloning. *Proc. Natl. Acad. Sci. U.S.A.* **91,** 247–251.

CHAPTER 6

Role of G Proteins in the Regulation of Apical Membrane Sodium Permeability by Aldosterone in Epithelia

Sarah Sariban-Sohraby
Laboratoire de Physiologie, CP 604, Université Libre de Bruxelles, 1070 Brussels, Belgium

I. INTRODUCTION

Aldosterone is the primary steroid hormone for long-term regulation of sodium balance by epithelial tissues, mainly the kidney. The mechanisms of this regulation can be studied in high electrical resistance model epithelia such as the A6 cultured cells derived from toad kidney (Handler *et al.*, 1981, Benos *et al.*, 1986).

Na^+ entry into epithelia occurs via channels characterized by high Na^+ selectivity, low electrical conductance, and sensitivity to the inhibitor amiloride. The increase in apical membrane conductance after aldosterone is attributed to either activation of quiescent channels preexisting in the membrane or an increase of the channel open probability (Kemendy *et al.*,

1063-5823/99 $30.00

1993; Palmer *et al.,* 1982; Schafer and Hawk, 1992). In contrast, neither increased single-channel conductance nor synthesis of new channels has been demonstrated (Kipnowski *et al.,* 1983; Garty and Edelman, 1983; Kleyman *et al.,* 1989; Tousson *et al.,* 1989). In this regard, while the model of the hormonal control of gene transcription proposed more than 30 years ago remains valid (Edelman *et al.,* 1963), there is no evidence that any of the aldosterone-induced proteins represent apical Na^+ channels per se or subunits of these channels.

A number of distinct biochemical pathways are involved in aldosterone's action on Na^+ permeability, namely, the activation of phospholipid fatty acid metabolism and phospholipase A_2 (Goodman *et al.,* 1975; Yorio and Bentley, 1978), the methylation of apical proteins and lipids (Sariban-Sohraby *et al.,* 1984, 1993; Wiesmann *et al.,* 1985), and the activation of GTP-binding proteins (Cantiello *et al.,* 1989; Garty *et al.,* 1989; Ohara *et al.,* 1993; Sariban-Sohraby *et al.,* 1995). Some or possibly all of these pathways, not mutually exclusive, ultimately increase the transepithelial Na^+ transport rate. In this chapter, we focus on GTP-dependent carboxymethylation and on activation of GTPase activity, two pathways which are tightly linked, with regard to aldosterone stimulation of Na^+ transport.

II. CONTROL OF BASAL Na^+ TRANSPORT BY G PROTEINS

Guanine nucleotides are important regulators of a number of ion channels (Bubien *et al.,* 1994; Wickman and Clapham, 1995; Yatani *et al.,* 1987). In epithelia, the G-protein-mediated pathways controlling Na^+ channels are different from the ones involved with membrane-bound G-protein-coupled receptors and diffusible second messengers. Activation of Na^+ transport by guanine nucleotides was first observed in membrane vesicles prepared from toad urinary bladder, where GTPγS, a nonhydrolyzable derivative of GTP, increases amiloride-sensitive Na^+ uptake (Garty *et al.,* 1989). In studies of single channels, whether in isolated membrane patches of A_6 cells or reconstituted into lipid bilayer membranes, GTPγS modulates Na^+ permeability (Ismailov *et al.,* 1994b; Ohara *et al.,* 1993). Interestingly, this modulation varies with the state in which the channels reside and/or with their membrane environment. For example, when A6 cells are grown on plastic supports in the absence of aldosterone, a condition under which Na^+ transport is low and GTP-dependent protein methylation is not observed (Sariban-Sohraby *et al.,* 1993), GTPγS increases the open probability of "high conductance" Na^+ channels (Cantiello *et al.,* 1989). However, when these cells are grown on porous supports and in the presence of aldosterone,

GTPγS decreases the open probability of "low conductance" Na^+ channels (Duchatelle *et al.*, 1992; Ohara *et al.*, 1993).

III. GTP-DEPENDENT METHYLATION OF MEMBRANE PROTEINS

Guanine nucleotides stimulate protein carboxymethylation in a variety of mammalian cell membranes such as macrophages, brain, liver, and buffy-coat cells (Backlund and Aksamit, 1988). Several classes of carboxyl methyltransferase catalyse methyl esterification at different amino acid residues (Clarke, 1985). Because transmethylation of membrane proteins is rapidly reversible, this would seem to represent an ideal locus for regulating cellular functions (Clarke, 1992; Yamane and Fung, 1993). Similar to bacteria, where carboxymethylation of membrane proteins is involved in chemoreceptor regulation (Clarke, 1985; Springer *et al.*, 1979), Ras-related proteins in human neutrophils are carboxymethylated in a GTP-dependent fashion in response to chemotactic factors (Philips *et al.*,1993). Similarly, G proteins of the *Ras* and *Rho* subfamilies with the carboxy-terminal C-A-A-X consensus sequence are targeted to membranes in response to prenylcysteine carboxymethylation, possibly because of increased hydrophobicity (Yamane and Fung, 1993).

In the mammalian kidney proximal tubule and collecting duct, carboxymethylation of low-molecular-weight G proteins is reported (Gingras *et al.*, 1993; Gupta *et al.*, 1991). In membranes isolated from A6 cultured cells, GTPγS stimulates the transmethylation of a 90-kDa protein. However, when cells are first exposed to aldosterone, a situation in which the level of apical methylation is elevated, GTPγS does not further increase methylation of this protein (Sariban-Sohraby *et al.*, 1993).

IV. CONTROL OF BASAL Na^+ PERMEABILITY BY METHYLATION

Several observations have been reported after *in vitro* methylation by *S*-adenosylmethionine (Adomet) of either isolated A6 membranes or single Na^+ channels:

1. Amiloride-sensitive Na^+ permeability is increased by Adomet in membrane vesicles in the absence of exogenous addition of methyltransferase indicating that the enzyme is membrane-associated (Sariban-Sohraby *et al.*, 1984). This correlates with an increase in whole cell lipid and protein methylation as well as with the methylation of a 90-kDa membrane protein, which is further increased by GTPγS (Sariban-Sohraby *et al.*, 1993).

2. In excised patches of A6 cells, inhibition of methylation by deazaadenosine leads to decreased single-channel P_o, while Adomet prevents the time-dependent decay of channels usually observed in patch-clamp experiments (Kemendy and Eaton, 1990; Duchatelle *et al.,* 1992; Bechetti *et al.,* 1994).

3. When Na^+ channels are reconstituted in planar lipid bilayers, single-channel P_o is increased after Adomet in the presence of purified methyltransferase. This effect is enhanced in the presence of GTPγS (Ismailov *et al.,* 1994b).

V. ALDOSTERONE-DEPENDENT METHYLATION OF MEMBRANE PROTEINS

Aldosterone stimulates protein methylation in apical membranes from A6 epithelia. The substrate of this reaction again is a polypeptide of M_r 90 kDa when methylation is initiated in cultured cells in the presence of [^{3}H]methionine (Sariban-Sohraby *et al.,* 1993). Contrary to what is observed in control membranes, this methylation reaction is not further stimulated by *in vitro* addition of GTPγS, indicating that the GTP-dependent methylation pathway is already activated by aldosterone.

VI. ALDOSTERONE-DEPENDENT MEMBRANE GTPase ACTIVITY

Apically enriched A6 membrane preparations hydrolyze GTP and this reaction is specifically stimulated by aldosterone, which causes a doubling in the maximal rate of the GTPase activity. The specificity of the correlation between increased Na^+ transport and GTPase activity after aldosterone is underscored by several observations:

1. Spironolactone, a competitive inhibitor of aldosterone (Porter, 1968), decreases both transepithelial Na^+ current in A6 cells and the aldosterone stimulation of GTPase activity by the isolated membranes (Sariban-Sohraby *et al.,* 1995).

2. When A6 cells are exposed to antidiuretic hormone, which also increases the Na^+ transport rate, no increase in the rate of GTP hydrolysis by the membranes is detected (Sariban-Sohraby *et al.,* 1995).

3. Increasing GTP hydrolysis *per se* is insufficient to stimulate transepithelial Na^+ transport. This uncoupling of transport and GTP hydrolysis is observed after destruction of proteic synthetic pathways by cycloheximide (Sariban-Sohraby *et al.,* 1995).

GTP hydrolysis is linearly correlated with Na^+ transport rate whether stimulated by aldosterone alone or, in addition, partially inhibited by PTx (see below) or spironolactone or abolished by growing cells on plastic. In view of the likelihood that several intracellular biochemical pathways utilize GTP as energy sources and/or cofactors, the correlation of I_{Na+} and GTP hydrolysis demonstrates an intricate dependence of Na^+ transport on one or more GTPases as related to aldosterone's action.

VII. THE EFFECT OF PERTUSSIS TOXIN ON MEMBRANE Na^+ TRANSPORT AND GTPase ACTIVITY

A number of effector systems are sensitive to *Pertussis* toxin (PTx), which specifically ADP-ribosylates the α subunit of G_i/G_o-type heterotrimeric G proteins (Bourne *et al.,* 1991; Yamane and Fung, 1993). A6 epithelia contain G_i proteins in their membrane, closely associated with the Na^+ channels (Ausiello *et al.,* 1992; Sariban-Sohraby *et al.,* 1995).The effect of PTx is complex, the toxin causing either activation or inhibition of Na^+ permeability (Cantiello *et al.,* 1989; Duchatelle *et al.,* 1992; Ismailov *et al.,* 1994a, Ohara *et al.,* 1993; Sariban-Sohraby *et al.,* 1995). Thus, the uncoupling of one or more G_i proteins by ADP-ribosylation leads to either channel opening or closing. This dual regulation is likely to depend on the conformational state of the channels, which in turn, varies with protein kinase A-mediated phosphorylation, GTP-dependent methylation, and/or growth conditions of the cells (Bubien *et al.,* 1994; Fields and Casey, 1997; Sariban-Sohraby and Fisher, 1995). In A6 cells grown on porous supports, PTx decreases transepithelial Na^+ transport and membrane GTP hydrolysis only marginally, suggesting that steady-state, basal levels of Na^+ transport are not influenced significantly by G_i proteins. In contrast, aldosterone leads to increased toxin sensitivity of both Na^+ transport rate and GTPase activity. I_{Na+} and GTPase activity again appear to be linearly related when PTx is administered with aldosterone (Sariban-Sohraby *et al.,* 1995).

VIII. THE EFFECT OF ALDOSTERONE ON G-PROTEIN EXPRESSION

Expression of G_i/G_o-type G proteins, the putative target of the PTx-induced ADP-ribosylation, is not increased in A6 membranes after short-term aldosterone (Sariban-Sohraby *et al.,* 1995). Likewise, mRNA levels of heterotrimeric G-protein α subunits do not change with aldosterone (Spindler *et al.,* 1997). Therefore, stimulation of GTP hydrolysis by aldosterone must result from the activation of G proteins residing in the apical

TABLE I

Modulation of Membrane Functions by Aldosterone and G Protein-Dependent Mechanisms

	Na^+ transport	Protein methylation	GTP hydrolysis	Single Na^+ channel P_o
Aldosterone	⇑ (1,2)	⇑ (2,5)	⇑ (4)	⇑ (6)
GTPγS	⇑ (3)	⇑ (5)	NA	⇑ or ⇓ (7,8,9)
GDPβS	⇓ (3)	⇓ (5)	⇓ (4)	⇑ or ⇓ (7,8,9)
Spironolactone	⇓ (4)	⇓ (5)	⇓ (4)	ND
Pertussis toxin	⇓ (4)[a]	ND	⇓ (4)	⇑ or ⇓ (7,8,9,10)
$G\alpha_{1\text{-}3}$	ND	ND	ND	⇑ (7)
Methylation	⇑ (2)	NA	ND	⇑ (11)

Note. Numbers in parentheses refer to: (1) Handler *et al.*, 1981; (2) Sariban-Sohraby *et al.*, 1984; (3) Garty *et al.*, 1989; (4) Sariban-Sohraby *et al.*, 1995; (5) Sariban-Sohraby *et al.*, 1993; (6) Kemendy *et al.*, 1993; (7) Cantiello *et al.*, 1989; (8) Duchatelle *et al.*, 1992; (9) Ohara *et al.*, 1993; (10) Bubien *et al.*, 1994; (11) Ismailov *et al.*, 1994b. NA, not applicable; ND, not determined.

[a] PTx inhibits transepithelial Na^+ transport only when A6 cells are stimulated with aldosterone.

membrane, triggered either by a "GTPase-activating protein" (GAPs) or one of the "regulators of G-protein signaling" (RGSs) (Bourne *et al.*, 1991; Fields and Casey, 1997; Markby *et al.*, 1993).

IX. CONCLUSION

The effects of aldosterone on Na^+ transport, protein methylation, GTP hydrolysis, and single-channel open probability in A6 cells are summarized in Table I. Both GTPγS and exogenous methylation reproduce the stimulatory effects of aldosterone on the above parameters, while GDPβS, spironolactone, and PTx antagonize them. The inhibition observed with PTx of both aldosterone-stimulated Na^+ transport and GTP hydrolysis implies the role of one or more G_i-type G proteins. This role, however, remains to be defined further, especially since heterotrimeric G proteins are not induced by the hormone. The regulation by aldosterone of heterotrimeric G proteins other than G_i and of low-molecular-weight G proteins, shown to be associated with epithelial membranes, has not been reported. In this regard, the recent report by Spindler *et al.* (1997) points to an increase in the mRNA level of *Xenopus* K-*ras* 2 in A6 cells after aldosterone, but the function of the "aldosterone-induced proteins" of Edelman *et al.* (1963) remain elusive.

References

Ausiello, D. A., Stow, J. L., Cantiello, H. C., Bruno de Almeida, J., and Benos, D. J. (1992). Purified epithelial Na^+ channel complex contains the Pertussis toxin sensitive $G\alpha_{i\text{-}3}$ protein. *J. Biol. Chem.* **267,** 4759–4765.

Backlund, P. S., and Aksamit, R. R. (1988). Guanine nucleotide-dependent carboxyl methylation of mammalian membrane proteins. *J. Biol. Chem.* **263,** 15864–15867.

Bechetti, A. S., Sariban-Sohraby, S., and Eaton, D. C. (1994). Transmethylation increases highly selective Na channels in A6 epithelia. *Faseb J.* **8,** A339.

Benos, D. J., Saccomani, G., Brenner, B. M., and Sariban-Sohraby, S. (1986). Purification and characterization of the epithelial sodium channel from A6 cultured cells and bovine renal papilla. *Proc. Natl. Acad. Sci. U.S.A.* **83,** 8525–8529.

Bourne, H. R., Sanders, D. A., and McCormick, F. (1991). The GTPase superfamily: Conserved structure and molecular mechanism. *Nature (London)* **349,** 117–127.

Bubien, J. E., Jope, R. S., and Warnock, D. G. (1994). G proteins modulate amiloride-sensitive sodium channels. *J. Biol. Chem.* **269,** 17780–17788.

Cantiello, H. F., Patenaude, C. R., and Ausiello, D. A. (1989). G protein subunit, $\alpha_{i\text{-}3}$, activates a pertussis toxin-sensitive Na^+ channel from the epithelial cell line A6. *J. Biol. Chem.* **264,** 20867–20870.

Clarke, S. (1985). Protein carboxyl methyltransferases: Two distinct classes of enzymes. *Annu. Rev. Biochem.* **54,** 479–506.

Clarke, S. (1992). Protein isoprenylation and methylation at carboxyl-terminal cysteine residues. *Annu. Rev. Biochem.* **61,** 355–386.

Duchatelle, P., Ohara, A., Ling, B. N., Kemendy, A. E., Kokko, K. E., Matsumoto, P. S., and Eaton, D. C. (1992). Regulation of renal epithelial sodium channels. *Mol. Cell. Biochem.* **114,** 27–34.

Edelman, I. S., Bogoroch, R., and Porter, G. A. (1963). On the mechanism of action of aldosterone on sodium transport: The role of protein synthesis. *Proc. Natl. Acad. Sci. U.S.A.* **50,** 1169–1177.

Fields, T. A., and Casey, J. P. (1997). Signaling functions and biochemical properties of pertussis toxin-resistant G-proteins. *Biochem. J.* **321,** 561–571.

Garty, H., and Edelman, I. S. (1983). Amiloride-sensitive trypsinization of apical sodium channels. Analysis of hormonal regulation of sodium transport in toad bladder. *J. Gen. Physiol.* **81,** 785–803.

Garty, H., Yeger, O., Yanovsky, A., and Asher, C. (1989). Guanosine nucleotide-dependent activation of the amiloride-blockable Na^+ channel. *Am. J. Physiol.* **256,** F965–F969.

Gingras, D., Boivin, D., and Beliveau, R. (1993). Subcellular distribution and guanine nucleotide dependency of COOH-terminal methylation in kidney cortex. *Am. J. Physiol.* **265,** F316–F322.

Goodman, D. B. P., Wong, M., and Rasmussen, H. (1975). Aldosterone-induced membrane phospholipid fatty acid metabolism in toad urinary bladder. *Biochemistry* **14,** 2803–2809.

Gupta, A., Bastani, B., Chardin, P., and Hruska, K. A. (1991). Localization of *ral,* a small GTP-binding protein, to collecting duct cells in bovine and rat kidney. *Am. J. Physiol.* **261,** F1063–F1070.

Handler, J. S., Perkins, F. M., and Johnson, J. P. (1981). Studies of renal cell function using cell culture techniques. *Am. J. Physiol.* **240,** C103–C105.

Ismailov, I. I., McDuffie, J. H., and Benos, D. J. (1994a). Protein kinase A phosphorylation and G protein regulation of purified renal Na^+ channels in planar bilayer membranes. *J. Biol. Chem.* **269,** 10235–10241.

Ismailov, I. I., McDuffie, J. D., Sariban-Sohraby, S., and Benos, D. J. (1994b). Carboxymethylation activates purified renal amiloride-sensitive Na^+ channels in planar lipid bilayers. *J. Biol. Chem.* **269,** 22193–22197.

Kemendy, A. E., and Eaton, D. C. (1990). Aldosterone-induced Na^+ transport in A6 epithelia is blocked by 3-deazaadenosine, a methylation blocker. *FASEB J.* **4,** A445.

Kemendy, A. E., Kleyman, T. R., and Eaton, D. C. (1993). Aldosterone alters the open probability of amiloride-blockable sodium channels in A6 epithelia. *Am. J. Physiol.* **263,** C825–C837.

Kipnowski, J., Park, C. S., and Fanestil, D. D. (1983). Modification of carboxyl of Na^+ channel inhibits aldosterone action on Na^+ transport. *Am. J. Physiol.* **245,** F726–734.

Kleyman, T. R., Cragoe, E. J., and Kraehenbuhl, J.-P. (1989). The cellular pool of Na^+ channels in the amphibian cell line A6 is not altered by mineralocorticoids. *J. Biol. Chem.* **264,** 11995–12000.

Markby, D. W., Onrust, R., and Bourne, H. R. (1993). Separate GTP binding and GTPase activating domains in a G alpha subunit. *Science* **262,** 1895–1901.

Ohara, A., Matsunaga, H., and Eaton, D. C. (1993). G-protein activation inhibits amiloride-blockable highly selective sodium channels in A6 cells. *Am. J. Physiol.* **264,** C352–C360.

Palmer, L. G., Li, J. H., Lindemann, B., and Edelman I. S. (1982). Aldosterone control of the density of sodium channels in the toad urinary bladder. *J. Membr.Biol.* **64,** 91–102.

Philips, M. R., Pillinger, M. H., Staud, R., Volker, C., Rosenfeld, M. G., Weissmann, G., and Stock, J. B. (1993). Carboxyl methylation of ras-related proteins during signal transduction in neutrophils. *Science* **259,** 977–979.

Porter, G. A. (1968). In vitro inhibition of aldosterone-stimulated Na^+ transport by steroidal spirolactones. *Mol. Pharmacol.* **4,** 224–237.

Sariban-Sohraby, S., and Fisher, R. S. (1995). Guanine nucleotide-dependent carboxymethylation: A pathway for aldosterone modulation of apical Na^+ permeability in epithelia. *Kidney Int.* **48,** 965–969.

Sariban-Sohraby, S., Burg, M., Wiesmann, W. P., Chiang, P. K., and Johnson, J. P. (1984). Methylation increases sodium transport into A6 apical membrane vesicles. Possible mode of aldosterone action. *Science* **225,** 745–746.

Sariban-Sohraby, S., Fisher, R. S., and Abramow, M. (1993). Aldosterone-induced and GTP-stimulated methylation of a 90 kDa polypeptide in the apical membrane of A6 epithelial cells. *J. Biol. Chem.* **268,** 26613–26617.

Sariban-Sohraby, S., Mies, F., Abramow, M., and Fisher, R. S. (1995). Aldosterone stimulation of GTP hydrolysis in membranes from renal epithelia. *Am. J. Physiol.* **268,** C557–C562.

Schafer, J. A., and Hawk, C. T. (1992). Regulation of Na^+ channels in the cortical collecting duct by AVP and mineralocorticoids. *Kidney Int.* **41,** 255–268.

Spindler, B., Mastroberardino, M., Custer, M., and Verrey, F. (1997). Characterization of early aldosterone-induced RNAs identified in A6 kidney epithelia. *Pfluegers Arch.* **434,** 323–331.

Springer, M. S., Goy, M. F., and Adler, J. (1979). Protein methylation in behavioral control mechanisms and in signal transduction. *Nature.* (*London*) **280,** 279–281.

Tousson, A., Alley, C. D., Sorscher, E. J., Brinkley, B. R., and Benos, D. J.(1989). Immunochemical localization of amiloride-sensitive sodium channels in sodium-transporting epithelia. *J. Cell Sci.* **93,** 349–362.

Wickman, K., and Clapham, D. E. (1995). Ion channel regulation by G proteins. *Physiol. Rev.* **75,** 865–885.

Wiesmann, W. P., Johnson, J. P., Miura, G. A., and Chiang, P. K. (1985). Aldosterone-stimulated transmethylations are linked to sodium transport. *Am. J. Physiol.* **248,** F43–F47.

Yamane, K., and Fung, B. K. K. (1993). Covalent modifications of G-proteins. *Annu. Rev. Pharmacol. Toxicol.* **32,** 201–241.

Yatani, A., Codina, J., Brown, A. M., and Birnbaumer, L. (1987). Direct activation of mammalian atrial muscarinic potassium channels by GTP regulatory protein G_k. *Science.* **235,** 207–211.

Yorio, T., and Bentley, P. J. (1978). Phospholipase A and the mechanism of action of aldosterone. *Nature* (*London*) **271,** 79–81.

CHAPTER 7

The Role of Posttranslational Modifications of Proteins in the Cellular Mechanism of Action of Aldosterone

J. P. Johnson, J.-M. Wang, and R. S. Edinger
Laboratory of Epithelial Cell Biology, Renal-Electrolyte Division, Department of Medicine, University of Pittsburgh, Pittsburgh, Pennsylvania 15213

The cellular mechanism of action by which aldosterone stimulates Na^+ reabsorption in responsive epithelia has remained an elusive physiologic puzzle for decades (Edelman, 1978; Garty and Palmer, 1997; Johnson, 1992; Verrey, 1995). Since the early 1960s when Edelman and colleagues demonstrated that mineralocorticoids mediate their effects through transcriptional regulation (Edelman, 1978; Edelman *et al.,* 1963), physiologists and biochemists have worked within the classic paradigm indicated in Fig. 1. Acting through binding to specific cellular receptors, the hormone stimulates transcriptional activity, which results in the production of aldosterone-induced proteins (AIPs), which in turn induce the upregulation of Na^+ transport. Targets for regulation by these proteins could be the epithelial Na^+ channel, through which Na^+ enters the cell from the luminal surface; the enzyme Na^+,K^+-ATPase, which mediates an energy-dependent exit from the basolateral surface of the cell; or finally through an effect on the coupling of energy metabolism to transport. In fact, all these putative sites of action are probably affected by the hormone but the exact cellular mechanisms remain poorly defined.

The activities of the Na^+ channel and Na^+,K^+-ATPase have long been recognized as regulated by aldosterone in responsive epithelia, but this regulation is not simply induction of synthesis of these critical transport proteins. Studies employing such disparate approaches as apical channel inactivation by trypsin (Garty and Edelman, 1983), covalent modification (Kipnowski *et al.,* 1983; Park and Fanestil, 1983), and electrophysiologic

1063-5823/99 $30.00

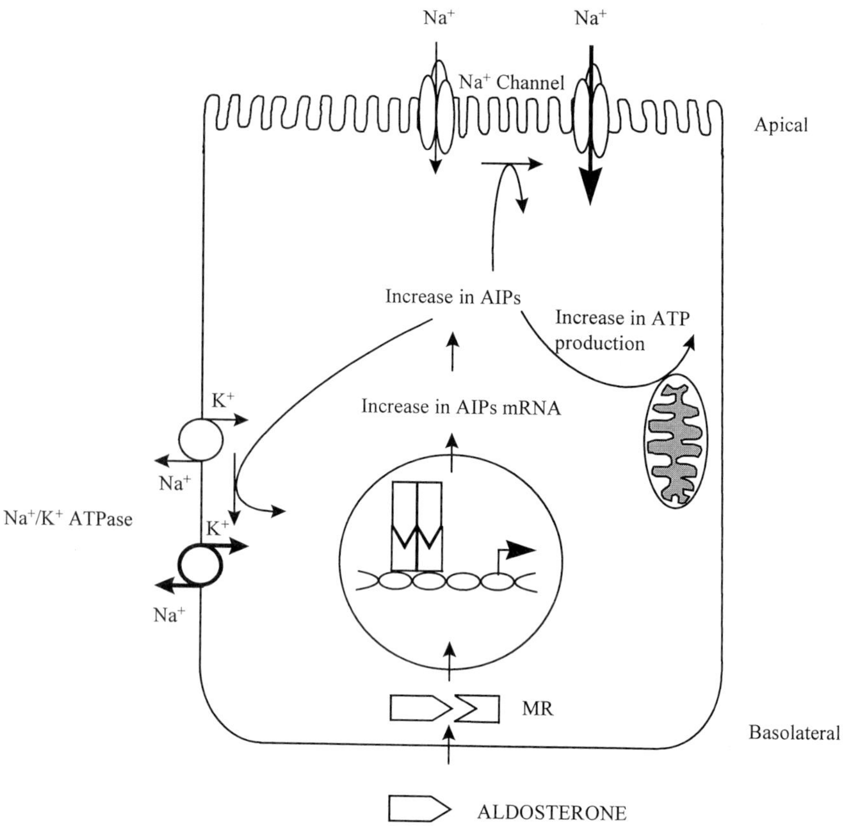

FIGURE 1 Proposed classic schema of the mechanism of action of aldosterone with sites of action at Na^+ channel, Na^+ pump, and mitochondrial energy supply.

noise analysis (Palmer *et al.,* 1982) and patch clamp (Kemendy *et al.,* 1992) have all suggested that aldosterone stimulates channel activity by increasing the number or open probability of active channels within the apical membrane without the insertion of new channel subunits, at least during the first 4–6 h of its action. These observations have been confirmed by direct and indirect (Kleyman *et al.,* 1989, 1992; May *et al.,* 1997) measures of channel proteins and mRNA for these proteins which have become possible following the remarkable studies which led to the identification of ENaC as the amiloride-sensitive Na^+ channel in aldosterone-responsive epithelia (Canessa *et al.,* 1993, 1994a,b; Cantiello *et al.,* 1989; Duc *et al.,* 1994). Conclusions from these studies have largely agreed that aldosterone leads to activation of preexisting channels through some as yet undetermined process involving either some biochemical modification of the channel (Asher and Garty, 1998; Benos *et al.,* 1995) or stimulation of a signaling cascade. Late in the course of aldosterone action there are modest increases in mRNA for channel subunits and possibly an increase in the biochemical pool of channels as well (May *et al.,* 1997).

Aldosterone-mediated effects on the Na^+ pump are also complex. In this case there is direct evidence for early regulation of mRNA for pump subunits (Verrey and Beron, 1996; Verrey *et al.,* 1987, 1989), but evidence of increased pump activity or number at the basolateral membrane cannot generally be detected until well after the initial transport stimulation occurs. It is speculated that pump activation may be dependent on the early stimulation of Na^+ entry through channel activation (Barlet-Bas *et al.,* 1990). Direct effects of aldosterone on synthesis of mitochondrial enzymes, particularly citrate synthase, have also been demonstrated (Kirsten *et al.,* 1968; Law and Edelman, 1978) but these effects may be dissociated from transport stimulation in some epithelia (Handler *et al.,* 1981; Johnson and Green, 1981). A number of other cellular effects of aldosterone have been noted, including stimulation of acylation/deacylation reactions with an increase in unsaturated long-chain fatty acids in cell membrane phospholipids (Goodman *et al.,* 1971, 1975; Lien *et al.,* 1975), stimulation of phosphytidylcholine biosynthesis and phospholipid methylation (Wiesmann *et al.,* 1985), and stimulation of phospholipase A_2 (Yorio and Bentley, 1978). In each case, inhibition of these aldosterone-stimulated processes of fatty acid metabolism has been associated with inhibition of the transport response, but a unifying hypothesis relating the activation of channels, insertion of pumps, and modification of cellular membranes has not emerged.

A number of attempts have been made to directly identify AIPs or transcriptional products of aldosterone employing classic techniques of protein biochemistry or molecular biology (Benjamin and Singer, 1974; Geheb *et al.,* 1981; Spindler *et al.,* 1997). Although providing interesting and sometimes provocative results, these studies have yet to shed further light on the complex process by which aldosterone produces its transport effects. This process is complex in that it induces responses at both cell membranes with differential actions over a prolonged time course. It may thus be seen as a problem in cellular signaling and coordination or orchestration of these responses to produce the simple physiologic end-point of increased Na^+ reabsorption. One way to approach this problem is to examine posttranslational effects on proteins, particularly as they relate to signaling. Verrey has referred to this as the "backwards" approach (Verrey, 1995), inferring transcriptional effects from biochemical or physiologic targets as they are modified by aldosterone, and this is an accurate description. Yet the approach may yield fruitful and interesting results. We review the process listed in Table 1 with reference to studies relating them to aldosterone action. As can be seen, these process primarily relate to posttranslational protein processing or signalling pathways.

Glycosylation is a posttranslational modification of proteins often associated with targeting to intracellular sites or membranes. Considerable evidence suggests that glycosylation is involved in processing of proteins regulating Na^+ transport (Asher and Garty, 1998; Duc *et al.,* 1994) and AIPs (Blazer-Yost and Cox, 1985; Szerlip *et al.,* 1988). The amiloride-sensitive Na^+ channel consists of three subunits which have extensive extracellular domains which are heavily glycosylated (Asher and Garty, 1998; Benos *et al.,* 1995; Duc *et al.,* 1994). Cox and colleagues, in a series of studies examining AIPs by two-dimensional protein separation techniques, have identified a cluster of acidic proteins, not apparently ENaC subunits, which are also glycosylated (Blazer-Yost and Cox, 1985; Szerlip *et al.,* 1988). The identity of these proteins remains unknown. Inhibition of glycosylation with tunicamycin results in a time-dependent decrease in amiloride-sensitive Na^+ transport and an inhibition of aldosterone-stimulated transport (Szerlip *et al.,* 1988). There is no evidence, however, that glycosylation per se is a regulated process with regard to Na^+ transport, and the effects of glycosylation inhibitors seem likely to be related to constitutive processing of channel subunits. This has not been studied in any detail and there remain some questions of this interpretation particularly with regard to time course. Studies of channel turnover in A6 suggest that the channel is a relatively short-lived protein in apical membranes (May *et al.,* 1997), while the time course of tunicamycin inhibition of transport is slow relative to this turnover (Szerlip *et al.,* 1988).

Protein acylation is another form of posttranslational modification associated with processing and targeting of proteins within the cell. A number

TABLE I

Posttranslational Systems Implicated in the Action of Aldosterone

Glycosylation
Acylation
Palmitoylation
Myristoylation
Isoprenylation
Protein carboxylmethylation
Activation of kinases or phosphatases
Stimulation of G-protein synthesis or activation
Direct alterations in lipid bilayer structure
Activation of phospholipase to produce second messengers

of fatty acid acyl groups have been described in this regard, including palmitate, myristate, and the isoprenoids. The possibility that such reactions might be involved in aldosterone regulation of Na^+ transport has recently been considered (Blazer-Yost *et al.,* 1997; Rokaw *et al.,* 1996a; Rokaw and Johnson, 1994) and is a provocative notion, consistent with earlier observations that inhibition of synthesis of fatty acid acyl groups is associated with attenuation of the aldosterone response in some epithelia (Goodman *et al.,* 1971, 1975; Lien *et al.,* 1975; Wiesmann *et al.,* 1985). There is no evidence that acyltransferase enzymes per se are regulated by steroids, so it is hypothesized that these pathways might be involved through increase in amount of a putatively acylated protein substrate. We have described aldosterone-stimulated palmitoylation of several G proteins in A6 cells (Rokaw *et al.,* 1996a), including the $G_{\alpha i3}$ subunit of the heterotrimeric G proteins which may be a channel-associated regulatory protein (Ausiello *et al.,* 1992; Cantiello *et al.,* 1989). Inhibition of palmitoylation markedly downregulated the aldosterone response without a marked effect on basal transport rates. The exact site of this effect is not entirely clear. While we have reported an apparent increase in palmitoylated $G_{\alpha i3}$ in association with immunoprecipitates of an Na^+ channel complex in A6 cells (Rokaw *et al.,* 1996a), Sariban-Sohraby was unable to demonstrate an increase in immunodetectable $G_{\alpha i3}$ in apical membrane preparations in response to aldosterone (Sariban-Sohraby *et al.,* 1995), and Verrey was unable to demonstrate an early increase in mRNA for this protein in response to aldosterone using differential display PCR (Spindler *et al.,* 1997). It may be, therefore, that a palmitoylation reaction critical to aldosterone activation of Na^+ transport has a different protein substrate. This possibility is particularly intriguing in view of recent evidence that a K-ras protein may be induced by aldosterone in A6 cells (Al-Baldawi *et al.,* 1998; Spindler *et al.,* 1997). This protein is known to be posttranslationally modified by palmitoylation as well as isoprenylation and carboxylmethylation (Dudler and Gelb, 1996).

Myristoylation of proteins in response to aldosterone has not been extensively studied. We were unable to detect a stimulation of protein myristoylation in response to aldosterone (Rokaw and Johnson, 1994) and inhibitors of myristoylation reactions produced an increase in basal and aldosterone-induced transport rates in A6 (Johnson *et al.*). Isoprenylation of proteins by either farnesyl or geranylgeranyl moieties has been studied indirectly by several groups with conflicting results (Blazer-Yost *et al.,* 1997; Rokaw *et al.,* 1996c). This pathway of posttranslational modification of proteins has emerged as a significant event in signaling pathways. Multiple small G proteins are activated by sequential acylation and C-terminal carboxylmethylation (Clark, 1992). While there is agreement that HMG-Coreductase inhibitors, which act early in the cascade of isoprenoid biosynthesis, result in downregulation of the aldosterone response (Blazer-Yost *et al.*, 1997; Rokaw *et al.*, 1995), these agents are extremely nonspecific. No protein that is isoprenylated in response to aldosterone has been identified by metabolic labeling techniques (Blazer-Yost *et al.*, 1997; Rokaw *et al.*, 1995, 1996c). Inhibitors of carboxylmethylation of farnesylated and geranylgeranylated proteins demonstrate that block of this processing pathway for geranylgeranylated proteins is associated with partial (Rokaw *et al.*, 1996c) or complete (Blazer-Yost *et al.*, 1997) inhibition of the aldosterone effect on Na^+ transport. Eaton and colleagues reported an increase in carboxylmethylation of K-ras (Al-Baldawi *et al.*, 1998), the only isoprenylated G protein known to be induced by aldosterone (Spindler *et al.*, 1997). While this would be consistent with the effects of palmitoylation inhibitors noted above, it is hard to integrate with the findings of the isoprenylation studies. The Ras proteins are known to be posttranslationally modified by farnesylation not by geranylgeranylation (Hancock *et al.*, 1989). This area, nonetheless, remains extremely interesting in view of the evidence that aldosterone may act, in part, through G-protein activation (Al-Baldawi *et al.*, 1998; Sariban-Sohraby and Fisher, 1995; Spindler *et al.*, 1997) and that there may be multiple effects of protein carboxylmethylation in the action of aldosterone (Al-Baldawi *et al.*, 1998; Blazer-Yost *et al.*, 1997; Rokaw *et al.*, 1995; Sariban-Sohraby and Fisher, 1995).

Protein carboxylmethylation (PCM) has been suggested as a posttranslational modification associated with aldosterone stimulation of Na^+ transport on the basis of a number of observations. Aldosterone stimulates PCM in responsive cells (Weismann *et al.*, 1985) and inhibition of methylation blocks the transport response. Sariban-Sohraby has identified aldosterone stimulation of PCM of a 95-kDa protein in the apical membrane of A6 cells (Sariban-Sohraby *et al.*, 1993) and demonstrated that *in vitro* methylation of apical membrane vesicles from A6 results in an increase in Na^+ uptake similar to that seen in vesicles prepared from cells pretreated with aldosterone (Sariban-Sohraby *et al.*, 1984). Ismailov has shown that carboxylmethylation of the 95-kDa subunit of the immunopurified Na^+ channel complex results in an increase in open probability (Ismailov *et al.*, 1994b), and we have demonstrated that aldosterone stimulates PCM of the β subunit of xENaC in A6 cells (Rokaw *et al.*, 1998). Carboxylmethylation of ENaC reconstituted in planar lipid bilayers also results in an increase in an open probability of the channel (Rokaw *et al.*, 1998), similar to the results obtained with the immunopurified complex.

Taken together, these results suggest that aldosterone may modify channel kinetics through direct modification of the β subunit by methylation. The site of the regulated step, however, remains unknown. In this regard, two groups have now reported stimulation of membrane-associated methyltransferase activity by aldosterone (Al-Baldawi *et al.,* 1998; Johnson *et al.,* 1999). This enzyme activity is also stimulated by GTP analogs, consistent with the GTP dependence of channel methylation noted in the above studies (Ismailov *et al.,* 1994b; Sariban-Sohraby and Fisher, 1995; Sariban-Sohraby *et al.,* 1993). It remains to be determined whether aldosterone stimulates activity of this methyltransferase directly or through stimulation of a G protein.

Other potential direct modifications of channel subunits have not been conclusively linked to aldosterone action. Channel activity is known to be regulated by the action of a number of kinases, including PKA (Asher and Garty, 1998; Garty and Palmer, 1997; Prat *et al.,* 1993a; Helman *et al.,* 1983), PKC (Ling and Eaton, 1989; Rokaw *et al.,* 1996b), and tyrosine kinases (Blazer-Yost *et al.,* 1998; Matsumoto *et al.,* 1993). While a number of studies have identified phosphorylations of the immunopurified channel complex reconstituted in planar lipid bilayers which convincingly modify channel kinetics (Ismailov *et al.,* 1994a; Oh *et al.,* 1993), it is not clear that ENaC subunits are gated directly by phosphorylation in cells in response to hormone stimulation. Kinase effects may be indirect, acting through cytoskeletal elements (Cantiello *et al.,* 1991; Prat *et al.,* 1993b), or through activation of signaling cascades resulting in alterations in channel number or activity) in apical membrane. Elegant studies by Canessa and colleagues (Skimkets *et al.,* 1998) have demonstrated hormonal modulation of channel subunit phosphorylation in MDCK cells stably transfected with ENaC. These studies demonstrated basal levels of phosphorylation of β and γ subunits, but not α. These phosphorylations are upregulated by aldosterone (as early as 3 h), insulin, PKA, and PKC. The sites appear to be serines or threonines in the carboxy termini of these two subunits. No regulated phosphorylation of α was detectable. These observations, coupled with the bilayer studies, support the notion that kinase activity may directly regulate channel function and may represent one of the early mechanisms of aldosterone action. Whether aldosterone-stimulated phosphorylations regulate channel kinetics or turnover is not currently known. The only currently available studies directly examining a role of aldosterone on kinase activity in epithelial systems are those of Liu and Greengard (1974, 1976; Liu *et al.,* 1981) which suggest activation of a protein phosphatase activity by aldosterone, and a serine/threonine kinase activity which is activated in many steroid-responsive tissues and does not appear to be mineralocorticoid specific.

Other pathways linked to aldosterone activation of Na^+ transport include G-protein activation and direct actions to modify the lipid bilayer. G-protein activation is addressed elsewhere in this volume (Sariban-Sohraby, Chapter 6) and will not be discussed in further detail here beyond noting that multiple G-protein-mediated effects on channel activity have been

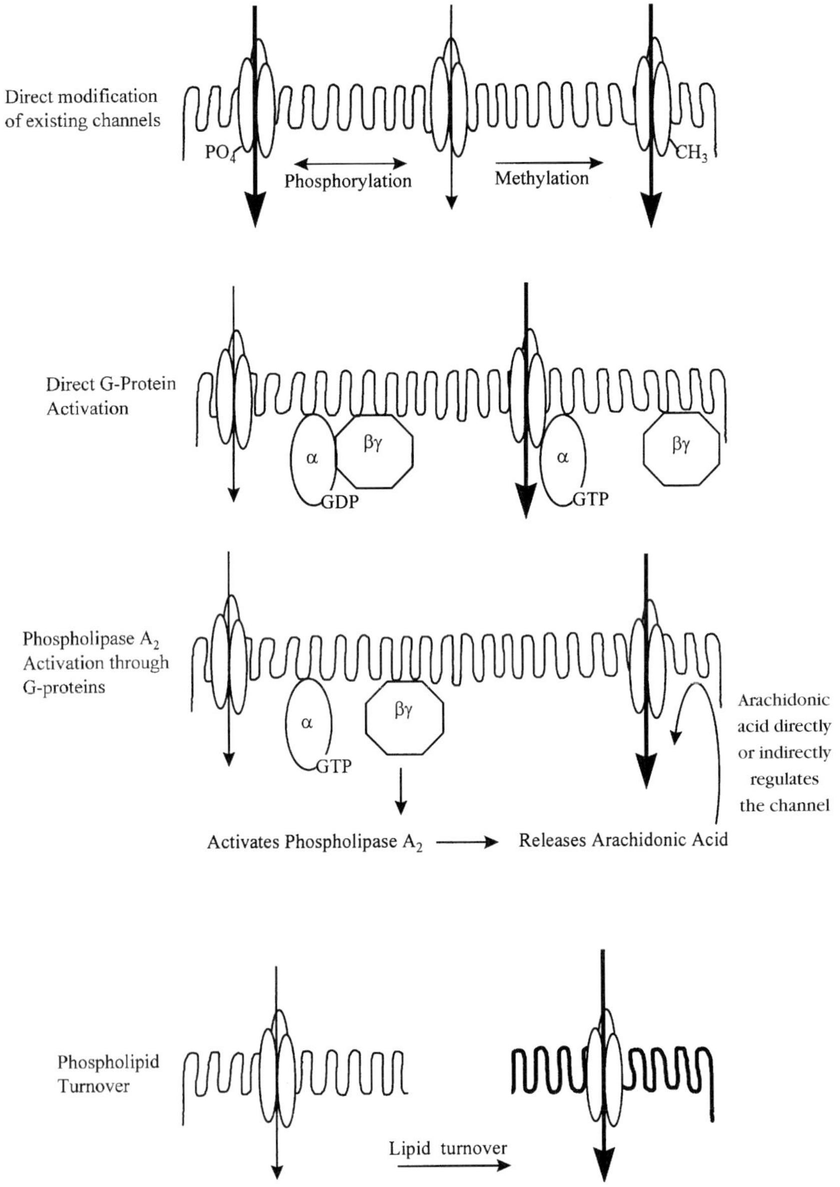

FIGURE 2 Schematization of potential sites of action of aldosterone in cell membranes which modify transporter activity. Evidence in support of each of these proposed mechanisms is discussed in the text. G proteins are depicted as heterotrimerics, but localization could involve activated small G proteins as well (see text).

noted (Garty and Palmer, 1997; Johnson, 1992; Sariban-Sohraby and Fisher, 1995) and may be linked to aldosterone action. Posttranslational processing steps associated with G-protein activation described above have also been implicated in aldosterone action so elucidation of G-protein pathways stimulated by aldosterone and coupled to channel activation remain potentially fruitful areas of research.

It has been known for years that steroids have direct effects on turnover of components of the lipid bilayer, including alterations in composition of acyl subgroups (Goodman *et al.,* 1971, 1975; Lien *et al.,* 1975) and alterations in turnover of polar head groups (Wiesmann *et al.,* 1985). No convincing studies relating lipid membrane composition or fluidity, potentially altered in response to steroid activation and directly affecting channel activity, are currently available. It remains possible that alterations in membrane lipids may produce transport regulation by activation of signaling cascades. Phospholipase A_2 has been shown to be stimulated by aldosterone in responsive epithelia, and this action is essential to transport stimulation by aldosterone (Yorio and Bentley, 1978). Whether this action is mediated through an effect on membrane remodeling or on the action of an arachidonate metabolite as originally proposed remains unknown. But evidence indicating that metabolites of PLA_2 activation directly stimulate channel activity (Cantiello *et al.,*) suggests that the original observation may have been prescient.

Figure 2 illustrates a number of pathways for which evidence exists in a suggestive but not conclusive manner linking them to aldosterone activation of Na^+ transport. Studies looking at posttranslational effects of steroids have implicated all these pathways at some point in the complex cellular action of aldosterone. One or all of these pathways may serve to modify channel activity through alterations in kinetic activity, activation of latent channels, or alterations in the turnover of channels in the apical membrane. It may be hoped that the results which produced this approach through the "back door" of posttranslational action will eventually meet with those currently underway through the "front door" of transcriptional products, producing a satisfying understanding of the mechanism of action of this fascinating hormone.

Acknowledgment

Work described in this report was supported in part by a grant from NIDDK, DK, 4787.

References

Al-Baldawi, N. F., Stockland, J. D., and Eaton, D. C. (1998) Protein methylation of Ras p21 induced by aldosterone. *FASEB J.* **12,** A421.

Asher, C. and Garty, H. (1998). Aldosterone increases the apical Na^+ permeability of toad bladder by two different mechanisms. *Proc. Natl. Acad. Sci. U.S.A.* **85,** 7413–7417.

Ausiello, D. A., Stow, J. L., Cantiello, H. F., DeAlmedia, and J. B., Benos, D. J. (1992). Purified epithelial Na^+ channel complex contains the pertussis toxin-sensitive $G_{\alpha i3}$ protein. *J. Biol. Chem.* **267,** 4756–4765.

Barlet-Bas, C., Khadouri, C., Marsy, S. and Doucet, A. (1990). Enhanced intracellular sodium concentration in kidney cells recruits a latent pool of Na-K-ATPase whose size is modulated by corticosteroids. *J. Biol. Chem.* **265,** 7799–7803.

Benjamin, W. B., and Singer, I. (1974). Aldosterone-induced proteins in toad urinary bladder. *Science* **186,** 69–272.

Benos, D. J., Awayda, M. S., Ismailov, I. I., and Johnson, J. P. (1995). Structure and function of amiloride-sensitive Na^+ channels. *J. Membr. Biol.* **143,** 1–18.

Blazer-Yost, B., and Cox, M. (1985). Aldosterone-induced proteins: Characterization using lectin-affinity chromatography. *Am. J. Physiol.* **249,** C215–C225.

Blazer-Yost, B., Hughes, C. L., and Nolan, P. L. (1997). Protein prenylation is required for aldosterone-stimulated Na^+ transport. *Am. J. Physiol.* **272,** C1928–C1935.

Blazer-Yost, Liu, X., and Helman, S. I. (1998). Hormonal regulation of ENaCs: Insulin and Aldosterone. *Am. J. Physiol.* **274,** C1373–C1379.

Canessa, C. M., Horisberger, J. D., and Rossier, B. C. (1993). Epithelial sodium channel related to proteins involved in neurodegeneration. *Nature* (*London*) **361,** 467–470.

Canessa, C. M., Merillat, A.-M., and Rossier, B. C., (1994a). Membrane topology of the epithelial sodium channel. *Am. J. Physiol.* **267,** C1682–C1690.

Canessa, C. M., Schild, L., Buell, G., Thorens, B., Gautschi, I., Horisberger, J.-D., and Rossier, B. C. (1994b). Amiloride-sensitive epithelial Na^+ channel is made of three homologous subunits. *Nature* (*London*), **367,** 463–467.

Cantiello, H. F., Patenaude, C. R., and Ausiello, D. A. (1989). G. protein subunit, α_{i3}, activates a pertussis toxin-sensitive Na^+ channel from the epithelial cell line, A6. *J. Biol. Chem.* **264,** 20867–20870.

Cantiello, H. F., Patenaude, C. R., Codina, J., Birnbaumer, L., and Ausiello, D. A., (1990). $G\alpha i3$ regulates epithelial Na^+ channels by activation of phospholipase A_2 and lipoxygenase pathways. *J. Biol. Chem.* **65,** 21624–21628.

Cantiello, H. F., Stow, F. J., Prat, A. G., and Ausiello, D. A. (1991). Actin filaments control epithelial Na^+ channel activity. *Am. J. Physiol.* **261,** C882–C888.

Clark, S. (1992). Protein isoprenylation and methylation at carboxyl-terminal cysteine residues. *Annu. Rev. Biochem.* **61,** 355–386.

Duc, C., Farman, N. Cannessa, C. M., Bonvalet, J.-P., and Rossier, B. C. (1994). Cell-specific expression of epithelial sodium channel α, β and γ subunits in aldosterone-responsive epithelia from the rat: Localization by in situ hybridization and immunocytochemistry. *J. Cell. Biol.* **127,** 1907-1921.

Dudler, T., and Gelb, M. H. (1996) Palmitoylation of H-ras facilities membrane binding, activation of downstream effectors and meiotic maturation in *Xenopus* oocytes. *J. Biol. Chem.* **271,** 11541–11547.

Edelman, I. S. (1978). Candidate mediators in the action of aldosterone on Na^+ transport. *In* "Membrane Transport Processes" (J. F. Hoffman, ed.), Vol. 1, pp. 125-140. Raven Press, New York.

Edelman, I. S., Bogoroch, R., and Porter, G. A. (1963). On the mechanism of action of aldosterone on sodium transport: The role of protein synthesis. *Proc. Natl. Acad. Sci. U.S.A.* **50,** 1169–1177.

Garty, H., and Edelman, I. S. (1983). Analysis of hormonal regulation of sodium transport in toad bladder. *J. Gen. Physiol.* **81,** 785–803, (1983).

Garty, H., Palmer L. G. (1997). Epithelial sodium channels: Function, structure and regulation. *Physiol. Rev.* **77,** 359–396.

Geheb, M., Huber, G., Hercker, E., and Cox, M. (1981). Aldosterone-induced proteins in toad urinary bladders. *J. Biol. Chem.* **256,** 11716–11723.

Goodman, D. B. P., Allen, J. E., and Rasmussen, H. (1971). Studies on the mechanism of action of aldosterone: Hormone-induced changes in lipid metabolism. *Biochemistry* **10,** 3825–3831.

Goodman, D. P. B., Wong, M., and Rasmussen, H. (1975). Aldosterone-induced membrane phospholipid fatty acid metabolism in the toad urinary bladder. *Biochemistry* **14,** 2803–2809.

Hancock, J. F., Magee, A. I., Childs, J. E., and Marshall, C. J. (1989). All ras proteins are polyisoprenylated but only some are palmitoylated. *Cell (Cambridge, Mass.)* **57,** 1167–1177.

Handler, J. S., Preston, A. S., Perkins, F. M., Matsamura, M., Johnson, J. P., and Watlington, C. O. (1981). The effect of adrenal steroids on epithelia formed in culture by A6 cells. *Ann. N.Y. Acad. Sci.* **372,** 442–454.

Helman, S. I., Cox, T. C., and Van Driessche, W. (1983). Hormonal control of apical membrane Na transport in epithelia. Studies with fluctuation analysis. *J. Gen. Physiol.* **82,** 201–220.

Ismailov, I. I., McDuffie, J. H., and Benos, D. J. (1994a). Protein kinase A phosphorylation and G protein regulation of purified renal Na^+ channels in planar bilayer membranes. *J. Biol. Chem.* **269,** 10235–10241.

Ismailov, I. I., McDuffie, J. D., Sariban-Sohraby, S., Johnson, J. P., and Benos, D. J. (1994b). Carboxymethylation activates purified renal amiloride-sensitive channels lipid bilayers. *J. Biol. Chem.* **269,** 22193–22197.

Johnson, J. P. (1992). Cellular mechanisms of action of mineralocorticoid hormones. *Pharmacol. Ther.* **53,** 1–29, 1992.

Johnson, J. P., and Green, S. W. (1981). Aldosterone stimulates Na^+ transport without affecting citrate synthase in cultured cells. *Biochim. Biophys. Acta.* **647,** 293–296.

Johnson, J. P., Gordon, R. K., Chiang, P., Wang, P.-M., Hui, D., Edinger, R. S., and Rokaw, M. D. (1999). Effect of methyl donors on the early and late aldosterone response in A6 cells: Induction of a methyltransferase. In review.

Kemendy, A. E., Kleyman, T. R., and Eaton, D. C. (1992) Aldosterone alters the open probability of amiloride-blockable sodium channels in A6 epithelia. *Am. J. Physiol.* **263,** C825–C83.

Kipnowski, J., Park, C. S., and Fanestil, D. D. (1983). Modification of carboxyl of Na^+ channels inhibits aldosterone action on Na^+ transport. *Am. J. Physiol.* **245**, F726–734.

Kirsten, E., Kirsten, R., Leaf, A., and Sharp, G. W. G. (1968). Increased activity of enzymes of the tricarboxylic acid cycle in response to aldosterone in toad bladder. *Pfluegers Arch.* **300,** 213-225.

Kleyman, T. R., Cragoe, E. J., and Kraehenbuhl, J. P. (1989) The cellular pool of Na^+ channels in amphibian cell line A6 is not altered by mineralocorticoids: Analysis using a new photoactive amiloride analog in combination with anti-amiloride antibodies. *J. Biol. Chem.* **264,** 11995–12000.

Kleyman, T. R., Coupaye-Gerard, B., and Ernst, S. A. (1992). Aldosterone does not alter apical cell-surface expression of epithelial Na^+ channels in the amphibian cell line A6. *J. Biol. Chem.* **267,** 9622–9628.

Law, P. L., and Edelman, E. S. (1978). Induction of citrate synthase by aldosterone in rat kidney. *J. Membr. Biol.* **41,** 15–40.

Lien, E. L., Goodman, D. P. B., and Rasmussen, H. (1975). Effects on an acetyl-coenzyme A carboxylase inhibitor and a sodium sparing diuretic on aldosterone-stimulated sodium transport, lipid synthesis and phospholipid fatty acid composition in the toad urinary bladder. *Biochemistry* **14,** 2749–2754.

Ling, B. N., and Eaton, D. C. (1989). Effects of luminal Na^+ on single Na^+ on single Na^+ channels in A6 cells, a regulatory role for protein kinase C. *Am. J. Physiol.* **256,** F1094–F1103.

Liu, A., and Greengard, P. (1974). Aldosterone-induced increase in protein phosphatase activity in toad bladder. *Proc. Natl. Acad. Sci. U.S.A.* **71,** 3869–3873.

Liu, A., and Greengard, P. (1976) Regulation by steroid hormones of phosphorylation of specific protein common to several target tissues. *Proc. Natl. Acad. Sci. U.S.A.,* **73,** 568–572.

Liu, A., Walter, U., and Greengard, P. (1981). Steroid hormones may regulate autophosphorylation of adenosine-3′,5′-monophosphate-dependent protein kinase in target tissues. *Eur. J. Biochem.* **114,** 539–548.

Matsumoto, P. S., Ohara, A., Duchatelle, P., and Eaton, D. C., (1993). Tyrosine kinase regulates epithelial sodium transport in A6 cells. *Am. J. Physiol.* **264,** C246–C250.

May, A., Puoti, A., Gsaeggeler, H.-P., Horisberger, J.-D., and Rossier, B. C. (1997). Early effect of aldosterone on the rate of synthesis of the epithelial sodium channel a subunit in A6 renal cells. *J. Am. Soc. Nephrol.* **8,** 1813–1822.

Oh, Y., Smith, P. R., Bradford, A. L., Keeton, D., and Benos, D. J. (1993). Regulation by phosphorylation of purified epithelial Na^+ channels in planar lipid bilayers. *Am. J. Physiol.* **265,** C85–C91.

Palmer, L. G., Li, J., Lindemann, B., and Edelman, I. S. (1982). Aldosterone control of the density of sodium channels in the toad urinary bladder. *J. Membr. Biol.* **64,** 91–102.

Park, C. S., and Fanestil, D. D. (1983). Functional groups of the Na^+ channel: Role of carboxyl and histidyl groups. *Am. J. Physiol.* **245,** F716–F725.

Prat, A. G., Ausiello, D. A., and Cantiello, H. F., (1993a). Vasopressin and protein kinase A activate G protein-sensitive epithelial Na^+ channels. *Am. J. Physiol.* **265,** C218–C223.

Prat, A. G., Bertorello, A. M., Ausiello, D. A., and Cantiello, H. G., (1993b). Activation of epithelial Na^+ channels by protein kinase A requires actin filaments. *Am. J. Physiol.* **265,** C224–C233.

Rokaw, M. D., and Johnson, J. P. (1994). Small GTP, binding proteins are synthesized and acylated in response to aldosterone. *Clin. Res.* **42,** 319.

Rokaw, M., West, M., and Johnson, J. P. (1995). Dual effect of aldosterone on methylation reactions in A6 cells. *J. Am. Soc. Nephrol.* **6,** 744.

Rokaw, M. D., Benos, D. J., Palevsky, P. M., Cunningham, S. A., West, M. E., and Johnson, J. P. (1996a). Regulation of a sodium channel-associated G-Protein by aldosterone. *J. Biol. Chem.* **271,** 4491–4496.

Rokaw, M. D., West, M., and Johnson, J. P. (1996b). Rapamycin inhibits protein kinase C activity and stimulates Na^+ transport in A6 cells. *J. Biol. Chem.* **271,** 32468–32473.

Rokaw, M. D., West, M., and Johnson, J. P. (1996c). Aldosterone inhibits carboxylmethltransferase activity towards small molecular weight GTP binding proteins in A6 cells *FASEB J.* **10,** 548.

Rokaw, M. D., Wang, J.-M., Edinger, R., Weisz, O., Middleton, P., Shylyonsky, V., Berdiev, B. K., Eaton, D. C., Benos, D. J., and Johnson, J. P. (1998). Methylation of β subunit of xENaC regulates channel activity. *J. Biol. Chem.* **273,** 28746–28751.

Sariban-Sohraby, S., and Fisher, R. S., (1995). Guanine nucleotide-dependent carboxymethylation: A pathway for aldosterone modulation of apical Na^+ permeability in epithelia. *Kidney Int.* **48,** 965–969.

Sariban-Sohraby, S., Burg, M., Wiesmann, W. P., Chiang, P., and Johnson, J. P. (1984). Methylation increases sodium transport into A6 apical membrane vesicles: Possible mode of aldosterone action. *Science* **225,** 745–746.

Sariban-Sohraby, S., Fisher, R. S., and Abranow, M. (1993). Aldosterone-induced and GTP stimulated methylation of a 90 kD polypeptide in the apical membrane of A6 epithelial cells. *J. Biol. Chem.* **268,** 26613–26617.

Sariban-Sohraby, S., Mies, F., Abramow, M., and Fisher, R. S. (1995). Aldosterone stimulation of GTP hydrolysis in membranes from renal epithelia. *Am. J. Physiol.* **268,** C557–c562.

Skimkets, R. A., Lifton, R., and Canessa, C. M., (1998). In vivo phosphorylation of the epithelial sodium channel. *Proc. Natl. Acad. Sci. U.S.A.* **95,** 3301–3305.

Spindler, B., Mastroberardino, L. Custer, M., and Verrey, F. (1997). Characterization of early aldosterone-induced RNAs identified in A6 kidney epithelia. *Pfluegers Arch.—Eur. J. Physiol.* **434,** 323–331.

Szerlip, H. M., Weisberg, L., Geering, K., Rossier, B. C., and Cox, M. (1988). Aldosterone induced glycoproteins: electrophysiological-biochemical correlatikon. *Biochim. Biophys. Acta.* **940,** 1–9.

Verrey, F. (1995). Transcriptional control of sodium transport in tight epithelia by adrenal steroids. *J. Membr. Biol.* **144,** 93–110.

Verrey, F., and Beron, J., (1996). Activation and supply of channels and pumps by aldosterone. *News Physiol. Sci.* **11,** 126–133.

Verrey, F., Schaerer, E., Zoerkler, P., Paccolat, M. P., Geering, K., Kraehenbuhl, J. P., and Rossier, B. C. (1987). Regulation by aldosterone of Na^+,K^+-ATPase mRNAs, protein synthesis, and sodium transport in cultured kidney cells. *J. Cell Biol.* **104,** 1231–1237.

Verrey, F., Kraehenbuhl, J. P., and Rossier, B. C. (1989). Aldosterone induces a rapid increase in the rate of Na, K-ATPase gene transcription in cultured kidney cells. *Mol. Endocrinol.* **3,** 1369–1376.

Wiesmann, W. P., Johnson, J. P., Miura, G. A., and Chiang, P. K. (1985). Aldosterone-stimulated transmethylations are linked to sodium transport. *Am. J. Physiol.* **248,** F43–F47.

Yorio, T., and Bentley, P. J. (1978). Phospholipase A and the mechanism of action of aldosterone. *Nature (London)* **271,** 79–81.

CHAPTER 8

Regulation of Amiloride-Sensitive Na^+ Channels in the Renal Collecting Duct

James A. Schafer, Li Li, Duo Sun, Ryan G. Morris, and Teresa W. Wilborn
Departments of Physiology and Biophysics, and Medicine, University of Alabama at Birmingham, Birmingham, Alabama 35294

I. INTRODUCTION

The mammalian renal collecting duct, in particular the cortical collecting duct (CCD), has been regarded for many years as the preeminent example of an epithelium that actively reabsorbs Na^+ via an amiloride-sensitive sodium channel (ASSC). In fact, ion transport across the principal cell of the CCD closely follows the classic frog skin model of Ussing and Zerhan (1951)—Na^+ enters the cell via an electrogenic transporter in the apical membrane, subsequently identified as an ASSC (see Garty and Benos, 1988), while K^+ that is actively transported into the cells via the basolat-

1063-5823/99 $30.00

eral Na^+,K^+-ATPase diffuses passively out of the cell through separate K^+-selective channels in the apical membrane (Koeppen *et al.,* 1983; O'Neil and Sansom, 1984a,b; Schafer and Hawk, 1992; Schlatter and Schafer, 1987). The subsequent cloning and sequencing of the three subunits of an epithelial Na^+ channel (ENaC)[1] from a rat colonic library by Canessa *et al.* (1993, 1994) have had tremendous impact on understanding the physiological and pathophysiological regulation of the CCD Na^+ channel.

Based on the mRNA sequence from the rat colon, all three rENaC subunits were also found in the distal convoluted tubule (DCT), connecting tubule, and CCD (Duc *et al.,* 1994). Volk *et al.* (1995) also showed that the three subunits are expressed at the mRNA level and by functional analysis in the inner medullary collecting duct (IMCD); however, it appears that a significant part of Na^+ reabsorption in the IMCD is mediated by a cation channel that is structurally and functionally different from ENaC (Ciampolillo *et al.,* 1996; Karlson *et al.,* 1995; McCoy *et al.,* 1995). Although the alternate IMCD channel is inhibited by amiloride at micromolar concentrations, it exhibits no selectivity among Na^+, K^+ and Ca^{2+}, whereas the ENaC channel is generally regarded to be highly selective for Na^+ and Li^+ (Canessa *et al.,* 1994). The IMCD channel is also inhibited by atrial natriuretic peptide (ANP) acting through guanosine 3′,5′-cyclic monophosphate (cGMP) and thus it has been referred to as the cGMP-inhibitable cation channel (Ciampolillo *et al.,* 1996). This channel is likely to be equally important in the regulation of renal Na^+ reabsorption, but in this chapter we focus on the ENaC channels because of recent observations linking genetic abnormalities in this channel to hypertension.

As discussed elsewhere in this volume by Benos (Chapter 19), Shimkets *et al.* (1994) have demonstrated that a mutation resulting in an early truncation in the C-terminal intracellular region of the β-ENaC subunit is linked to Liddle's syndrome. This syndrome is marked by excessive Na^+ retention, leading to a low-renin, low-aldosterone, salt-sensitive form of hypertension, referred to in general as pseudohyperaldosteronism. Additional kindreds with other β and γENaC gene defects have since been identified and exhibit similar pathology (e.g., Hansson *et al.,* 1995a,b). Interestingly, a mutation in the N-terminal region of the αENaC subunit has been linked to pseudohypoaldosteronism type I, which is characterized by renal salt-wasting and

[1] Throughout this chapter we distinguish between the ENaC subunits and channels that are known to be constituted of one or more subunits, and ASSC. ASSC represents the general class of electrogenic, amiloride-inhibitable, Na^+ channels that may or may not consist of ENaC subunits. In the view of many investigators in this field, these two terms are synonymous; however, in the pre-1993 work cited here, ENaC subunits had not as yet been identified and it is not certain that the transporter studied was the same.

hypotension. Thus loss of function or gain of function resulting from mutations in a subunit or subunits of ENaC produces either, respectively, salt-wasting and hypotension, or salt retention and hypertension, and these defects relate primarily to the regulation of Na^+ reabsorption by the CCD and possibly the IMCD.

In addition to mutant β and γENaC genes, abnormalities in the biosynthesis and metabolism of adrenal steroids can also lead to salt retention and hypertension. The distal nephron, i.e., the DCT, CCD, and IMCD, are the renal target tissues for aldosterone. However, the selectivity of the response of these tissues to aldosterone compared with the higher circulating levels of glucocorticoids depends not on the selectivity of the steroid receptors but on the ability of 11β-hydroxysteroid dehydrogenase (11β-HSD) to inactivate steroids with glucocorticoid activity (for references, see Giebisch *et al.,* 1996). Licorice and certain drugs inhibit CCD 11β-HSD activity, leading to the syndrome of apparent mineralocorticoid excess (AME) that has the same manifestations of salt retention and hypertension as Liddle's syndrome (for references, see Laragh and Blumenfeld, 1996; Whorwood *et al.,* 1993). In addition to these acquired forms of AME, mutations in the human 11β-hydroxysteroid dehydrogenase gene have now been linked to heritable forms of this syndrome (Mune *et al.,* 1995; Mune and White, 1996).

Although genetic defects in the ENaC subunits or in 11β-HSD are responsible for no more than a small fraction of the vast proportion of the world population with salt-dependent hypertension, they illustrate the importance of the regulation of Na^+ reabsorption in the collecting duct for the maintenance of normal body salt balance and blood pressure. *The important point in the context of the present discussion is that single gene defects that result in unregulated Na^+ reabsorption by the CCD are sufficient to produce severe hypertension.* These abnormalities can arise anywhere in the pathway between the Na^+ channel itself and the hormones that regulate it, including receptors, G-proteins, phospholipases, protein kinases, and other intracellular second messengers. This chapter examines the association between some of these other elements and the regulation of ENaC subunits in the CCD.

II. VASOPRESSIN CAN ACT AS AN ANTINATRIURETIC AS WELL AS AN ANTIDIURETIC HORMONE

The syndrome of AME underscores the well-known importance of aldosterone as the primary regulator of Na^+ reabsorption and of K^+ and H^+ secretion by the collecting duct. It has been less widely appreciated that arginine vasopressin (AVP), in addition to regulating osmotic water permeability (P_f) and thus water reabsorption by the collecting duct, is important in regulating Na^+ reabsorption and K^+ secretion by this segment in some species (see Farman *et al.*, this volume, Chapter 4; Schafer and Hawk, 1992). Na^+ reabsorption by the amphibian epithelia such as the frog skin and toad bladder that served as the early models of the collecting duct is stimulated not only by aldosterone but also by AVP or by the natural amphibian antidiuretic hormone arginine vasotocin (Leaf *et al.*, 1957). Given this precedent, it was somewhat puzzling that AVP, although raising P_f, had no significant effect on Na^+ transport in the isolated perfused rabbit CCD, which was the first AVP-responsive mammalian nephron segment to be examined. As shown in the classic study of Frindt and Burg (1972) and subsequently by others (Chen *et al.*, 1990; Holt and Lechene, 1981), AVP produces only a transient and modest increase in Na^+ reabsorption (J_{Na}) in the rabbit CCD. Furthermore, as shown by Chen *et al.* (1990) this response is absent in CCD in which J_{Na} has been elevated by the mineralocorticoid deoxycorticosterone (DOC).

In subsequent studies with the isolated perfused rat CCD , Tomita *et al.* (1985) showed that AVP augments J_{Na}, and we demonstrated that this stimulation is sustained and synergistic with the effects of aldosterone or DOC even at physiological concentrations (Chen *et al.*, 1990; Hawk and Schafer, 1996; Reif *et al.*, 1984, 1986; Schafer and Hawk, 1992). As shown in Fig. 1, the synergism between the two hormones in the rat CCD is so significant that, even with a high pharmacological dose of DOC used to treat one group of rats, net Na^+ reabsorption and K^+ secretion (estimated by unidirectional $^{86}Rb^+$ fluxes) by the CCD were less than 25% of the rates observed upon further stimulation with AVP (Chen *et al.*, 1990; Schafer and Troutman, 1987). However, as discussed in more detail below, the absence of a natriferic effect of AVP in the rabbit CCD appears to be due more to inhibitory feedback mechanisms (Chen *et al.*, 1991) than to the lack of an effect of AVP-dependent cAMP production on the Na^+ channel.

There are additional differences between the regulation of the ASSC in the rat and rabbit that may have physiologic significance. In the rat CCD α_2-adrenergic agonists inhibit AVP-dependent cAMP production, whereas prostaglandin E_2 (PGE_2) has no effect. In the rabbit CCD PGE_2 inhibits cAMP production while α_2-adrenergic agonists do not (Chabardès *et al.*, 1988). We examined the consequences of these phenomena on Na^+ and water transport in isolated perfused CCD of the rat and of the rabbit and found that although PGE_2 inhibited basal Na^+ transport and AVP-stimulated water transport in the rabbit (as had been observed in several early studies (Grantham and Orloff, 1968; Stokes and Kokko, 1977)), it had no inhibitory effect in the rat (Chen *et al.*, 1991). Also, as expected from the cAMP production results, 100 nM epinephrine and 1 μM clonidine

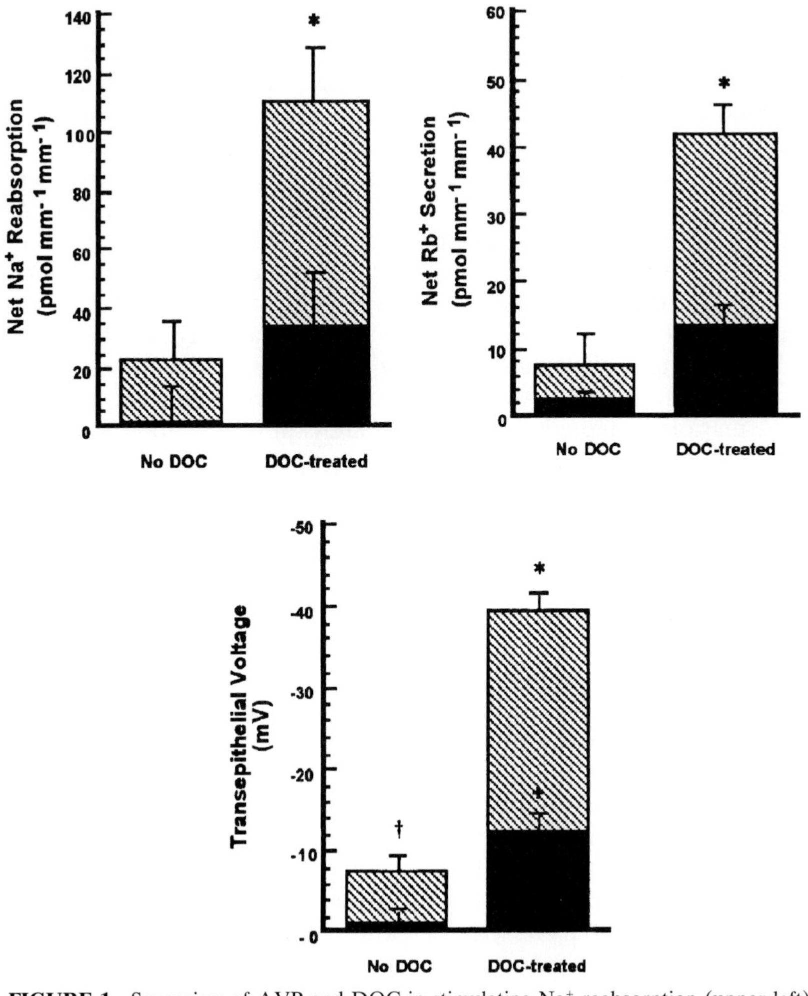

FIGURE 1 Synergism of AVP and DOC in stimulating Na^+ reabsorption (upper left), Rb^+ secretion (upper right), and transepithelial voltage (bottom) in the isolated perfused rat CCD. The net flux of Na^+ was measured as the difference in unidirectional fluxes of $^{22}Na^+$ and that of K^+ was approximated from the difference between unidirectional $^{86}Rb^+$ fluxes. The transepithelial voltage was lumen-negative in all cases. The black portion of the bar indicates the flux or transepithelial voltage observed in the absence of AVP and the cross-hatched portion of the bar shows increments produced by addition of 200 pM AVP. CCD were obtained from rats either implanted with a timed release pellet containing DOC (DOC-treated) or with no implanted pellet (No DOC). Data are from Reif *et al.* (1986) and Schafer and Troutman (1986). Statistically significant differences: *, compared with no AVP treatment in the DOC group or with No DOC group with or without AVP; †, effect of AVP treatment in the No DOC group; +, DOC-treated group compared with No DOC group in the absence of AVP.

were potent inhibitors of AVP-dependent Na^+ and water transport in the rat but not the rabbit CCD (Chen *et al.,* 1991; Hawk and Schafer, 1993b).

Of more importance, we have found that the regulatory phenotype of the rabbit CCD changes to resemble that of the rat CCD when immunodissected CCD cells are grown in culture. In these primary cultures grown on permeable supports, we measured the amiloride-inhibitable short-circuit current (I_{sc}) as an indicator of net Na^+ apical-to-basolateral flux. AVP produced a sustained increase in I_{sc} that was additive to the stimulatory effect of 10 nM aldosterone, which had been added to the culture medium 48 h before the experiment in one experimental group. Furthermore, I_{sc} was not inhibited by prostaglandin E_2 (Canessa and Schafer, 1992).

In the context of a potential role for AVP in regulating salt excretion, it is important to note that AVP is required for the production of experimental hypertension by the administration of DOC in combination with normal or elevated dietary salt—so-called "DOC-salt hypertension" (Berecek and Brody, 1982; Crofton *et al.,* 1993; Liang *et al.,* 1997), which is quite similar to Liddle's syndrome and AME. These findings together with the observed natriferic effect of AVP in the rat CCD and primary cultures of rabbit CCD led us to question the possible involvement of defective regulation of CCD Na^+ transport in certain forms of hypertension. Plasma AVP levels are markedly elevated in edematous states such as congestive heart failure and cirrhosis in which the kidney conserves salt despite extreme volume expansion (Robertson and Berl, 1996). The extent to which the natriferic actions of AVP may contribute to the effects of aldosterone in driving Na^+ reabsorption (and K^+ and H^+ secretion) may depend on the influence of other autacoids that modulate the action of AVP. Among the autacoids known to have such a function in the rat or rabbit CCD or IMCD are atrial (and related) natriuretic peptides, bradykinin, prostaglandins, in particular prostaglandin E_2 (PGE_2), and catecholamines. We have concentrated our research on the latter two systems because of the above-mentioned differences between their effects in the rat and rabbit CCD (Schafer, 1994; Schafer and Hawk, 1992). We have tried to exploit the differences between the regulatory phenotypes in these two species as a key to determining the critical signaling mechanism(s) that can either augment or suppress the natriferic actions of AVP.

III. AUTACOIDS THAT LIMIT THE ACTIONS OF ALDOSTERONE AND VASOPRESSIN IN THE CCD

A. Prostaglandin E_2 Is an Inhibitory Regulator

The inhibitory effect of PGE_2 in the rabbit CCD has largely been ascribed to the EP_1 receptor linked to the phospholipase C (PLC) system, resulting in increased intracellular Ca^{2+} ($[Ca^{2+}]_i$) and PKC activation (Breyer *et al.*, 1996). The EP_1 receptor has been localized to the CCD in the rabbit and not to other nephron segments or renal cells (Zhang *et al.*, 1997). The EP_3 receptor, which is coupled primarily to the inhibition of adenylate cyclase via G_i, is also found in the CCD as well as the medullary thick ascending limb of the rabbit (Breyer *et al.*, 1996). The absence of an inhibitory effect of PGE_2 in the rat CCD or in primary cultures of rabbit CCD cells suggests that neither EP_1 nor EP_3 receptors are expressed, or that their coupling to the inhibitory effectors is interrupted, perhaps by the absence of a necessary PKC isoform. The absence of the PGE_2 effect assumes particular importance when one considers that it may be acting as an autocrine inhibitor of AVP action in the rabbit CCD.

B. Ca^{2+} and Protein Kinase C Isoforms as Intracellular Mediators of Inhibitory Regulation

In the rabbit CCD elevation of intracellular calcium ($[Ca^{2+}]_i$) and activation of PKC have been shown to inhibit both basal Na^+ transport and the P_f produced by AVP as would be expected if they mediated the inhibitory effect of PGE_2 (for references, see Breyer and Fredin, 1991). However, as shown in Fig. 2 in the isolated perfused rat CCD we have found that neither the elevation of $[Ca^{2+}]_i$ by ionomycin or thapsigargin nor the activation of PKC by phorbol myristate acetate (PMA) or oleoyl-acetylglycerol (OAG) had any inhibitory effect on the lumen-to-bath $^{22}Na^+$ flux (J_{Na}). Not shown in Fig. 2, there was also no effect of any of these agents on the transepithelial voltage or the osmotic water permeability (P_f) in the presence of AVP (Rouch *et al.*, 1993). Even when OAG was combined with ionomycin, there was no significant inhibition of J_{Na} or P_f.

Because of the difference in the response to PKC activation between the rat and the rabbit CCD, we have conducted studies to determine if there are differences in the expression of PKC isoform(s) in CCD from the two species. In particular we considered the possibility that PKC in the rat CCD might be of the "atypical" class that shows no activation by diacyl glycerol, PMA, or Ca^{2+}. In these studies, we used reverse transcription and polymerase chain reaction (RT-PCR) of RNA and immunoblotting with isoform-specific antibodies to identify PKC isoforms in the rat and rabbit CCD. Contrary to our expectations, the initial pattern of PKC isoform expression was largely the same in the rat and the rabbit CCD: both expressed the α, δ, ε, and ζ, but not the β or γ isozymes, with no obvious differences in relative levels (Wilborn and Schafer, 1996). However, in the same studies we also obtained RT-PCR evidence for the presence of the PKC-θ and η isoforms in the rabbit CCD, and this was confirmed by

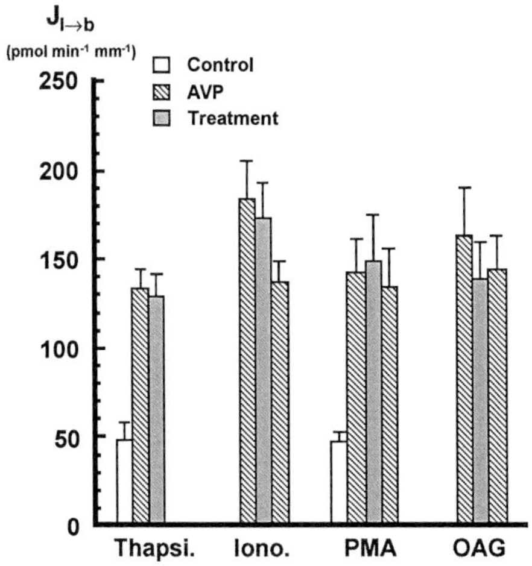

FIGURE 2 Lack of inhibition by elevation of $[Ca^{2+}]_i$ and PKC activation on Na^+ transport in the rat CCD. The lumen-to-bath flux of $^{22}Na^+$ ($J_{l\to b}$) was measured in isolated perfused CCD from DOC-treated rats. In some protocols there was an initial period with no AVP present (open bars). AVP was added to the bathing solution at 200 pM (cross-hatched bars before and in some protocols after the treatment period). In the treatment period one of four agents was added to the bathing solution: to raise $[Ca^{2+}]_i$, 1–2 μM thapsigargin (Thapsi.) or 0.5–1.0 μM ionomycin (Iono.); to activate PKC, 300 nM PMA or 100 μM OAG. There was no significant change in $J_{l\to b}$ from AVP period(s) with any of the treatments. Data of Rouch *et al.* (1993). Figure reprinted from Schafer (1994) with permission of the publisher.

immunoblotting with isoform specific antibodies. Of more interest in the context of the regulatory changes observed after the growth of rabbit CCD cells in culture, as shown in Fig. 3, we found that, whereas both the PKC-θ and η isoforms were expressed in freshly immunodissected rabbit CCD cells, after culture to confluency the protein expression of the η isoform increased while the θ isoform disappeared almost completely (Wilborn and Schafer, 1996). In more recent preliminary studies we have found that the largest change in expression of PKC-θ at the mRNA and protein levels occurs within the first 24 h in culture. Thus PKC-θ virtually disappears from the rabbit cells long before the cells have become confluent, and it remains barely detectable after the cells have become confluent and exhibit transepithelial Na^+ transport. The expression of other PKC isoforms also falls within 24 h of culturing, but as for the ε and ζ isoforms, expression is restored when the cells have repolarized (Wilborn and Schafer, 1997a,b).

Despite several attempts using RT-PCR we have been unable to obtain evidence for expression of PKC-θ in the rat CCD. In these attempts we have used the degenerate primer sets used to amplify the previously unidentified

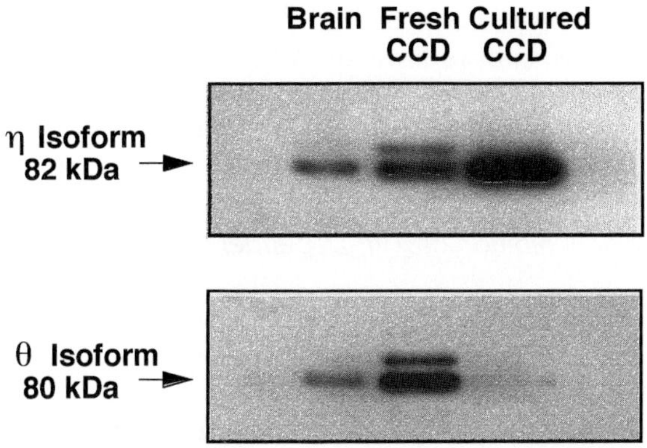

FIGURE 3 Immunoblots of PKC isoforms η and θ in immunodissected rat CCD cells. Proteins were extracted from rat brain tissue (as a positive control) and from "fresh cells" immediately after isolation of CCD. The cultured cells were immunodissected and then grown in culture until they were confluent with a high transepithelial resistance and a voltage more negative than −10 mV at which time the proteins were extracted. Western blotting was conducted using isoform specific antibodies. Figure from Wilborn and Schafer (1996) with permission of the publisher.

isoforms in the rabbit CCD, specific primers for mouse and human PKC-θ, and various combinations of degenerate and specific primers. The absence of PKC-θ in the rat and its disappearance from rabbit CCD cells after growth in culture make it a candidate effector that could explain the difference in AVP response between the rat and rabbit CCD, and between freshly immunodissected and cultured rabbit CCD cells.

It should be noted that our conclusions regarding the PKCs involved in the AVP response phenotype differ from those of DeCoy *et al.* (1995). Based on their experiments, they proposed that the expression of PKC-ε was associated with the transient Na^+ transport response in the rabbit CCD. They observed that when this isoform was down-regulated in cultured cells by PMA or an antisense oligonucleotide to the PKC-ε mRNA, AVP and cAMP analogs produced a stable stimulation of V_T. However, upon further examination of the data of DeCoy *et al.* (1995) it is seen that the *amiloride-sensitive* portion of the I_{sc} in their experiments did not differ significantly between AVP-treated control cells and those treated chronically with PMA or the antisense oligo (Wilborn and Schafer, 1996). We believe that there was a transient activation of a Cl^- conductance superimposed on the AVP-induced stimulation of Na^+ transport in the studies of DeCoy *et al.* (1995), and that inhibition of PKC-ε expression prevented activation of the Cl^- conductance resulting in a monotonic increase in I_{sc} (Wilborn and Schafer, 1996). As discussed below, we and others have seen the parallel activation of the amiloride-sensitive Na^+ channel (ASSC) and a transient CFTR Cl^- channel in A6 cells, and we have also observed a reversed I_{sc} in response to AVP in some of our rabbit CCD cultures that could be due to activation of a Cl^- channel allowing active electrogenic Cl^- reabsorption.

C. α-Adrenergic Inhibition of AVP-Dependent Transport in the CCD

In the rat CCD the only autacoids that have been shown to modulate the effect of AVP on Na^+ transport and P_f are catecholamines that act via an inhibitory G protein (G_i) to decrease AVP-dependent cAMP generation. These include epinephrine or norepinephrine acting through α_2-adrenoceptors (α_2-AR) and dopamine acting primarily via the D_4 receptor, and possibly also through the D_{1A} receptor. It has frequently been suggested that abnormalities in renal adrenoceptors may contribute to the development of hypertension, and renal denervation has been found to attenuate the development of hypertension in the SHR, DOC-salt, and renoprival ("one kidney, one clip") models of hypertension (DiBona, 1992). Three α_1-adrenoceptor (AR) isoforms, now referred to as α_{1A}, α_{1B}, and α_{1D}, are present in the rat and human. Three α_2-AR isoforms, α_{2A}, α_{2B}, and α_{2C} have also been identified. Both α_{2A}- and α_{2B}-ARs are found in the kidney, with α_{2A} dominating in the human and α_{2B} dominating in the rat (Handy *et al.*, 1993; Michel *et al.*, 1989). Importantly, the expression levels of α_2-ARs in diverse tissues changes with physiological stimuli as a consequence of regulated gene transcription (Handy *et al.*, 1993, 1995; Handy and Gavras, 1996).

Numerous studies have shown that the numbers of both α_1 and α_2 renal ARs are increased in rat models of genetic hypertension; however, it has been problematic to demonstrate that changes in AR density are associated with the cause, rather than being a result, of the hypertension (Jackson and Insel, 1993). Most of this increased α_2-AR density is found in the CCD where circulating catecholamines could inhibit Na^+ reabsorption and AVP-dependent cAMP production. It has also been hypothesized that the increase in α_2-receptor density in the CCD may be a consequence of a deficiency in the downregulation of α_1-receptors in the proximal tubule in hypertensive rat models (Gellai, 1990; Pettinger *et al.*, 1987).

Using the isolated perfused rat CCD preparation, we have shown that clonidine and epinephrine inhibit AVP-dependent Na^+ transport and P_f in the CCD both with and without DOC treatment, and that the decrease in Na^+ transport is a consequence of a decreased Na^+ conductance of the apical membrane (Chen *et al.*, 1991; Hawk *et al.*, 1993; Hawk and Schafer, 1993a). The effects of both agonists appear to be mediated primarily through an α_2-AR because when J_{Na} and P_f were stimulated by the cAMP analog 8-bromo-cAMP (Br-cAMP) plus isobutyl-methylxanthine (IBMX) or by forskolin, the subsequent inhibitory effect of epinephrine was less than 50% of that observed in the presence of AVP alone; however, the residual inhibitory effect was variably reversed by yohimbine (Hawk *et al.,* 1993). Thus we concluded that the primary effect of epinephrine was exerted through an α_2 adrenoceptor linked to a G_i protein which inhibited AVP-dependent cAMP production. However, there was also a variable component of inhibition that might be mediated through coupling of the α_2-adrenoceptor to G_q and the phospholipase C system or by an α_1-receptor. Rather than pursuing a pharmacological evaluation of the α-AR isoforms in the difficult isolated perfused tubule preparation, we opted to identify the α-adrenoceptor(s) isoforms present in the rat CCD at the mRNA level by RT-PCR.

We identified the presence of α_{2A} and α_{2B} but not α_{2C} in the rat CCD and proximal tubule, and we found that all three α_1-AR isoforms are present in the rat CCD, while only α_{1A} and α_{1B} are present in proximal tubules (Wilborn and Schafer, 1998). Although all three α_1-ARs are expressed in the CCD, it appears that their functional effect is minimal in comparison with the effects of α_2-AR because we have been unable to produce any effects using α_1-AR agonists either in the bathing solution or in the luminal perfusate in isolated perfused CCD experiments. We have also been unable to block the inhibitory effects of epinephrine, norepinephrine, or clonidine on Na^+ transport and AVP-dependent P_f in the isolated perfused rat CCD with α_1-AR antagonists (Wilborn and Schafer, 1998).

We made numerous attempts to demonstrate the presence of α_1- and α_2-ARs in the CCD by binding studies, but the amount of membrane protein that could be dissected from collagenase-treated kidney cortex was insufficient to obtain reliable results. Unfortunately, studies of catecholamine receptors has also been very much constrained by the paucity of reliable antibodies to identify receptor proteins by Western blotting or immunohistochemistry. We are currently unaware of reliable antibodies for the identification of α_1-AR subtypes. However, antibodies with high specificity for the renal α_2-AR have been produced by the laboratories of Drs. K. Lynch, D. Rosin, and M. Okusa at the University of Virginia. We have used an α_{2B} antibody from this laboratory (Huang *et al.,* 1996) to demonstrate the presence of the α_{2B}-AR by Western blotting of proteins from the rat CCD. As shown in Fig. 4, the antibody reveals an expected band at ~48 kDa in extracts from whole rat kidney cortex and proximal tubules as well as a larger band at ~62 kDa, which was the only band seen in the CCD extract. The higher MW band was also observed in Dr. Okusa's laboratory (Huang *et al.,* 1996) and was absent when antibody was preadsorbed to the corresponding fusion protein, thus it appears to be an alter-

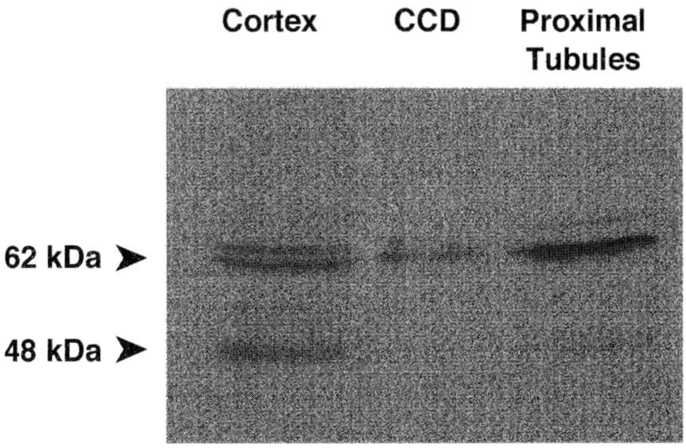

FIGURE 4 Immunoblots of α_{2B}-AR in rat CCD. Protein was extracted from whole rat kidney cortex, microdissected CCD, and proximal tubule segments. Western blotting was conducted using an α_{2B}-AR isoform-specific antibody (Huang *et al.*, 1996). Figure from Wilborn and Schafer (1998) with permission of the publisher.

nately processed form of the α_{2B}-AR (Wilborn and Schafer, 1998). Given the relative prevalence of the α_{2B}-AR in whole kidney homogenates and the relative abundance of the RT-PCR product, it is likely that this is the primary adrenoceptor coupled to the inhibition of AVP-dependent cAMP generation.

D. The Dopamine D_4 Receptor Is Coupled to the Primary Inhibitory Effect in the Rat CCD

Dopamine has a marked natriuretic and diuretic action in humans, rats, and other species and is clinically useful because of these properties (Jose *et al.*, 1992). Although the hemodynamic effects of dopamine are a determinant of excretory changes, direct actions of exogenous and endogenous dopamine on the nephron appear to be equally important. Dopamine excretion increases in proportion to the natriuresis produced by a salt load, and both the natriuresis and dopamine excretion are diminished by inhibition of renal dopamine synthesis, which occurs primarily in proximal tubules and acts as an autocrine or paracrine regulator of Na^+ reabsorption (Jose *et al.*, 1992).

There are two groups of dopamine receptors: the D_1-type receptors that include the D_{1A} and the D_{1B} (D_5 is the human equivalent of rat D_{1B}) receptors and the D_2-type receptors that include the D_2 (which has both long and short splice transcripts), D_3, and D_4 receptors (Gingrich and Caron, 1993). D_1-type receptors are primarily coupled to activation of adenylate cyclase through the stimulatory G protein (G_s), but they may also be coupled to phospholipases A_2 (PLA_2) and C (PLC) and hence to activation of protein kinase C (PKC) (Gingrich and Caron, 1993; Jose *et al.,* 1992). The D_2-type receptors, through coupling to an inhibitory G_i protein, act primarily to decrease adenylate cyclase activity (Gingrich and Caron, 1993), but they have also been shown to be coupled to PLC and PLA_2, and result in prostaglandin E_2 production in some systems (Jose *et al.,* 1992).

Abnormalities in the renal response to dopamine have been associated with certain forms of genetic hypertension in the rat. Although dopamine production in response to salt loading in the Dahl salt-sensitive rat is normal, D_1 agonists do not have the usual inhibitory effect on Na^+ reabsorption in the proximal tubule (Ohbu and Felder, 1991). In the spontaneously hypertensive (SHR) rat, although renal dopamine excretion is increased, there is a defective coupling of the D_{1A} receptor to the intracellular mediators of transport in the proximal tubule (Albrecht *et al.,* 1996).

There have been conflicting reports on the action of dopamine in the CCD. On the one hand, Muto *et al.* (1985) observed that dopamine inhibited the P_f response to AVP in the isolated perfused rabbit CCD and that this inhibition was reversed by metoclopramide, which, at that time, was regarded to be a specific inhibitor of D_2-dopaminergic receptors. In contrast, other groups have shown that the CCD has a D_{1A} receptor (Ohbu and Felder, 1991; Takemoto *et al.,* 1991) and that both dopamine and the D_{1A}-specific agonist fenoldopam increase intracellular cAMP in the rat CCD and thereby inhibit Na^+,K^+-ATPase, thus contributing to dopamine-induced natriuresis and diuresis (Takemoto *et al.,* 1992). However, such an action of dopamine seems incompatible with the known effects of cAMP production in the rat CCD to increase not only P_f but also transepithelial Na^+ transport in the isolated perfused rat CCD (Chen *et al.,* 1990; Schafer and Troutman, 1990; Schlatter and Schafer, 1987; Sun and Schafer, 1996). For that reason we examined the effects of dopamine on Na^+ transport and AVP-dependent P_f in the isolated perfused rat CCD.

In the CCD from DOC-treated rats in the presence of 22 pM AVP, 10 μM dopamine in the bathing solution inhibited Na^+ transport (measured by J_{Na} and V_T) and P_f by 40–60% (Sun and Schafer, 1996). The effects of dopamine were not reversed by the D_1-specific antagonist SCH-23390, and no inhibition was produced by the D_1-specific agonists fenoldopam or chloro-PB (SKF-81297). When Na^+ transport and P_f were stimulated with 8-(4-chlorophenylthio)-cAMP (cpt-cAMP) plus isobutyl-methylxanthine (IBMX), dopamine was not inhibitory, suggesting a "D_2-type" receptor. A variety of D_2 and D_3 agonists and antagonists were without effect—only clozapine, a specific D_4 antagonist, reversed the effects of dopamine (Sun and Schafer, 1996). These data strongly suggested that the effects of dopamine on transepithelial transport were produced via a D_2-type receptor coupled via a G_i protein to inhibition of adenylate cyclase (rather than a D_1-type receptor coupled to G_s) and that this receptor was at least similar to the unique clozapine-sensitive D_4 receptor (Sun and Schafer, 1996). Using a D_4 receptor-specific antibody to perform immunohistochemistry on the rabbit CCD, we have found that the receptor protein is distributed exclusively in both the cortical and medullary regions of the collecting duct (Sun *et al.,* 1998).

We have also examined the effects of various dopamine agonists and antagonists on cAMP production in microdissected CCD segments. As shown in Fig. 5 in the absence of AVP, dopamine produced no significant stimulation of cAMP generation indicating there was no significant D_1-type receptor response. The D_{1A}-specific agonist fenoldopam also produced no stimulation of cAMP production (see Li and Schafer, 1998). As expected, AVP significantly stimulated cAMP production, and this stimulation was significantly reduced by 10 μM dopamine. The inhibitory effect of dopamine was completely reversed by 10 μM clozapine, a specific D_4 antagonist (Fig. 5). To further rule out the participation of D_2-type receptors other than D_4, we tested the ability of other antagonists to reverse the effects of dopamine, but only clozapine was effective (Li and Schafer, 1998).

Despite the functional effects supporting an action of dopamine by a D_4 receptor, the distal convoluted tubule and to a lesser extent the cortical collecting duct (CCD) also have D_{1A} receptors (Ohbu and Felder, 1991; Takemoto *et al.,* 1991), and Satoh *et al.* (1993) have shown that dopamine decreases Na^+,K^+-ATPase activity in the CCD and the medullary thick ascending limb. Using RT-PCR, we have confirmed the expression of both the D_{1A} and D_4 receptors at the mRNA level (Sun *et al.,* 1998). Nevertheless, at least in rats of the age used in these studies (4 to 6 weeks), dopamine appears to inhibit Na^+ and water reabsorption in the CCD primarily through the D_4 receptor.

IV. TRAFFICKING AND THE REGULATION OF THE AMILORIDE-SENSITIVE Na^+ CHANNEL

Ultimately, the effect of AVP, modulated by autacoids such as the α_2-AR, dopamine, and PGE_2, must be exerted on ENaC. Presumably, this effect is mediated by cAMP-dependent protein kinase A (PKA) phosphorylation of the channel itself, or of proteins that are responsible for activation of the channel. As reviewed in the chapter by Smith in this volume (Chapter 9), it is presently debated whether such phosphorylation results in regulated exocytosis or activation of channel subunits that are already present in the

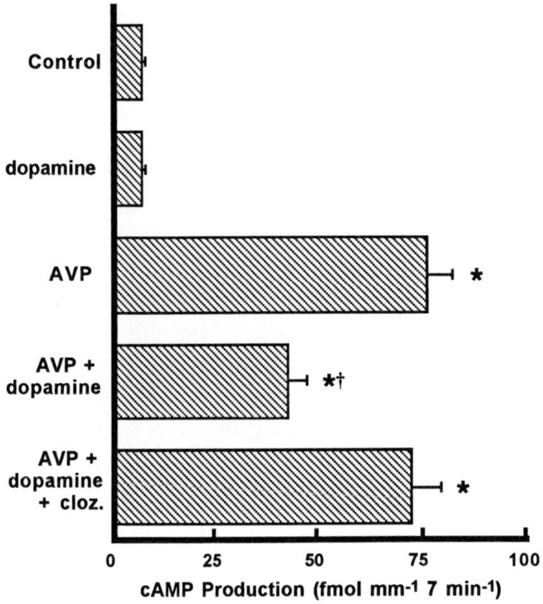

FIGURE 5 Effect of dopamine on cAMP production by intact (nonpermeabilized), microdissected rat CCD. Agents added to the bathing solution were as follows: control, no additions; 10 μM dopamine; 200 pM AVP; 200 pM AVP + 10 μM dopamine; and 200 pM AVP + 10 μM dopamine + 10 μM clozapine (cloz.) Results were from 17 experiments. Statistically significant differences: *, compared with control; †, compared with preceding and following groups. Reprinted from Li and Schafer (1998) with permission of the publisher.

apical membrane. In other words, stimulation of ENaC activity may involve both trafficking of channel subunits to the luminal membrane and/or their activation once they are present in the membrane, or both processes may occur.

We have recently examined the possibility that microtubules are involved in the regulation of ASSC in A6 cells (Morris *et al.,* 1998). In these studies, we, as well as Verrey *et al.* (1995), determined that the increase in short-circuit current across monolayers of A6 cells in response to arginine vasotocin (AVT) had two components: an amiloride-sensitive component and a CFTR-mediated Cl^- conductance (Morris *et al.,* 1998). In our functional studies of these cells, we found that disruption of the microtubules by nocodazole inhibited the CFTR-dependent component of the response to AVT but not the amiloride-sensitive component. To examine the structural correlates of these effects of microtubule disruption, we used immunohistochemistry to examine the localization of ASSC and CFTR in the A6 cells and the effects of AVT and nocodazole. As shown in Fig. 6A, immunohistochemistry with a CFTR-specific antibody revealed that, in the absence of AVT, this transporter was primarily located in the deeper part of the cytoplasm surrounding the nuclei, but it almost completely relocated to the apical surface after AVT treatment (Morris *et al.*, 1998). On the other hand, in agreement with the previous studies of Tousson *et al.* (1989), an antibody against the multimeric ASSC complex isolated from bovine renal medulla showed that ASSC was localized exclusively in or immediately adjacent to the apical membrane in the presence or absence of AVT (Morris *et al.*, 1998). We then examined the effect of AVT on the localization of both ASSC and CFTR in the presence of the inhibitor of nocodazole. As shown in Fig.6B, nocodazole decreased the apical labeling of ASSC and increased its appearance in the cytoplasm. It also prevented the redistribution of CFTR from the cytoplasm to the apical membrane upon the addition of AVT (seen in Fig. 6A). In parallel short-circuit current experiments we found that nocodazole completely inhibited the AVT-induced increase in CFTR-mediated Cl^- current (as determined by its sensitivity of glibenclamide), but it did not prevent the increase in amiloride-sensitive Na^+ current produced by AVT. From these results we concluded that AVT caused activation of CFTR by microtubule-dependent trafficking from deep cytoplasmic regions to the apical membrane. Although there was no evidence for any dependence of the Na^+ channel activation by AVT on the integrity of the microtubules, we could not rule out the possibility that Na^+ channels in vesicles immediately adjacent to the plasma membrane were docking and inserting after AVT stimulation. In other words, our experiments, as those of Tousson *et al.* (1989), demonstrated that the AVT-dependent activation of Na^+ conductance in the A6 cell did not involve trafficking from deeper intracellular regions as was the case for CFTR. However, because of the limited resolution of immunohistochemistry at the light level, our studies could not rule out the possibility that AVT caused docking and exocytosis of Na^+ channels already localized in the subapical region, as occurs in the case of AVP-induced activation of aquaporin 2 (AQ2) water channels (Morris *et al.*, 1998).

V. A CHALLENGE FOR INTEGRATIVE PHYSIOLOGY—THE LINK BETWEEN Na^+ RETENTION AND HYPERTENSION

Our recent knowledge of the molecular and genetic details of the Na^+ channel in the CCD, and its regulation by hormones and autacoids, has established a clear link between abnormal channel regulation and hypertension. It is now widely accepted that the kidney is involved in the genesis

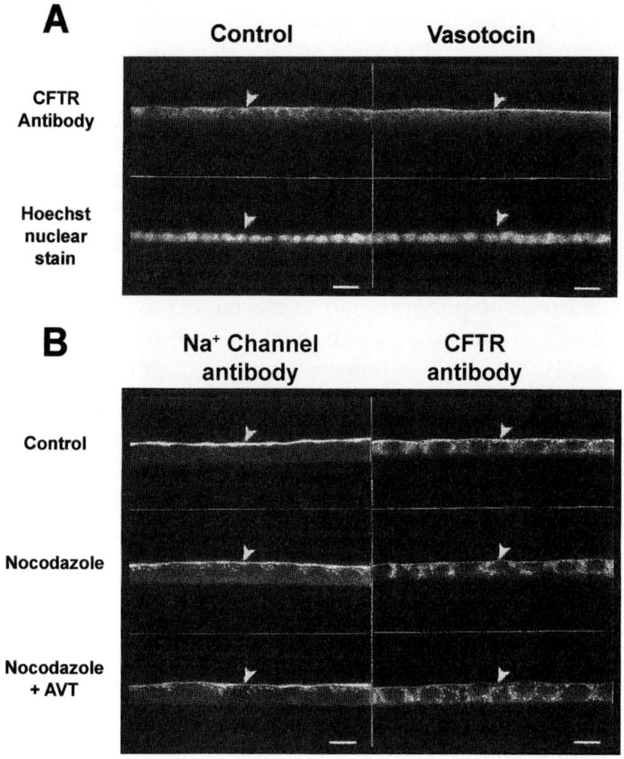

FIGURE 6 (A) Immunohistochemical localization of CFTR protein in A6 cells using a CFTR-specific antibody (Genzyme, Boston, MA). The cells are viewed in cross-section with the apical surface marked by an arrowhead. In the lower two panels Hoechst stain is used to localize the nuclei with respect to the apical membrane. A6 cells in the left two panels received no hormone treatment, while those on the right were treated with 0.1 μM arginine vasotocin (AVT). The scale bar is 20 μm in length. Revised from Morris *et al.* (1998) with permission of the publisher. (B) Immunohistochemical localization of CFTR and ASSC protein in A6 cells. Panels on the left were stained with the antibody of Tousson *et al.* (1989) which was raised against the Na^+ channel protein complex from bovine renal medulla. Panels on the left were stained with the CFTR antibody (see Fig. 6). The cells are viewed in cross-section with the apical surface, whose location is marked by an arrowhead, uppermost. In the middle two panels the cells had been treated with 33 μM nocodazole for 30 min at 4°C followed by 3 h at 26°C. In the lower two panels, the A6 cells were treated with nocodazole as described, and subsequently with 0.1 μM arginine vasotocin (AVT) for 15 min at 26°C. The scale bar is 20 μm in length. Revised from Morris *et al.* (1998) with permission of the publisher.

and maintenance of certain forms of human hypertension as well as in the most widely studied animal models of hypertension. This association has been further suggested for humans with essential hypertension in whom removal of nephrosclerotic kidneys and transplantation with a normal kidney causes remission of the hypertension (Curtis *et al.,* 1983). In animal models of hypertension ample evidence now exists that there is a diminished natriuretic response to a salt-load even before hypertension develops. The accepted "explanation" for this linkage is that the Na^+ retentiveness of the kidney results in a shift in the relationship between renal perfusion pressure and Na^+ excretion (the pressure-natriuresis curve) toward higher systemic blood pressures in order to maintain Na^+ balance (e.g., Woolfson and de Wardener, 1996). Thus Na^+-retentive kidneys cause extracellular volume expansion that is counteracted by an increase in blood pressure sufficient to maintain normal Na^+ excretion by pressure natriuresis. The missing link is how an increase in extracellular volume causes the necessary increase in blood pressure and resulting pressure natriuresis that restores Na^+ balance. Numerous studies have shown that the acute infusion of large amounts of isotonic saline or isoncotic plasma or plasma substitutes results in rapid natriuresis and diuresis without any increase in blood pressure. Thus when the kidneys are operating properly, volume expansion per se does not produce hypertension, at least acutely (Schafer, 1994). We need to know what component of the normal feedback regulation of extracellular volume by the kidneys is impaired in salt-dependent hypertension. As discussed above, numerous studies suggest that, in addition to aldosterone, AVP, PGE_2, norepinephrine, and dopamine are essential hormones in the regulation of salt excretion, and abnormalities in their effects on salt reabsorption are likely to be associated with salt-sensitive hypertension. The challenge is now to determine which regulatory systems are disturbed in which individuals with salt-dependent hypertension. It is quite likely that currently available modes of therapy and their immediate derivatives will be more effective in controlling hypertension if targeted to a particular regulatory system that is known to be defective.

Acknowledgments

We gratefully acknowledge the superb technical support of Ms. Mary Lou Watkins and the administrative support of Ms. Carole D. Marks. Data from our laboratory was supported by NIH Grants DK-25519 and DK-45768, and the Department of Physiology and Biophysics at UAB.

References

Albrecht, F. E., Drago, J., Felder, R. A., Printz, M. P., Eisner, G. M., Robillard, J. E., Sibley, D. R., Westphal, H. J., and Jose, P. A. (1996). Role of the D1A dopamine receptor in the pathogenesis of genetic hypertension. *J. Clin. Invest.* **97,** 2283–2288.

Berecek, K. H., and Brody, M. J. (1982). Vasopressin and deoxycorticosterone hypertension in Brattleboro rats. *Ann. N.Y. Acad. Sci.* **394,** 319–329.

Breyer, M. D., and Fredin, D. (1991). Feedback inhibition of cyclic adenosine monophosphate-stimulated Na^+ transport in the rabbit cortical collecting duct via Na^+-dependent basolateral Ca^{2+} entry. *J. Clin. Invest.* **88,** 1502–1510.

Breyer, M. D., Jacobson, H. R., and Breyer, R. M. (1996). Functional and molecular aspects of renal prostaglandin receptors. *J. Am. Soc. Nephrol.* **7,** 8–17.

Canessa, C. M., and Schafer, J. A. (1992). AVP stimulates Na^+ transport in primary cultures of rabbit cortical collecting duct cells. *Am. J. Physiol.* **262,** F454–F461.

Canessa, C. M., Horisberger, J. D., and Rossier, B. C. (1993). Epithelial sodium channel related to proteins involved in neurodegeneration. *Nature* (*London*) **361,** 467–470.

Canessa, C. M., Shild, L., Buell, G., Thorens, B., Gautschi, I., Horisberger, J.-D., and Rossier, B. C. (1994). Amiloride-sensitive epithelial Na^+ channel is made of three homologous subunits. *Nature* (*London*) **367,** 463–467.

Chabardès, D., Brick, G. C., Montégut, M., and Siaume-Perez, S. (1988). Effect of PGE_2 and α-adrenergic agonists on AVP-dependent cAMP levels in rabbit and rat CCT. *Am. J. Physiol.* **255,** F43–F48.

Chen, L., Williams, S. K., and Schafer, J. A. (1990). Differences in synergistic actions of vasopressin and deoxycorticosterone in rat and rabbit CCD. *Am. J. Physiol.* **259,** F147–F156.

Chen, L., Reif, M. C., and Schafer, J. A. (1991). Clonidine and PGE_2 have different effects on Na^+ and water transport in the rat and rabbit CCD. *Am. J. Physiol.* **261,** F126–F136.

Ciampolillo, F., McCoy, D. E., Green, R. B., Karlson, K. H., Dagenais, A., Molday, R. S., and Stanton, B. A. (1996). Cell-specific expression of amiloride-sensitive, Na^+-conducting ion channels in the kidney. *Am. J. Physiol.* **271,** C1303–C1315.

Crofton, J. T., Ota, M., and Share, L. (1993). Role of vasopressin, the renin-angiotensin system and sex in Dahl salt-sensitive hypertension. *J. Hypertens.* **11,** 1031–1038.

Curtis, J. J., Luke, R. G., Dustan, H. P., Kashgarian, M., Welchel, J. D., Jones, P., and Diethelm, A. (1983). Remission of hypertension after renal transplantation. *N. Engl. J. Med.* **309,** 1009–1015.

DeCoy, D. L., Snapper, J. R., and Breyer, M. D. (1995). Anti-sense DNA down regulates protein kinase C-e and enhances vasopressin stimulated Na^+ absorption in rabbit cortical collecting duct. *J. Clin. Invest.* **95,** 2749–2756.

DiBona, G. F. (1992). Sympathetic neural control of the kidney in hypertension. *Hypertension* **19,** 128–135.

Duc, C., Farman, N., Canessa, C. M., Bonvalet, J. P., and Rossier, B. C. (1994). Cell-specific expression of epithelial sodium channel alpha, beta, and gamma subunits in aldosterone-responsive epithelia from the rat: Localization by in situ hybridization and immunocytochemistry. *J. Cell Biol.* **127,** 1907–1921.

Frindt, G., and Burg, M. B. (1972). Effect of vasopressin on sodium transport in renal cortical collecting tubules. *Kidney Int.* **1,** 224–231.

Garty, H., and Benos, D. J. (1988). Characteristics and regulatory mechanisms of the amiloride-blockable Na^+ channel. *Physiol. Rev.* **68,** 309–73.

Gellai, M. (1990). Modulation of vasopressin antidiuretic action by renal α_2-adrenoceptors. *Am. J. Physiol.* **259,** F1–F8.

Giebisch, G., Malnic, G., and Berliner, R. W. (1996). Control of renal potassium excretion. *In* "The Kidney" (B. M. Brenner, ed.), 5th ed., pp. 371–407. Saunders, Philadelphia.

Gingrich, J. A., and Caron, M. G. (1993). Recent advances in the molecular pharmacology of dopamine receptors. *Annu. Rev. Neurosci.* **16,** 299–321.

Grantham, J. J., and Orloff, J. (1968). Effect of prostaglandin E_1 on the permeability response of the isolated collecting tubule to vasopressin, adenosine 3',5'-monophosphate, and theophylline. *J. Clin. Invest.* **47,** 1154–1161.

Handy, D. E., and Gavras, H. (1996). Evidence for cell-specific regulation of transcription of the rat α_{2A}-adrenergic receptor gene. *Hypertension* **27,** 1018–1024.

Handy, D. E., Flordellis, C. S., Bogdanova, N. N., Bresnahan, M. R., and Gavras, H. (1993). Diverse tissue expression of rat α_2-adrenergic receptor genes. *Hypertension* **21,** 861–865.

Handy, D. E., Zanella, M. T., Kanemaru, A., Tavares, A., Flordellis, C., and Gavras, H. (1995). A negative regulatory element in the promoter region of the rat alpha 2A-adrenergic receptor gene overlaps an SP1 consensus binding site. *Biochem. J.* **311,** 541–547.

Hansson, J. H., Nelson-Williams, C., Suzuki, H., Schild, L., Shimkets, R., Yin, L. U., Canessa, C., Iwasaki, T., Rossier, B., and Lifton, R. P. (1995a). Hypertension caused by a truncated epithelial sodium channel γ subunit: Genetic heterogeneity of Liddle syndrome. *Nat. Genet.* **11,** 76–82.

Hansson, J. H., Schild, L., Lu, Y., Wilson, T. A., Gautschi, I., Shimkets, R., Nelson-Williams, C., Rossier, B. C., and Lifton, R. P. (1995b). A de novo missense mutation of the beta subunit of the epithelial sodium channel causes hypertension and Liddle syndrome, identifying a proline-rich segment critical for regulation of channel activity. *Proc. Natl. Acad. Sci. U.S.A.* **92,** 11495–11499.

Hawk, C. T., and Schafer, J. A. (1993a). Clonidine, but not bradykinin or ANP, inhibits Na^+ and water transport in Dahl SS rat CCD. *Kidney Int.* **40,** 30–35.

Hawk, C. T., and Schafer, J. A. (1993b). Dose-response to AVP, and synergism with aldosterone in CCD from Sprague-Dawley and Dahl salt-sensitive rats. *J. Am. Soc. Nephrol.* **4,** 854 (abstr.).

Hawk, C. T., and Schafer, J. A. (1996). AVP and aldosterone at physiologic concentrations have synergistic effects on Na^+ transport in rat CCD. *Kidney Int.* **57,** S35–S41.

Hawk, C. T., Kudo, L. H., Rouch, A. J., and Schafer, J. A. (1993). Inhibition by epinephrine of AVP- and cAMP-stimulated Na^+ and water transport in Dahl rat CCD. *Am. J. Physiol.* **265,** F449–F460.

Holt, W. F., and Lechene, C. (1981). ADH-PGE_2 interactions in cortical collecting tubule. I. Depression of sodium transport. *Am. J. Physiol.* **241,** F452–F460.

Huang, L., Wei, Y. Y., Momonse-Hotokezaka, A., Dickey, J., and Okusa, M. D. (1996). α_{2B} Adrenergic receptors: Immunolocalization and regulation by potassium depletion in rat kidney. *Am. J. Physiol.* **270,** F1015–F1026.

Jackson, C. A., and Insel, P. A. (1993). Renal a-adrenergic receptors and genetic hypertension. *Pediatr. Nephrol.* **7,** 853–858.

Jose, P. A., Raymond, J. R., Bates, M. D., Aperia, A., Felder, R. A., and Carey, R. M. (1992). The renal dopamine receptors. *J. Am. Soc. Nephrol.* **2,** 1265–1278.

Karlson, K. H., Ciampolillo-Bates, F., McCoy, D. E., Kizer, N. L., and Stanton, B. A. (1995). Cloning of a cGMP-gated cation channel from mouse kidney inner medullary collecting duct. *Biochim. Biophys. Acta, Biomembr.* **1236,** 197–200.

Koeppen, B. M., Biagi, B. A., and Giebisch, G. H. (1983). Intracellular microelectrode characterization of the rabbit cortical collecting duct. *Am. J. Physiol.* **244,** F35–F47.

Laragh, J. H., and Blumenfeld, J. D. (1996). Essential hypertension. *In* "The Kidney" (B. M. Brenner, ed.), 5th ed., pp. 2071–2105. Saunders, Philadelphia.

Leaf, A., Anderson, J., and Page, L. B. (1957). Active sodium transport by the isolated toad bladder. *J. Gen. Physiol.* **41,** 657–668.

Li, L., and Schafer, J. A. (1998). Dopamine inhibits vasopressin-dependent cAMP production in the rat cortical collecting duct. *Am. J. Physiol.* **275,** F62–F67.

Liang, J., Toba, K., Ouchi, Y., Nagano, K., Akishita, M., Kozaki, K., Ishikawa, M., Eto, M., and Orimo, H. (1997). Central vasopressin is required for the complete development of deoxycorticosterone-salt hypertension in rats with hereditary diabetes insipidus. *J. Auton. Nerv. Syst.* **62,** 33–39.

McCoy, D. E., Guggino, S. E., and Stanton, B. A. (1995). The renal cGMP-gated cation channel: its molecular structure and physiological role. *Kidney Int.* **48,** 1125–1133.

Michel, M. C., Insel, P. A., and Brodde, O.-E. (1989). Renal a-adrenergic receptor alterations: A cause of essential hypertension? *FASEB J.* **3,** 139–144.

Morris, R. G., Tousson, A., Benos, D. J., and Schafer, J. A. (1998). Microtubule disruption inhibits an AVT-stimulated chloride secretion but not sodium reabsorption in A6 cells. *Am. J. Physiol.* **274,** F300–F314.

Mune, T., and White, P. C. (1996). Apparent mineralocorticoid excess: Genotype is correlated with biochemical phenotype. *Hypertension* **27,** 1193–1199.

Mune, T., Rogerson, F. M., Nikkilä, H., Agarwal, A. K., and White, P. C. (1995). Human hypertension caused by mutations in the kidney isozyme of 11b-hydroxysteroid dehydrogenase. *Nat. Genet.* **10,** 394–399.

Muto, S., Tabei, K., Asano, Y., and Imai, M. (1985). Dopamine inhibition of the action of vasopressin on the cortical collecting tubule. *Eur. J. Pharmacol.* **114,** 393–397.

Ohbu, K., and Felder, R. A. (1991). DA_1 dopamine receptors in renal cortical collecting duct. *Am. J. Physiol.* **261,** F890–F895.

O'Neil, R. G., and Sansom, S. C. (1984a). Characterization of apical cell membrane Na^+ and K^+ conductances of cortical collecting duct using microelectrode techniques. *Am. J. Physiol.* **247,** F14–F24.

O'Neil, R. G., and Sansom, S. C. (1984b). Electrophysiological properties of cellular and paracellular conductive pathways of the rabbit cortical collecting duct. *J. Membr. Biol.* **82,** 281–295.

Pettinger, W. A., Umemura, S., Smyth, D. D., and Jeffries, W. B. (1987). Renal α_2-adrenoceptors and the adenylate cyclase-cAMP system: Biochemical and physiological interactions. *Am. J. Physiol.* **252,** F199–F208.

Reif, M. C., Troutman, S. L., and Schafer, J. A. (1984). Sustained response to vasopressin in isolated rat cortical collecting tubule. *Kidney Int* **26,** 725–732.

Reif, M. C., Troutman, S. L., and Schafer, J. A. (1986). Sodium transport by rat cortical collecting tubule. Effects of vasopressin and desoxycorticosterone. *J. Clin. Invest.* **77,** 1291–1298.

Robertson, G. L., and Berl, T. (1996). Pathophysiology of water metabolism. *In* "The Kidney (B. M. Brenner, ed.), 5th ed., pp. 873–928. Saunders, Philadelphia.

Rouch, A. J., Chen, L., Kudo, L. H., Bell, P. D., Fowler, B. C., Corbitt, B. D., and Schafer, J. A. (1993). Intracellular Ca^{2+} and PKC activation do not inhibit Na^+ and water transport in rat CCD. *Am. J. Physiol.* **265,** F569–F577.

Satoh, T., Cohen, H. T., and Katz, A. I. (1993). Different mechanisms of renal Na-K-ATPase regulation by protein kinases in proximal and distal nephron. *Am. J. Physiol.* **265,** F399–F405.

Schafer, J. A. (1994). Salt and water homeostasis—is it just a matter of good bookkeeping? *J. Am. Soc. Nephrol.* **4,** 1933–1950.

Schafer, J. A., and Hawk, C. T. (1992). Regulation of Na^+ channels in the cortical collecting duct by AVP and mineralocorticoids. *Kidney Int.* **41,** 255–268.

Schafer, J. A., and Troutman, S. L. (1986). Effect of ADH on rubidium transport in isolated perfused rat cortical collecting tubules. *Am. J. Physiol.* **250,** F1063–F1072.

Schafer, J. A., and Troutman, S. L. (1987). Potassium transport in cortical collecting tubules from mineralocorticoid-treated rat. *Am. J. Physiol.* **253,** F76–F88.

Schafer, J. A., and Troutman, S. L. (1990). cAMP mediates the increase in apical membrane Na^+ conductance produced in the rat CCD by vasopressin. *Am. J. Physiol.* **259,** F823–F831.

Schlatter, E., and Schafer, J. A. (1987). Electrophysiological studies in principal cells of rat cortical collecting tubules. ADH increases the apical membrane Na^+-conductance. *Pfluegers Arch.* **409,** 81–92.

Shimkets, R. A., Warnock, D. G., Bositis, C. M., Nelson-Williams, C., Hansson, J. H., Schambelan, M., Gill, J., Jr., Ulick, S., Milora, R. V., Findling, J. W., Canessa, C. M., Rossier, B. C., and Lifton, R. P. (1994). Liddle's syndrome: Heritable human hypertension caused by mutations in the beta subunit of the epithelial sodium channel. *Cell (Cambridge, Mass.)* **79,** 407–414.

Stokes, J. B., and Kokko, J. P. (1977). Inhibition of sodium transport by prostaglandin E_2 across the isolated, perfused rabbit collecting tubule. *J. Clin. Invest.* **59,** 1099–1104.

Sun, D., and Schafer, J. A. (1996). Dopamine inhibits AVP-dependent Na^+ transport and water permeability in rat CCD via a D_4-like receptor. *Am. J. Physiol.* **271,** F391–F400.

Sun, D., Wilborn, T. W., and Schafer, J. A. (1998). Dopamine D_4 receptor isoform mRNA and protein is expressed in the rat cortical collecting duct. *Am. J. Physiol.* **275,** F742–F751.

Takemoto, F., Satoh, T., Cohen, H. T., and Katz, A. I. (1991). Localization of dopamine-1 receptors along the microdissected rat nephron. *Pfluegers Arch.* **419,** 243–248.

Takemoto, F., Cohen, H. T., Satoh, T., and Katz, A. I. (1992). Dopamine inhibits Na/K-ATPase in single tubules and cultured cells from distal nephron. *Pfluegers Arch.* **421,** 302–306.

Tomita, K., Pisano, J. J., and Knepper, M. A. (1985). Control of sodium and potassium transport in the cortical collecting duct of the rat. Effects of bradykinin, vasopressin, and deoxycorticosterone. *J. Clin. Invest.* **76,** 132–136.

Tousson, A., Alley, C. D., Sorscher, E. J., Brinkley, B. R., and Benos, D. J. (1989). Immunochemical localization of amiloride sensitive sodium channels in sodium-transporting epithelia. *J. Cell Sci.* **93,** 349–362.

Ussing, H. H., and Zerhan, K. (1951). Active transport of sodium as the source of electric current in the short-circuited frog skin. *Acta Physiol. Scand.* **23,** 110–127.

Verrey, F., Groscurth, P., and Bolliger, U. (1995). Cytoskeletal disruption in A6 kidney cells: Impact on endo/exocytosis and NaCl transport regulation by antidiuretic hormone. *J. Membr. Biol.* **145,** 193–204.

Volk, K. A., Sigmund, R. D., Snyder, P. M., McDonald, J., Welsh, M. J., and Stokes, J. B. (1995). rENaC is the predominant Na^+ channel in the apical membrane of the rat renal inner medullary collecting duct. *J. Clin. Invest.* **96,** 2748–2757.

Whorwood, C. B., Sheppard, M. C., and Steward, P. M. (1993). Licorice inhibits 11β-hydroxysteroid dehydrogenase ribonucleic acid level and potentiates glucocorticoid action. *Endocrinology (Baltimore)* **132,** 2287–2296.

Wilborn, T. W., and Schafer, J. A. (1996). Protein kinase C isoforms and Na^+ transport response to AVP in cultured rabbit CCD cells. *Am. J. Physiol.* **270,** F766–F775.

Wilborn, T. W., and Schafer, J. A. (1997a). Expression of mRNA for δ, η, and θ PKC isoforms in the rabbit cortical collecting duct. *FASEB J.* **11,** A248.

Wilborn, T. W., and Schafer, J. A. (1997b). Rabbit cortical collecting duct cell growth in culture produces a rapid fall in protein kinase C q isoform expression. *J. Am. Soc. Nephrol.* **8,** 47A.

Wilborn, T. W., and Schafer, J. A. (1998). Expression of multiple a-adrenoceptor isoforms in rat CCD. *Am. J. Physiol.* **275,** F111–F118.

Woolfson, R. G., and de Wardener, J. E. (1996). Primary renal abnormalities in hereditary hypertension. *Kidney Int.* **50,** 717–731.

Zhang, Y.-H., Guan, Y.-F., Davis, L., Hebert, R. L., Breyer, R. M., and Breyer, M. D. (1997). Tissue distribution and function of the rabbit prostaglandin EPA receptor. *J. Am. Soc. Nephrol.* **8,** 419A. (abstr.).

CHAPTER 9

cAMP-Mediated Regulation of Amiloride-Sensitive Sodium Channels: Channel Activation or Channel Recruitment?

Peter R. Smith
Department of Physiology, MCP–Hahnemann University, Philadelphia, Pennsylvania 19129

I. INTRODUCTION

Sodium-reabsorbing epithelia contain Na^+ channels situated within their apical membranes that are inhibited by the diuretic amiloride and its analogs (Benos *et al.*, 1995; Garty and Palmer, 1997). These amiloride-sensitive Na^+ channels are rate limiting for net Na^+ reabsorption because they mediate entry of Na^+ from the luminal fluid into the cells during the first stage of electrogenic Na^+ transport (Benos *et al.*, 1995; Garty and Palmer, 1997). Amiloride-sensitive Na^+ channels thus play a key role in the regulation of whole body fluid and electrolyte balance and their activity is highly regulated. Although these channels are commonly refereed to as epithelial Na^+ channels, they are also expressed in nonepithelial cells such as lymphocytes (Bubien and Warnock, 1993), endothelial cells (Vigne *et al.*, 1989), vascular smooth muscle cells (Van Renterghem and Lazdunski, 1991), and oocytes (Kupitz and Atlas, 1993).

1063-5823/99 $30.00

The physiological regulation of amiloride-sensitive Na^+ channels is complex. Activity of these channels has been demonstrated to be regulated by proteases, G proteins, methylation, arachidonic acid metabolites, tyrosine kinases, the membrane cytoskeleton, the cystic fibrosis transmembrane conductance regulator (CFTR), and protein kinase C (PKC) and cyclic adenosine monophosphate (cAMP)-dependent protein kinase A (PKA)-mediated phosphorylation (see Benos *et al.,* 1995; Eaton *et al.,* 1995; Barbry and Hofman, 1997; Garty and Palmer, 1997). In addition, studies have indicated that the number of active channels expressed at the cell surface is regulated by the recruitment of channels from an intracellular pool (see Benos *et al.,* 1995; Eaton *et al.,* 1995; Barbry and Hofman, 1997; Garty and Palmer, 1997).

In this chapter we focus on the mechanisms whereby intracellular cAMP increases amiloride-sensitive Na^+ channel activity. However, it should be noted that in some tissues, such as the rat distal colon (Bridges *et al.,* 1984) and human nasal epithelia (Boucher *et al.,* 1986; Rückes *et al.,*1997), amiloride-sensitive Na^+ channel activity is not stimulated by cAMP. The peptide hormone vasopressin and its analogs, vasotocin and oxytocin, increase amiloride-sensitive Na^+ transport across mammalian collecting ducts and amphibian epithelia (i.e., toad urinary bladder, frog skin and colon, and *Xenopus* A6 renal epithelial cells) two- to fourfold through an increase in intracellular cAMP (see Garty and Palmer, 1997). This increase in Na^+ transport develops rapidly over a period of 5 to 30 min and is initiated by the binding of vasopressin to V_2 receptors situated in the basolateral membrane. This in turn leads to the stimulation of adenylate cyclase via the heterotrimeric G protein, $G\alpha_s$, and a concomitant increase in intracellular cAMP (Breyer and Ando, 1994; Eaton *et al.,*1995), which results in increased Na^+ channel activity. The fact that membrane-permeable cAMP analogs, inhibitors of phosphodiesterase, and activators of adenylate cyclase all mimic the vasopressin induced activation of amiloride sensitive Na^+ channels clearly indicates that the action of vasopressin is mediated by cAMP.

Chronic (10–50 min) exposure to prostaglandin E_2 (PGE_2) has been shown to stimulate amiloride-sensitive Na^+ channels in A6 renal epithelial cells (Kokko *et al.,* 1994; Matsumoto *et al.,* 1997) and frog skin (Els and Helman, 1997). Like vasopressin, PGE_2 increases intracellular cAMP through its binding to a basolateral receptor which activates adenylate cyclase through a coupled stimulatory G protein (Sonnenberg and Smith, 1988; Kokko *et al.,* 1994). cAMP has also been implicated in the regulation of amiloride-sensitive Na^+ channels in the lung by catecholamines and other β_2-adrenergic receptor agonists which stimulate adenylate cyclase activity via a G_s coupled pathway (see O'Brodovich, 1991; Matalon *et al.,* 1996). β_2-Receptor agonists, such as epinephrine, induce epithelial Na^+ transport in the fetal and perinatal lung during labor and delivery, resulting in the clearance of fluid from the alveolar spaces as the lung changes to a gas conducting system (see Matalon *et al.,* 1996). β_2-Receptor agonists have also been implicated in the upregulation of amiloride-sensitive Na^+ transport in the adult lung (see Matalon *et al.,* 1996).

Although it is widely accepted that cAMP-mediated increases in amiloride-sensitive Na^+ reabsorption are the result of PKA activation and a concomitant phosphorylation event (Lester *et al.,* 1988; Gary and Palmer, 1997), the molecular identity of the proteins that are phosphorylated and how their phosphorylation translates into an increase in Na^+ channel activity remains to be clarified. Two alternative, but not mutually exclusive, hypotheses have been proposed to explain cAMP-mediated upregulation of amiloride-sensitive Na^+ channel activity. The first hypothesis proposes that there is a direct phosphorylation of the channel or an associated regulatory protein which activates quiescent channels preexisting in the membrane (Fig. 1A). The alternative hypothesis proposes that PKA mediated phosphorylation of a vescile trafficking or cytoskeletal protein(s) results in the insertion of new channels into the cell membrane from an intracellular pool (Fig. 1B). In this chapter we first summarize the data supporting activation of quiescent channels through a PKA mediated phosphorylation of a channel subunit or an associated regulatory protein. We then turn to a discussion of the role of channel recruitment in the cAMP-mediated upregulation of amiloride-sensitive Na^+ channel activity. We conclude with perspectives for further research in this area.

II. EVIDENCE FOR cAMP-MEDIATED ACTIVATION OF AMILORIDE-SENSITIVE SODIUM CHANNELS

The laboratories of Rossier (Canessa *et al.,* 1993, 1994) and Barbry (Lingueglia *et al.,* 1993, 1994) have recently isolated the cDNAs encoding for an epithelial Na channel (ENaC) from corticosteroid-stimulated rat distal colon using expression cloning techniques. This channel is a heteromeric complex consisting of at least three homologous subunits (α, β, and γ) that share 34–37% identity at the amino acid level (Canessa *et al.,* 1993; 1994; Lingueglia *et al.,* 1993, 1994). When expressed in *Xenopus* oocytes, this channel has a single-channel conductance of 4–5 pS and exhibits ion selectivity, gating kinetics, and an amiloride pharmacological profile similar to that of the 4–5 pS, highly selective Na^+ channel expressed in native Na^+-reabsorbing epithelia (Canessa *et al.,* 1994). Analysis of the deduced amino acid sequences of each of the three ENaC subunits shows no conserved intracellular PKA phosphorylation sites. Shimkets and co-workers (1998)

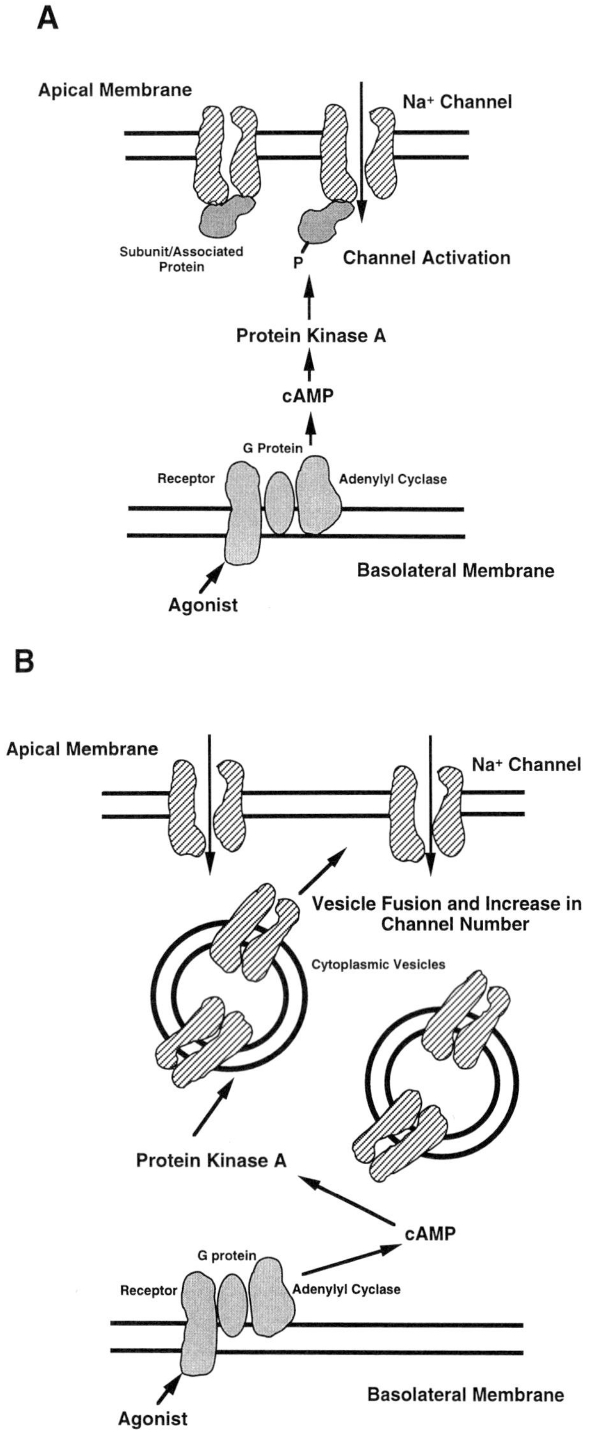
A
Apical Membrane
Na+ Channel
Subunit/Associated Protein
P
Channel Activation
Protein Kinase A
cAMP
G Protein
Receptor
Adenylyl Cyclase
Basolateral Membrane
Agonist
B
Apical Membrane
Na+ Channel
Vesicle Fusion and Increase in Channel Number
Cytoplasmic Vesicles
Protein Kinase A
cAMP
G protein
Receptor
Adenylyl Cyclase
Basolateral Membrane
Agonist

have revealed that both β and γ, but not α, rENaC are phosphorylated in their basal state when stably expressed in MDCK cells and PKA increases the phosphorylation of the β and γ subunits in their C termini. They have proposed that phosphorylation of channels in their basal state may be required to maintain basal Na^+ transport or, alternatively, it may serve to chronically down regulate basal Na^+ transport (Shimkets *et al.*, 1998).

Benos and co-workers (Awayda *et al.*, 1996) have examined whether heterologously expressed α, β, and γENaC cloned from rat distal colon can be directly activated by cAMP. Stimulation of *Xenopus* oocytes expressing $\alpha\beta\gamma$rENaC with forskolin and isobutylxanthine (IBMX) failed to increase amiloride-sensitive Na^+ currents above control (unstimulated) levels. Analysis of the current and *I/V* relationships revealed they were similar to unstimulated, $\alpha\beta\gamma$ENaC expressing oocytes. Similar findings have been reported by Kunzelmann and associates (Mall *et al.*, 1996; Kunzelmann *et al.*, 1997) for $\alpha\beta\gamma$rENaC expressed in *Xenopus* oocytes following stimulation with IBMX. The direct effects of PKA on $\alpha\beta\gamma$ENaC activity were also examined by incorporating ENaC into planar lipid bilayers. The ENaC used in these experiments was derived either by *in vitro* translation or by heterologous expression in *Xenopus* oocytes followed by isolation of the ENaC containing plasma membranes. Neither αrENaC nor $\alpha\beta\gamma$rENaC incorporated into lipid bilayers was activated by the addition of PKA and ATP to either side of the bilayer (Awayda *et al.*, 1996). This lack of cAMP activation of ENaC when expressed in *Xenopus* oocytes or incorporated into planar lipid bilayers is consistent with amiloride-sensitive Na^+ transport in the rat distal colon being insensitive to cAMP (Bridges *et al.*, 1984).

Liebold and co-workers (1996) have demonstrated cAMP sensitivity of ENaC in the guinea pig distal colon, in contrast to the rat colon. The membrane-permeable cAMP analog 8-chlorophenyl-thio-cAMP (8-cpt-cAMP) together with IBMX was shown to significantly increase the amiloride-sensitive current induced in *Xenopus* oocytes following the injection of poly(A^+) RNA isolated from the distal colon of guinea pigs maintained on a low salt diet. The amiloride-sensitive current was ~125% greater

---

FIGURE 1 Schematic diagram summarizing the two mechanisms proposed to explain the increase in amiloride-sensitive Na^+ channel activity in response to agonists that stimulate adenylate cyclase activity and increase intracellular cAMP. (A) Inactive channels situated within the plasma membrane are activated through a protein kinase A-mediated phosphorylation of a channel subunit or an associated regulatory protein. (B) Protein kinase A-mediated phosphorylation of a vesicle trafficking and/or cytoskeletal protein(s) stimulates the insertion of new functional channels into the plasma membrane from a cytoplasmic pool of channel-containing vesicles.

than the current from unstimulated, poly(A^+) RNA-injected oocytes. Patch-clamp analysis demonstrated that poly(A^+) RNA injection induced the expression of a highly Na^+-selective, amiloride-inhibitable 4-pS channel which was identical in its biophysical characteristics to ENaC (Liebold *et al.,* 1996). In light of the data of Benos and co-workers demonstrating a lack of effect of cAMP on ENaC when expressed in oocytes (Awayda *et al.,* 1996), it was proposed that the cAMP sensitivity observed in their own study was due to the injected mRNA encoding for associated proteins which confer cAMP sensitivity on ENaC (Liebold *et al.,* 1996).

Stutts and co-workers (1997) have reached similar conclusions for the effects of cAMP and PKA on the activity of $\alpha\beta\gamma$ENaC stably expressed in NIH 3T3 fibroblasts. Using whole-cell voltage-clamp techniques, these authors demonstrated that expression of $\alpha\beta\gamma$ENaC in NIH 3T3 cells induced amiloride-sensitive currents which were stimulated by the cAMP analog 8-cpt-cAMP. Rundown of ENaC activity in excised patches could be partially reversed or prevented by exposure of the cytoplasmic face of the patch to the catalytic subunit of PKA and ATP (Stutts *et al.,* 1997). A peptide inhibitor of PKA (pKI), which had been myristoylated to promote its association with membranes, reversed the effects of PKA and ATP by inhibiting channel open probability (P_o) when added to excised membrane patches (Stutts *et al.,* 1997). Furthermore, forskolin plus 8-cpt-cAMP was shown to significantly increase the P_o of ENaC in cell attached patches when compared to basal (unstimulated) conditions. To explain the effects of cAMP and PKA on ENaC activity, it was proposed that PKA induced the phosphorylation of an ENaC associated protein(s) which regulates channel gating (Stutts *et al.,* 1997).

Studies examining the effect of PKA on ENaC activity in native epithelial cells have produced conflicting results. Using the patch-clamp technique, Frings *et al.* (1988) reported that PKA induced activation of a highly Na^+-selective, 4.8-pS channel, resembling the cloned ENaC in its biophysical characteristics, in the toad urinary bladder. In contrast, Lester *et al* (1988) found that incorporation of PKA, cAMP, and ATP into apical membrane vesicles prepared from toad urinary bladder did not increase the amiloride-blockable uptake of ^{22}Na into the vesicles. Chalfant *et al.* (1996) were unable to detect a consistent change in the P_o, unitary current, or kinetics of the 4-pS, highly selective Na^+ channel following the addition of PKA and ATP to excised apical patches from the mouse M-1 cortical collecting duct cell line. Alkaline phosphatase pretreatment did not alter the lack of PKA sensitivity of the channel, suggesting that the channel had not been stably phosphorylated prior to the application of PKA (Chalfant *et al.,* 1996).

In an elegant study Bubien and co-workers (1994) presented data that offer an explanation for such contradictory findings on the effects of PKA on ENaC activity. They showed that the cAMP-mediated pathway and pertussis toxin (PTX)-sensitive G proteins act synergistically to regulate amiloride-sensitive Na^+ channel activity in Epstein–Barr-transformed human B lymphocytes. Expression of α, β, and γENaC has been identified in these cells using the reverse transcriptase polymerase chain reaction (RT-PCR) (Fuller *et al.,* 1997), indicating that the amiloride-sensitive Na^+ channel activity examined by Bubien and co-workers (1994) is due to ENaC expression. Cholera toxin, which increases intracellular cAMP through the activation of $G\alpha_s$, and PTX, which ADP-ribosylates the GTP-binding proteins $G\alpha_i$ and $G\alpha_o$, independently increased the lymphocyte amiloride sensitive Na^+ conductance (Bubien *et al.,* 1994). When the toxins were applied simultaneously, the sodium conductance failed to increase, indicating that the same Na^+ channel was affected by both toxins. Addition of 8-cpt-cAMP, following activation of Na^+ currents with PTX, reduced Na^+ currents to the basal state (Bubien *et al.,* 1994). Comparable results were obtained when the lymphocytes were activated by 8-cpt-cAMP and then treated with the GTP analog GDPβS (Bubien *et al.,* 1994). Taken together, these data indicate that cAMP can either activate or inactivate Na^+ channels, depending on whether or not the channel had previously been activated by PTX-sensitive G proteins. This synergistic regulation of amiloride-sensitive Na^+ channels by PTX-sensitive G proteins and the cAMP pathway has been corroborated using purified renal epithelial Na^+ channels incorporated into planar bilayer membranes (Ismailov *et al.,* 1994).

In addition to PTX-sensitive G proteins, the coexpression of CFTR with amiloride-sensitive Na^+ channels in specific cell types provides an explanation for the conflicting effects of cAMP on Na^+ channel activation. Recent data indicate that CFTR can regulate the cAMP sensitivity of amiloride-sensitive Na^+ channels. Stutts and associates (1995, 1997) demonstrated that when $\alpha\beta\gamma$ENaC were expressed in MDCK cells or NIH 3T3 cells either alone (see discussion above) or together with the Δ508 mutation of CFTR, Na^+ transport increased in response to agonists which elevated intracellular cAMP. In contrast, the coexpression of $\alpha\beta\gamma$ENaC with wild-type CFTR in these cells inhibited the cAMP-mediated increase in Na^+ transport (Stutts *et al.,* 1995, 1997). These findings have been corroborated using ENaC and CFTR coexpressed in *Xenopus* oocytes (Mall *et al.,* 1996). Ismailov *et al.* (1996a) have shown that co-incorporation of CFTR and immunopurified bovine renal epithelial Na^+ channels into planar lipid bilayers inhibits the PKA induced activation of the Na^+ channel, whereas the nonconductive CFTR mutant G551D has no effect on the PKA sensitivity of this Na^+ channel.

Regulation of the cAMP sensitivity of ENaC by CFTR has also been demonstrated in native Na^+-reasborbing epithelia. Forskolin decreases an amiloride-sensitive Na^+ conductance attributable to ENaC and stimulates an apical CFTR-like Cl^- conductance in the M-1 mouse cortical collecting duct cell line (Letz and Korbmacher, 1997). Similarly, the amiloride-inhibitable Na^+ conductance is reduced and the Cl^- conductance is enhanced when isolated rat colonic crypt cells, which express both ENaC and CFTR, are exposed to either forskolin or IBMX (Ecke *et al.,* 1996). In addition, inhibition of CFTR expression in A6 renal epithelial cells by antisense oligonucleotides has recently been shown to significantly increase both forksolin-stimulated amiloride-sensitive I_{SC} and Na^+ channel P_o when compared to forskolin-stimulated A6 cells exposed to control oligonucleotides (Ling *et al.,* 1997).

The mechanism whereby CFTR regulates the cAMP sensitivity of amiloride-sensitive Na^+ channels is unclear. The data of Ismailov *et al.* (1996a) suggest that ENaC and CFTR physically interact. In addition, Kunzelmann *et al.* (1997) have presented data using the yeast two hybrid system indicating that the C-termini of α ENaC and CFTR can directly interact. Nevertheless, biochemical evidence for an *in vivo* interaction between these two proteins is lacking. The fact that many Na^+-transporting epithelia express both Na^+ channels and CFTR and still exhibit cAMP-mediated activation of amiloride-sensitive Na^+ transport has lead Ismailov and co-workers (1996a) to propose that the interactions between CFTR and amiloride-sensitive Na^+ channels may involve additional factors such as extracellular mediators released through the activation of CFTR or associated proteins.

Increasing data indicate that proteins associated with the amiloride-sensitive Na^+ channels are the target for cAMP-mediated phosphorylation and that their phosphorylation regulates ENaC gating. Biochemical evidence for a role of associated proteins in conferring cAMP sensitivity on amiloride-sensitive Na^+ channels was first provided by Benos and associates who purified an amiloride-sensitive Na^+ channel complex from A6 renal epithelial cells and bovine papillary collecting ducts (Benos *et al.* 1986, 1987). This channel was biochemically isolated as a heterooligomeric complex consisting of at least 6 associated polypeptides which formed a Na^+-selective, highly amiloride-sensitive channel when incorporated into planar lipid bilayers (Benos *et al.,* 1986, 1987). Although the exact relationship of this channel to the cloned ENaC remains to be clarified, recent data have demonstrated that antibodies directed against αENaC cross react with two polypeptides in this channel, indicating that ENaC comprises the conductive element of this Na^+ channel (Ismailov *et al.,* 1996b). A 300- to 315-kDa polypeptide in the Na^+ channel complex was found to be specifically phosphorylated in filter grown A6 cells following stimulation with vasopressin and its phosphorylation correlated with an increase in transepithelial Na^+

transport (Sariban-Sohraby *et al.,* 1988). Furthermore, this polypeptide was phosphorylated *in vitro* when the purified channel complex was incubated with the catalytic subunit of PKA plus [γ-^{32}P]ATP (Sariban-Sohraby *et al.,* 1988; Oh *et al.,* 1993). Planar lipid bilayer studies demonstrated that PKA plus ATP significantly increased the P_o of this channel from values of <0.20 to 0.60–0.95, with no detectable change in the single channel conductance or ion selectivity (Ismailov *et al.,* 1994). PKA phosphorylation of this channel conferred a voltage dependence to the channel activity, with the channel P_o being greater at negative voltages (Ismailov *et al.,* 1994). Treatment of the channel with low concentrations of dithiothreitol (DTT), an agent which reduces disulfide bonds, uncoupled PKA regulation (Ismailov *et al.,* 1996b), suggesting that the PKA sensitivity of this renal Na^+ channel is conferred by an associated protein, namely the 300-kDa polypeptide.

Amiloride-sensitive Na^+ channels biochemically purified from rat lymphocytes (Bradford *et al.,* 1995) and rabbit alveolar type II cells (Berdiev *et al.,* 1997) using an antibody generated against the renal epithelial Na^+ channel also exhibited a PKA-stimulated increase in P_o when incorporated into planar lipid bilayers. These data are in agreement with patch-clamp studies which have demonstrated cAMP sensitivity of these channels in native cells (Bubien *et al.,* 1994; Yue *et al.,* 1994). Biochemical analysis revealed that the 300- to 315-kDa polypeptide is not a component of either the lymphocyte (Bradford *et al.,* 1995) or alveolar type II cell Na^+ channel (Oh *et al.,* 1992; Senyk *et al.,* 1995), suggesting that the proteins conferring PKA sensitivity on the Na^+ channels are cell specific.

Data from Cantiello and co-workers (Prat *et al.,* 1993a,b) have indicated a role for actin in the vasopressin and cAMP-mediated upregulation of a poorly selective, 9-pS amiloride-blockable Na^+ channel expressed in A6 cells grown on nonpermeable supports. It is unclear whether the 9-pS channel and the 4-pS ENaC channel are distinct entities or if they represent different manifestations of the same channel under different growth conditions. This channel has previously been shown to be activated by both cytochalasin D, an agent that depolymerizes actin filaments, and short actin filaments (Cantiello *et al.,* 1991; Cantiello, 1995). Vasopressin induces a transient depolymerization of an apical submembrane pool of actin in both toad urinary bladder and mammalian collecting duct that may permit the fusion of vesicles containing water channels with the apical membrane, thereby facilitating transepithelial water movement (Ding *et al.,* 1991; Simon *et al.,* 1993). This actin pool is also sensitive to cytochalasin D (Franki *et al.,* 1992). In light of the vasopressin-induced depolymerization of actin in renal epithelial cells, Cantiello and co-workers investigated the role of changes in actin filament length on the vasopressin and PKA-mediated activation of the 9-pS channel using the patch clamp technique in the excised inside-out configuration (Prat *et al.,* 1993a,b). Pretreatment of A6 cells with cytochalasin D inhibited channel activation by PKA; however, subsequent addition of short actin filaments to the patches induced channel activity. These data were interpreted to indicate that actin filaments are required for the PKA-mediated activation of this channel and that actin may be a target for PKA (Prat *et al.,* 1993b). To test whether actin is a substrate for PKA-mediated phosphorylation, actin was incubated with PKA and ATP in varying molar ratios of actin to ATP. Both G- and F-actin were phosphorylated by PKA under physiologically relevant conditions, and phosphorylation could be inhibited by a specific inhibitor of PKA, 5- to 24-amide (Prat *et al.,* 1993b, Cantiello, 1995). When the effects of phosphorylated G- and F-actin on channel activity in excised patches were examined, only phosphorylated F-actin filaments were capable of activating Na^+ channels. It was therefore proposed, based on these data, that PKA-dependent effects of actin on Na^+ channel activity are mediated either through a direct molecular interaction between actin and the channel protein or through actin-binding proteins associated with the channel (Prat *et al.,* 1993b; Cantiello, 1995).

Benos and associates (Berdiev *et al.,* 1996) recently extended these observations to ENaC using the planar lipid bilayer system. In their work, α, β, and γrENaC cDNAs were individually *in vitro* translated in the presence of pancreatic microsomes and subsequently co-reconstituted into planar lipid bilayers. Although the catalytic subunit of PKA had no effect on channel activity when added to either side or both sides of the bilayer membrane, the presence of ATP alone was found to increase channel P_o (Berdiev *et al.,* 1996). When PKA and ATP were added in the presence of G-actin, the P_o of the channel increased by a factor of ~ 2 over the P_o in the presence of ATP. Cytochalasin D, added after run down of ENaC activity, produced a second peak of channel activation (Berdiev *et al.,* 1996). Subsequently, this group presented data indicating that actin interacts only with αENaC (Ismailov *et al.,* 1997). It was concluded from these data that one mechanism whereby cAMP activates ENaC is through the direct interaction between actin and αENaC and that actin is the target for PKA-mediated phosphorylation (Ismailov *et al.,* 1997).

In addition to actin and the 300-kDa polypeptide associated with the biochemically characterized renal epithelial Na^+ channel, McDonald and Welsch (1995) identified a 134-kDa αENaC associated polypeptide which is a substrate for PKA-mediated phosphorylation in H441 cells, a human lung adenocarcinoma cell line. The 134-kDa polypeptide specifically bound to the proline-rich domain of the C terminus of human αENaC expressed as a GST fusion protein and was phosphorylated *in vitro* following incubation with [γ-^{32}P]dATP and PKA. Although the molecular identity of this

ERRATA

Due to a printing error, the legend for Figure 1 in the Preface was omitted. For the reader's convenience, the correct versions of pages xix and xx are given here.

Preface

Since the publication of the first subunit of the cloned epithelial Na^+ channel (ENaC) in 1993, many laboratories throughout the world have focused attention on this important ion channel. Progress has accelerated, and a plethora of novel discoveries concerning the channel's structure–function relationships and regulatory influences have been forthcoming. For example, Fig. 1 highlights the sheer number of new members of the ENaC/degenerin superfamily that have been identified in four years. Moreover, the direct involvement of ENaC in important physiological processes and human disease has been demonstrated.

This volume is a compilation of current work in the general area of amiloride-sensitive Na^+ channels. Researchers from many of the leading laboratories working in this area have contributed chapters. The volume is organized into five main sections: ENaC structure–function relationships, regulation of amiloride-sensitive Na^+ channels, sodium channels in the lung, sensory and mechanical transduction, and involvement of Na^+ channels in disease. These sections deal with the following questions:

1. What is the molecular basis for the functional and biochemical diversity of amiloride-sensitive Na^+ channels?
2. What are the biophysical properties of the ENaCs?
3. What is the stoichiometry of a functional ENaC?

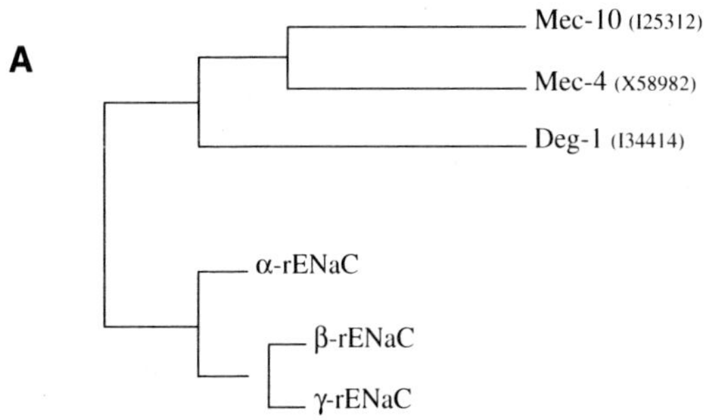

FIGURE 1A Phylogenetic trees of the DEG/ENaC supergene family. Tree circa 1994.

Current Topics in Membranes, Volume 47 (Benos, Ed.)
0-12-153347-6 (case)
0-12-089030-5 (paperback)

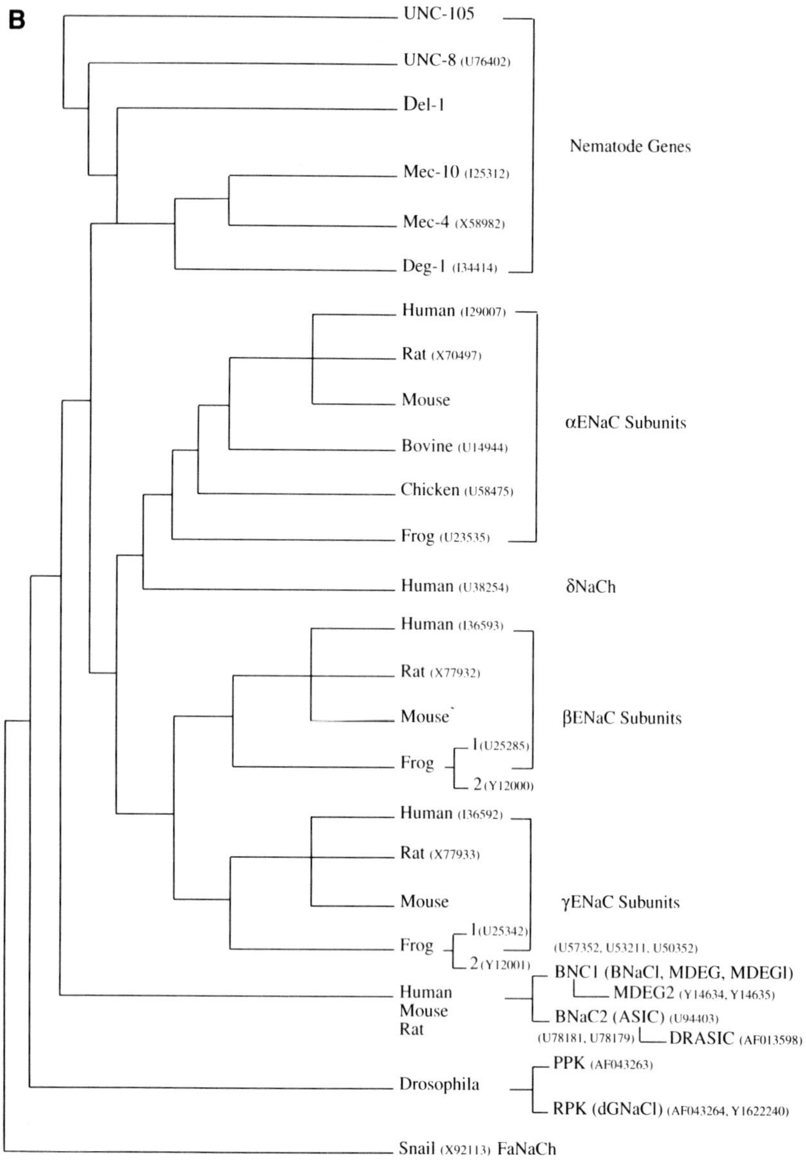

FIGURE 1B Phylogenetic trees of the DEG/ENaC supergene family. Tree circa 1998. Sequence analysis indicates that this superfamily contains five subfamilies, namely, the *C. elegans* members, the ENaCs, the mammalian neuronal channels (BNaC1 and BNaC2, MDEG2, and DRASIC), and *Drosphila* pickpocket (PPK, neuronal), ripped pocket (RPK, embryonic) channels, and the snail FaNaCh. Relevant GenBank accession numbers are given in parentheses.

Current Topics in Membranes, Volume 47 (Benos, Ed.)
0-12-153347-6 (case)
0-12-089030-5 (paperback)

polypeptide is unknown, McDonald and Welsch (1995) speculated that it contains a SH3 domain because of its capacity to bind to the proline-rich region of αENaC. Furthermore, it was proposed that this polypeptide may function in the regulation of ENaC activity by coupling ENaC to a ligand-dependent signaling pathway which activates PKA (McDonald and Welsch, 1995).

III. EVIDENCE FOR cAMP-MEDIATED RECRUITMENT OF AMILORIDE-SENSITIVE SODIUM CHANNELS

A role for the recruitment of Na^+ channels to the cell surface in response to cAMP stems from the work of Garty and Edelman (1983) on the toad urinary bladder. They showed that exposure of the apical surface of the toad bladder to trypsin, which blocks most of the baseline Na^+ transport, does not impair the vasopressin-induced increase in Na^+ absorption, whereas trypsinization inhibits the response to aldosterone. Because the actions of vasopressin and aldosterone on amiloride-sensitive Na^+ transport have been shown to be synergistic in the toad urinary bladder (Handler *et al.*, 1969; Fanestil *et al.*, 1967) their observations suggested that vasopressin stimulates the recruitment of Na^+ channels from an intracellular pool which was not accessible to trypsin. However, Park and Fanestil (1980) showed that when the apical surface of the toad urinary bladder was pretreated with tyrosine-specific protein-modifying reagents, such as 7-chloro-4-nitrobenzo-2-oxa-1,3 diazole or diazolsulfonic acid, baseline and vasopressin stimulated Na^+ currents were inhibited to the same extent. From the cloning of the ENaC subunits it is now known that there are multiple tyrosine residues within their extracellular loops (Canessa *et al.*, 1994) which would be modified by these agents. The most parsimonious explanation for these findings is that the channels may be continuously present in the apical membrane in a conformational state that is resistant to trypsin proteolysis, but are susceptible to other protein-modifying reagents. Such an explanation would be in keeping with studies indicating (1) that vasopressin-induced changes in capacitance of the apical membranes of the toad urinary bladder (Stetson *et al.*, 1982; Palmer and Lorenzen, 1983) and the amphibian colon (Krattenmacher and Clauss, 1988; Krattenmacher *et al.*, 1988), which also shows a marked increase in sodium transport in response to vasopressin, are not correlated with increased Na^+ transport; and (2) agents which disrupt microtubules and microfilaments, such as colchicine and cytochalasin, inhibit vasopressin-induced changes in membrane capacitance and water transport but do not effect vasopressin stimulated Na^+ transport across the toad urinary bladder (e.g., Palmer and Lorenzen, 1983; Taylor *et al.,* 1978). Nevertheless, as discussed below, multiple lines of evidence indicate that there is recruitment of Na^+ channels to the cell surface in response to cAMP.

Immunocytochemical studies have revealed the presence of intracellular pools of amiloride-sensitive Na^+ channels in Na^+-reabsorbing epithelial cells. Duc *et al.* (1994) demonstrated using ENaC subunit-specific antibodies that α, β, and γENaC are coexpressed not only at the apical cell surface but also intracellularly in the kidney, colon, sweat, and salivary ducts of rats maintained on a low salt diet. Similar observations for ENaC have been reported by Barbry and co-workers (Renard *et al.,* 1995; Barbry and Hofman, 1997). However, because the animals used in these studies were maintained on a low salt diet, it is unclear what proportion of the intracellular labeling is the result of steroid-hormone-induced upregulation of ENaC expression. Antibodies directed against the biochemically purified renal epithelial Na^+ channel were found to label the apical membrane and subapical cytoplasm of filter grown A6 cells (Morris *et al.,* 1998). Comparable results were obtained in A6 cells using an anti-idiotypic antibody directed against the amiloride binding site on the epithelial Na^+ channel; however, labeling was distributed throughout the cytoplasm rather than being restricted to the subapical cytoplasm (Kleyman *et al.,* 1991).

The C termini of α, β and γENaC contain two adjacent tyrosine-based internalization signals for clathrin-mediated endocytosis, NPXY and YXXL (where X is any polar amino acid) (Shimkets *et al.,* 1997). Internalization motifs are present in transport proteins recruited to the cell surface from a recycling intracellular storage pool in response to physiologic stimuli, such as GLUT-4 (Rea and James, 1997) and the gastric H^+,K^+-ATPase β subunit (Caplan, 1997). Shimkets and co-workers (1997) have shown that mutagensis of the tyrosine to an alanine within these motifs in β and γENaC increased the retention time of ENaC at the cell surface by reducing endocytosis, whereas this mutation in αENaC had no effect. In patients with Liddle's syndrome, a hereditary form of hypertension characterized by hyperactivity of the amiloride-sensitive Na^+ channel, mutations within the C terminus of β and γENaC led to the deletion of these internalization motifs and increased retention of ENaC at the surface (Awayda *et al.,* 1997; Firsov *et al.,*1996; Snyder *et al.,* 1995; Staub *et al.,* 1996, 1997). Although ENaC expression at the cell surface is regulated by clathrin-mediated endocytosis (Shimkets *et al.,* 1997), it remains to be determined if the channel is recycled to the cell surface.

Brefeldin A (BFA) is a fungal metabolite that has recently been employed in electrophysiological studies designed to examine if the cAMP-mediated increase in amiloride-sensitive Na^+ transport results from the recruitment of epithelial Na^+ channels in response to increases in intracellular cAMP.

BFA acts by reversibly disrupting the Golgi apparatus, thereby interrupting the posttranslational processing of proteins and intracellular trafficking and secretion (Fujiwara *et al.,* 1988; Lipponcott-Schwartz *et al.,* 1990, 1991; Strous *et al.,* 1993; Pelham, 1991). In addition, BFA has been shown to effect endosome function and the movement of membrane proteins from an intracellular pool to the cell surface (Low *et al.,* 1992; Miller *et al.,* 1992). The use of BFA to examine recruitment of Na^+ channels in response to forskolin stimulation has produced conflicting results in A6 renal epithelial cells. Treatment of A6 cells with BFA prior to stimulation with forskolin was shown by Kleyman and co-workers (1994) to significantly inhibit the forskolin induced in amiloride-sensitive I_{SC}. In contrast, another group has presented data indicating that under these conditions BFA has no effect on amiloride-sensitive I_{SC} (Coupaye-Gerard *et al.,* 1994). However, this may reflect differences in the culture conditions. The presence of aldosterone in the culture media has been shown to significantly enhance both vasotocin/forskolin-stimulated increases in intracellular cAMP and apical endocytosis/exocytosis (Verrey *et al.,*1993, 1995). Recently, Niisato and Marunaka (1997) demonstrated that BFA treatment has no affect on the vasopressin-stimulated increase in I_{SC} across A6 cell monolayers cultured in the absence of aldosterone, whereas when A6 cells are cultured in the presence of aldosterone, the vasopressin stimulated increase in I_{SC} is sensitive to BFA. Taken together, these data strongly suggest that in A6 cells grown in the absence of aldosterone, vasopressin either activates preexisting channels in the apical membrane or stimulates translocation of channels via a BFA insensitive pathway, whereas in cells grown in the presence of aldosterone, vasopressin stimulates the recruitment of Na^+ channels from a BFA-sensitive pool.

In the toad urinary bladder, Weng and Wade (1994) observed that BFA treatment did not alter the initial vasopressin induced increase in I_{SC}. This is in agreement with earlier data indicating that vasopressin activates quiescent channels already present in the apical membrane of the toad urinary bladder (Stetson *et al.,* 1982; Palmer and Lorenzen, 1983). However, subsequent I_{SC} responses to repeated vasopressin challenge were markedly reduced (Weng and Wade, 1994). Similar results were obtained in response to repeated treatment with forskolin. Because BFA has been shown to inhibit the apical delivery of membrane proteins from endosomes as well as newly synthesized proteins, it remains to be clarified if the BFA effects in the toad bladder are due to an inhibition of the recruitment of newly synthesized Na^+ channels, recycled Na^+ channels, or a combination of the two.

Ito and colleagues (1997) used BFA to show that the β_2-adrenergic agonist terbutaline increases Na^+ transport across primary cultures of rat fetal distal lung epithelium by stimulating the intracellular trafficking of a nonselective, 27-pS amiloride-sensitive cation channel to the apical cell surface. This channel corresponds to the high (H-type)-amiloride-affinity Na^+ channel biochemically characterized from rat fetal distal lung epithelia (Matalon *et al.,* 1993). Basolateral application of terbutaline stimulated amiloride-sensitive short circuit current 2.5-fold within 50 min of application and this response was inhibited by a 30-min pretreatment of the monolayers with brefeldin A (Ito *et al.,* 1997). Single-channel patch-clamp experiments demonstrated that terbutaline increased the number of nonselective cation channels in the apical membrane from 1.6 ± 0.2 in untreated cells to 4.3 ± 0.5 in treated cells and this increase in channel number was inhibited by brefeldin A (Ito *et al.,* 1997). In contrast, terbutaline was not found to alter the apical cell surface expression of the 12-pS highly Na^+-selective channel expressed in these cells in either the presence or the absence of brefeldin A. Taken together, these data were interpreted to indicate that a β_2-agonist induced increase in Na^+ transport across the fetal distal lung epithelium is due to a cAMP-mediated recruitment of nonselective 27-pS cation channels to the apical cell surface (Ito *et al.,* 1997).

Compelling data supporting the cAMP-mediated recruitment of Na^+ channels from an intracellular pool to the apical surface in A6 cells comes from the laboratories of Kleyman (Kleyman *et al.,* 1994) and Eaton (Marunaka and Eaton, 1991). Kleyman and co-workers (1994) assessed changes in the levels of Na^+ channel expression at apical cell surface following the addition of vasopressin or forskolin by immunoprecipitation of Na^+ channels from A6 cell monolayers which had been radioiodinated apically. Immunoprecipitation was performed using the anti-idiotypic antibody directed against the amiloride-binding site of the Na^+ channel. Vasopressin or forskolin induced a 1.7- to 2.9-fold increase in A6 cell amiloride-sensitive Na^+ transport which was accompanied by a 1.6- to 2.9-fold increase in apical cell surface expression of amiloride-sensitive Na^+ channels (Kleyman *et al.,* 1994). Marunaka and Eaton (1991) were able to show using patch clamp techniques that both vasopressin and the cAMP analog N^6,2′-*O*-dibutyryladenosine 3′,5′-cyclic monophosphate (DBcAMP) increase the number of Na^+ channels per patch from 2.0 ± 1.5 (unstimulated) to 9.2 ± 1.5 without a change in channel P_o in filter-grown A6 cells. These data imply that the cAMP-mediated increase in amiloride-sensitive Na^+ transport in these cells is due to the insertion of channels into the apical membrane from an intracellular pool and is in keeping with the effect of BFA on this process observed by Niisato and Marunaka (1997) and Kleyman *et al.* (1994). Further evidence corroborating recruitment of channels from an intracellular pool in A6 cells comes from the fact that agents which disrupt microtubules and microfilaments have been shown to decrease the vasopressin-stimulated increase in Na^+ transport across A6 cell monolayers (Verrey *et al.*, 1995).

In contrast, data obtained by Morris *et al.* (1998) using the antibody directed against the renal epithelial Na^+ channel argue against a cAMP-mediated recruitment of Na^+ channels from an intracellular to the cell surface in A6 cell monolayers. Quantitative immunofluorescence microscopy did not reveal a significant difference in cell surface labeling between either vasopressin- or vasotocin-stimulated monolayers and control monolayers despite an ~2-fold increase in amiloride-sensitive I_{SC} (Smith *et al.*, 1995; Morris *et al.*, 1998). Furthermore, microtubule disruption had no effect on the increase in amiloride-sensitive I_{SC}. However, as noted by Morris and co-workers (1998), immunofluorescence microscopy may not provide the resolution necessary to distinguish between channels at the cell surface and channels lying in close proximity to the apical membrane which are recruited by vasopressin. On the other hand, when the anti-Na^+ channel antibody was used together with cell surface biotinylation to examine if vasopressin stimulates recruitment of [^{35}S]methionine-labeled Na^+ channels in A6 cell monolayers, changes in apical Na^+ channel expression were not detected (Oh *et al.*, 1993). However, as discussed above, these contrasting data on Na^+ channel recruitment in response to cAMP may reflect differences in the conditions used to culture the A6 cells.

Frindt *et al.* (1995) have examined cAMP-mediated activation of ENaC in isolated rat cortical collecting tubules. When the permeant cAMP analog cpt-cAMP was added to the superfusate during patch clamp recording from cell attached patches, neither channel activation nor an increase in channel density was observed. However, when cpt-cAMP was added to the bath before formation of the patch, the conducting channels increased from 10 ± 2 to 37 ± 6 per patch, suggesting that the increased channel density was due to the recruitment of channels to the apical membrane from an intracellular pool. They speculated that their inability to detect an increase in channel density when cpt-cAMP was added to cell attached patches was due to the plasma membrane pulling away from vesicles or cytoskeletal elements during patch formation (Frindt *et al.*, 1995).

Chronic (10–50 min) exposure to prostaglandin E_2 (PGE_2) increases intracellular cAMP levels and stimulates transepithelial Na^+ transport in A6 cells grown on permeable supports (Kokko *et al.*, 1994; Matsumoto *et al.*, 1997) and frog skin (Els and Helman, 1997). Kokko *et al.* (1994), using patch clamp techniques, have shown that the increase in Na^+ transport is the result of an increase in the mean number of open ENaC in the apical membrane (NP_o, 0.98 ± 0.14 vs 0.56 ± 0.08 (control)) rather than an increase in ENaC P_o (Kokko *et al.*, 1994). Analysis of the distribution of the number of functional channels per patch in unstimulated cells revealed that they could be fitted with a single Gaussian with a mean of 2.1 channels per patch. In contrast, following stimulation by PGE_2, the distribution of the number of channels was best fit by two Gaussians, with mean values of 2.7 and 10 channel per patch, indicating that there was activation of clusters of channels in the apical membrane (Kokko *et al.,* 1994). The most parsimonious explanation for these data was that there was recruitment of clusters of channels into the apical membrane from a vesicular pool in response to PGE_2-stimulated increase in intracellular cAMP.

IV. PERSPECTIVES

It is evident from the preceding discussion that the mechanisms whereby cAMP regulates amilioride-sensitive Na^+ channels are cell specific and involve channel activation, channel recruitment, or possibly both. Critical to further understanding the role channel-associated proteins play in the cAMP-mediated activation of Na^+ channels will be the molecular identification of proteins which are substrates for PKA-mediated phosphorylation and which interact, either directly or indirectly, with the channel. The cloning of ENaC subunits from a variety of species facilitates the use of solid-phase binding assays (i.e., McDonald and Welsch, 1995) as well as the yeast two hybrid system and functional complementation expression cloning assays for identifying tissue- and cell-specific ENaC-associated proteins which are substrates for PKA-mediated phosphorylation. Staub and co-workers (1996, 1997) used the yeast two hybrid system to identify an association between ENaC and the ubiquitin ligase Nedd4, and Vallet and colleagues (1997) used a functional complementation assay to identify the ENaC regulatory protein CAP1. The identification of ENaC-associated proteins will allow their role in mediating channel activity to be examined using a combination of molecular, biochemical, and biophysical techniques. It will also be important to determine the molecular mechanisms whereby CFTR regulates the cAMP sensitivity of ENaC in a cell-specific manner.

Key to elucidating how cAMP mediates the recruitment of amiloride-sensitive Na^+ channels from an intracellular pool to the cell surface will be the identification of specialized proteins involved in the targeting (i.e., v-SNARES , t-SNARES) and trafficking (i.e., cytoskeleton-associated proteins and molecular motors) of Na^+-channel-containing vesicles to the apical membrane and how these proteins are modulated by cAMP. Progress in the area of protein trafficking has revealed that the processes and molecules involved are essentially the same not only between diverse cell types, such as myocytes and neurons, but also between diverse organisms, such as yeast and mammals. SNARE proteins have been implicated in the trafficking of

H^+, K^+-ATPase (Peng *et al.,* 1997), GLUT-4 (Rea and James, 1997), and aquaporin 2 (Knepper and Inoue, 1997) from an intracellular pool to the cell surface in response to physiologic stimuli and a preliminary report from Peters *et al.* (1998) suggests that the t-SNARE syntaxin 1A is involved in the trafficking of ENaC.

Electrophysiological studies have revealed that amiloride sensitive Na^+ channels are also expressed in invertebrate epithelia, such as the dorsal skin of the leech *Hirudo medicinalis* (Weber *et al.,* 1993) and gills of the Chinese crab *Eriocheir sinensis* (Zeiske *et al.,* 1992). cAMP has been shown to increase the density of active Na^+ channels in the leech dorsal skin (Weber *et al.,* 1993). Further analysis of the mechanism(s) whereby by cAMP regulates invertebrate amiloride-sensitive Na^+ channels will provide insight into the evolution of Na^+ channel regulatory mechanisms.

In this chapter, we have presented an overview of the current understanding of the mechanism by which cAMP mediates the activity of amiloride-sensitive Na^+ channels. Although it is clear that both channel activation and channel recruitment are mechanisms whereby cAMP mediates the activity of amiloride-sensitive Na^+ channels, our understanding of the protein–protein interactions underlying these mechanisms and how they are regulated by cAMP is only beginning.

Acknowledgment

This work was supported by National Institute of Diabetes and Digestive and Kidney Diseases Grant DK 46705.

References

Awayda, M. S., Ismailov, I. I., Berdiev, B. K., Fuller, C. M., and Benos, D. J. (1996). Protein kinase regulation of a cloned epithelial Na^+ channel. *J. Gen. Physiol.* **108,** 49–65.

Awayda, M. S., Tousson, A. and Benos, D. J. (1997). Regulation of a cloned epithelial Na^+ channel by its beta and gamma subunits. *Am. J. Physiol.* **273,** C1889—C1899.

Barbry, P., and Hoffman, P. (1997). Molecular biology of Na^+ absorption. *Am. J. Physiol.* **273,** G571–G585.

Benos, D. J., Saccomani, G., Brenner, B. M., and Sariban-Sohraby, S. (1986). Purification and characterization of the amiloride-sensitive sodium channel from A6 cultured cells and bovine renal papilla. *Proc. Natl. Acad. Sci. U.S.A.* **83,** 8525–8529.

Benos, D. J., Saccomani, G., and Sariban-Sohraby, S. (1987). The epithelial sodium channel: Subunit number and location of amiloride-binding site. *J. Biol. Chem.* **262,** 10613–10618.

Benos, D. J., Awayda, M. S., Ismailov, I. I., and Johnson, J. P. (1995). Structure and function of amiloride-sensitive Na^+ channels. *J. Membr. Biol.* **143,** 1–18.

Berdiev, B. K., Prat, A. G., Cantiello, H. F., Ausiello, D. A., Fuller, C. M., Jovov, B., Benos, D. J., and Ismailov, I. I. (1996). Regulation of epithelial sodium channels by short actin filaments. *J. Biol. Chem.* **271,** 17704–17710.

Berdiev, B. K., Shylonsky, V. G., Senyk, O., Keeton, D, Guo, Y., Matalon, S., Cantiello, H. F., Prat, A. G., Ausiello, D. A., Ismailov, I. I., and Benos, D. J. (1997). Protein kinase

A phosphorylation and G protein regulation of type II pneumocyte Na^+ channels in lipid bilayers. *Am. J. Physiol.* **272,** C1262–C1270.

Boucher, R. C., Stutts, M. J., Knowles, M. R., Cantley, L., and Gatzy, J. T. (1986). Na^+ transport in cystic fibrosis respiratory epithelia. Abnormal basal rate and response to adenylate cyclase activation. *J. Clin. Invest.* **78,** 1245–1252.

Bradford, A. L., Ismailov, I. I., Achard, J.-M., Warnock, D. G., Bubien, J. K., and Benos, D. J. (1995). Immunopurification and functional reconstitution of a Na^+ channel complex from rat lymphocytes. *Am. J. Physiol.* **269,** C601–C611.

Breyer, M. D., and Ando, Y. (1994). Hormonal signaling and regulation of salt and water transport in the collecting duct. *Physiol. Rev.* **56,** 711–739.

Bridges, R. J., Rummel, W., and Wollenberg, P. (1984). Effects of vasopressin on electrolyte transport across isolated colon from normal and dexamethasone-treated rats. *J. Physiol. (London)* **355,** 11–23.

Bubien, J. K., and Warnock, D. G. (1993). Amiloride-sensitive sodium conductance in human B lymphoid cells. *Am. J. Physiol.* **265,** C1175–C1183.

Bubien, J. K., Jope, R. S., and Warnock, D. G. (1994). G-proteins modulate amiloride-sensitive sodium channels. *J. Biol. Chem.* **269,** 17780–17783.

Canessa, C. M., Horisberger, J.-D., and Rossier, B. C. (1993). Epithelial sodium channel related to proteins to proteins involved in neurodegeneration. *Nature (London)* **361,** 467–470.

Canessa, C. M., Schild, L., Buell, G., Thorens, B., Gautschi, I., Horisberger, J.-D., and Rossier, B. C. (1994). Amiloride-sensitive epithelial Na^+ channel is made up of three homologous subunits. *Nature (London)* **367,** 463–467.

Cantiello, H. F. (1995). Role of the actin cytoskeleton on epithelial Na^+ channel regulation. *Kidney Int.* **48,** 9700–9984.

Cantiello, H. F., Stow, J. L., Prat, A. G., and Ausiello, D. A. (1991). Actin filaments regulate epithelial Na^+ channel activity. *Am. J. Physiol.* **261,** C882–C888.

Caplan, M. J. (1997). Ion pumps in epithelial cells: Sorting, stabilization, and polarity. *Am. J. Physiol.* **272,** G1304–G1313.

Chalfant, M. L., Peterson-Yantorno, K., O'Brien, T. G., and Civan, M. M. (1996). Regulation of epithelial Na^+ channels from M-1 cortical collecting duct cells. *Am. J. Physiol.* **271,** F861–F870.

Coupaye-Gerard, B., Kim, H. J., Singh, A., and Blazer-Yost, B. L. (1994). Differential effects of brefeldin A on hormonally regulated Na^+ transport in a model renal epithelial cell line. *Biochim. Biophys. Acta* **1190,** 449–456

Ding, G., Franki, N., Condeelis, L., and Hays, R. M. (1991). Vasopressin depolymerizes F-actin in toad urinary bladder epithelial cells. *Am. J. Physiol.* **260,** C9–C16.

Duc, C., Farman, N., Canessa, C. M., Bonvalet, J.-P., and Rossier, B. C. (1994). Cell-specific expression of epithelial sodium channel α,β and γ subunits in aldosterone-responsive epithelia from rat: Localization by *in situ* hybridization and immunocytochemistry. *J. Cell Biol.* **127,** 1907–1921.

Eaton, D. C., Becchetti, A., Ma, H., and Ling, B. N. (1995). Renal sodium channels: regulation and single channel properties. *Kidney Int.* **48,** 941–949.

Ecke, D., Bleich, M., and Greger, R. (1996). The amiloride inhibitable Na^+ conductance of rat colonic crypt cells is suppressed by forskolin. *Pfluegers Arch.* **431,** 984–986.

Els, W. J., and Helman, S. I. (1997) Dual role of prostaglandins (PGE_2) in regulation of channel density an open probability of epithelial Na^+ channels in frog skin *(R. pipiens)*. *J. Membrane Biol.* **155,** 75–87.

Fanestil, D. D., Porter, G. A., and Edelman, I. S. (1967). Aldosterone stimulation of sodium transport. *Biochim. Biophys. Acta* **135,** 74–88.

Firsov, D., Schild, L., Gautschi, I., Mérillat, A-M., Schneeberger, E., and Rossier, B. C. (1996). Cell surface expression of the epithelial Na channel and a mutant causing Liddle syndrome: A quantitative approach. *Proc. Natl. Acad. Sci. U.S.A.* **93,** 15370–15375.

Franki, N., Ding, G., Gao, Y., and Hays, R. M. (1992). Effect of cytochalasin D on the actin cytoskeleton of the toad urinary bladder epithelial cell. *Am. J. Physiol.* **263,** C995–C1000.

Frindt, G., Silver, R. B., Windhager, E. E., and Palmer, L. G. (1995). Feedback regulation of Na channels in rat CCT III. Response to cAMP. *Am. J. Physiol.* **268,** F480–F489.

Frings, S., Purves, R. D., and MacKnight, A. D. C. (1988). Single-channel recordings from the apical membrane of the toad urinary bladder epithelial cell. *J. Membr. Biol.* **106,** 157–172.

Fujiwara, T., Oda, K., Yokata, S., Takatsuki, A., and Ikehara, Y. (1988). Brefeldin A causes disassembly of the Golgi complex and accumulation of secretory proteins in the endoplasmic reticulum. *J. Biol. Chem.* **263,** 18545–18522.

Fuller, C. M., Benos, D. J., and Bubien, J. K. (1997). Identification of α, β, and γ hENaC in human lymphocytes. *Physiologist* **40,** A5.

Garty, H., and Edelman, I. S. (1983). Amiloride-sensitive trypsinization of apical sodium channels. Analysis of hormonal regulation of sodium transport in toad bladder. *J. Gen. Physiol.* **81,** 785–803.

Garty, H., and Palmer, L. G. (1997). Epithelial sodium channels: Function, structure, and regulation. *Physiol. Rev.* **77,** 359–396.

Handler, J. S., Preston, A. S., and Orloff, J. (1969). Effect of adrenal steroid hormones on the response of the toad's urinary bladder to vasopressin. *J. Clin. Invest.* **48,** 823–833.

Ismailov, I. I., McDuffie, J. H., and Benos, D. J. (1994). Protein kinase A phosphorylation and G protein regulation of purified renal Na^+ channels in planar bilayer membranes. *J. Biol. Chem.* **269,** 10235–10241.

Ismailov, I. I., Awayda, M. S., Jovov, B., Berdiev, B. K., Fuller, C. M., Dedman, J. R., Kaetzel, M. A., and Benos, D. J. (1996a). Regulation of epithelial sodium channels by the cystic fibrosis transmembrane conductance regulator. *J. Biol. Chem.* **271,** 4725–4732.

Ismailov, I. I., Berdiev, B. K., Bradford, A. L., Awayda, M. S., Fuller, C. M., and Benos, D. J. (1996b). Associated proteins and renal epithelial Na^+ channel function. *J. Membr. Biol.* **149,** 123–132.

Ismailov, I. I., Berdiev, B. K., Shlyonsky, V. G., Fuller, C. M., Prat, A. G., Jovov, B., Cantiello, H. F., Ausiello, D. A., and Benos, D. J. (1997). Role of actin in regulation of epithelial sodium channels by CFTR. *Am. J. Physiol.* **272,** C1077–1086.

Ito, Y., Niisato, N., O'Brodovich, H., and Marunaka, Y. (1997). The effect of brefeldin A on terbutaline-induced sodium absorption in fetal rat distal lung epithelium. *Pfluegers Arch.* **434,** 492–494.

Kleyman, T. R., Kraehenbuhl, J.-P., and Ernst, S. A. (1991). Characterization and cellular localization of the epithelial Na^+ channel. Studies using an anti-Na^+ channel antibody raised by an antiidiotypic route. *J. Biol. Chem.* **266,** 3907–3915.

Kleyman, T. R., Ernst, S. A., and Coupaye-Gerard, B. (1994). Arginine vasopressin and forskolin regulate apical cell surface expression of epithelial Na^+ channels in A6 cells. *Am. J. Physiol.* **266,** F506–F511.

Knepper, M. A., and Inoue, T. (1997). Regulation of aquaporin-2 water channel trafficking by vasopressin. *Curr. Opin. Cell Biol.* **9,** 560–564.

Kokko, K. E., Matsumoto, P. S., Ling, B. N., and Eaton, D. C. (1994). Effects of prostaglandin E_2 on amiloride-blockabale Na^+ channels in a distal nephron cell line (A6). *Am. J. Physiol.* **267,** C1414–C1425.

Krattenmacher, R. and Clauss, W. (1988). Electrophysiological analysis of sodium-transport in the colon of the frog (*Rana esculenta*). *Pfluegers Arch.* **411,** 606–612.

Krattenmacher, R., Fischer, H., van Driessche, W., and Clauss, W. (1988). Noise analysis of cAMP-stimulated Na current in frog colon. *Pfluegers Arch.* **412,** 568–573.

Kunzelmann, K., Kiser, G. L., Schreiber, R., and Riordan, J. R. (1997). Inhibition of epithelial Na^+ currents by intracellular domains of the cystic fibrosis transmembrane conductance regulator. *FEBS Lett.* **400,** 341–344.

Kupitz, Y., and Atlas, D. (1993). A putative ATP-activated Na^+ channel involved in sperm-induced fertilization. *Science* **261,** 484–486.

Lester, D. S., Asher, C., and Garty, H. (1988). Characterization of cAMP-induced activation of epithelial sodium channels. *Am. J. Physiol.* **254,** C802–C808.

Letz, B., and Korbmacher, C. (1997). cAMP stimulates CFTR-like Cl^- channels and inhibits amiloride-sensitive Na^+ channels in mouse CCD cells. *Am. J. Physiol.* **272,** C657–C666.

Liebold, K. M., Reifarth, F. W., Clauss, W., and Weber, W.-M. (1996). cAMP-activation of amiloride-sensitive Na^+ channels from guinea-pig colon expressed in *Xenopus* oocytes. *Pfluegers Arch.* **431,** 913–922.

Ling, B. N., Zuckerman, J. B., Lin, C., Harte, B. J., McNulty, K. A., Smith, P. R., Gomez, L. M., Worrell, R. T., Eaton, D. C., and Kleyman, T. R. (1997). Expression of the cystic fibrosis phenotype in a renal amphibian epithelial cell line. *J. Biol. Chem.* **272,** 594–600.

Lingueglia, E., Voilley, N., Waldmann, R., Lazdunski, M., and Barbry, P. (1993). Expression cloning of an epithelial amiloride-sensitive Na^+ channel. *FEBS Lett.* **318,** 95–99.

Lingueglia, E., Renard, S., Waldmann, R., Voilley, N., Champigney, G., Plass, H., Lazdunski, M., and Barbry, P. (1994). Different homologous subunits of the amiloride sensitive Na^+ channel are differently regulated by aldosterone. *J. Biol. Chem.* **269,** 13736–13739.

Lipponcott-Schwartz, J., Donaldson, J. G., Schweizer, A., Berger, E. G., Hauri, H. P., Yaun, L. C., and Klausner, R. D. (1990). Microtubule-dependent retrograde transport of proteins into ER in the presence of brefeldin A suggests an ER recycling pathway. *Cell* (*Cambridge, Mass.*) **60,** 821–836.

Lipponcott-Schwartz, J., Yuan, L. C., Tipper, C., Amherdt, M., Orci, L., and Klausner, R. D. (1991). Brefeldin A's effects on endosomes, lysosomes, and TGN suggest a general mechanism for regulating organelle structure and membrane function. *Cell* (*Cambridge. Mass.*) **67,** 601–616.

Low, S. H., Tang, B. L., Wong, S. H., and Hong, W. (1992). Selective inhibition of protein targeting to the apical domain of MDCK cells by brefeldin A. *J. Cell Biol.* **118,** 51–52.

Mall, M., Hipper, A., Greger, R. and Kunzelmann, K. (1996). Wild type but not ΔF508 inhibits Na^+ conductance when coexpressed in *Xenopus* oocytes. *FEBS Lett.* **381,** 47–52.

Marunaka, Y., and Eaton, D. C. (1991). Effects of vasopressin and cAMP on single amiloride-blockable Na^+ channels. *Am. J. Physiol.* **260,** C1071–C1084.

Matalon, S., Bauer, M. L., Benos, D. J., Kleyman, T. R., Lin, C., Cragoe, E. J., and O'Brodovich, H. (1993). Fetal lung epithelial cells contain two populations of amiloride-sensitive Na^+ channels. *Am. J. Physiol.* **264,** L357–L364.

Matalon, S., Benos, D. J., and Jackson, R. M. (1996). Biophysical and molecular properties of amiloride-inhibitable Na^+ channels in alveolar epithelial cells. *Am. J. Physiol.* **271,** L1–L22.

Matsumoto, P. S., Mo, L., and Wills, N. K. (1997). Osmotic regulation of Na^+ transport across A6 epithelium: Interactions with prostalgandin E_2 and cyclic AMP. *J. Membr. Biol.* **160,** 27–38.

McDonald, F. J., and Welsch, M. J. (1995). The proline-rich region of the epithelial Na^+ channel binds SH3 domains and associates with specific cellular proteins. *Biochem. J.* **312,** 491–497.

Miller, S. G., Carnell, L., and Moore, H.-P. H. (1992). Post Golgi membrane traffic: Brefeldin A inhibits export of distal Golgi compartments to the cell surface but not recycling. *J. Cell Biol.* **118,** 267–283.

Morris, R. G., Tousson, A., Benos, D. J., and Schafer, J. A. (1998). Microtubule disruption inhibits AVT-stimulated Cl^- secretion but not Na^+ reabsorption in A6 cells. *Am. J. Physiol.* **274,** F300–F314.

Niisato, N., and Marunaka, Y. (1997). Hyposmolality-induced enhancement of ADH action on amiloride-sensitive Isc in renal epithelial A6 cells. *Jpn. J. Physiol.* **47,** 131–137.

O'Brodovich, H. (1991). Epithelial ion transport in the fetal and perinatal lung. *Am. J. Physiol.* **261,** C555–C564.

Oh, Y., Matalon, S., Kleyman, T. R. and Benos, D. J. (1992). Biochemical evidence for the presence of an binding protein in adult alveolar type II pneumocytes. *J. Biol. Chem.* **267,** 18498–18504.

Oh, Y, Smith, P. R., Bradford, A. L., Keeton, D., and Benos, D. J. (1993). Regulation by phosphorylation of purified epithelial Na^+ channels in planar lipid bilayers. *Am. J. Physiol.* **265,** C85–C91.

Palmer, L. G. and Lorenzen, M (1983). Antidiuretic hormone-dependent membrane capacitance and water permeability in the toad urinary bladder. *Am. J. Physiol.* **244,** F195–F204.

Park, C. S., and Fanestil, D. D. (1980). Covalent modification and inhibition of an epithelial sodium channel by tyrosine-reactive reagents. *Am. J. Physiol.* **239,** F299–F306.

Pelham, H. R. B. (1991). Multiple targets for brefeldin A. *Cell (Cambridge, Mass.)* **67,** 449–451.

Peng, X. R., Vao, X., Chow, D. C., Forte, G., and Bennett, M. K. (1997). Association of syntaxin 3 and vesicle associated membrane protein (VAMP) with H^+/K^+-ATPase containing tubulovesicles in gastric parietal cells. *Mol. Biol. Cell* **8,** 399–407.

Peters, K. W., Qi, J.-J., Liu, C., Watkins, S. C., Venable, D. F., and Frizzell, R. A. (1998). Syntaxin 1A inhibits the functional expression of the amiloride-sensitive epithelial sodium channel (ENaC). *FASEB J.* **12,** A981.

Prat, A. G., Ausiello, D. A., and Cantiello, H. F. (1993a). Vasopressin and protein kinase A activate G protein-sensitive epithelial Na^+ channels. *Am. J. Physiol.* **265,** C218–C223.

Prat, A. G., Bortorello, A. M., Ausiello, D. A., and Cantiello, H. F. (1993b). Activation of epithelial Na^+ channels by protein kinase A requires actin filaments. *Am. J. Physiol.* **265,** C224–C233.

Rea, S., and James, D. E. (1997). Moving GLUT4. The biogenesis and trafficking of GLUT4 storage vesicles. *Diabetes* **46,** 1667–1677.

Renard, S., Voilley, N., Bassilana, F., Lazdunski, M., and Barbry, P. (1995). Localization and regulation by steroids of the α, β and γ subunits of the amiloride-sensitive Na^+ channel in colon, lung and kidney. *Pfluegers Arch.* **430,** 299–307.

Rückes, C., Blank, U., Möller, K., Rieboldt, J., Lindemann, H., Münker, G., Clauss, W., and Weber, W.-M. (1997). Amiloride-sensitive Na^+ channels in human nasal epithelium are different from classical epithelial Na^+ channels. *Biochem. Biophys. Res. Commun.* **237,** 488–491.

Sariban-Sohraby, S., Sorscher, E. J., Brenner, B. M., and Benos, D. J. (1988). Phosphorylation of a single subunit of the epithelial Na^+ channel protein following vasopressin treatment of A6 cells. *J. Biol. Chem.* **263,** 13875–13879.

Senyk, O., Ismailov, I., Bradford, A. L, Baker, R. R., Matalon, S., and Benos, D. J. (1995). Reconstitution of immunopurified alveolar type II cell Na^+ channel protein into planar lipid bilayers. *Am. J. Physiol.* **268,** C1148–C1156.

Shimkets, R. A., Lifton, R. P., and Canessa, C. M. (1997). The activity of the epithelial sodium channel is regulated by clathrin-mediated endocytosis. *J. Biol. Chem.* **272,** 25537–25541.

Shimkets, R. A., Lifton, R., and Canessa, C. M. (1998). *In vivo* phosphorylation of the epithelial sodium channel. *Proc. Natl. Acad. Sci. U.S.A.* **95,** 3301–3305.

Simon, H., Gao, Y., Franki, N., and Hays, R. M. (1993). Vasopressin depolymerizes apical F-actin in rat inner medullary collecting duct. *Am. J. Physiol.* **265,** C757–C762.

Smith, P. R., Stoner, L. C., Viggiano, S. C., Angelides, K. J., and Benos, D. J. (1995). Effects of vasopressin and aldosterone on the lateral mobility of epithelial Na^+ channels in A6 renal epithelial cells. *J. Membr. Biol.* **147,** 195–205.

Snyder, P. M., Price, M. P., McDonald, F. J., Adams, C. M., Volk, K. A., Zeiher, B. G., Stokes, J. B., and Welsh, M. J. (1995). Mechanism by which Liddle's syndrome mutations increase activity of a human epithelial Na^+ channel. *Cell (Cambrige, Mass.)* **83,** 969–978.

Sonnenburg, W. K., and Smith, W. L. (1988). Regulation of cyclic AMP metabolism in rabbit cortical collecting tubule cells by prostaglandins. *J. Biol. Chem.* **263,** 6155–6160.

Staub, O., Dho, S., Henry, P. C., Correa, J., Ishikawa, T., McGlade, J., and Rotin, D. (1996). WW domains of Nedd4 bind to the proline rich PY motifs in the epithelial Na^+ channel deleted in Liddle's syndrome. *EMBO J.* **15,** 2371–2380.

Staub, O., Gautschi, I., Ishikawa, T., Breitschopf, K., Ceichanover, A., Schild, L., and Rotin, D. (1997). Regulation of stability and function of the epithelial Na^+ channel (ENaC) by ubiquitination. *EMBO J.* **16,** 6325–6336.

Stetson, D. L., Lewis, S. A., Alles, W., and Wade, J. B. (1982). Evaluation by capacitance measurements of antidiuretic hormone induced area changes in toad bladder. *Biochim. Biophys. Acta* **689,** 267–274.

Strous, G. J., Kerkhof, P. V., Meer, G. V., Rijnboutt, S., and Stoorvogel, W. (1993). Differential effects of brefeldin A in transport of secretory and lysosomal proteins. *J. Biol. Chem.* **268,** 2341–2347.

Stutts, M. J., Canessa, C. M., Olsen, J. C., Hamrick, M., Cohn, J. A., Rossier, B. C., and Boucher, R. C. (1995). CFTR as a cAMP-dependent regulator of sodium channels. *Science* **269,** 847–850.

Stutts, M. J., Rossier, B. C., and Boucher, R. C. (1997). Cystic fibrosis transmembrane conductance regulator inverts protein kinase A-mediated regulation of epithelial sodium channel single channel kinetics. *J. Biol. Chem.* **272,** 14037–14040.

Taylor, A., Mamelak, M., Golbetz, H., and Maffly, R. (1978). Evidence for the involvement of microtubules in the action of vasopressin in toad urinary bladder. *J. Membr. Biol.* **40,** 213–235.

Vallet, V., Chraibi, A., Gaeggler, H. P., Horisberger, J.-D., and Rossier, B. C. (1997). An epithelial serine protease activates the amiloride-sensitive sodium channel. *Nature (London)* **389,** 607–610.

Van Renterghem, C., and Lazdunski, M. (1991). A new non voltage-dependent epithelial like Na^+ channel in vascular smooth muscle cells. *Pfluegers Arch.* **419,** 401–408.

Verrey, F., Digicaylioglu, M., and Bolliger, U. (1993). Polarized membrane movements in A6 kidney cells are regulated by aldosterone and vasopressin/vasotocin. *J. Membrane Biol.* **133,** 213–226.

Verrey, F., Groscurth, P., and Bolliger, U. (1995). Cytoskeletal disruption in A6 kidney cells: Impact on endo/exocytosis and NaCl transport regulation by antidiuretic hormone. *J. Membr. Biol.* **145,** 193–204.

Vigne, B., Champigny, G., Marsault, R., Barbry, P., Frelin, C., and Lazdunski, M. (1989). A new type of amiloride-sensitive cationic channel in endothelial cells of brain microvessels. *J. Biol. Chem.* **264,** 7663–7668.

Weber, W.-M., Dannenmaier, B., and Clauss, W. (1993). Ion transport across leech integument. I. Electrogenic Na^+ transport and current fluctuation analysis of the apical Na^+ channel. *J. Comp. Physiol. B* **163,** 153–159.

Weng, K., and Wade, J. B. (1994). Effect of brefeldin A on ADH-induced transport responses of toad bladder. *Am. J. Physiol.* **266,** C1069–C1076.

Yue, G., Shoemaker, R. L, and Matalon, S. (1994). Regulation of low-amiloride-affinity sodium channels in alveolar type II cells. *Am. J. Physiol.* **267,** L94–L100.

Zeiske, W., Onken, H., Schwarz, H. J., and Graszynski, K. (1992). Invertebrate epithelial Na^+ channels: Amiloride-induced current-noise in crab gill. *Biochim. Biophys. Acta* **1105,** 245–252.

CHAPTER 10

Human Lymphocyte Ionic Conductance

James K. Bubien
Department of Physiology and Biophysics, University of Alabama at Birmingham, Birmingham, Alabama 35294

I. INTRODUCTION

Human lymphocytes express a variety of ion channels. Expression of at least two types of potassium channels has been documented electrophysiologically (Brent *et al.,* 1990; Chandy *et al.,* 1984; DeCoursey *et al.,* 1984). These channels appear to play a role in activation and cell cycle progression, because K^+ channel blockers inhibit lymphocyte cell cycle progression at G_1-S (Brent *et al.,* 1990; Chandy *et al.,* 1984; DeCoursey *et al.,* 1984). Lymphocytes express CFTR, which is also regulated in a cell-cycle-dependent fashion (Bubien *et al.,* 1990; Chen *et al.,* 1989). The cell surface antigen CD20 has been shown to be a unique type of calcium channel that is not sensitive to dihydropyridine inhibition. CD20 channels appear to have the structure of a homotetramer, that allows calcium conduction when either four monomers or two homodimers are brought together in the plasma membrane (Bubien *et al.,* 1993). Human lymphocytes also express a highly regulated, amiloride-sensitive sodium conductance (ASSC) (Bubien and Warnock, 1993).

It is the purpose of this review to summarize the electrophysiological studies on human lymphocytes that have investigated each of these types

1063-5823/99 $30.00

of ionic conductors, because our current understanding of the role of these channels, their regulation, and their relationship to each other is relatively rudimentary. Some of these channels have been linked to human disease (cystic fibrosis, CFTR, Liddle's syndrome, ASSC). The examination of the properties, drug sensitivity, and regulation of these channels in human tissue makes lymphocytes an ideal model for an electrophysiological system. Thus, while little is known about the role of these channels in the functioning of lymphocytes, the expression of these different channel types in these cells allows for the direct electrophysiological examination of abnormally functioning channels from individuals with channel-related diseases.

The sections are arranged approximately chronologically, starting with potassium channels, moving to CFTR chloride channels, followed by a short section on CD20 calcium channels. A summary of the amiloride-sensitive sodium conductance completes this short review, with special attention given to abnormalities in the regulation of this conductance in disease states such as Liddle's syndrome.

II. LYMPHOCYTE POTASSIUM CHANNELS AND CELL CYCLE REGULATION

Shortly after it was proven that high-resistance electrical seals were possible between cells and low-resistance microelectrodes (i.e., patch clamp), the technique was applied to human lymphocytes. Prior to these initial studies, the ionic conductance properties of human lymphocyte membranes were virtually unknown. Early flow cytometry (Amigorena *et al.,* 1984) and patch-clamp studies showed that ion channel activity was altered when B and T lymphocytes were stimulated with mitogens (DeCoursey *et al.,* 1984). It was also shown that these potassium channels were required for T-lymphocyte activation (Chandy *et al.,* 1984), and that these potassium channels played a role in T-lymphocyte volume regulation (Cahalan and Lewis, 1989). Simultaneously, the same roles for potassium channels were established for B lymphocytes (Choquet and Korn, 1988). Further studies showed that potassium channel inhibition specifically arrested B lymphocytes at G_1-S of the cell cycle (Amigorena *et al.,* 1984; Brent *et al.,* 1990). Figure 1 shows whole-cell potassium currents recorded from a normal resting (G_0) human peripheral blood lymphocyte.

Figure 2 shows that treatment with potassium channel blockers (4-aminopyridine, tetraethylammonium) does not inhibit expression of the early proliferative marker 4F2, indicating that mitogen-stimulated lymphocytes enter G_1 of the cell cycle. However, K^+ channel inhibition completely prevented the expression of the late G_1 activation antigen FC6. The effects

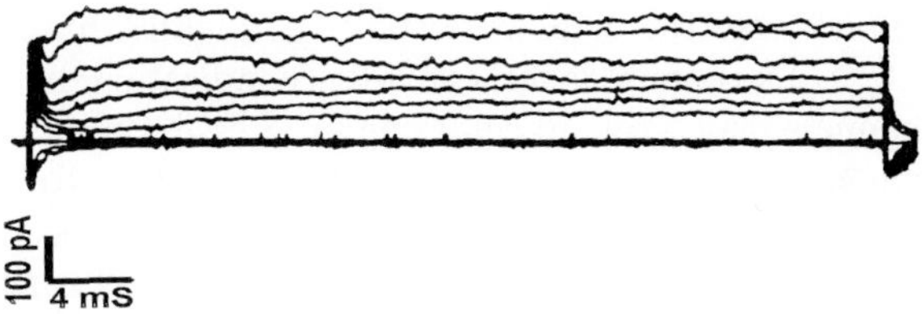

FIGURE 1 Whole-cell potassium currents from a resting (G_0) human peripheral blood lymphocyte. The cell was held at a voltage of −60 mV and test clamped to voltages from −100 mV to +100 mV in 20 mV increments. (Reprinted with permission from Brent *et al.* (1990), *J. Immunol.* **145,** 2381–2389. Copyright 1990. The American Association of Immunologists.)

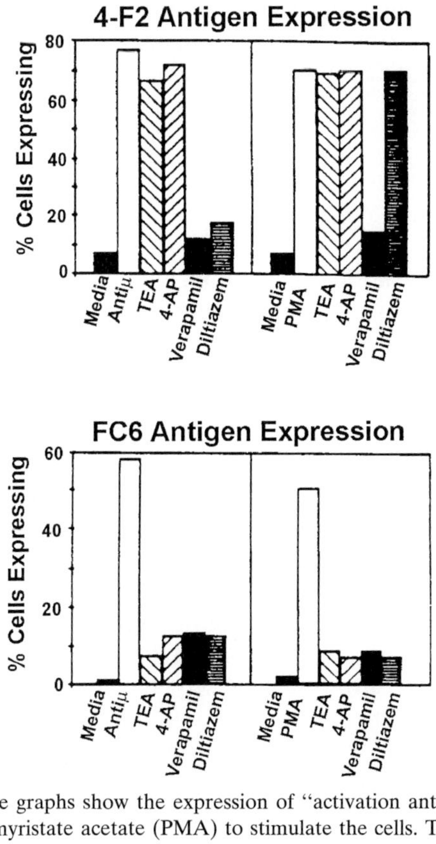

FIGURE 2 These graphs show the expression of "activation antigens" using anti-IgM (antiμ) and phorbol myristate acetate (PMA) to stimulate the cells. The potassium channel blockers do not inhibit expression of the early antigen (4F2), but completely inhibit the expression of the antigen expressed late in G_1 (FC6). (Reprinted with permission from Brent *et al.* (1990), *J. Immunol.* **145,** 2381–2389. Copyright 1990. The American Association of Immunologists.)

were the same whether the cells were stimulated with Anti Ig or phorbol myristate acetate. Thus, there is direct evidence indicating that lymphocyte potassium channels play a role in cell cycle progression because K^+ inhibition arrests the cells in mid-G_1, and prevents progression to S phase.

The implication from these early studies is that changes in membrane potential or abnormal depolarizations (induced by K^+ channel blockade) inhibit functions or enzymes necessary for cell cycle progression, but that the initial activation process is refractory to the K^+ channel blockade. Another possibility is that by depolarizing lymphocytes with K^+ channel blockers, the driving force for Na^+-dependent transport is reduced. Thus, cell cycle progression inhibition by the K^+ channel blockade may result from an indirect metabolic deficit created by reduced nutrient influx by Na^+-dependent cotransport. Even though the precise mechanism of the effect of the K^+ channel blockade has not been firmly established, it is likely that the expression of these channels by lymphocytes plays a role in cellular housekeeping, by maintaining and regulating a hyperpolarized membrane potential.

III. CELL-CYCLE-DEPENDENT EXPRESSION OF CHLORIDE CHANNELS BY HUMAN LYMPHOCYTES

The expression of chloride channels by lymphocytes was initially demonstrated by single-channel studies on "normal" and cystic fibrosis lymphocytes (Chen *et al.,* 1989). Subsequently, fluorescent analysis and whole-cell patch clamp demonstrated directly that two chloride conductances were present in normal human B lymphocytes (Bubien *et al.,* 1990; Krauss *et al.,* 1992b). One conductance was activated by cAMP, and another chloride conductive pathway was activated by increases in intracellular calcium. These currents were carried by CFTR or regulated by CFTR because they were not present in lymphocytes that were homozygous for the Δ_{508} mutation that produces clinical cystic fibrosis. Chloride currents through these conductive pathways are shown in Fig. 3.

The cell cycle dependence of lymphocyte chloride conductance was further investigated by synchronizing transformed B lymphocytes at G_1 and S phases of the cell cycle with hydroxyurea. Subsequently G_1 cells and cells in S phase were whole-cell-clamped under conditions exclusive for observation of the chloride conductance. Figure 4 shows examples of representative electrophysiological experiments and the corresponding flow cytometry findings of DNA stained cells in G_1 and S phases. Cells synchronized to S phase were released from hydroxyurea block and cultured an additional 24 h. The findings showed that G_1 phase cells expressed a cAMP-activated chloride conductance that was not present in S phase cells (Krauss *et al.,* 1992a).

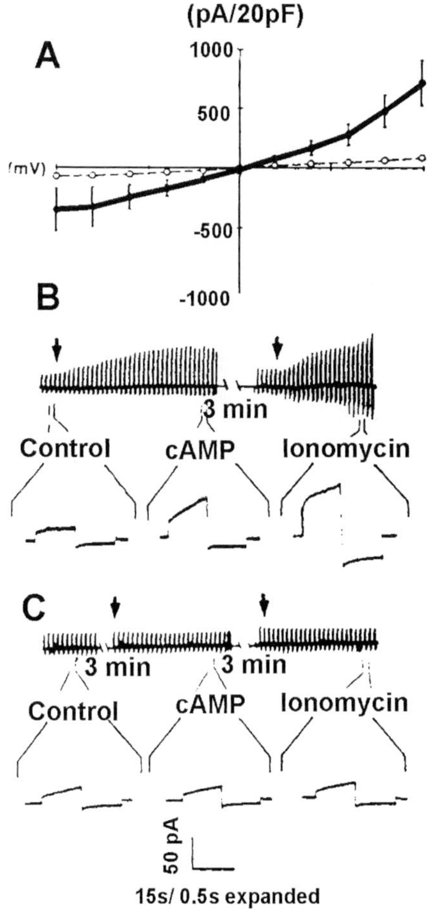

FIGURE 3 (A) The average cAMP-activated chloride current-voltage relations for normal ($n = 5$) and CF ($n = 6$) B lymphocytes. (B) Real time pulse protocols showing the time course of cAMP-mediated chloride current activation and Ca^{2+}-mediated chloride current activation in a whole-cell clamped normal human B lymphocyte. It is possible that these distinct currents are conducted via different channels. (C) The same protocols as shown in B fail to activate any chloride current in cystic fiborsis lymphocytes. (Reprinted with permission from Bubien *et al.* (1990). Cell cycle dependence of chloride permeability in normal and cystic fibrosis lymphocytes. *Science* **248,** 1416–1419. Copyright 1990 American Association for the Advancement of Science.)

These findings demonstrate directly that lymphocytes in G_1 have a regulated chloride conductance that is completely absent in cells in S phase. The fact that the conductance will not respond to cAMP stimulation sug-

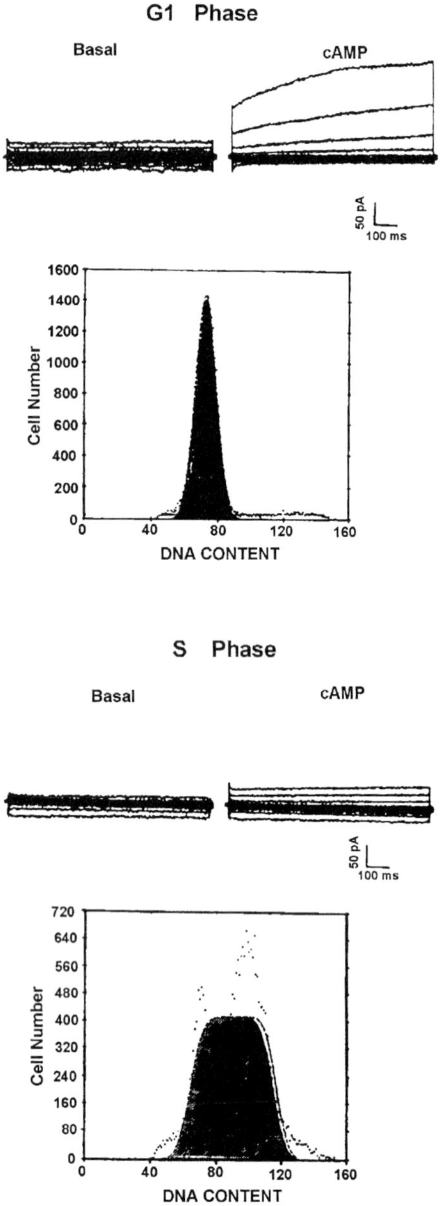

FIGURE 4 The upper panels show whole-cell chloride currents in G_1 and S phase human lymphocytes. The bottom panels show the DNA content (left, $2n$; right, $>2n<4n$) of samples from cell cycle phased lymphocyte cultures. The whole-cell clamped cells were taken from these populations. (Reprinted from Krauss *et al.* (1992), *EMBO J.* **11,** 875–883, by permission of Oxford University Press.)

gests that the channels are not present in the plasma membrane. However, it is possible that a component of the regulatory mechanism downregulates during DNA synthesis. Certainly, the cytosolic biochemistry is radically altered during this phase. Since the chloride permeability plays only a minor role in the regulation of membrane potential, it is unlikely that alterations in chloride conductance are required for membrane potential changes during cell cycling. It is more likely that membrane trafficking is reduced during S phase, and that the CFTR membrane trafficking function is downregulated during this cell cycle phase.

IV. CD20: A B-LYMPHOCYTE-SPECIFIC UNIQUE CALCIUM CONDUCTOR

CD20 was among the first lineage-specific (B-cell lineage) antigens to be identified (Stashenko *et al.,* 1981). It has now been demonstrated directly that CD20 can function as a calcium channel that is not inhibited by dihydropyridines. The conductance is not voltage-sensitive; however, it does appear to inactivate quickly after a hyperpolarizing voltage clamp pulse (Fig. 5).

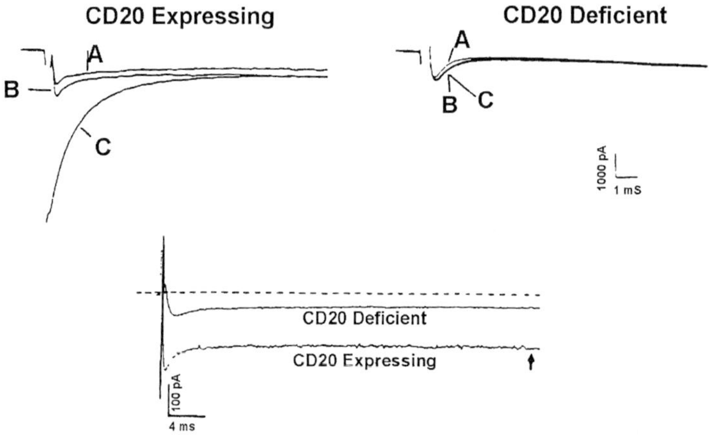

FIGURE 5 Whole-cell currents from Jurkat T lymphocytes that do not express CD20 endogenously. The upper left panel shows three currents elicited from the same CD20 transfected Jurkat cell. Current b was elicited with a bat of RPMI and demonstrates the endogenous current magnitude in response to a hyperpolarizine voltage clamp to −100 mV from a holding potential of 0 mV. Current a was elicited in response to the same voltage clamp protocol in an RPMI bath supplemented with EGTA (2 mM), thereby eliminating the bath Ca^{2+}. Current c was in a bat supplemented with 2 mM Ba^{2+}. The upper right are corresponding records from a Jurkat T lymphocyte transfected with the vector alone. The lower records are the Ba^{2+} currents form the upper records placed on the same scale to show the increased steady state current in the CD20 transfected cells. (Reproduced from Bubien *et al.* (1993), *The Journal of Cell Biology* **121,** 1121–1132, by copyright permission of The Rockefeller University Press.)

These CD20-associated calcium currents can result in enough calcium entry to substantially increase the cytosolic calcium (Fig. 6). However, it is not known whether the entire increase in cytosolic calcium is from influx across the plasma membrane, or whether Ca^{2+}- activated Ca^{2+} release from cytosolic stores is involved in the process.

Like the potassium and chloride conductances, the CD20 calcium conductance appears to be associated with cell cycle progression and B-lymphocyte differentiation because monoclonal antibodies to CD20 inhibit cell cycle progression from S phase to G_2/M phase following mitogen stimulation (Forsgren *et al.,* 1987. CD20 becomes phosphorylated upon mitogen stimulation of B lymphocytes, and this modification apparently regulates CD20 function. Transfection of CD20 and the subsequent expression of a new Ca^{2+} conductance leave little doubt that CD20 can function as a Ca^{2+} channel (Bubien *et al.,* 1993). Also, activation of a calcium whole-cell conductance by CD20-specific monoclonal antibodies in cells that express this protein (Fig. 7) leaves little doubt that CD20 is involved in plasma membrane calcium conductance.

Because it is possible to immunoprecipitate and crosslink CD20, some unique aspects of this protein's configuration have been elucidated, leading to interesting speculation of how the channel function of CD20 may be regulated. Membrane protein crosslinking experiments with DSP followed by immunoprecipitation with CD20-specific monoclonal antibodies revealed three

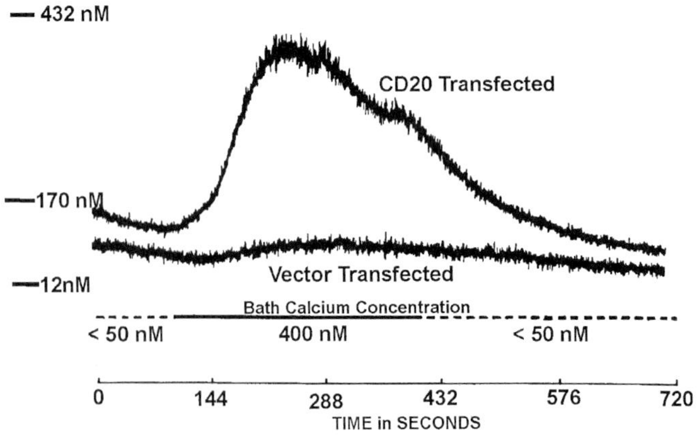

FIGURE 6 Fura-2 fluorescence measurements of the changes in the cytosolic calcium concentration in response to an increase in extracellular calcium. These are the same transfected Jurkat cells as used for the electrophysiological studies shown in Fig. 5. The increase in calcium in the $CD20^+$ cells is obvious, and must at least in part result from a transmembrane influx via CD20 calcium channels. (Reproduced from Bubien *et al.* (1993), *The Journal of Cell Biology* **121,** 1121–1132, by copyright permission of The Rockefeller University Press.)

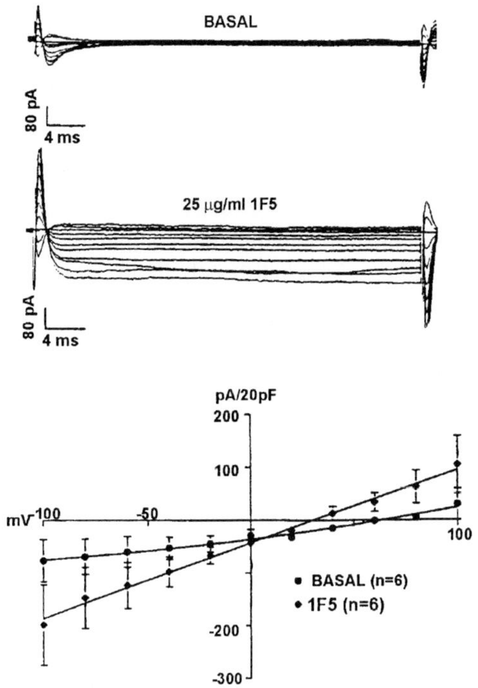

FIGURE 7 The whole-cell voltage clamps from a Daudi B lymphocyte (Daudi cell endogenously express CD20) show that the CD20-specific monoclonal antibody specifically activates an inward calcium current. The only charge carrier available for these experiments was calcium. All other pipette and bath solution components were substituted with NMDG-CL. (Reproduced from Bubien *et al.* (1993), *The Journal of Cell Biology* **121,** 1121–1132, by copyright permission of The Rockefeller University Press.)

configurations of the protein, monomers, dimers, and tetramers, when separated under nonreducing conditions. When the immunoprecipitated products were subsequently reduced, all of the product was observed at the 35-kDa position for CD20. These configiurations suggest a multimeric structure for the mature channel and also suggest that multimer formation and deformation may be the means of channel regulation (Bubien *et al.*, 1993).

V. LYMPHOCYTE AMILORIDE-SENSITIVE SODIUM CONDUCTANCE

When human lymphocytes are whole-cell voltage-clamped using normal ionic gradients the predominant conductance is outward, indicating that the cells are relatively more permeable to potassium that sodium (Fig. 8, upper left panel). However, when the cells are superfused with the membrane-permeant analog of cyclic AMP (8-CPT-cAMP, 40 μM) there is a rapid increase in the inward current and a shift to the right of the reversal (i.e., membrane) potential, indicating a specific cAMP-stimulated increase in the sodium conductance (Fig. 8, upper right). These activated inward Na^+ currents are completely inhibited by 2 μM amiloride (Fig. 8, lower left). Dose–efficacy experiments revealed an IC_{50} for amiloride of 75 nM. The IC_{50} for benzamil was approximately 7 nM and for phenamil it was greater that 1 μM (Bubien and Warnock, 1993). The inhibition of this conductance by amiloride is completely reversible (Fig. 8, lower right). The mean current–voltage relations for these sequential treatment effects on the ensemble whole-cell current are shown in Fig. 9. The reversal potential shifted right 30 mV with cAMP treatment and 30 mV left when the conductance was blocked with amiloride. The amiloride and basal mean current–voltage relations are indistinguishable, indicating the complete inhibition of this conductance by amiloride.

Cholera toxin activates adenylate cyclase specifically, thereby raising cellular cAMP. Pertussis toxin ADP-ribosylated certain GTP-binding proteins. When human lymphocytes are treated with these toxins, they respond with a specific activation of the amiloride-sensitive inward conductance. Figure 10 shows the effects of these toxins on whole-cell clamped human lymphocytes. The conclusion from these findings is that two distinct signal

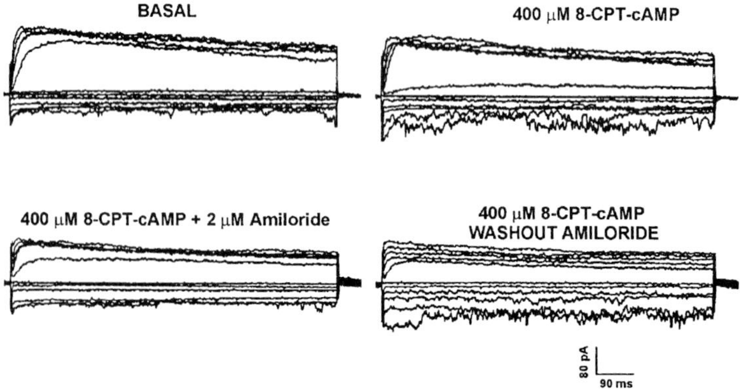

FIGURE 8 Ensemble whole-cell currents from a human lymphocyte using "normal ionic gradients for Na^+, K^+, Cl^-, and Ca^{2+}. Typically the cells have prominent outward currents and very little inward current. Inward current can be activated with 8-CPT-cAMP, and the activated current can be completely inhibited by 2 μM amiloride. (Reprinted with permission from Bubien and Warnock, 1993.)

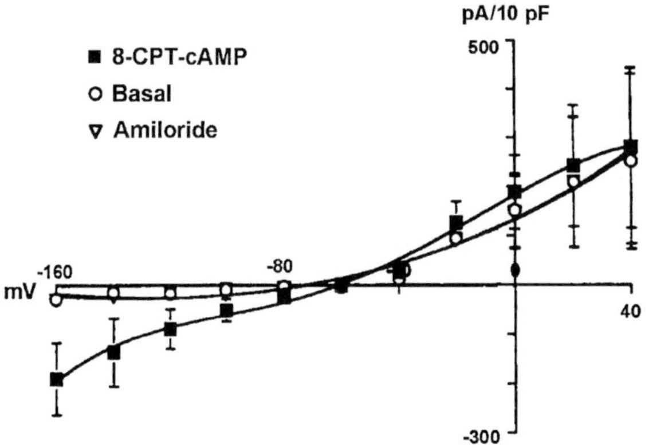

FIGURE 9 Current–voltage relation showing the specific activation of the inward current by 8-CPT-cAMP and the complete inhibition of the activated current by 2 μM amiloride. (Reprinted with permission from Bubien and Warnock, 1993.)

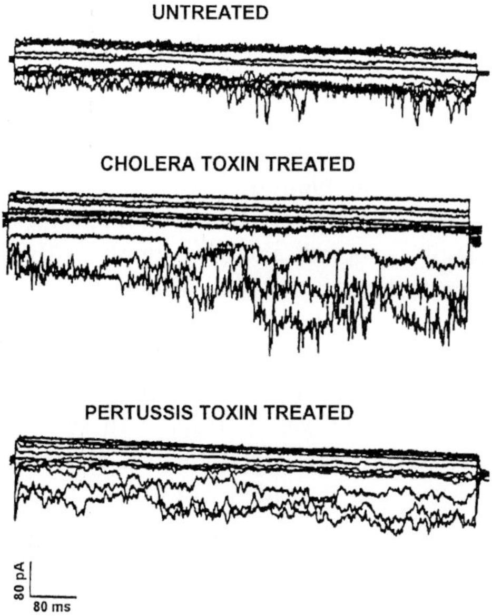

FIGURE 10 Lymphocyte whole-cell inward Na^+ currents activated by cholera toxin and pertussis toxin, respectively. (Reprinted with permission from Bubien *et al.,* 1994.)

transduction pathways are capable of activating amiloride-sensitive sodium channels in human lymphocytes.

One of the interesting features of the activated current is the "ragged" appearance of the currents. While the outward currents are generally smooth, the inward currents are typically uneven, tending to increase and decrease rapidly during hyperpolarizing voltage clamps. This atypical characteristic was present in whole-cell clamped lymphocytes as well as renal cortical collecting duct principal cells (Bubien *et al.*, 1994) and appeared to be a specific "signature" for amiloride-sensitive sodium channels. Since the hyperpolarizing pulses were relatively large (as negative as −160 mV), it was possible that the membrane was being altered by the voltage clamp. However, amiloride completely inhibited all of the "ragged" inward current. This finding is inconsistent with a breakdown of the membrane or its components due to the voltage clamp. To further examine this phenomenon, single channels were recorded in the cell-attached configuration. Typically cell-attached patches with Na^+-filled pipettes did not have single-channel activity that carried inward current. However, such activity was observed consistently after the cells were superfused with 40 μM 8-CPT-cAMP (the same treatment used to activate the whole-cell current). Since whole-cell voltage clamps lasted 800 ms, a stable single-channel recording lasting 30 s was obtained. This record was broken into 30 segments and all of the segments were digitally added together to produce a single "virtual" 800-ms-long whole-cell record. After correcting for the number of channels in the patch and the projected number of channels in the whole membrane (approximately 800/cell, by H^3 bromobenzamil binding), and assuming a P_{open} of 0.25, a virtual record was constructed and compared to an actual whole-cell record at the same membrane potential (-70 mV). Both records had the same "ragged" appearance. These findings show that the whole-cell current accurately reflected, and was simply the sum of, the single-channel activity. The findings are shown in Fig. 11.

The question remained whether cholera toxin and pertussis toxin activated the same or different sets of channels. To answer this question human lymphocytes were incubated with cholera toxin and pertussis toxin simultaneously. The surprising result was that under these conditions there was no activation of the inward whole-cell current. After the possibility that the toxins were ineffective was ruled out, experiments were performed whereby the cells were incubated in one toxin, whole-cell clamped, and superfused with the other toxin. In both cases (i.e., incubated with cholera toxin and superfused with pertussis toxin, or vice versa) activated current was observed initially and was subsequently inhibited upon superfusion with the other toxin. An example of one such experiment is shown in Fig. 12.

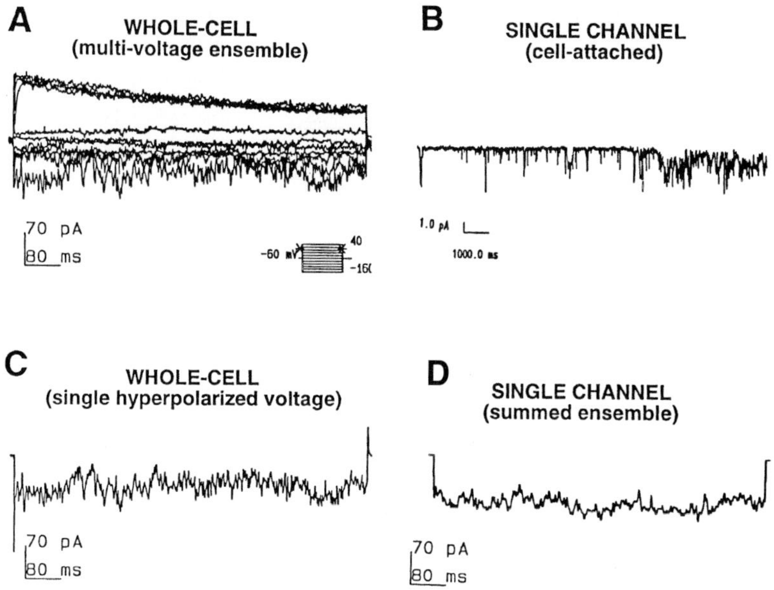

FIGURE 11 These records show the connection between single-channel properties and the whole-cell inward Na^+ currents in human lymphocytes. The virtual whole-cell current matches the actual current precisely. In particular, the current produced from the summed single-channel records shows the same "ragged" appearance as the actual whole-cell current, lending support to the hypothesis that the single-channel properties account for the whole-cell current properties.

The conclusion from these experiments was that both second-messenger pathways had the ability to regulate the function of the same set of lymphocyte amiloride-sensitive sodium channels. These experiments led to a model for the dual regulation of these channels shown in Fig. 13.

Table I shows inward sodium conductance values that summarize the multitreatment experiments used to verify the main features of the dual regulation model.

Usually, cytosolic signal transduction pathways link plasma membrane receptors to a cytosolic process, a nuclear process, or another plasma membrane process, such as an ion channel. Since there are two independent signal transduction pathways that regulate lymphocyte amiloride-sensitive sodium channels, the logical hypothesis is that there are two independent receptors that are capable of regulating these channels. To test this hypothesis whole-cell clamped lymphocytes were superfused with norepinephrine and isoproterenol (Bubien *et al.*, 1998). It was found that norepinephrine

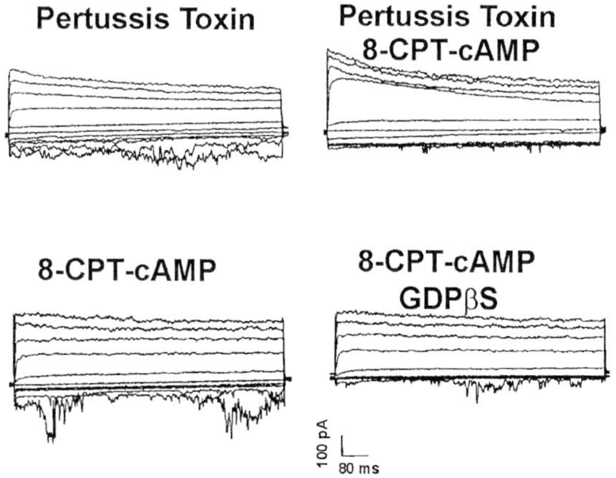

FIGURE 12 Upper panels: Cells were pretreated with 200 ng/ml pertussis holotoxin, and consequently had activated inward currents at hyperpolarized membrane potentials. Subsequent treatment with cAMP completely and specifically inhibited these currents. Lower panels: Cells with inward currents activated by 8-CTP-cAMP were subsequently treated with GDPβS (in the pipette solution), which completely and specifically inactivated the inward currents. (Reprinted with permission from Bubien *et al.*, 1994.)

but not isoproterenol specifically activated lymphocyte amiloride-sensitive sodium channels. Examples of these experiments are shown in Fig. 14.

These findings indicate that α-adrenergic but not β-adrenergic receptors play a role in the regulation of the channels. There are two distinct classes of α-adrenergic receptors, α_1 and α_2. Yohimbine is a specific α_2 receptor inhibitor, and terazosin is a specific α_1-adrenergic receptor ligand. Thus experiments were performed to discern which of these receptors regulated amiloride-sensitive sodium channels. Also, cellular cyclic AMP was measured in response to stimulation with norepinephrine. It was found that norepinephrine increased cellular cAMP 10-fold within 5 min of exposure. This provided indirect evidence that norepinephrine-mediated activation of amiloride-sensitive sodium channels was accomplished via cAMP-mediated signal transduction. This hypothesis was confirmed by inhibiting the agonist effect of norepinephrine with the specific cAMP inhibitor Rp-CPT-cAMP. It was found that norepinephrine-mediated ASSC activation could be reversed with Rp-CPT-cAMP and that cells pretreated with Rp-CPT-cAMP were refractory to stimulation with norepinephrine. The agonist effect was also inhibited by the α_2 antagonist yohimbine and the α_1 antagonist terazosin. Thus, by simple blockade of the activation it was not possible to determine whether α_1 or α_2 receptors were responsible for lymphocyte ASSC activation. Interestingly, however, terazosin was also able to inhibit currents activated by 8-CPT-cAMP, which acts downstream from the recep-

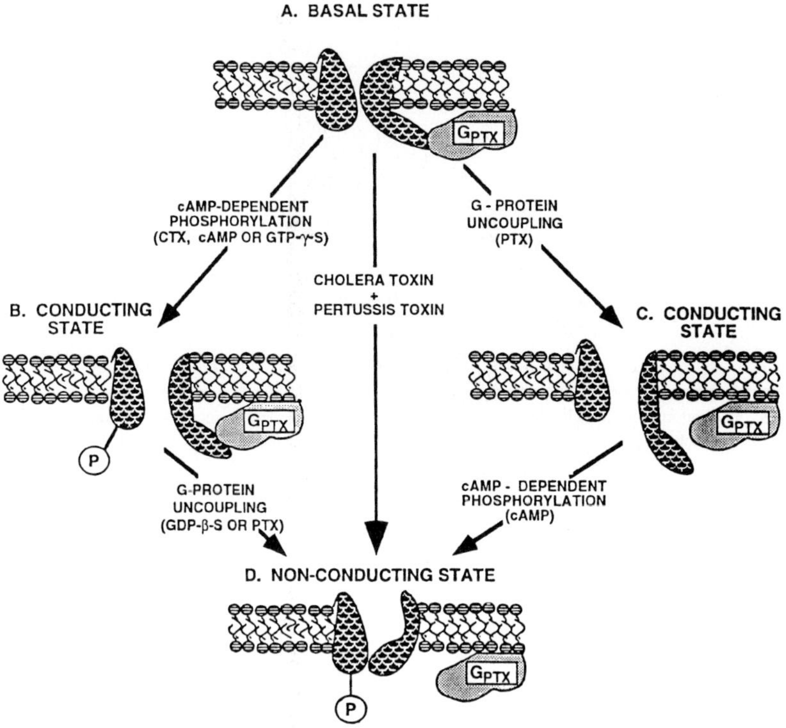

FIGURE 13 Model for the regulation of lymphocyte amiloride-sensitive sodium channels. Since the channels could be activated and inactivated by sequential application of treatments, the channels are closed basally and application of either cAMP or pertussis toxin activates the channels. Application of both causes the channels to close. (Reprinted with permission from Bubien *et al.*, 1994.)

tor in the signal transduction pathway. Figure 15 shows terazosin-mediated inhibition of cAMP-activated lymphocyte ASSCs.

Doxazosin is another closely related member of this class of α_1 ligand compounds, and it has the same inhibitory effect as terazosin. This finding is important because there is no mechanism whereby these drugs could antagonize a direct stimulation by cyclic AMP. One possible explanation for the effect would be a direct inhibition of the channels by doxazosin and terazosin. However, it was found that these drugs had no effect on α, β, γ ENaC channels incorporated into planar lipid bilayers and expressed in *Xenopus* oocytes, thereby ruling out this hypothesis. The other possibility was that terazosin and doxazosin could be acting as agonists on the α_1 receptor. If this hypothesis is correct, a regulatory mechanism consistent with that shown by the model in Fig. 13 could be responsible for the

TABLE I

8-CPT-CAMP	−	+	−	−	−	−	−	−	−	−	+	+
Rp-CPT-cAMP	−	−	−	−	+	+	+	−	+	−	−	−
Cholera toxin	−	−	−	−	+	+	−	−	−	+	−	−
Pertussis toxin	−	−	−	−	−	−	−	+	+	+	+	−
GTPγS	−	−	+	−	−	−	−	−	−	−	−	−
GDT βS	−	−	−	+	−	−	−	−	−	−	−	+
Mean	588	1833	1409	298	1507	581	672	1428	1558	566	658	411
SEM	81	232	296	100	244	243	56	163	324	79	76	84
n	17	5	8	4	14	3	4	16	4	6	6	6

Note. The values reported in this table are chord conductances between −140 mV and the reversal potential for each cell. The + indicates the agent listed in the left column was present, a − indicates it was not. (Reprinted with permission from Bubien *et al.*, 1994.)

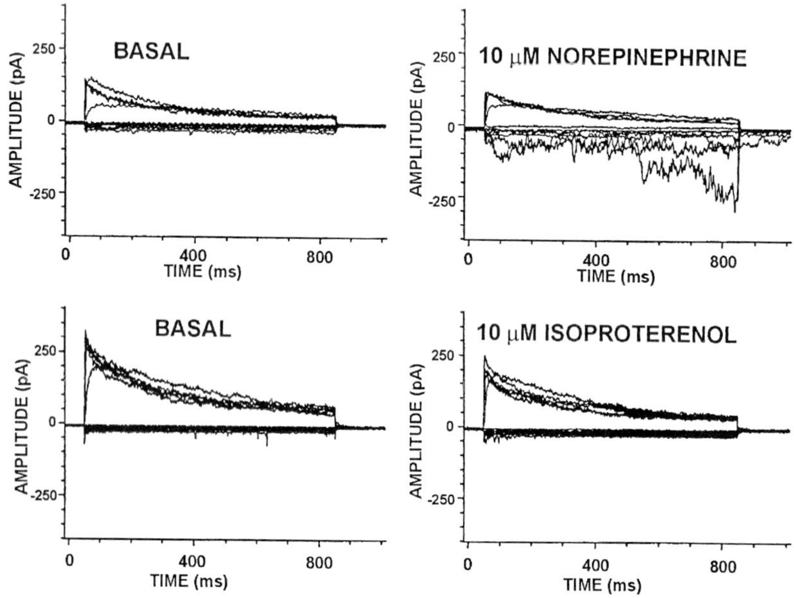

FIGURE 14 Whole-cell clamp current records showing the specific activation of the inward ASSC by norepinephrine and the failure of isoproterenol to alter human lymphocyte whole-cell currents. (Reprinted with permission from Bubien *et al.*, 1998.)

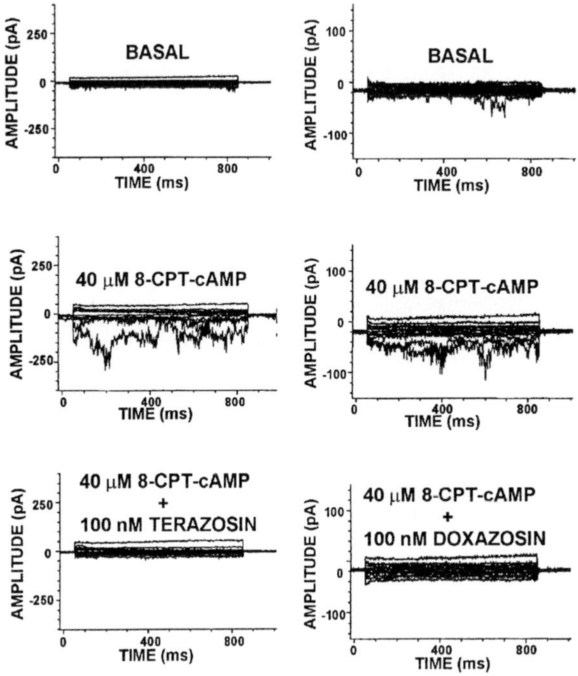

FIGURE 15 Whole-cell clamped lymphocytes show dual regulation of amiloride-sensitive sodium channels by cAMP and α_1-specific ligands. (Reprinted with permission from Bubien *et al.*, 1998.)

observation. A prediction from that model is that terazosin alone would activate the channels. This prediction was tested by superfusing lymphocytes with 100 nM terazosin. It was found that without preactivation, terazosin alone acted as an agonist and specifically activated lymphocyte amiloride-sensitive sodium channels. An example of these experiments is shown in Fig. 16.

As shown in Fig. 16, the α_2-specific ligand yohimbine has no direct effects on lymphocyte ion channels. Also, this ligand does not interfere with cAMP-mediated ASSC activation. However, yohimbine does prevent activation mediated by norepinephrine. Thus, yohimbine appears to act solely as a passive antagonist to α_2-adrenergic receptor activation in lymphocytes with respect to regulation of amiloride-sensitive sodium channels. The findings with the α_1-adrenergic receptor ligand terazosin were completely different. Direct exposure to this compound activated the ASSC current. Subsequent treatment with cAMP inactivated it. These results are consistent with the model for lymphocyte ASSC regulation depicted in Fig. 13 and consistent

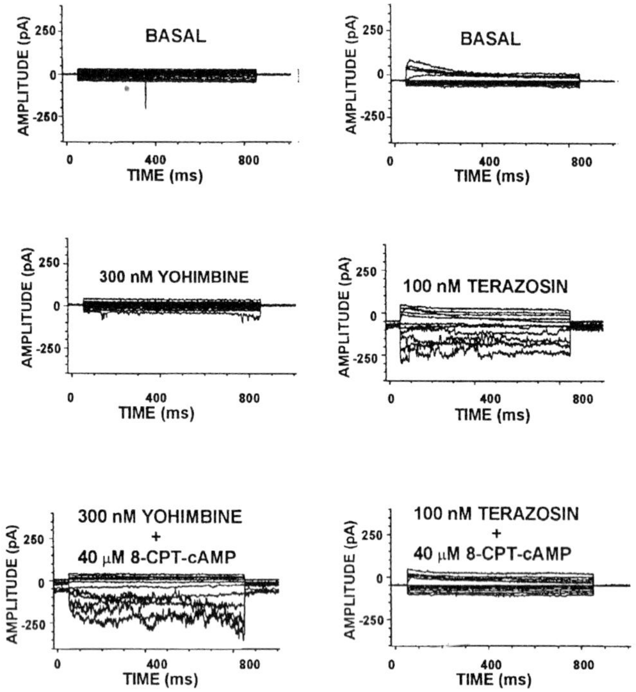

FIGURE 16 The α_2-adrenergic ligand yohimbine has no direct agonist effect on lymphocyte whole-cell currents and does not antagonize amiloride-sensitive sodium channel activation by cAMP. In contrast, terazosin directly activates lymphocyte ASSCs and this activation is specifically inhibited by cAMP. (Reprinted with permission from Bubien *et al.,* 1998.)

with the hypothesis that two independent sets of receptors and signal transduction pathways play roles in the regulation of lymphocyte amiloride-sensitive sodium channels. These findings show the complexity and sophistication of lymphocyte amiuloride-sensitive sodium channel regulation. Liddle's syndrome provides an example of how specific alterations of the β subunit of the amiloride-sensitive sodium channel are manifested in functional and regulatory alterations of the current carried by these channels.

Liddel's syndrome is caused by a premature stop mutation of the β subunit of the cloned ENaC polypeptide (Skimkets *et al.,* 1994). This disease is characterized by severe hypertension, with potassium wasting. The pathophysiology of the disease is restricted to the kidney, because renal transplantation completely eliminates the disease. It is now known that the excessive constitutive salt and water reabsorption occurs in the collecting duct and is mediated by principal cells. These cells express amiloride-sensitive sodium channels in thier apical membrane. These channels constitute the rate-limiting step in the the vectorial transport of salt and water by the collecting duct. The channels are highly regulated by the peptide hormone vasopressin and the steroid hormone aldosterone. In the absence of a need to conserve salt and water the channels remain closed, thereby limiting salt and water reabsorption by the collecting duct. The pathophysiology of Liddle's syndrome suggests that there is inappropriate basal reabsorption of salt and water and an inappropriate removal of potassium as a by-product of the reabsorptive process. Since the mutation is found in a polypeptide expressed by renal collecting duct principal cells, it is reasonable to hypothesize that these cells mediated the pathophysiology of the disease. Human renal principal cells are not available for electrophysiological study, and it is not possible to obtain these cells from individuals with Liddle's syndrome. However, blood samples containing lymphocytes are readily available. Thus, the properties of lymphocyte amiloride-sensitive sodium channels were examined from individuals with Liddle's syndrome and unaffected genetic relatives. It was found that cells from unaffected individuals had quiescent channels and very little basal inward sodium current. In contrast, cells from affected individuals had constitutively active channels and the resultant large inward sodium currents when examined by whole-cell voltage clamp. Figure 17 shows the difference.

The first abnormality that is apparent in the records from the affected individual is that the basal current is constitutively active in the absence

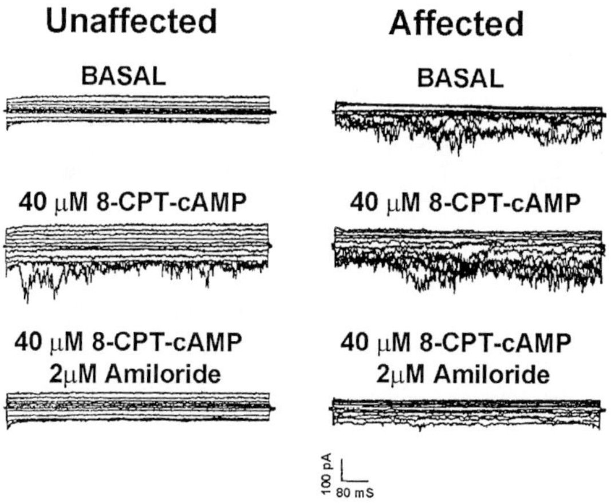

FIGURE 17 Lymphocyte whole-cell clamp records from an unaffected (left) individual and an individual with Liddle's syndrome (right). (Reprinted with permission from Bubien *et al.*, 1996.)

of any stimulus. The second abnormality is that superfusion with 8-CPT-cAMP has no effect on the currents. This means that the channels have been altered by the mutation such that they are unresponsive to regulation via this signal transduction mechanism. In further experiments, it was found that the constitutively active current in cells from affected individuals was inhibited by the specific inhibitor of cyclic AMP, Rp-CPT-cAMP. This second finding suggested that the cells may produce excessive cyclic AMP. However, direct measurement showed that the cAMP levels were the same in affected and unaffected cells. Thus, the mutation to the β subunit of the channels induced the channels to function as if they were being stimulated by cAMP, in the absence of the second messenger. This abnormal function/regulation was also present when affected cells were treated with the specific cAMP inhibitor Rp-CPT-cAMP. Figure 18 shows whole-cell voltage-clamp records from affected lymphocytes with constitutively activated currents in the basal state that are completely inhibited by the cAMP antagonist. The inward sodium currents could be reactivated in cells treated with Rp-CPT-cAMP by further treatment with pertussis toxin (Fig. 18). This finding is consistent with the model for ASSC regulation shown in Fig. 13 and also shows that this independent regulatory pathway remains intact in lymphocytes from individuals with Liddle's syndrome.

Amiloride-sensitive sodium channels were purified from lymphocytes of individuals with Liddle's syndrome and unaffected genetic relatives. The purified protein complexes were inserted into planar lipid bilayers and single channels were examined electrophysiologically. It was found that channels from unaffected individuals had a relatively low open probability, while channels purified from lymphocytes of affected individuals had a high

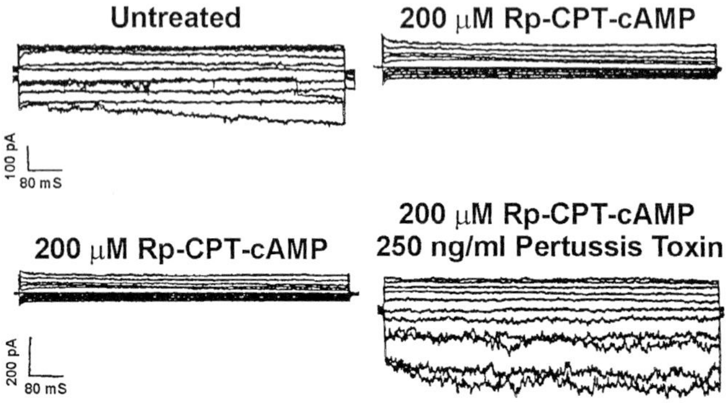

FIGURE 18 Whole-cell voltage clamp records from cells expressing the mutation for Liddle's syndrome, showing abnormal responses to compounds used to study second-messenger regulation. (Reprinted with permission from Bubien *et al.*, 1996.)

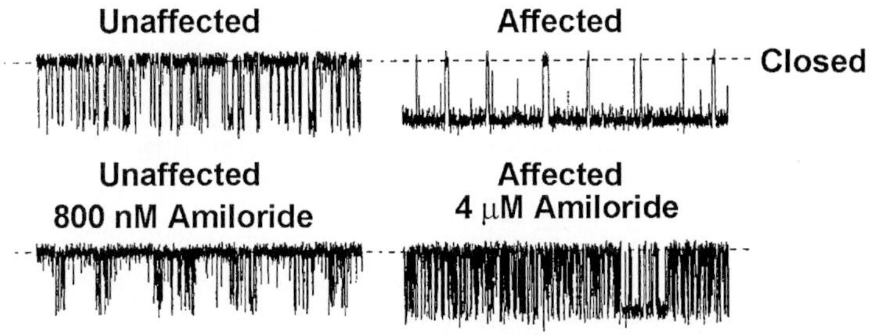

FIGURE 19 Single-channel records from amiloride-sensitive sodium channels purified from lymphocytes of individuals affected with Liddle's syndrome and unaffected genetic relatives. (Reprinted with permission from Ismailov *et al.*, 1996.)

open probability (Fig. 19). Amiloride inhibited the reconstituted normal channels with high affinity and the abnormal channels with a somewhat lower affinity (Ismailov *et al.*, 1996). These findings demonstrate directly that the mutation responsible for Liddle's syndrome is present in the amiloride-sensitive sodium channels expressed by lymphocytes and that the mutation alters basal channel function in a manner consistent with the pathophysiology of the disease.

The research findings presented in this short review show that human lymphocytes express a variety of ion channels. There is some evidence to indicate that these channels play a role in activation and cell cycle progression. However, to date the most important aspect of ion channel expression by lymphocytes is that they appear to faithfully express channels that have been implicated in human disease states. This has been demonstrated directly for CFTR and cystic fibrosis, and now ENaC and hypertension. These findings are important because they mean that cells, and expressed channels, can be obtained from individuals affected with these diseases for more detailed analysis with relative ease and little discomfort. These findings also hold the promise of specific diagnostic potential for individuals with these diseases and provide material to test hypotheses directed at alleviating the abnormal pathophysiology.

References

Amigorena, S., Choquet, D., Teillaud, J-L., Korn, H., and Fridman, W. H. (1984). Ion channel blockers inhibit B cell activation at a precise stage of the G1 phase of the cell cycle. *J. Immunol.* **144,** 2038–2045.

Brent, L. H., Butler, J. I., Woods, W. T., and Bubien, J. K. (1990). Transmembrane ion conductance in human B lymphocyte activation. *J. Immunol.* **145,** 2381–2389.

Bubien, J. K., and Warnock, D. G. (1993). Amiloride-sensitive sodium conductance in human B lymphoid cells. *Am. J. Physiol.* **265,** C1175–C1183.

Bubien, J. K., Kirk, K. L., Rado, T. A., and Frizzell, R. A. (1990). Cell cycle dependence of chloride permeability in normal and cystic fibrosis lymphocytes. *Science* **248,** 1416–1419.

Bubien, J. K., Zhou, L.-J., Bell, P. D., Frizzell, R. A., and Tedder, T. F. (1993). Transfection of the CD20 cell surface molecule into ectopic cell types generates a Ca^{++} conductance found constitutively in B lymphocytes. *J. Cell Biol.* **121,** 1121–1132.

Bubien, J. K., Jope, R. S., and Warnock, D. G. (1994). G-proteins modulate amiloride-sensitive sodium channels. *J. Biol. Chem.* **269,** 17780–17783.

Bubien, J. K., Ismailov, I. I., Berdiev, B. K., Cornwell, T., Lifton, R. P., Fuller, C. M., Achard, J.-M., Benos, D. J., and Warnock, D. G. (1996). Liddle's disease: Abnormal regulation of amiloride-sensitive sodium channels by β subunit mutation. *Am. J. Physiol.* **270,** C208-C213.

Bubien J. K., Cornwell, T., Bradford A. L., Fuller, C. M., DuVall, M. D., and Benos, D. J. (1998). Alpha adrenergic receptors regulate human lymphocyte amiloride-sensitive sodium channels. *Am. J. Physiol.* **275,** C702-C710

Cahalan, M. D., and Lewis, R. S. (1989). Role of potassium and chloride channels in volume regulation by T lymphocytes. *In* "Cell Physiology of Blood." Rockefeller Press, New York.

Chandy, K. G., DeCoursey, T. E., Cahalan, M. D., McLaughlin, C., and Gupta, S. (1984). Voltage-gated potassium channels are required for human T lymphocyte activation. *J. Exp. Med.* **160,** 369–374.

Chen, J. H., Shulman, H., and Gardner, P. (1989). A cAMP-regulated chloride channel in lymphocytes that is affected in cystic fibrosis. *Science* **243,** 657–660.

Choquet, D., and Korn, H. (1988). Modulation of voltage-dependent potassium channels in B lymphocytes. *Biochem. Pharmacol.* **37,** 3797–3803.

DeCoursey, T. E., Chandy, K. G., Gupta, S., and Cahalan, M. D. (1984). Voltage-gated K^+ channels in human T lymphocytes: A role for mitogenesis. *Nature London* **307,** 465–467.

Forsgren, A., Penta, A., Schlossman, S. F., and Tedder, T. F. (1987). Regulation of B cell function through the CD20 molecule. *In* "Leukocyte Typing III." Oxford University Press, Oxford.

Ismailov, I. I., Berdiev, B. K., Fuller, C. M., Bradford, A. L., Lifton, R. P., Warnock, D. G., Bubien, J. K., and Benos, D. J. (1996). Peptide block of constitutively activated Na^+ channels in Liddle's Disease. *Am. J. Physiol.* **270,** C214–C223.

Krauss, R. D., Bubien, J. K., Drumm, M. L., Zheng, T., Peiper, S. C., Collins, F. S., Kirk, K. L., Frizzell, R. A., and Rado, T. A. Transfection of wild type CFTR into cystic fibrosis lymphocytes restores chloride conductance at G1 of the cell cycle. *EMBO J.* **11,** 875–883. (1992a).

Krauss, R. D., Berta, G., Rado, T. A., and Bubien, J. K. (1992b). Antisense oligonucleotides to CFTR confer cystic fibrosis phenotype on B lymphocytes. *Am. J. Physiol.* **263,** C1147–C1151.

Shimkets, R. A., Warnock, D. G., Bositis, C. M., Nelson-Williams, C., Hansson, J. H., Schambelan, M., Gills, J. R. J., Ulick, S., Milora, R. V., Findling, J. W., Canessa, C. M., Rossier, B. C., and Lifton, R. P. (1994) Liddle's syndrome: Inheritable human hypertension caused by mutations in the β subunit of the epithelial sodium channel. *Cell (Cambridge, Mass.)* **79,** 407–414.

Stashenko, P. L., Nadler, L. M., Hardy, R., and Schlossman, S. F. (1981). Expression of cell surface markers after human B lymphocyte activation. *Proc. Natl. Acad. Sci. U.S.A.* **78,** 3848–3852.

CHAPTER 11

Regulatory Aspects of Apx, a Novel Na^+ Channel with Connections to the Cytoskeleton

Horacio F. Cantiello
The Renal Unit, Massachusetts General Hospital East, Charlestown, Massachusetts 02129, and the Department of Medicine, Harvard Medical School, Boston, Massachusetts

I. SODIUM CHANNELS OF A6 EPITHELIAL CELLS

The 9-pS Epithelial Na^+ Channel, a Novel Ion Channel Structure. Molecular information about the protein(s) responsible for the vasopressin-sensitive and protein kinase A (PKA)-regulated Na^+ transport in epithelial cells is still controversial. At least three major "types" of apical Na^+ channels can be found in epithelial cells (Benos *et al.*, 1995; Garty, 1994; Garty and Palmer, 1997). These Na^+ channel types include a small conductance (~3–5 pS) ENaC type (see below), a highly Na^+ selective channel with a $Na^+:K^+$ perm-selectivity ratio higher than 10 (Hamilton and Eaton, 1986; Marunaka and Eaton, 1991), a higher mean conductance (7–15 pS) Na^+ channel (that we refer to as the 9-pS channel) with a lower perm-selectivity ratio (3–5) (Cantiello *et al.*, 1989, 1990, 1991; Hamilton and Eaton, 1985), and a nonselective cation channel with an even higher (23–28 pS) single-channel conductance (Light *et al.*, 1988). The only common feature among these epithelial channel fingerprints is their ability to be blocked by the

1063-5823/99 $30.00

Cl-pirazine carboxylguanidine amiloride. No molecular information is yet available on the structural identity of different functional channel fingerprints, and the possibility that these Na^+ channel phenotypes may be, at least partly, a reflection of different functional conformations of a single channel complex, has not been completely ruled out. Nevertheless, a solid body of evidence has mounted to suggest that the small channel type may represent the heterotrimeric α,β,γ-ENaC (*e*pithelial *Na*$^+$ *c*hannel), originally cloned from the rat distal colon by two independent groups (Canessa *et al.,* 1993; Lingueglia *et al.,* 1993). Functional expression of ENaC homologs (Canessa *et al.,* 1994) is largely consistent with the presence of a small conductance, highly Na^+-selective ion channel(s) previously described (for a comprehensive review, see Garty and Palmer, 1997). On the other hand, the nonselective cation channel present in the rat IMCD, (Na(28)), previously included among the functionally characterized epithelial Na^+ channels, was originally observed to be a channel functionally regulated by atrial natriuretic peptide and the second messenger cGMP (Light *et al.,* 1989a,b). This channel seems to correspond structurally to an entirely different family, whose closest homolog is the regulated cation channel of the eye (Karlson *et al.,* 1995).

Perhaps the most elusive of the amiloride-sensitive apical epithelial Na^+ channels is the 9-pS channel, which is a target for vasopressin regulation in renal epithelia (Prat *et al.,* 1993a). The features of the Na(9) channel (to use a recently coined, albeit confusing nomenclature) (Garty and Palmer, 1997), include its presence in cells grown on plastic (Hamilton and Eaton, 1986) and glass supports (Cantiello *et al.,* 1989, 1991), where the Na^+ channels can be localized in single cells in all stages of growth, and confluent monolayers. Chronic aldosterone treatment is not required for Na(9) expression (Cantiello *et al.,* 1989; Hamilton and Eaton, 1986). This Na^+ channel has been primarily studied in the renal tubular A6 amphibian cells (Cantiello *et al.,* 1989; Hamilton and Eaton, 1986; Prat *et al.,* 1992), although functional mammalian homologs have been determined in LLC-PK_1 proximal tubular cells (Moran and Moran, 1984), and the rabbit straight proximal tubule (Gogelein and Greger, 1986).

The Na(9) channels of A6 cells are modulated by a regulatory pathway involving the G protein-mediated activation of phospholipase A_2 (Cantiello *et al.,* 1989, 1990). In this regulatory pathway, the pertussis toxin-mediated ADP-ribosylation of $G_{\alpha i\text{-}3}$, seems to be a strong inhibitory signal for the Na(9) channel, in contrast to the opposite effect observed on the Na(5) channel function. The most relevant regulatory pathway of Na(9) channels, however, seems to involve the mediation of ion channel function by dynamic changes in actin filament organization (Cantiello *et al.,* 1991; Prat *et al.,* 1993a,b). Interactions between the G protein and the actin-mediated signaling pathways may also exist, linking the various regulatory steps associated with Na^+ channel activation. Incubation of A6 cells with the actin filament disrupter cytochalasin D for more than 2 h resulted in the cells being unresponsive to either AVP and/or PKA activation (Prat *et al.,* 1993b). Under these conditions, neither GTPγS nor the direct addition of the purified, activated G protein subunit $G_{\alpha i\text{-}3}$ effected apical Na^+ channel activation, in contrast to previous results in control cells displaying an organized actin cytoskeleton (Cantiello *et al.,* 1989). Further, addition of arachidonic acid, the hydrolytic byproduct and parental substrate of the phospholipase A_2 regulatory pathway, was also without effect. However, addition of actin plus ATP, readily activated Na(9) channel activity in cytoskeletaly deranged A6 cells (Prat *et al.,* 1993b). Thus, actin filament organization is required to effect the proper regulatory pathway(s) involving G proteins and the phospholipase A_2 production of lipid metabolites, linking the apical regulatory pathways of Na^+ channel regulation, including the cAMP-mediated hormonal response to AVP. Actin, therefore, seems to be the more proximal regulatory effector system controlling Na^+ channel activity. Although the molecular structure underlying the 9-pS epithelial Na^+ channel was apparently unavailable until recently, studies from our laboratory have progressed, elucidating the molecular structure of Na(9), Apx (see below), one of the main targets for vasopressin-sensitive Na^+ reabsorption in the amphibian kidney (Prat *et al.,* 1996).

II. Apx, AN ACTIN-REGULATED SODIUM CHANNEL

Pioneering work by Benos and collaborators initiated the purification and characterization of an amiloride-sensitive Na^+ channel complex from A6 cells and the bovine renal medulla by specific binding to amiloride derivatives (Benos *et al.,* 1987). This 700-kDa molecular weight Na^+ channel is a multimeric structure containing polypeptides varying from 320 to 35 kDa, with an apparent pore subunit of 150 kDa, which binds amiloride and analogs (Benos *et al.,* 1987). The 150-kDa subunit of the A6 Na^+ channel complex displays Na^+ channel activity in lipid bilayers (Sariban-Sohraby and Fisher, 1992; Sariban-Sohraby *et al.,* 1984). Thus, Apx and the 150-kDa subunit may represent the same transmembrane protein. This is further suggested by the fact that antibodies raised against the 150-kDa subunit immunoprecipitate Apx (Staub *et al.,* 1992) and conversely, anti-idiotypic antibodies to Apx immunoprecipitate the 150-kDa subunit of the Na^+ channel complex (Staub *et al.,* 1992). Although the simplest encompassing possibility for the previous data was that Apx may be the pore-bearing component of the renal epithelial Na^+ channel, its expression in *Xenopus* oocytes failed to elicit amiloride-sensitive Na^+ channel activity (Staub *et al.*, 1992). However, the possibility that Apx could be tonically inhibited by regulatory proteins was not contemplated in that study. Under resting conditions little or no Na^+ channel activity is present in A6 cells in the absence of vasopressin stimulation. However, any disruption of the endogenous actin cytoskeleton, or direct addition of actin readily activated Na(9) channels. This phenomenon is rapidly reversed by addition of monomeric actin binding proteins such as DNase I or actin filament cross-linking proteins such as filamin/ABP-280 (Cantiello *et al.*, 1991). Further, the ABP-280 homolog filamin (Cantiello *et al.*, 1991) inhibits the PKA- and actin-induced Na^+ channel activity in A6 cells (Cantiello *et al.*, 1991; Prat *et al.*, 1993b). Thus, actin filament cross-linking may be an endogenous mechanism for keeping an organized actin cytoskeleton and thus an inactive channel. This has been confirmed by the inhibitory effect of filamin/ABP-280 on the anion channel CFTR (Prat *et al.*, 1995) and the cell volume-regulated cation channels in human melanoma cells (Cantiello *et al.*, 1993).

Thus, the issue of whether Apx is implicated directly in epithelial Na^+ channel activity was addressed by stable transfection of the Apx gene into ABP-280-deficient human melanoma cells (M2) expressing a deranged actin cytoskeleton (Prat *et al.*, 1996). The central hypothesis hinged upon the possibility that Apx was inhibited tonically in cells expressing an organized cytoskeleton.

Functional expression of Apx using the whole-cell patch-clamp technique was assessed in different clones (Apx-2 through 8) produced by stable transfection of M2 (ABP(−)) melanoma cells with the full-length cDNA for Apx. The presence of Apx was determined by immunolocalization of Apx with immunopurified anti-Apx antibodies raised against a fusion protein containing the amino acid sequence 1194 to 1395 from Apx (Prat *et al.*, 1996). Immunocytochemical analysis indicated that all clones expressed various amounts of intracellular Apx. No labeling of Apx was detected in the parent, nontransfected M2 human melanoma cells. Quantification of Apx labeling indicated that immunofluorescence labeling of the Apx-7 clone (the most thoroughly studied) was comparable to that of A6 cells, from which Apx was originally described (Staub *et al.*, 1992). It is important to note that not only was Apx immunolabeling undetectable in the M2 cells, but also that no Na^+-selective currents were observed under any experimental conditions, including voltage activation, osmotic shock, and PKA and PKC additions, further suggesting that no endogenous Na^+ conductance is present in M2 cells prior to Apx expression. This was further supported by failure to immunolocalize the Na(5) channel ENaC (Canessa *et al.*, 1994), in either M2 or Apx-7 cells.

Nontransfected M2 cells had a basal whole-cell conductance of 1.2 ± 0.4 nS/cell (n = 13) and were insensitive to amiloride (10 μM). Six Apx-expressing clones were studied whose basal whole-cell conductance was higher than that of M2 cells (58–892% higher). Clone Apx-2 (58% higher), however, did not reach statistical difference as compared to M2 cells. The whole-cell conductance of clone Apx-6 (3.4 ± 1.2 nS/cell) was further increased (129%, $P<0.01$) after cell depolarization. The voltage-activated Apx-6 whole-cell currents were inhibited by amiloride. The whole-cell conductance of Apx-7 cells ranged from 3 to 16 nS/cell, and in average was 625% higher than M2 cells (8.6 ± 0.7 nS/cell, $n = 29$, $P< 0.001$; Fig. 1). Apx-7 whole-cell currents were highly selective for Na^+ over Cl^- and had a $Na^+:K^+$ perm-selectivity ratio of 4.3. The dose–response decrease of the whole-cell currents by amiloride indicated an affinity of 3.3 μM (Fig. 2). However, the best fitting of the experimental results was optimal with two amiloride binding sites (H. F. Cantiello, unpublished), indicating that more than one affinity site for the drug may coexist in a single cell. The higher affinity binding site was lower than 10^{-7} M amiloride and accounted for >25% of the amiloride-sensitive currents (data not shown). The whole-cell currents of Apx-7 cells were also inhibited by thc amiloride analog benzamil (<1 μM) but not by *N*-(ethyl-*N*-isopropyl)-amiloride (EIPA, 100 μM; data not shown). Interestingly, although most clones had a higher cationic conductance compared to M2 cells, no correlation was found between the whole-cell and the amiloride-sensitive conductances, thus most clones had largely variable responses to amiloride (Fig. 2). Cyclic AMP-stimulation of Apx-7 cells induced a 420 ± 264% ($n = 4$) increase of the whole-cell currents in 4 out of 9 experiments ($P < 0.001$). No effect was observed on M2 cells (− 47 ± 11%, $n = 6$, NS).

To confirm the presence of Apx-mediated single-channel currents, two techniques were applied. First, the cell-attached and inside-out patch-clamp configurations determined that spontaneous Na^+ channel activity was present in 50% of patches tested on Apx-7, but not M2 cells ($n = 28$) under cell-attached conditions. The single-channel current-voltage relationship (Fig. 3) was highly similar to that of Na^+ channels previously reported in A6 epithelial cclls (but not Na(5)) (Hamilton and Eaton, 1985; Prat *et al.*, 1993a). After excision, Na^+ channel currents of the Apx-7 clone had a single channel conductance of 6.2 ± 0.6 pS ($n = 5$) in symmetrical Na^+ (140 mM NaCl or Na isethionate). A smaller subconductance state of approximately 3 pS was also observed. Either addition of PKA plus ATP (1 mM) or polymerizing concentrations of monomeric actin (1 mg/ml) to excised, inside-out patches of Apx-7 cells also induced and/or increased Na^+ channel activity in 78% of the experiments tested ($n = 9$, Fig. 4). Actin-induced Na^+ channels had a linear single channel conductance of 6.2 ± 0.3 pS ($n = 6$) and a perm-selectivity sequence of $Na^+ > Li^+ > K^+$ with a 4:1 perm-selectivity ratio between Na^+ and K^+.

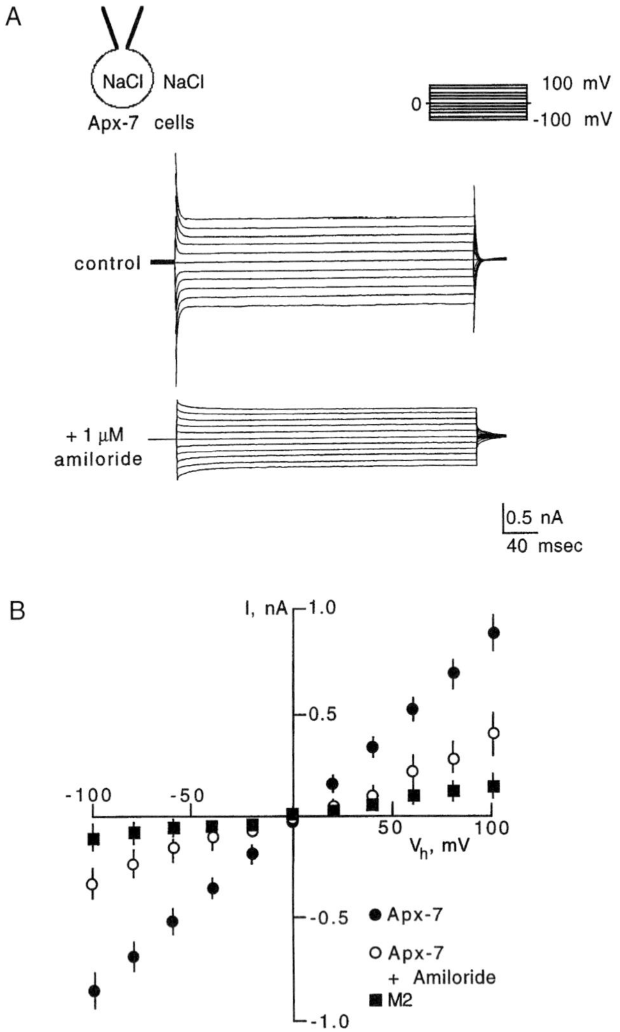

FIGURE 1 Effect of amiloride on basal Na^+ currents of Apx-expressing human melanoma cells. (A) Representative whole-cell currents of M2 human melanoma cells stably transfected with Apx (Apx-7) before (top) and after addition of amiloride (1 μM, bottom). (B) Current-voltage relationship of M2 (filled squares, n = 13), and Apx-7 cells, before (filled circles, n = 29) and after addition of 1 μM amiloride (open circles, n = 6). Data are the mean ± SEM. Whole-cell conductance of Apx-7 cells was not affected by replacement of Cl^- by isethionate. Reproduced from Prat *et al.* (1996), with permission.

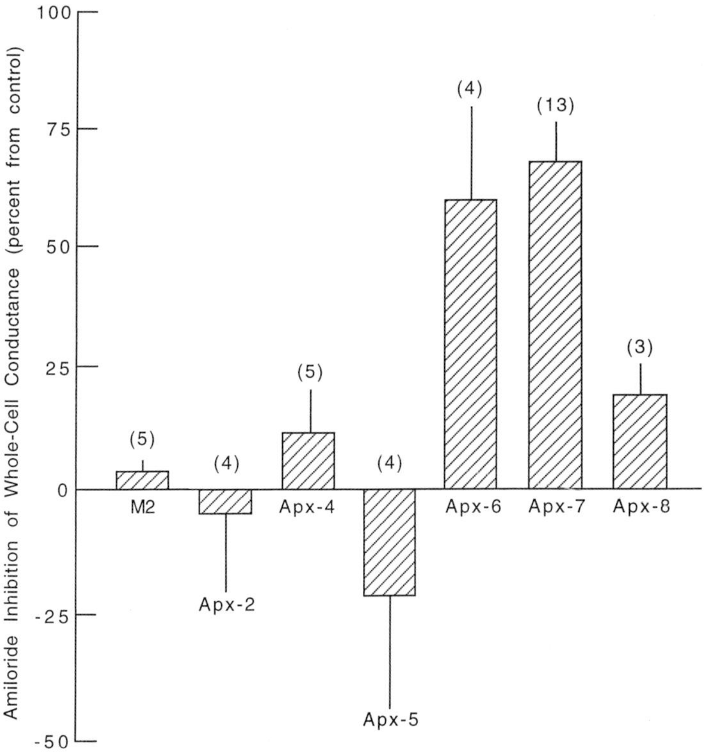

FIGURE 2 Comparison of amiloride effect on the various Apx-expressing clones. Percent inhibition by amiloride (10–100 μM) of whole-cell conductance on the various Apx-expressing clones. The numbers in parentheses indicate the number of whole-cell experiments analyzed. No correlation was observed between the Apx-induced expression of a cation-selective conductance and the sensitivity to amiloride. Reproduced from Prat *et al.* (1996), with permission.

The second technique that allowed the determination of the functional characterization of Apx channels was lipid bilayer reconstitution of membrane-enriched fractions from Apx-7 cells (Fig. 5). Spontaneous Na^+-selective channel activity in asymmetrical NaCl (600 : 50 mM NaCl, *cis:trans,* respectively) was only observed in 1/14 experiments but was readily activated in 9/14 experiments by addition of actin to the *cis* side of the chamber (data not shown). Actin-induced Na^+-selective channel activity was also observed in symmetrical 300 mM NaCl (Fig. 5). Reconstituted channels had a single channel conductance of 10.9 $\pm$ 1.0 pS (n = 3), and were completely inhibited by addition of amiloride added to the *trans* but not to the *cis* side of the chamber (Fig. 5). A smaller conductance substate was also observed displaying the same characteristics. However, membrane

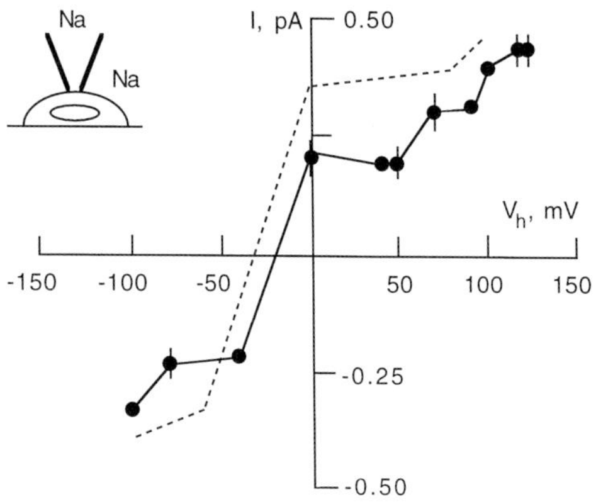

FIGURE 3 Na^+ channel activity of Apx-expressing melanoma cells. Current-voltage relationship of spontaneous Na^+ channel activity of Apx-7 cells under cell-attached conditions. Data are the mean ± SEM of 3 experiments. The dashed line represents the current-voltage relationship of "9-pS" Na^+ channel activity previously reported in epithelial A6 cells (Prat *et al.*, 1993a). Reproduced from Prat *et al.*, (1996), with permission.

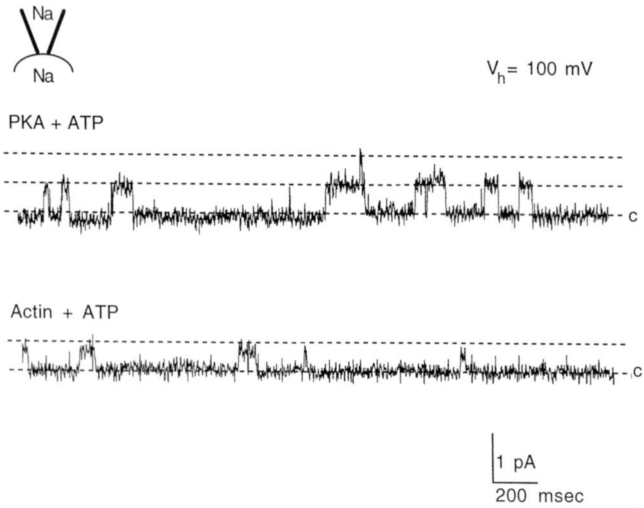

FIGURE 4 Effect of protein kinase A and actin on Na^+ channel activity of Apx-expressing cells. Top, addition of PKA (10 μg/ml) plus ATP (1 mM), to the cytoplasmic side of quiescent excised, inside-out patches from Apx-7 cells, induced and/or increased Na^+ channel activity within 2 min. Data are representative of 3 experiments obtained in symmetrical Na^+. Bottom, addition of actin (1 mg/ml) to the cytoplasmic side of quiescent excised, inside-out patches from Apx-7 cells, induced Na^+ channel activity within 2 min. Data are representative of 7 experiments obtained in symmetrical Na^+ conditions. Reproduced from Prat *et al.* (1996), with permission.

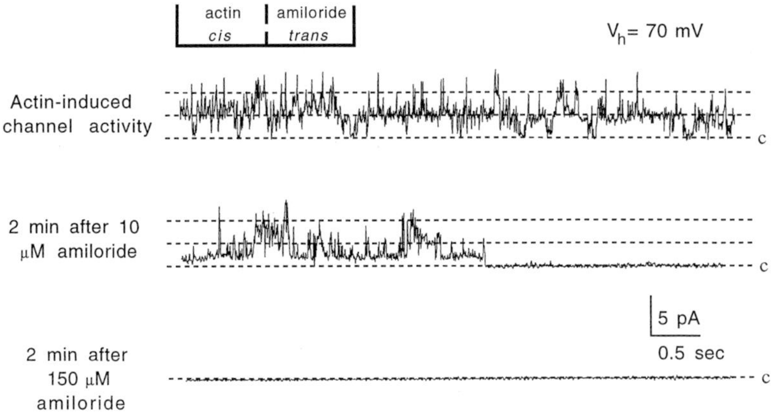

FIGURE 5 Reconstitution studies. Membrane-enriched preparations from Apx-7 cells were reconstituted into lipid bilayers using either asymmetrical (650/50 *cis/trans,* data not shown) or symmetrical 300 mM NaCl. Cation-selective ion channel activity was readily activated by addition of actin to the *cis* side of the chamber (top tracing). Reconstituted channels showed a single channel conductance of 10.9 pS. Channel activity was completely inhibited by addition of amiloride (bottom tracing) to the *trans* but not to the *cis* chamber. M2 membrane preparations showed no channel activity under the same conditions. Experimental conditions were identical to those reported in (Prat *et al.,* 1996).

preparations from M2 cells showed no channel activity in the presence or absence of actin.

Thus, Apx-mediated single Na^+ channel currents had similar functional and regulatory properties to those of Na^+ channels originally reported in A6 epithelial cells (Hamilton and Eaton, 1985; Prat *et al.,* 1993a). These included single channel conductances and kinetics, inhibition by amiloride, and regulation by either PKA plus ATP, or polymerizing concentrations of actin. The encompassed data are most consistent with the contention that Apx is associated with the 9-pS apical Na^+ channel (Cantiello *et al.,* 1989, 1990, 1991; Hamilton and Eaton, 1985; Prat *et al.,* 1993a).

III. SODIUM TRANSPORT IN PROXIMAL TUBULAR CELLS

Electrodiffusional Na^+ Transport in LLC-PK_1 Cells. The mammalian renal proximal tubule reabsorbs approximately 80% of the salt and water load which clears through the glomerulus (Maude, 1974). It is accepted that most water reabsorption in this renal section is driven by Na^+ reabsorption in renal tubular cells. However, to a large extent, only carrier mechanisms have been described as relevant Na^+ entry pathways into renal proximal cells. Few studies to date have been able to assess Na^+ transport rates under physiological electrochemical gradients for Na^+ in isolated cells, and the question remains largely unanswered as to whether the proximal segments of the renal nephron, in addition to other Na^+-facilitated transport mechanisms (Cantiello *et al.,* 1986; Rabito and Karish, 1982), may also contain ion channels responsible for Na^+ reabsorption. The pig kidney tubular epithelial cell line, LLC-PK_1 (Hull *et al.,* 1976), possesses multiple differentiated characteristics of the straight segment (S_3) of the proximal tubule (Amsler and Cook, 1982; Cantiello *et al.,* 1986; Rabito and Karish, 1982). These cells, grown on a permeable support, provide a useful experimental model representing a single cell-type epithelium in which Na^+ transport can be assessed under experimental conditions difficult to obtain with conventional preparations (Cantiello and Ausiello, 1986; Cantiello *et al.,* 1986, 1987). Although the apical Na^+ entry step in tight epithelia occurs passively through a simple electrodiffusional transport system, a similar finding in the leaky epithelia, such as the renal proximal tubule, has been difficult to demonstrate. In the absence of any major cotransportable solutes, including sugars, amino acids, or phosphate, Na^+ movement across LLC-PK_1 cells is thought to be mediated via an electroneutral apical Na^+/H^+-exchanger present in these cells (Cantiello *et al.,* 1986).

To assess the presence of Na^+ entry pathways in the absence of cotransported substrates, confluent monolayers of LLC-PK_1 cells grown on permeable support (Cantiello *et al.,* 1987) were first Na^+-depleted in a solution containing low Na^+ (15 mM) and high K^+ (135 mM), following a 15 min incubation in tissue culture medium containing 10^{-3} M ouabain to block the Na^+,K^+-ATPase. The $^{22}Na^+$ uptake (Fig. 6) was then followed by transferring the LLC-PK_1 monolayers into the uptake solutions containing 143 mM Na^+ (and other salts, see Methods in Cantiello *et al.,* 1987). The data indicate that intracellular Na^+ reached equilibrium in approximately 10–15 min, following a single exponential uptake kinetics. The $^{22}Na^+$ accumulated during the first 2-min incubation times (Fig. 6, inset) was linear. The Na^+ influx in the absence of a H^+ gradient was amiloride sensitive (Fig. 7), but not associated with the apical Na^+/H^+ exchanger present in confluent LLC-PK_1 cells, since this pathway requires strong intracellular acidification (Cantiello *et al.,* 1986), and the amiloride effect on cells in which both pathways were active was not additive. Furthermore, amiloride inhibited the Na^+ influx in the absence of a proton gradient with an affinity higher than 10^{-7} M, most consistent with the inhibitory effect on Na^+ channels in tight epithelia. The localization of this high affinity Na^+ transport pathway was entirely apical (Table I).

To determine the electrodiffusional nature of the $^{22}Na^+$ transport mechanism, the net Na^+ influx in the absence of any exchange mechanisms should

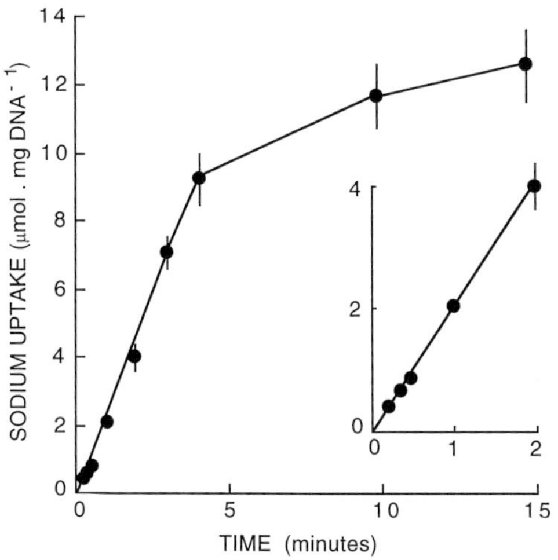

FIGURE 6 Radioactive Na^+ uptake into confluent monolayers of LLC-PK_1 monolayers. LLC-PK_1 monolayers were first Na^+-depleted in a Na^+-free solution also in the presence of 10^{-3} M ouabain to inhibit the Na^+,K^+-ATPase. Cells were incubated for 15 min in tissue culture medium containing 10^{-3} M ouabain and then transferred to a low Na^+ solution containing 130 mM K^+, 15 mM Na^+, and 10 mM HEPES, pH 7.40, also in the presence of 10^{-3} M ouabain (Cantiello *et al.,* 1987). Experiments were performed in a modified Earle's balanced salt solution (EBSS) containing, in mM; 143 Na^+, 5.36 K^+ 0.8 Mg^{2+}, 1.8 Ca^{2+}, 125 Cl^-, 0.8 SO_4^{2-}, and 10 mM *N*-2-hydroxyethylpiperazine-N′,-2-ethanesulfonic acid (HEPES) adjusted to pH 7.4 with 1N NaOH. Uptake experiments were performed in monolayers preincubated at 37°C under two different conditions. The $^{22}Na^+$ uptake was conducted by transferring the LLC-PK_1 monolayers from the preincubation medium into the uptake solutions containing $^{22}Na^+$ in the form of NaCl (1μCi/ml). Inset shows uptake of $^{22}Na^+$ recorded in first 3-min incubation. Results are means ± SEM of 6–9 determinations. Reproduced from Cantiello *et al.* (1987), with permission.

depend on a permeable counterion (to maintain electroneutrality) and should also generate a change in electrical potential (while transferring net charges). The Na^+ influx measured in the absence of a pH gradient in LLC-PK_1 cells was completely blocked by Cl replacement with the impermeable anion isethionate (Fig. 4 in Cantiello *et al.,* 1987). An electrical coupling to the electrodiffusional pathway should also apply. Thus, dissipation of the cell membrane potential by increasing the extracellular K^+ concentration in the presence of valinomycin reduced Na^+ influx measured in the absence of a proton gradient (Table II), while having no effect on the Na^+ influx induced by the presence of an H^+ gradient (Table II). The electrodiffusional nature of this Na^+ influx could also be observed by the membrane hyperpolarization, as expected by net charge displacement. Addition of 1 mM amiloride to cells uptaking extracellular Na under the above conditions

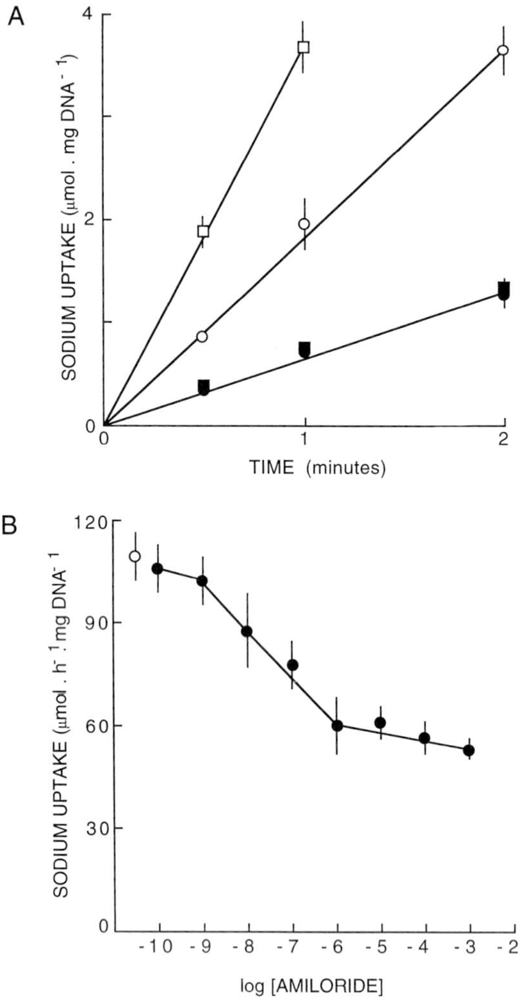

FIGURE 7 (A) Effect of 10^{-3} M amiloride on $^{22}Na^{+}$ influx. Net Na^{+} uptake was measured in absence (circles) or presence (squares) of a hydrogen ion gradient (pH_i 6.0, pH_o 7.4) without (open symbols) or with (closed symbols) 10^{-3} M amiloride. Results are means ± SEM of 6–9 determinations. Reproduced from Cantiello *et al.* (1987) with permission. (B) Effect of amiloride concentration on Na^{+}. Closed circles represent means ± SEM of 8 determinations performed at each concentration of amiloride. The open circle indicates the control value in the absence of amiloride. Reproduced from Cantiello *et al.* (1987), with permission.

TABLE I

Effect of Amiloride (1 mM) on Unidirectional Na^+ Influx in LLC-PK_1 Confluent Monolayers

	Na^+ influx $\mu mol \cdot H^{-1} \cdot mg\ DNA^{-1}$	
Unidirectional Na^+ flux	Control	Amiloride
Apical to basolateral	72.0 ± 3.0	49.2 ± 2.0
Basolateral to apical	73.0 ± 7.8	73.1 ± 9.8

Note. Unidirectional Na^+ fluxes of confluent LLC-PK_1 cells. The $^{22}Na^+$ uptake assays were performed by transferring the monolayers from the preincubation medium into the uptake solutions containing $^{22}Na^+$ in the form of NaCl (1μCi/ml). Transepithelial $^{22}Na^+$ transport from either the apical or basolateral side of the confluent monolayer was determined on monolayers of LLC-PK_1 cells mounted as a flat sheet between two hemichambers as described previously (Cantiello *et al.,* 1986). The solutions bathing both sides of the monolayers were of identical composition. The $^{22}Na^+$, however, was added only to the apical or basolateral solution. After a 2-min uptake period, both solutions were removed and the membrane was washed in ice-cold 0.1 M $MgCl_2$ solution. The radioactivity of the samples was determined by liquid scintillation counting techniques. During the 2-min uptake, no radioactivity was detected in the contralateral side. The results were normalized by the DNA content of the samples. Results are means ± SEM of 4–8 determinations. Amiloride concentration was 10^{-3} M. Experiments were conducted in the absence of intracellular acidification. Reproduced from Cantiello *et al.* (1987), with permission.

TABLE II

Effect of Cell Membrane Potential on Na^+ Influx Measured in the Presence or Absence of Proton Gradient

	Na^+ influx, $\mu mol \cdot h^{-1} \cdot mg\ DNA^{-1}$	
Condition	$pH_o = pH_i$	$pH_o < pH_i$
Control	69.3 ± 1.8	105.3 ± 6.6
Val + high $[K]_o$	57.0 ± 1.8	99.0 ± 5.7
Val + high $[K]_o$ + Amil	42.6 ± 1.2	46.0 ± 5.4

Note. Effect of amiloride on transepithelial Na^+ flux in LLC-PK_1 confluent monolayers. Results are means ± SEM of 5–6 determinations. Na^+ influx was measured at a Na^+ concentration of 70 mM. Part of the Na^+ (75 mM) was replaced by 75 mM choline (control) or 75 mM K^+. High K^+ medium contains, in addition, 4 mg/ml valinomycin (Val) with or without 10^{-3} M amiloride (Amil). Reproduced from Cantiello *et al.* (1987), with permission.

hyperpolarized the membrane potential (inside negative) by 10 mV, a value similar to external chloride replacement (6 mV).

Thus, confluent monolayers of LLC-PK_1 cells express an electrodiffusional, amiloride-sensitive Na^+ transport system which is independent of the Na^+/H^+ exchanger. As indicated by its localization, however, as well as the affinity for amiloride, this Na^+ transport pathway is most likely consistent with the presence of a Na^+-channel, as it has been previously reported by patch-clamping techniques (Moran and Moran, 1984) and observed in whole-cell currents (H. F. Cantiello, unpublished).

Immunochemical Localization of Apx in LLC-PK_1 Cells. To assess the possibility that Apx may be responsible for the electrodiffusional, amiloride-sensitive apical Na^+ transport in confluent LLC-PK_1 monolayers, immunocytochemistry of Apx was conducted in LLC-PK_1 cells grown on glass coverslips to partial confluency. The presence of both apical and subapical (intracellular) Apx was determined in LLC-PK_1 cells under control conditions (Fig. 8, left, see color insert). Interestingly, subconfluent cells where the actin cytoskeleton was heavily disrupted by a 20-min incubation with cytochalasin D (15 μg/ml) showed a dramatic increase in subapical labeling, consistent with strong plasma membrane redistribution of Apx (Fig. 8, right). Cells were, however, largely detached from the glass.

IV. CONCLUSION AND PERSPECTIVE

The above data indicate that Apx and perhaps related proteins underlie the molecular structure(s) associated with the functional group of channel species previously known as the "9-pS" channel. This is a distinct ion channel structure found, as indicated by the previous studies from our and other laboratories, in the bovine kidney, and most prominently in the apical domain of A6 cells. The biophysical features of Apx-Na(9) also make it a distinct functional structure as compared to the Na(5)-ENaC phenotypes. Most importantly their differences are further strenghthened by distinct renal localizations and regulatory features. To name a few relevant aspects of these differences, Apx and associated functional phenotypes are found in renal sections where neither ENaC mRNA nor protein has been detected, such as the proximal tubule (Duc *et al.,* 1994). This is further supported by our findings (this report) of a rather large amount of Apx in the proximal tubular LLC-PK_1 cell line, which is functionally consistent with the apical electrodiffusional Na^+ movement and the presence of functional Na^+ channels of this type on both LLC-PK_1 cells (Moran and Moran, 1984), and the straight segment of the proximal tubule (Gogelein and Greger, 1986).

It is important to note, therefore, that the previous "categorizing" of epithelial Na^+ channels on the basis of single ion conductances may be misleading, as a single ion channel structure may undergo several different functional states depending on the surrounding environments. Thus, although ENaC is hardly the molecular structure underlying the Na^+ channels in the renal proximal tubule/LLC-PK_1 cells, nevertheless it may have functional states "apparently similar" to those expected from Apx and related proteins. Conversely, "smaller" single-channel subconductance states (~4 pS) have been found in parallel to the higher (~7–10 pS) conductances in Apx (Prat *et al.,* 1996a), which do not necessarily imply the presence of ENaC.

In this regard, a more felicitous nomenclature would be to refer to Apx as the proximal (pNa, i.e., renal proximal tubule where Apx has been found; H. F. Cantiello unpublished), as opposed to the more distal Na^+ channel (dNa, i.e., distal colon, distal nephron ENaCs). Nevertheless, combined expression of these channels has been previously detected, and thus more studies will be required to assess the functional role of either channel, and in particular their preferred renal (or other organ) location.

Concerning the regulatory aspects of Apx and ENaCs, it is also important to establish the various signaling pathways associated with expression/regulation of either protein. ENaCs have been paradigmatically associated with chronic aldosterone treatment and/or Na^+ depletion (Garty and Palmer, 1997), while the 9-pS channel is spontaneusly expressed in A6 cells (Cantiello *et al.,* 1989). Furthermore, the regulatory role of vasopressin and cAMP activation may also be different between these two channels. Nevertheless, mutual regulatory pathways may combine functional efforts to avoid Na^+ overload, although this will require further investigation. Actin and associated cytoskeletal proteins, for example, control both Apx and ENaC function although the regulatory features of the cytoskeleton on either ion channel may be different. Polymerizing concentrations of actin modify the activation of Apx, while the same "ligands" modify both the open probability as well as the single-channel conductance of ENaCs (Berdiev *et al.,* 1996).

Finally, the possibility that Apx plays a relevant role in renal Na^+ reabsorption is warranted by the magnitude of the Na^+ uptake in confluent monolayers of A6 cells grown on permeable supports, a condition that highly resembles its functional role of equivalent epithelial sheets *in vivo* (Ausiello *et al.,* 1992). Although the nature of the underlying Na^+ channel structures was not determined in that study, the absence of aldosterone stimulation, and the fact that the amiloride-sensitive Na^+ uptake was also inhibited by pertussis toxin, would suggest this to be mediated by the 9-pS Apx channel. This also offers the possibility of interesting comparisons with other renal preparations. As indicated above, the amiloride-sensitive Na^+ influx into confluent monolayers of LLC-PK_1 cells grown on permeable supports was 1.20 μmol· mg DNA^{-1}· min^{-1} (Cantiello *et al.,* 1987), threefold higher than those reported for A6 cells (0.4 μmol· mg DNA^{-1}· min^{-1}, Table I in Ausiello *et al.,* 1992), grown and treated under identical conditions. Interestingly, this is in close agreement with a comparably higher Apx labeling in the LLC-PK_1 cells with respect to A6 cells (compare Fig. 8 with Fig. 1 of Prat *et al.,* 1996). Further, the present data would suggest that Apx expression and perhaps regulation as well is intimately related to the dynamic state of the cytoskeleton, the paradigm that allowed us to both discern the regulatory pathway of vasopressin action of the 9-pS channel in A6 cells and determine the presence of a functional Apx.

To conclude, a novel ion channel has been determined, which may be implicated in the characterization of an entire group of Na^+ channel phenotypes in regions which may or may not overlap with those where ENaCs are most expected. Further understanding of the regulatory features of Apx will not only help elucidate the renal role of these channels, but will be an invaluable tool for assessing the mandatory requirement of cytoskeletal organization on ion channel function. Most of the work is ahead of us.

Acknowledgments

The author gratefully acknowledges the constant technical contributions of Mr. G. Robert Jackson, Jr., and Dr. Adriana G. Prat. The author also thanks the American Physiological Society and the Journal of Biological Chemistry having made available previously published materials. The studies were supported in part by NIH Grant DK48040.

References

Amsler, K., and Cook, J. S. (1982). Development of Na^+-dependent hexose transport in a cultured line of porcine kidney cells. *Am. J. Physiol.* **242,** C94–C101.

Ausiello, D. A., Stow, J. L., Cantiello, H. F., de Almeida, J. B., and Benos, D. J. (1992). Purified epithelial Na^+ channel complex contains the pertussis toxin-sensitive $G_{\alpha i\text{-}3}$ protein. *J. Biol. Chem.* **267,** 4759–4765.

Benos, D. J., Saccomani, G., and Sariban-Sohraby, S. (1987). The epithelial sodium channel: Subunit number and location of the amiloride binding site. *J. Biol. Chem.* **262,** 10613–10618.

Benos, D. J., Awayada, M. S., Ismailov, I. I., and Johnson, J. P. (1995). Structure and function of amiloride-sensitive Na^+ channels. *J. Membr. Biol.* **143,** 1–18.

Berdiev, B., Prat, A., Cantiello, H., Ausiello, D., Fuller, C., Jovov, B., Benos, D., and Ismailov, I. (1996). Regulation of epithelial sodium channels by short actin filaments. *J. Biol. Chem.* **271,** 17704–17710.

Canessa, C., Horisberger, J.-D., and Rossier, B. (1993). Epithelial sodium channel related to proteins involved in neurodegeneration. *Nature (London)* **361,** 467–470.

Canessa, C., Schild, L., Buell, G., Thorens, B., Gautschl, I., Horisberger, J.-D., and Rossier, B. (1994). Amiloride-sensitive epithelial Na^+ channel is made of three homologous subunits. *Nature (London)* **367,** 463–467.

Cantiello, H. F., and Ausiello, D. A. (1986). Atrial natriuretic factor and cGMP inhibit amiloride-sensitive Na^+ transport system in the cultured renal epithelial cell line, LLC-PK_1. *Biochem. Biophys. Res. Commun.* **134,** 852–860.

Cantiello, H. F., Scott, J. A., and Rabito, C. A. (1986). Polarized distribution of the Na^+/H^+ exchange system in a renal cell line (LLC-PK_1). *J. Biol. Chem.* **261,** 3252–3258.

Cantiello, H. F., Scott, J. A., and Rabito, C. A. (1987). Conductive Na^+ transport in an epithelial cell line (LLC-PK_1) with characteristics of proximal tubular cells. *Am. J. Physiol. (Renal Fluid Electrolyte Physiol.)* **252,** F590–F597.

Cantiello, H. F., Patenaude, C. R., and Ausiello, D. A. (1989). G protein subunit, alpha i–3, activates a pertussis toxin-sensitive Na^+ channel from the epithelial cell line, A6. *J. Biol. Chem.* **264,** 20867–20870.

Cantiello, H. F., Patenaude, C. R., Codina, J., Birnbaumer, L., and Ausiello, D.A. (1990). $G_{\alpha i3}$ regulates epithelial Na^+ channels by activation of phospholipase A_2 and lipoxygenase pathways. *J. Biol. Chem.* **265,** 21624–21628.

Cantiello, H. F., Stow, J., Prat, A. G., and Ausiello, D. A. (1991). Actin filaments control epithelial Na^+ channel activity. *Am. J. Physiol.* **261,** C882–C888.

Cantiello, H. F., Prat, A. G., Bonventre, J. V., Cunningham, C. C., Hartwig, J., and Ausiello, D. A. (1993). Actin-binding protein contributes to cell volume regulatory ion channel activation in melanoma cells. *J. Biol. Chem.* **268,** 4596–4599.

Duc, C., Farman, N., Canessa, C., Bonvalet, J.-P., and Rossier, B. (1994). Cell-specific expression of epithelial sodium channel α, β, and γ subunits in aldosterone-responsive epithelia from the rat: Localization by in situ hybridization and immunocytochemistry. *J. Cell Biol.* **127,** 1907–1921.

Garty, H. (1994). Molecular properties of epithelial, amiloride-blockable Na^+ channels. *FASEB J.* **8,** 522–528.

Garty, H., and Palmer, L. G. (1997). Epithelial sodium channels: Function, structure, amd regulation. *Physiol. Rev.* **77,** 359–396.

Gogelein, H., and Greger, R. (1986). Na selective channels in the apical membrane of late proximal tubules (pars recta). *Pflugers Arch.* **406,** 198–203.

Hamilton, K. L., and Eaton, D. C. (1985). Single-channel recordings from amiloride-sensitive epithelial sodium channel. *Am. J. Physiol.* **249,** C200–C207.

Hamilton, K. L., and Eaton, D. C. (1986). Single channel recordings from two types of amiloride-sensitive epithelial Na^+ channels. *Membr. Biochem.* **6,** 149–171.

Hull, R. N., Cherry, W. R., and Weaver, G. W. (1976). The origin and characteristics of a pig kidney cell strain LLC-PK_1. *In Vitro Cell Dev. Biol.* **12,** 670–677.

Karlson, K., Ciampolillo-Bates, F., McCoy, D., Kizer, N., and Stanton, B. (1995). Cloning of a cAMP-gated cation channel from mouse kidney inner medullary collecting duct. *Biochim. Biophys. Acta* **1236,** 197–200.

Katsura, T., Verbavatz, J.-M., Farinas, J., Ma, T., Ausiello, D. A., Verkman, A. S., and Brown, D. (1995). Constitutive and regulated membrane expression of aquaporin 1 and aquaporin 2 water channels in stably transfected LLC-PK_1 epithelial cells. *Proc. Natl. Acad. Sci. U.S.A.* **92,** 7212–7216.

Light, D. B., McCann, F. V., Keller, T. M., and Stanton, B. A. (1988). Amiloride-sensitive cation channel in apical membrane of inner medullary collecting duct. *Am. J. Physiol.* **255,** F278–F286.

Light, D. B., Ausiello, D. A., and Stanton, B. A. (1989a). Guanine nucleotide-binding protein, αi-3, directly activates a cation channel in rat renal inner medullary collecting duct cells. *J. Clin. Invest.* **84,** 352–356.

Light, D. B., Schwiebert, E. M., Karlson, K. H., and Stanton, B. A. (1989b). Atrial natriuretic peptide inhibits a cation channel in renal inner medullary collecting duct cells. *Science* **243,** 383–385.

Linguegia, E., Voilley, N., Waldmann, R., Lazdunski, M., and Barbry, P. (1993). Expression cloning of an epithelial amiloride-sensitive Na^+ channel. A new channel type with homologies to *Caenorhabditis elegans* degenerins. *FEBS Letters* **318,** 95–99.

Marunaka, Y., and Eaton, D. (1991). Effects of vasopressin and cAMP on single amiloride-blockable Na^+ channels. *Am. J. Physiol.* **260,** C1071–C1084.

Maude, D. L. (1974). Mechanisms of tubular transport of salt and water. *In* "Kidney and Urinary Tract Physiology" (A. C. Guyton and K. Thurau, eds.), pp. 39–78. Butterworths, London.

Moran, A., and Moran, N. (1984). Amiloride-sensitive channels in LLC-PK_1 apical membranes. *Fed. Proc., Fed. Am. Soc. Exp. Biol.* **43,** 447 (abstract).

Prat, A. G., Ausiello, D. A., and Cantiello, H. F. (1993a). Vasopressin and protein kinase A activate G protein-sensitive Na channels. *Am. J. Physiol.* **265,** C218–C223.

Prat, A. G., Bertorello, A. M., Ausiello, D. A., and Cantiello, H. F. (1993b). Activation of epithelial Na^+ channels by protein kinase A requires actin filaments. *Am. J. Physiol.* **265,** C224–C233.

Prat, A. G., Xiao, Y.-F., Ausiello, D. A., and Cantiello, H. F. (1995). cAMP-independent regulation of CFTR by the actin cytoskeleton. *Am. J. Physiol.* **268,** C1552–C1561.

Prat, A., Holtzman, E., Brown, D., Cunningham, C., Reisin, I., Kleyman, T., McLaughlin, M., Jackson, G. R., Jr., Lydon, J., and Cantiello, H. F. (1996). Renal epithelial protein (Apx) is an actin cytoskeleton-regulated Na^+ channel. *J. Biol. Chem.* **271,** 18045–18053.

Rabito, C. A., and Karish, M. V. (1982). Polarized amino acid transport by an epithelial cell line of renal origin (LLC-PK_1). *J. Biol. Chem.* **257,** 6802–6808.

Sariban-Sohraby, S., and Fisher, R. S. (1992). Single channel activity by the amiloride binding subunit of the epithelial Na^+ channel. *Am. J. Physiol.* **263,** C1111–C1117.

Sariban-Sohraby, S., Latorre, R., Burg, M., and Benos, D. (1984). Amiloride-sensitive epithelial Na^+ channels reconstituted into planar lipid bilayer membranes. *Nature* (*London*) **308,** 80–82.

Staub, O., Verrey, F., Kleyman, T. R., Benos, D. J., Rossier, B. C., and Kraehenbuhl, J.-P. (1992). Primary structure of an apical protein from *Xenopus laevis* that participates in amiloride-sensitive sodium channel activity. *J. Cell Biol.* **119,** 1497–1506.

PART III

Sodium Channels in the Lung

CHAPTER 12

Species-Specific Variations in ENaC Expression and Localization in Mammalian Respiratory Epithelium

Colleen R. Talbot
Department of Biology, California State University, San Bernardino, California 92407

I. INTRODUCTION

A large body of literature exists on ion transport across airway epithelia; studies have been performed on a number of different species, from different regions within the respiratory tree, and at different developmental stages. These studies have led to a greater understanding of ion transport across airway epithelia and have suggested potential regulatory pathways. However, because of the complex morphology of the respiratory tract, a good understanding of the *in vivo* physiology of Na^+ and Cl^- transport and associated water flow and how these processes affect the composition and volume of the airway surface liquid layer is poorly understood. Additionally, electrophysiological studies done on airway and alveolar epithelia have

1063-5823/99 $30.00

suggested some differences in these pulmonary tissues relative to aldosterone-sensitive Na^+ reabsorbing epithelia of the kidney and colon. Different tissues have been shown to vary in ion selectivity, open and closed time, amiloride inhibition, and single-channel conductances (Smith and Benos, 1991; Palmer, 1992; Rossier *et al.,* 1994); for example, patch-clamp studies have shown single-channel Na^+ conductances for proximal airway and alveolar epithelia to be approximately 19 and 25 pS, respectively (Chinet *et al.,* 1993; Matalon *et al.,* 1992); whereas the highly selective Na^+ channels common to kidney and colon show single-channel conductances around 5 pS (Rossier *et al.,* 1994). The cloning of the amiloride-sensitive epithelial sodium channel (ENaC) may lead to an increased understanding of airway Na^+ transport physiology by permitting the mRNA of the ENaC subunits to be detected and quantified within the airway epithelia, allowing localization within the respiratory tree and, perhaps, to specific cell types. Knowledge of where ENaC is expressed and how expression of mRNA for the subunits varies within the airway could help estimate the contribution of specific regions to the establishment and maintenance of the airway surface liquid layer. Understanding variations in ENaC expression between species might also lead to an explanation of observed physiological differences in ion transport between species.

II. STRUCTURE/FUNCTION OF THE RESPIRATORY TRACT

Airways of the respiratory tract conduct air to and from the alveolar region, the site of gas exchange, and defend against inhaled particles and microbes. These defense functions depend, in part, on mucociliary transport, which moves mucous entrapped particles from the distal lung towards the mouth. Ciliated nasal epithelia is also involved in a similar defense function. The effective functioning of the mucociliary transport system depends on coordinated ciliary beating, mucous secretions, and regulation of the volume and composition of the airway surface liquid (ASL) layer that coats the luminal airway surfaces (Boucher, 1994a,b). The volume and depth of the ASL is maintained by active ion transport as solute transport creates local concentration gradients across the epithelium, resulting in osmotically coupled fluid movement (Diamond, 1979). Airway epithelia have the capacity for either fluid secretion or absorption, depending upon the gestational age, neurohumoral environment, airway region, and species examined (Widdicombe, 1994). Fluid secretion is affected by active secretion of Cl^- through the cystic fibrosis transmembrane conductance regulator (CFTR) and/or an "alternative" Cl^- channel; fluid absorption is mediated by electrogenic absorption of Na^+. Na^+ entry into the cell is thought to occur through the apical membrane amiloride-blockable epithelial Na^+ channel, ENaC, originally cloned from rat colon (Canessa *et al.*, 1993, 1994). Volume regulation of the airway surface liquid layer by these two processes is believed to play a critical role in mucociliary clearance. The ASL is composed of two discrete layers: the periciliary sol layer and the mucous layer resting on top of the cilia of the airway epithelia. Mucociliary clearance is mediated by the forward strokes of the cilia driving the mucous layer forward. Mucociliary clearance can be affected by altering the depth and composition of the periciliary sol layer or the hydration state of the mucous gel (Al Bazzaz, 1986). The importance of regulating the depth and composition of the ASL can be clearly seen in airways that have a defective CFTR, producing the disease state cystic fibrosis (CF). In CF, the combination of defective Cl^- secretion along with an associated increase in the rate of Na^+ absorption leads to increased fluid absorption and generates a dehydrated ASL. As a result, the efficiency of the normal mucociliary clearance defense mechanisms may be impaired and the lungs may become more susceptible to bacterial infection.

Respiratory epithelia are broadly defined by anatomic region and morphology and include the nasal epithelia lining the turbinates, the proximal airway epithelium (*i.e.*, the cartilaginous airways), the distal (bronchiolar) airway epithelium, and the alveolar epithelium. The respiratory nasal epithelium and the epithelium lining the proximal airways is pseudostratified and comprised of three major cell types: ciliated cells, secretory (goblet) cells, and basal cells. The role of the superficial epithelium of this region in maintaining the ASL may be supplemented by secretions from submucosal glands. These glands are comprised of both serous and mucous cells within the acinar epithelium and cuboidal and ciliated cells in the ductal epithelium. Bronchiolar airway epithelium is comprised primarily of ciliated and nonciliated (Clara) cells and is broken down into several regions including proximal bronchioles, terminal bronchioles, and respiratory bronchioles. The alveolar region is comprised of two cell types: Type I and Type II pneumatocytes. Type I cells are spread out thin over the alveolar surface and comprise the primary gas exchange surface whereas Type II cells are much more compact and generally confined to the corners of the alveoli. The roles of these regions are characterized to varying degrees with regards to their role in maintaining airway surface liquid and ion transport.

Chloride secretion, by either alveolar cells or the epithelia lining the small distal airways (bronchioles), is thought to generate net fluid flow into the airway, thus providing a source for the ASL (Fig. 1). However, there is contradictory evidence as to which of these regions is the primary site of fluid secretion. Direct measurements of ion concentrations in alveolar hypophase found that Cl^- is above equilibrium and suggest that the alveoli

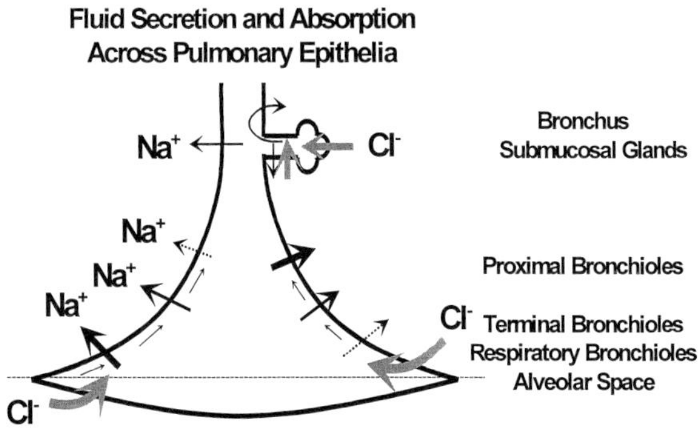

FIGURE 1 Possible sites of Cl^- secretion and Na^+ absorption across pulmonary epithelia as related to their role in the formation and maintenance of the airway surface liquid layer. The ASL may originate in either the alveolar or bronchial regions. A gradient of gene expression of the relevant transport proteins may exist in respiratory, terminal and proximal bronchioles, and bronchial epithelium. Submucosal gland serous cells probably secrete fluid and ducts may have either dominant secretory or absorptive functions.

may function in secretion of liquids (Nielson, 1986). However, cultures of Type II cells suggest the opposite, that the alveoli are involved in Na^+ absorption, and hence, fluid absorption (Mason *et al.*, 1982; O'Brodovich *et al.*, 1991). This is presumed to function in maintaining alveolar subphase and preventing pulmonary edema in adult, nonfetal lung. Additionally, these cultures fail to show significant Cl^- secretory activity and β-adrenergic stimulation does not increase Cl^- secretion but does increase Na^+ absorption. Others have hypothesized that the distal airways secrete Cl^- and fluid into the airway lumen to replenish fluid swept proximally by ciliary movement (Kilburn, 1968). However, Ballard and Taylor (1994) found that in microcannulated pig proximal bronchioles, active Na^+ absorption accounts for most of the resting short circuit current and little active Cl^- secretion occurs under basal conditions. Intact sheep bronchioles also have a dominant Na^+ absorptive capacity, as well as apical Cl^- channels that may mediate active secretion (Al Bazzaz, 1994). Cultured rabbit Clara cells have a dominant Na^+ absorptive function (Van Scott *et al.*, 1989), but stimulation of apical purinergic receptors with extracellular ATP induces a transient electrogenic Cl^- and HCO_3^- secretion (Van Scott *et al.*, 1995).

The role of the proximal airway submucosal glands in forming/maintaining the ASL is not fully understood. The submucosal glands develop as invaginations of surface epithelium in cartilagenous airways during gestation and are comprised of a proximal ciliated duct, collecting ducts, mucous tubules, and distal serous tubules (Meyrick *et al.*, 1969). Location within

```
---------------------------------------------MKGDKPEEPGPGPEP 15  beNaC
-------------------------MEGNKLEEQDSSPPQSTPGLMKGNKREEQGLGPEP 35  hENaC
MLDHTRAPELNIDLDLHASNSPKGSMKGNQFKEQDPCPPQPMQGLGKGDKREEQGLGPEP 60  rENaC
--------------------------------------------MPEGEKTRQCKQETE- 15  cENaC
-----------------------------------------------------------


SGPPPPTEEEEALLEFHRSYRELFEFFCNNTTIHGAIRLVCSQHNRMKTVFWAVLWLCTF 75  bENaC
AAPQQPTAEEEALIEFHRSYRELFEFFCNNTTIHGAIRLVCSQHNRMKTAFWAVLWLCTF 95  hENaC
SAPRQPTEEEEALIEFHRSYRELFQFFCNNTTIHGAIRLVCSKHNRMKTAFWAVLWLCTF 120 rENaC
--QQQKEDEREGLIEFYGSYQELFQFFCSNTTIHGAIRLVCSKKNKMKTAFWSVLFILTF 73  cENaC
-MTKEEKNEKEALIEFFSSYRELFEFFCSNTTIHGAIRLVCSRRNRMKTAFWLVLFLVTF 59  xENaC

GMMYWQFGQLFGEYFSYPVSLNINLNSDKLVFPAVSICTLNPYRYKEIQEELEELDRITE 135 bENaC
GMMYWQFGLLFGEYFSYPVSLNINLNSDKLVFPAVTICTLNPYRYPEIKEELEELDRITE 155 hENaC
GMMYWQFALLFEEYLSYPVSLNINLNSDKLVFPAVTVCTLNPYRYTEIKEELEELDRITE 180 rENaC
GLMYWQFGILYREYFSYPVNLNLNLNSDRLTFPAVTLCTLNPYRYSAIRKKLDELDQITH 133 cENaC
GLMYWQFGLLFGQYFSYPVSINLNVNSDKLPFPAVTVCTLNPYRYKAIQNDLQELDKETQ 119 xENaC

QTLFDLYKYNSSKTLVAHARSR--RDLREPLPHPLQRLPVPAPPHAARGVRRAGSSMRDN 193 bENaC
QTLFDLYKYSSFTTLVAGSRSR--RDLRGTLPHPLQRLRVPPPPHGARRARSVASSLRDN 213 hENaC
QTLFDLYKYNSSYTRQAGARRRSSRDLLGAFPHPLQRLRTPPPPYSGRTARSGSSSVRDN 240 rENaC
QTLLDLYDYNMSLARSDGSAQFSHRRTSRSLLHHVQRHPLR------RQKRDNLVSLPEN 187 cENaC
RTLYELYKYNSTGVQGWIPNNQRVKRDRAGLPYLLELLPPG------SETHRVSRSVIEE 173 xENaC

NPQVNRKDWKIGFQLCNQNKSDCFYQTYSSGVDAVREWYRFHYINILSRRRQDTSPSLEE 253 bENaC
NPQVDWKDWKIGFQLCNQNKSDCFYQTYSSGVDAVREWYRFHYINILSR-LPETLPSLEE 272 hENaC
NPQVDRKDWKIGFQLCNQNKSDCFYQTYSSGVDAVREWYRFHYINILSR-LSDTSPALEE 299 rENaC
SPSVDKNDWKIGFVLCSENNEDCFHQTYSSGVDAVREWYSFHYINILAQ--MPDAKDLDE 245 cENaC
ELQVKRREWNIGFKLCNETGGDCFYQTYTSGVDAIREWYRFHYINILAR--VPQEAAIDG 231 xENaC

DVLGKFIFTCRFNQDSCNEANYSHFHHPMYGNCYTFNDK--NSSNLWMSSMPGVNNGLSL 311 bENaC
DTLGNFIFACRFNQVSCNQANYSHFHHPMYGNCYTFNDK--NNSNLWMSSMPGINNGLSL 330 hENaC
EALGNFIFTCRFNQAPCNQANYSKFHHPMYGNCYTFNDK--NNSNLWMSSMPGVNNGLSL 357 rENaC
SDFENFIYACRFNEATCDKANYTHFHHPLYGNCYTFND---NSSSLWTSSLPGINNGLSL 302 cENaC
EQLENFIFACRFNEESCTKANYSSFHHAIYGNCYTFNQNQSDQSNLWSSSMPGIKNGLTL 291 xENaC

TLRTEQNDFIPLLSTVTGARVMVHERDEPAFMDDAGFNLRPGVETSISMSKEAVDRLGGD 371 bENaC
MLRAEQNDFIPLLSTVTGARVMVHGQDEPAFMDDGGFNLRPGVETSISMRKETLDRLGGD 390 hENaC
TLRTEQNDFIPLLSTVTGARVMVHGQDEPAFMDDGGFNLRPGVETSISMRKEALDSLGGN 417 rENaC
VVRTEQNDFIPLLSTVTGARVMVHDQNEPAFMDDGGFNVRPGIETSISMRKEMTERLGGS 362 cENaC
VLRTEQHDYIPLLSSVAGARVLVHGHKEPAFMDDNGFNIPPGMETSIGMKKETINRLGGK 351 xENaC

YGDCTKNGSEVPVENLYNTKYTQQVCIHSCFQESMIKECGCAYIFYPRPDGVEFCDYRKH 431 bENaC
YGDCTKNGSDVPVENLYPSKYTQQVCIHSCFQESMIKECGCAYIFYPRPQNVEYCDYRKH 450 hENaC
YGDCTENGSDVPVKNLYPSKYTQQVCIHSCFQENMIKKCGCAYIFYPKPKGVEFCDYRKQ 477 rENaC
YSDCTEDGSDVPVQNLYSSRYTEQVCIRSCFQLNMVKRCSCAYYFYPLPDGAEYCDYTKH 422 cENaC
YSDCSEDGSDVDVKNLFQSEYTEQVCVRSCFQAAMVARCGCGYAFYPLSPGDQYCDYNKH 411 xENaC

NSWGYCYYKLQDAFSSDRLGCFTKCRKPCSVTIYKLSASYSQWPSATSQDWVFQMLSRQN 491 bENaC
SSWGYCYYKLQVDFSSDHLGCFTKCRKPCSVTSYQLSAGYSRWPSVTSQEWVFQMLSRQN 510 hENaC
SSWGYCYYKLQGAFSLDSLGCFSKCRKPCSVINYKLSAGYSRWPSVKSQDWIFEMLSLQN 537 rENaC
VAWGYCYYKLLAEFKADVLGCFHKCRKPCKMTEYQLSAGYSRWPSAVSEDWVFYMLSQQN 482 cENaC
KSWGHCYYKLIIEFTSNKLGCFTKCRKPCLVSEYQLTAGYSKWPNRVSQDWVLHTLSRQ- 470 xENaC
```

CHAPTER 1 FIGURE 1 Alignment of αbENaC with the corresponding human, rat, chicken, and *Xenopus laevis* α subunit isoforms. αbENaC exhibits considerable homology with its counterparts from other species. Small and hydrophobic amino acids are shown in red, acidic residues in blue, basic residues in magenta, and all others in green. *(Continues)*

```
NYTIKNKRDGVAKLNIFFKELNYKSNSESPSVTMVTLLSNLGSQWSLWFGSSVLSVVEMA 551  bENaC
NYTVNNKRNGVAKVNIFFKELNYKTNSESPSVTMVTLLSNLGSQWSLWFGSSVLSVVEMA 570  hENaC
NYTINNKRNGVAKLNIFFKELNYKTNSESPSVTMVSLLSNLGSQWSLWFGSSVLSVVEMA 597  rENaC
KYNITSKRNGVAKVNIFFEEWNYKTNGESPAFTVVTLLSQLGNQWSLWFGSSVLSVMELA 542  cENaC
-YNLTD-RNGIAKLNIYFEELNYKTILESPTINMAMLLSLLGSQWSLWFGSSVLSVVEML 528  xENaC

ELIIDLLVITFLMLLRRFRSRYW--SPGRGGKGTQEVASTPAASLPSSFCPHPAFFSSSP 609  bENaC
ELVFDLLVIMFLMLLRRFRSRYW--SPGRGGRGAQEVASTLASSPPSHFCPHPMSLSLSQ 628  hENaC
DVIFDLLVITLLMLLRRFRSRYW--SPGRGARGAREVASTPASSFPSRFCPHPTSPPPSL 655  rENaC
ELILDFTVITFILAFRWFRSKRWH--------------SSPAPPPNSHDNTAFQDEASGL 597  cENaC
ELVIDFVIIGVMILLHRYYYKKANEGE------ETTVVPTPAPAFADLEQQVPHIPRGDL 584  xENaC

PDPAISP--ALSAPPPAYATLGPH-PAPSGLAEASTSAHAPGEP---------------- 650  bENaC
PGPAPSP--ALTAPPPAYATLGPR-PSPGGSAGASSSTCPLGGP---------------- 669  hENaC
PQQGMTPPLALTAPPPAYATLGPS-APPLDSAAPDCSACALAAL---------------- 698  rENaC
DAPHRFTVEAVVTTLPSYNSLEPCGPSKDGETGLE------------------------- 623  cENaC
SQRQISVV-ADITPPPAYESLDLRSVGTLSSRSSSMRSNRSYYEENGGRRN--------- 632  xENaC
```

CHAPTER 1 FIGURE 1 (Continued)

610 MLLRRFRSRYWSPGRGARGAREVASTPASSFPSRFCPHPT αrENaC

564 MLLRRFRSRYWSPGRGGKGTREVASTPAASLPSSFCPHPA αbENaC

650 SPPPSLPQQGMTPPLALTAPPPAYATLGPSAPPLDSAAPD αrENaC

604 FFSSSPPDPAI SP......ALSAPPPAYATLGPHPAPSGLAEAS αbENaC

690 CSACALAAL αrENaC
643 TSAHAPGEP αbENaC

CHAPTER 1 FIGURE 5 C-terminal alignments of αbENaC and αrENaC. The locations of nonconserved amino acids are shown in red. The site of the stop codon mutation in both αbENaC and αrENaC is shown in blue.

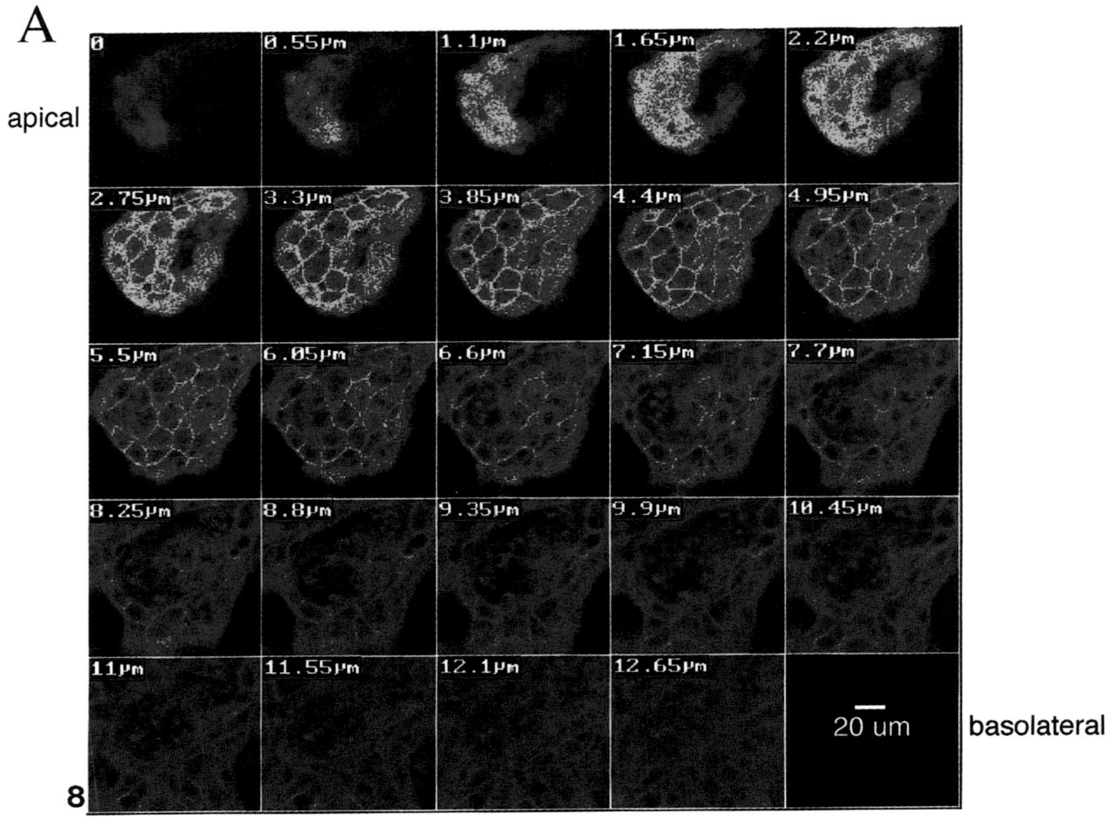

CHAPTER 5 FIGURE 8 (Continues)

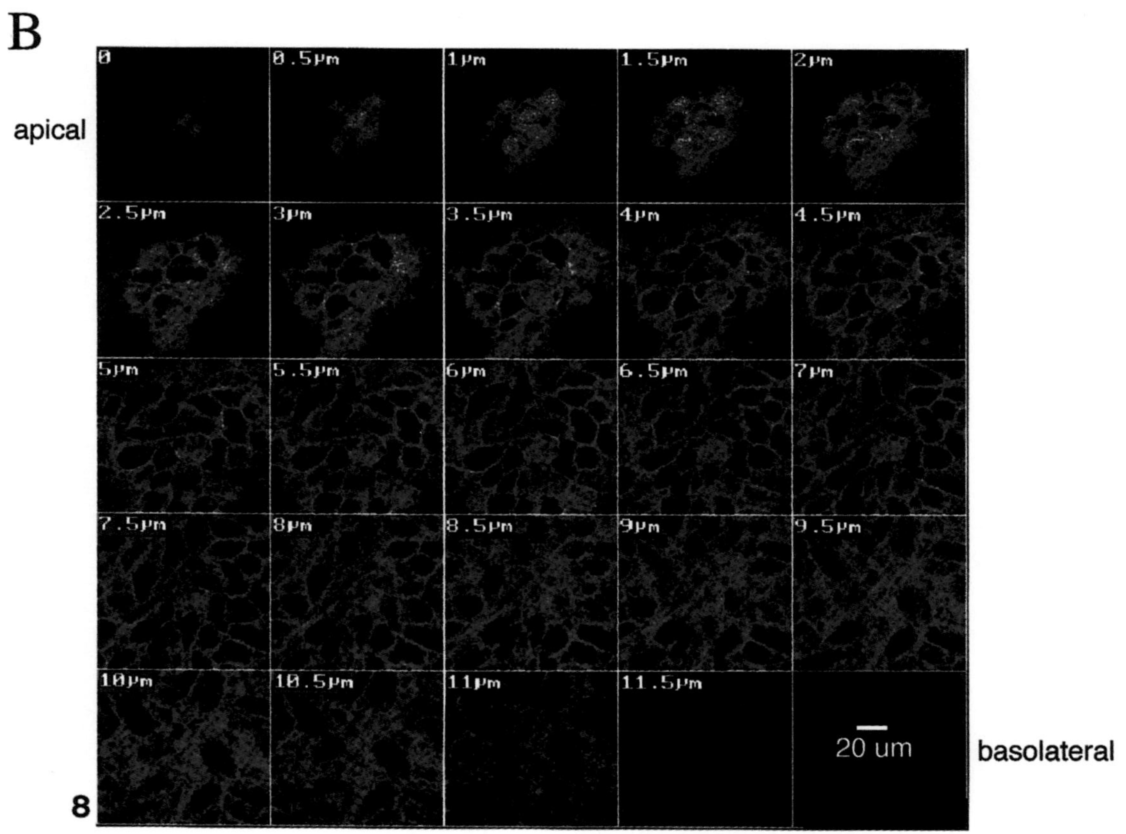

CHAPTER 5 FIGURE 8 (Continues)

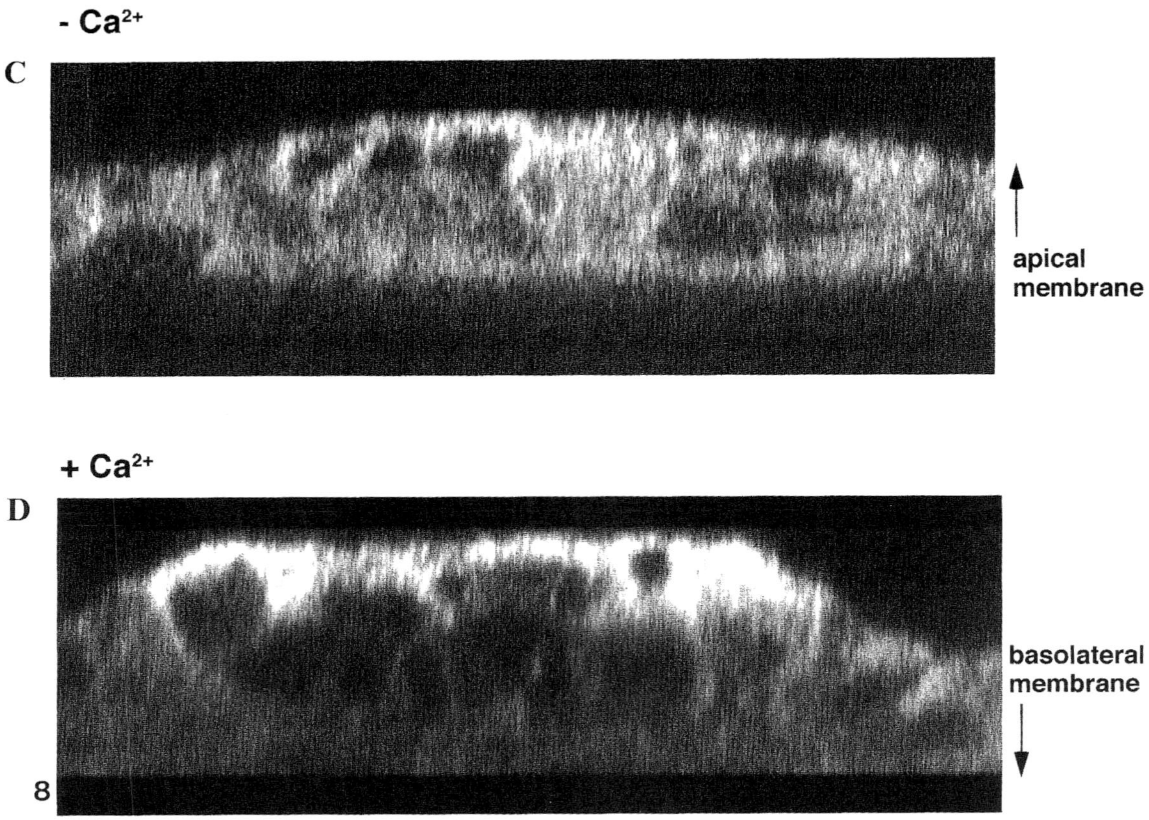

CHAPTER 5 FIGURE 8 (Continued)

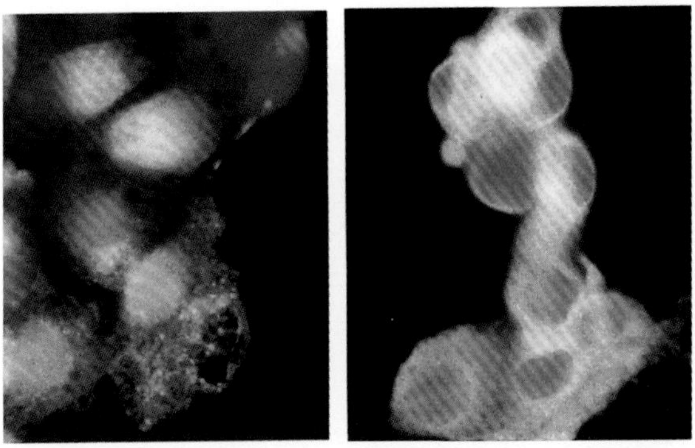

CHAPTER 11 FIGURE 8 Immunocytochemistry of LLC-PK_1 cells. The presence of Apx was assessed in subconfluent monolayers of LLC-PK_1 cells before (left) and after cytoskeletal disruption by a 20 min incubation in the presence of cytochalasin D (15 μg/ml). Control cells (left) were double-labeled with an anti-Apx antibody (FITC, green) and Evans blue (CY3, red) to counterstain with a contrast dye to enhance cell morphology. Both apical and subapical labeling of Apx were observed. In contrast, cytochalasin D incubated cells (right) showed a dramatic increase in plasma membrane distribution of Apx. Immunocytochemistry was performed as previously described for cultured cells (Katsura *et al.*, 1995; Prat *et al.*, 1996).

CHAPTER 5 FIGURE 8 Confocal analysis of Nedd4 distribution following Ca^{2+} plus ionomyc in treatment of polarized MDCK cells. Polarized MDCK cells were treated (A) or not (B) with Ca^{2+} (1 mM) plus ionomycin (1 μM) for 5 min. Cells were then fixed, permeabilized, and stained with affinity pure anti-Nedd4 (WWII) antibodies followed by anti-rabbit secondary antibodies conjugated to Texas Red, and serial confocal sections at 5 μm apart were taken starting from the apical (top) surface. Panels C and D represent X-Z reconstruction of the images of treated and untreated cells, respectively. Reproduced with permission from Plant *et al.* (1997).

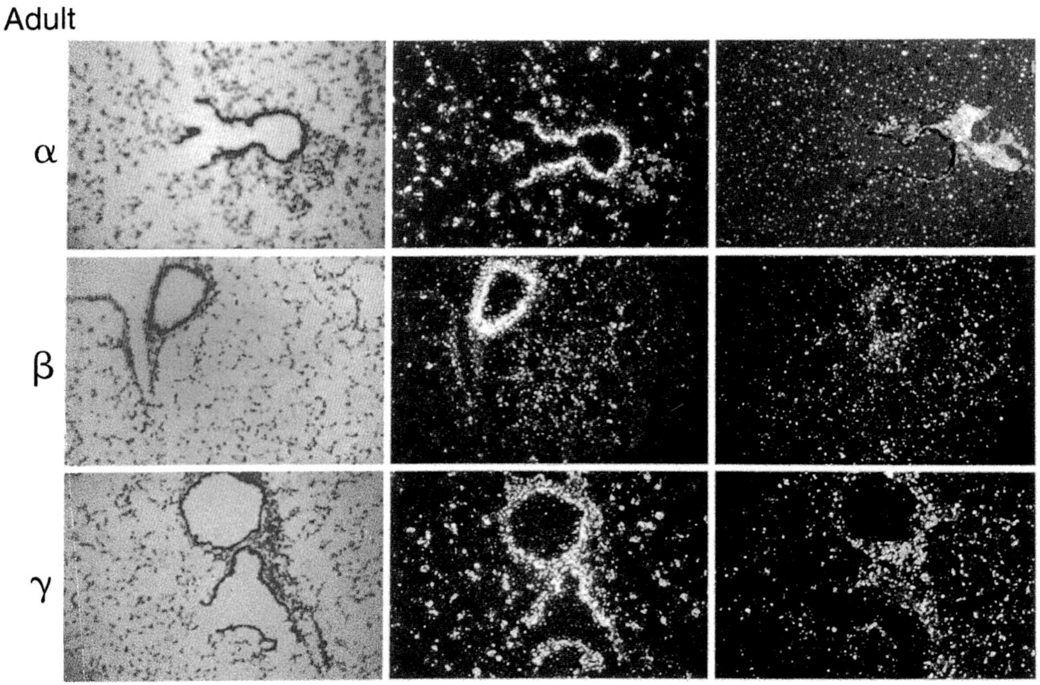

CHAPTER 12 FIGURE 3 *In situ* hybridization of αENaC (first row), βENaC (middle row), and γENaC (bottom row) in normal adult mouse lung. Bright field micrographs (first column) show H&E stained section of antisense labeled sections; dark field micrographs of same fields (middle column) demonstrate hybridization of all three subunit probes to bronchioles and punctate labeling in alveolar region for α and γENaC. Sense probes (right column) show only nonspecific hybridization in adjacent serial sections. (Modified with permission from Talbot *et al.* (1999). Quantitation and localization of ENaC subunit expression in fetal, newborn and adult mouse lung. *Am. J. Resp. Crit. Care Med.* **20,** 398–406. Official Journal of the American Thoracic Society. ©American Lung Association.)

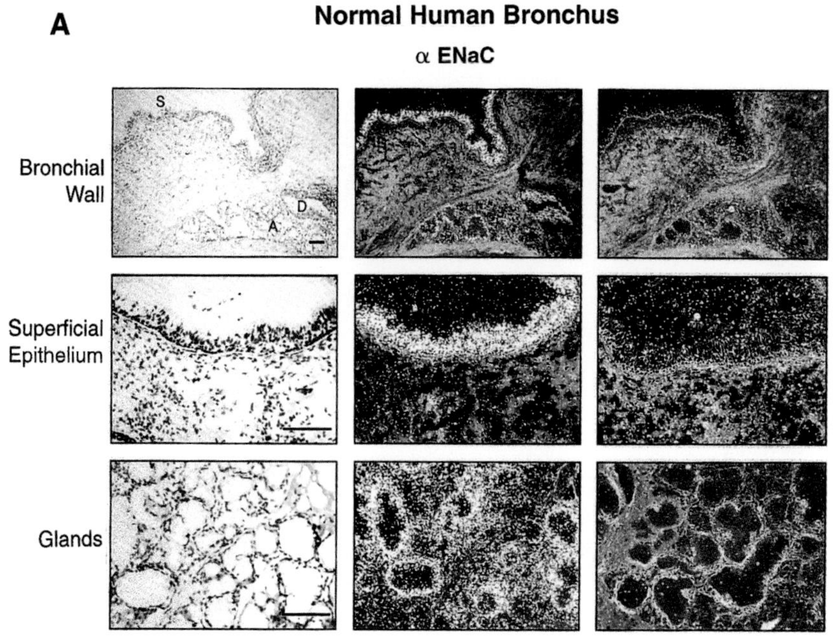

CHAPTER 12 FIGURE 4 *In situ* hybridization of normal human bronchus hybridized to probes for (A) αENaC or (B) β and γENaC. Sections were hybridized with antisense (middle column) or sense (right column) ^{35}S-labeled riboprobes and stained with hematoxylin and cosin (H&E). Bright field (left column) micrographs of same fields are shown for each antisense dark field image. S, superficial epithelium; D, gland duct; A, mucous and serous gland acini. Bar = 100 μm in all sections. (Modified from Burch *et al.*, 1995, with permission). *(Continues)*

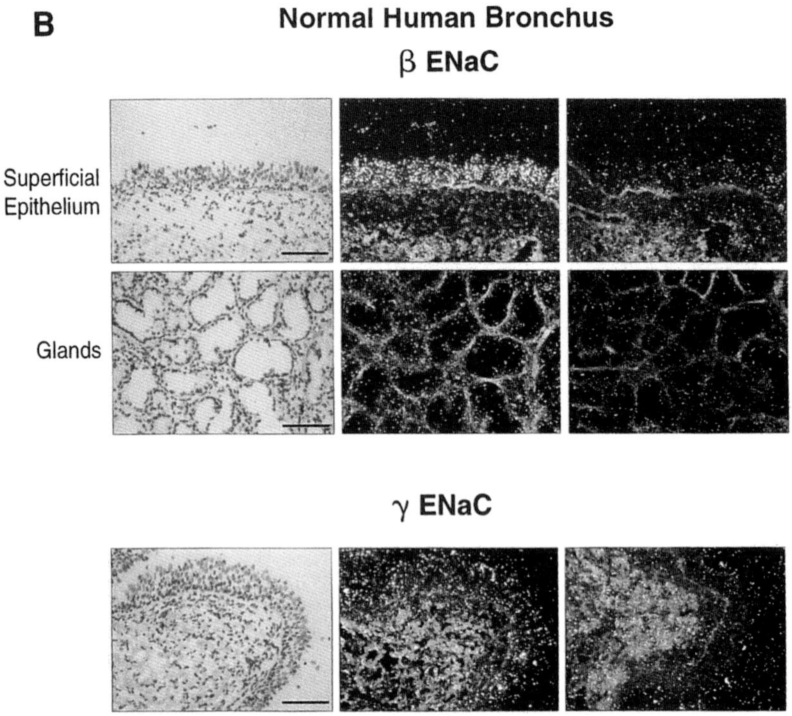

CHAPTER 12 FIGURE 4 (Continued)

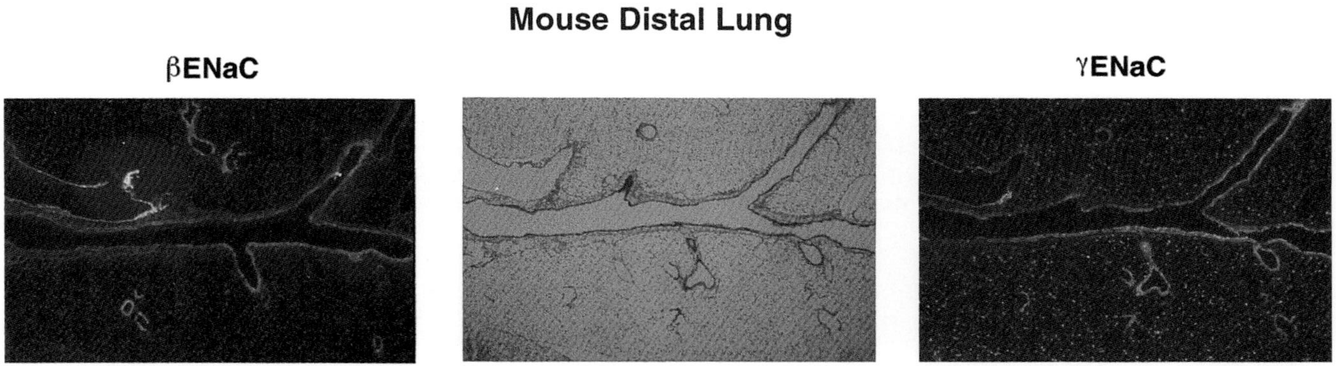

CHAPTER 12 FIGURE 6 Serial cross sections of an adult mouse lung hybridized with β (antisense dark field, left) or γENaC (antisense dark field, right). Bright field image (center) is the same field as shown for γENaC. Proximal, bronchial airways are to the left; distal, bronchiolar airways toward the right. A decrease in signal intensity is evident along the distal-to-proximal axis for both subunits. A large vein (in the upper left of each section) does not hybridize with either subunit. Magnification ×20. Probes and protocol as in Talbot *et al.* (1998, C. Talbot and R. Boucher, unpublished).

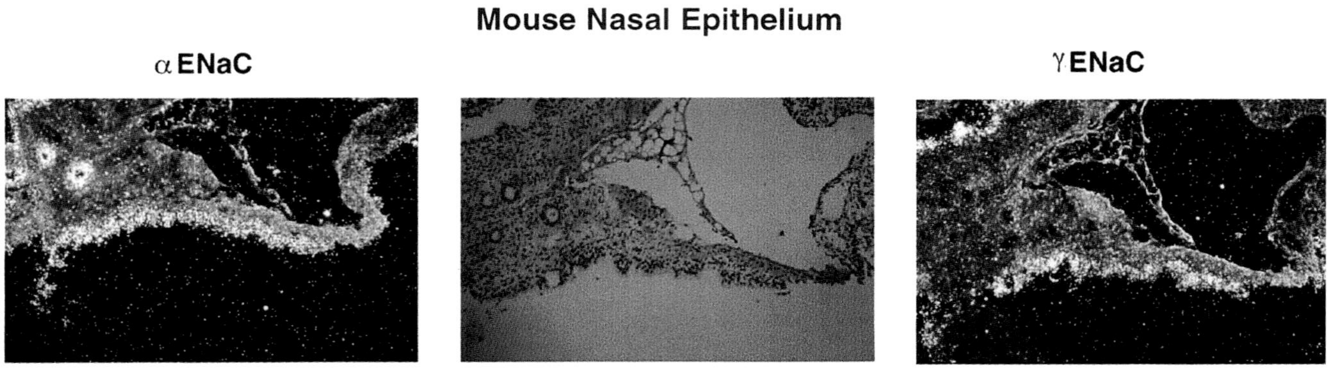

CHAPTER 12 FIGURE 7 Mouse nasal epithelium hybridized with α (antisense dark field, left) or γENaC (antisense dark field, right). Bright field image (center) is the same field as shown for γENaC. A strong signal for αENaC is evident for ciliated surface epithelium (bottom of each image) and gland ducts (left side of micrographs): γENaC displays a strong but uneven signal along the surface epithelium and no signal in ductular epithelium. Magnification ×40. Probes and protocol as in Talbot *et al.* (1998, C. Talbot and R. Boucher, unpublished).

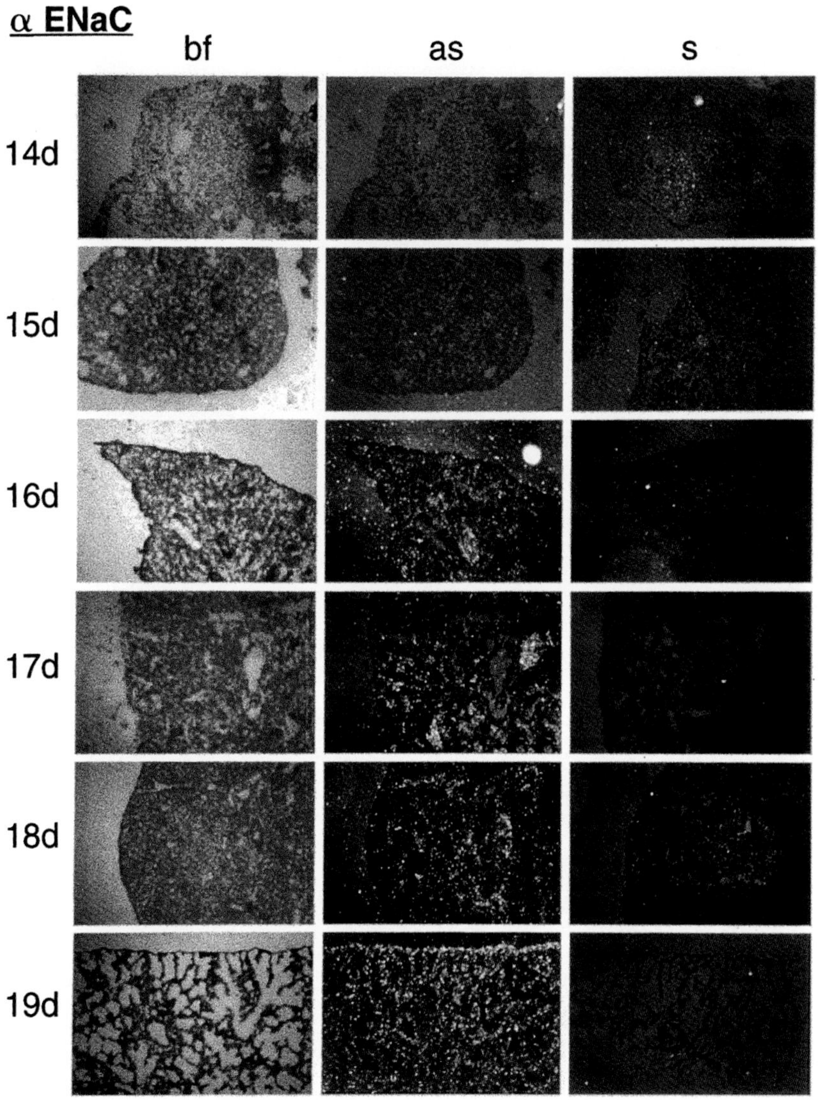

CHAPTER 12 FIGURE 9 *In situ* hybridization study of αENaC expression in fetal mouse lung. Lung sections from gestational ages 14 d through 19 d are shown. Sections were hybridized with antisense (middle column) or sense (right column) ^{35}S-labeled riboprobes and stained with H&E. Bright field (left column) micrographs of same fields shown for each antisense dark field image. Magnification ×40. (Reprinted with permission from Talbot *et al.* (1999). Quantitation and localization of ENaC subunit expression in fetal, newborn and adult mouse lung. *Am. J. Resp. Crit. Care Med.* **20,** 398–406. Official Journal of the American Thoracic Society. © American Lung Association.)

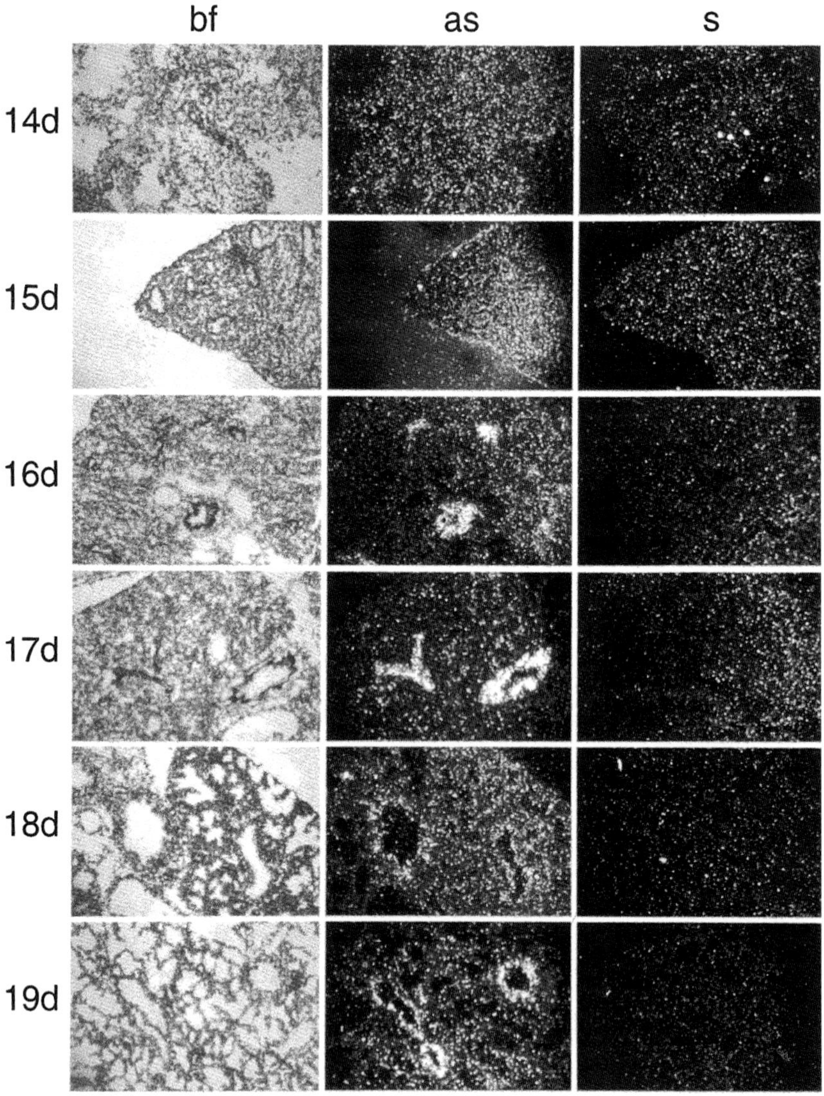

CHAPTER 12 FIGURE 10 *In situ* hybridization study of βENaC expression in fetal mouse lung. Lung sections from gestational ages 14 d through 19 d are shown. Sections were hybridized with antisense (middle column) or sense (right column) ^{35}S-labeled riboprobes and stained with H&E. Bright field (left column) micrographs of same fields are shown for each antisense dark field image. Magnification ×40. (Reprinted with permission from Talbot *et al.* (1999). Quantitation and localization of ENaC subunit expression in fetal, newborn and adult mouse lung. *Am. J. Resp. Crit. Care Med.* **20,** 398–406. Official Journal of the American Thoracic Society. © American Lung Association.)

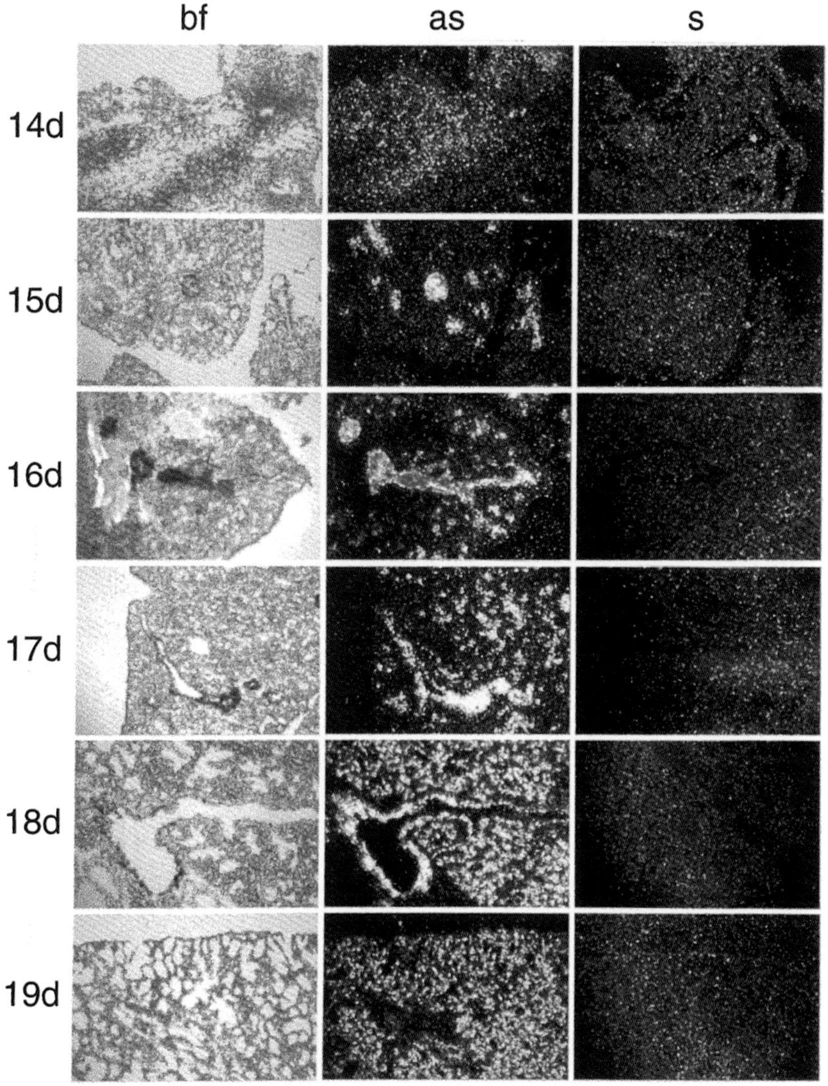

CHAPTER 12 FIGURE 11 *In situ* hybridization study of γENaC expression in fetal mouse lung. Lung sections from gestational ages 14 d through 19 d are shown. Sections were hybridized with antisense (middle column) or sense (right column) ^{35}S-labeled riboprobes and stained with H&E. Bright field (left column) micrographs of same fields are shown for each antisense dark field image. Magnification ×40. (Reprinted with permission from Talbot *et al.*(1999). Quantitation and localization of ENaC subunit expression in fetal, newborn and adult mouse lung. *Am. J. Resp. Crit. Care Med.* **20,** 398–406. Official Journal of the American Thoracic Society. © American Lung Association.)

WILD-TYPE CARBOXY TERMINUS OF β-hENaC

...^{559}KSLRQRRAQASYAGPPTVAELVEAHTNFGFQPDTAPRSPNTGPYPSEQALPIPGTPPPNYDSLRLQPLDVIESDSEGDAI $^{640\ (637)}$*

CARBOXY TERMINUS β-hENaC TRUNCATION MUTATIONS

K100 ...KSLRQR $^{565\ (564)}$

K175 ...KSLRQRRAQASYAGPPTVAELVEAHTNFGI $^{590\ (589)}$*

K176 ...KSLRQRRAQASYAGPPTVAELVEAHTNFGFQPDT $^{594\ (592)}$gppqpqphwalpg*

K101 ...KSLRQRRAQASYAGPPTVAELVEAHTNFGFQPDTAP $^{596\ (594)}$aaptlgptpvsrpcpsqaprpptmtpcvcsrwtsssltvrbmps*

β579del32...KSLRQRRAQASYAGPPTV ^{578}-----------LPA ^{581}*

CARBOXY TERMINUS β-hENaC POINT MUTATIONS

P616L...KSLRQRRAQASYAGPPTVAELVEAHTNFGFQPDTAPRSPNTGPYPSEQALPIPGTPP**L**$^{618\ (616)}$NYDSLRLQPLDVIESDSEGDAI $^{640\ (637)}$*

Y618H...KSLRQRRAQASYAGPPTVAELVEAHTNFGFQPDTAPRSPNTGPYPSEQALPIPGTPPPN**H** $^{620\ (618)}$DSLRLQPLDVIESDSEGDAI $^{640\ (637)}$*

T594M...KSLRQRRAQASYAGPPTVAELVEAHTNFGFQPD**M** ^{594}APRSPNTGPYPSEQALPIPGTPPPNYDSLRLQPLDVIESDSEGDAI $^{640\ (637)}$*

CHAPTER 19 FIGURE 1 Carboxy-terminal domains of the β subunit of human ENaC in Liddle's syndrome. The wild-type sequence from amino acid positions 559–640 (numbered according to the GenBank protein database) is shown at the top of the figure. A highly conserved Nedd4 (PY) binding domain is highlighted in red. Numbers at the beginning of each sequence, or in parentheses, indicate the numbering sequence employed in the original publication describing the mutation. Two classes of mutant producing Liddle's syndrome are shown: the truncation mutations (K100 [original Liddle's mutation], K175, K176, K101, and β579dc132), and point or missense mutations (P616L, Y618H, and T594M). The truncation mutations all have premature stop codons introduced into the translated sequence. Pedigrees K176 and K101 have frameshift mutations followed by new stop codons (indicated by asterisks). β579del32 has a deletion of 32 nucleotides (1735–1766) of the complementary DNA, resulting in a loss of 11 amino acids (Δ579–589), and the introduction of a new stop codon following the translation of three different amino acids (LPA). The point mutations P616L, Y618H, and T594M are indicated in bold. Two of these mutations are located within the PY motif. Data taken from Skimkets *et al.* (1994), Tamura *et al.* (1996), Su *et al.* (1996), and Jeunemaitre *et al.* (1997).

the airways of the submucosal glands is species specific. Humans have submucosal glands in trachea and bronchi; rat and mouse submucosal glands are confined to the proximal most regions of their tracheas. Additionally, there are numerous glands within the epithelium lining the nasal passages. Human submucosal glands are known to have a subpopulation of cells that strongly express CFTR, both within the gland acinar (i.e., the serous cells) and ductular epithelium (Englehardt *et al.,* 1992). Primary cultures of mixed seromucous cells, but not mucous cells, secrete Cl^- in response to stimulation by appropriate agonists (Yamaya *et al.,* 1991, Finkbeiner *et al.,* 1994). Additionally, studies utilizing porcine small bronchi, which contain numerous submucosal glands, support the hypothesis that the glands are a major source of Cl^- and fluid secretion (Ballard *et al.,* 1995). That study suggests that a direct relationship may exist between the magnitude of active Cl^- and fluid secretion and the presence of submucosal glands within an airway region. However, the gland acinar and ductal epithelium may be involved in Na^+ and fluid absorption as well. Mixed seromucous cultures show an approximately 40% decrease in short circuit current in the presence of the Na^+ channel blocker amiloride (Yamaya *et al.,* 1991) and there is evidence for the expression of mRNA for some of the ENaC subunits in gland ductular and acinar epithelium as well (Burch *et al.,* 1995; Farman *et al.,* 1997).

III. REGIONAL ENaC EXPRESSION IN THE RESPIRATORY TRACT

It is reasonable to expect that in tissues which express varying levels of a given mRNA that the relative expression level of that mRNA species would be proportional to function in various cells or tissues that express that mRNA. Thus, the physiological roles of specific airway regions in maintaining the composition and volume of the ASL may be estimated from the relative levels of expression of the mRNAs encoding ENaC subunits (and other proteins involved in ion and fluid balance). The underlying assumption would be that the greater the level of expression of the three ENaC subunits, the greater the rate of Na^+ and, hence, H_2O absorption in that region. While it is known that various regions of the respiratory tract show a basal amiloride-sensitive Na^+ absorption *in vitro,* the contribution of the various regions to *in vivo* ion transport physiology and its relationship to ASL homeostasis are not known. An examination of the relative expression of the ENaC subunits along the respiratory tract may help distinguish between alternative hypotheses on where airway liquid may originate and how the lung deals with this liquid (see Fig. 1). Strong, equivalent regional expression of the three ENaC subunits would suggest a role in fluid absorption across the airway; whereas high regional expression of CFTR (or other Cl^- transporters/channels) and reciprocally less ENaC expression might suggest fluid secretion. A gradient of expression of ENaC mRNAs may be detected along the distal to proximal axis of the airway tree corresponding to a specific role in regional fluid balance. The ASL may originate in the alveolar or bronchiolar regions, either by active or passive (leak) mechanisms. If the alveolar region is the primary site of fluid secretion into the lung, Na^+ absorption (and hence, ENaC expression) would likely be greatest in the distal bronchioles. However, if fluid secretion primarily occurs in the brochioles, fluid might accumulate at a greater rate in the proximal regions of the airway as it is swept up the respiratory tract into areas with a smaller overall surface area and thus Na^+ absorption and ENaC expression might be expected to be greatest in the proximal regions of the respiratory tract. *In situ* hybridization studies on ENaC expression in three different mammalian species allow comparison of these hypotheses.

A. Distal Lung ENaC Expression

In situ hybridization studies of distal lung produce two distinct patterns of ENaC mRNA expression (Table I). ENaC expression in bronchioles is similar to that observed in aldosterone-sensitive tissues with approximately equivalent levels of expression of all three subunits (Duc *et al.,* 1994); in contrast, alveolar epithelium has a noncoordinate pattern of expression of the ENaC subunits that is characteristic of many other regions in the respiratory tract (see Subsections III.A.1 and III.B).

1. Alveolar Epithelium

Both rat and mouse alveolar epithelium express ENaC in a noncoordinate fashion (Matsushita *et al.,* 1996; Farman *et al.,* 1997; Talbot *et al.,* 1999). Within the alveolar epithelium the relative level of expression of the three subunits is distinctly different. In rats, Farman *et al.* (1997) observed distinct expression of both α and γ subunits (although γ was expressed at a slightly lower level than the α subunit); the β subunit was not detectable by *in situ* hybridization. Matsushita *et al.* (1996) also found good evidence for expression of αENaC in rat alveolar epithelium, but with little or no β and γ expression (Fig. 2). In contrast, cultured alveolar type II cells derived from rats do show mRNA expression of all three ENaC subunits by RT-PCR or RNase protection assays (O'Brodovich *et al.,* 1993; Planès *et al.,* 1997) and the level of mRNA expression changes in parallel with amiloride-sensitive ^{22}Na flux (Planès *et al.,* 1997). Adult mouse lung alveolar epithelium produces a pattern that was similar to that observed by Farman and

TABLE I

Relative Regional Expression Levels of α, β, and γ ENaC in Human, Rat, and Mouse Respiratory Tract as detected by *in situ* Hydridization

Tissue	α	β	γ
Nasel epithelia			
Human[a]	++++	++	+
Rat	++	−	+
Mouse	+++	ND	+++
Nasal gland ducts			
Human	+++	ND	ND
Rat	+++	+++	+++
Mouse	+++	ND	+
Nasal gland acini			
Human	++	ND	ND
Rat	++	+/−	+
Mouse	+/−	ND	−
Trachea/bronchus			
Human	++++	++	+/−
Rat	+++	+/−	++
Mouse	++	++	++
Airway gland ducts			
Human	++++	++	−
Rat	+++	+/−	++
Mouse	ND	ND	ND
Airway gland acini			
Human	+++	+/−	−
Rat	+++	+/−	++
Mouse	ND	ND	ND
Bronchiole			
Human	ND	ND	ND
Rat	++	++	++
Mouse	++++	+++	++++
Alveolar acini			
Human	ND	ND	ND
Rat[b]	+++	+/−	++
Mouse[b]	++	+	+++

Note. ND, tissue not examined by *in situ* hybridzation.

[a] Comparison derived from expression levels observed on a Northern blot.

[b] Patchy distribution suggestive of alveolar type II cell distribution.

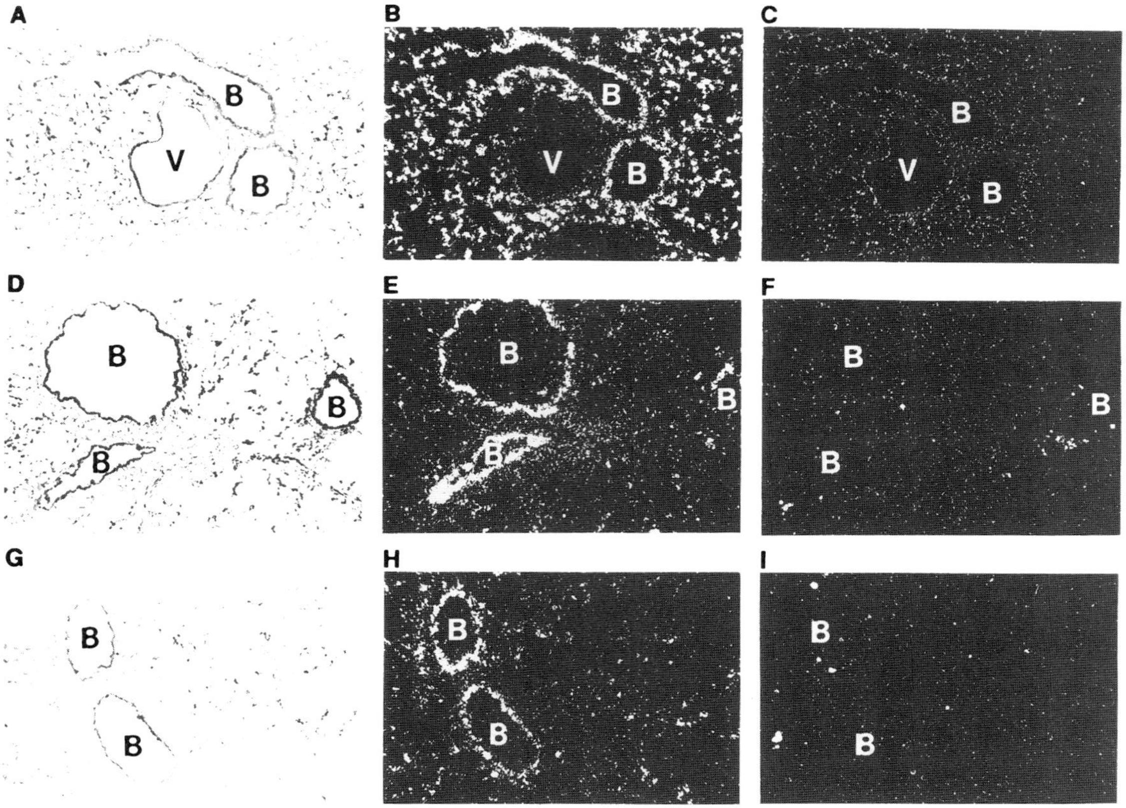

A
B
V
B
B
B
V
B
C
B
V
B
D
B
B
B
E
B
B
B
F
B
B
B
G
B
B
H
B
B
I
B
B

colleagues in rats: strong, but punctate expression of the α and γ subunits with low, diffuse β subunit expression (Talbot *et al.*, 1999; see Fig. 3).

In the species studied, when expression for a given ENaC subunit was detected, it was punctate, suggestive of localization to alveolar type II cells (Figs. 2 and 3). Yue *et al.* (1995) found that in rat alveoli all αrENaC positive cells (by *in situ* hybridization) were cuboidal and located in the alveolar corners, consistent with the localization of alveolar type II cells. Additionally, αrENaC has a pattern of expression similar to SP-C (Matsushita *et al.*, 1996), a protein known to be expressed only in type II cells in adult rat lung (Kalina *et al.*, 1992); however, Matsushita and his colleagues found that αrENaC mRNA expression is more diffuse than that of SP-C mRNA and may suggest that ENaC expression also occurs in type I cells.

2. Bronchiolar Epithelium

Bronchiolar epithelial expression of ENaC subunit mRNAs contrasts sharply relative to ENaC expression in other regions of the airways in that it is one of the very few airway regions that expresses approximately equivalent amounts of the three ENaC subunits (see Table I). This type of expression pattern is common in aldosterone-sensitive tissues, such as the distal nephron and colon (Duc *et al.*, 1994) and would be expected for maximal Na^+ current, based on the physiological evidence of *Xenopus* oocytes studies (Canessa *et al.*, 1994).

In rat (Fig. 2) and mouse (Fig. 3, see color insert) bronchiolar epithelium the levels of the α, β and γ ENaC subunits are all expressed in approximately equivalent amounts (Matsushita *et al.*, 1996; Farman *et al.*, 1997; Talbot *et al.*, 1999). Human bronchiolar epithelium shows the same pattern of equivalent levels of ENaC subunit expression (Rochelle *et al.*, 1998a). Farman and colleagues (1994) noted that the relative levels of expression in the bronchioles are similar to those observed in rat renal distal nephron and salivary gland ducts using the same 3' untranslated region probes (Duc *et al.*, 1994).

Equivalent levels of expression do not, however, imply homogenous expression throughout the bronchiolar epithelium. Small bronchioles in rat

---

FIGURE 2 *In situ* hybridization of αENaC (A–C), βENaC (D–F), and γENaC (G–I) in normal adult rat lung. Bright field micrographs (A, D, G) show bronchioles (B) and blood vessels (V). Dark field micrographs of the same fields (B, E, H) demonstrate hybridization of antisense probes for all three subunits to bronchioles. Sense probes (C, F, I) show only nonspecific hybridization in adjacent serial sections. αrENaC from paraffin-embedded sections; β and γrENaC from frozen sections. (Reprinted from Matsushita *et al.*, 1996, with permission.)

airways showed cellular heterogeneity of expression (Farman *et al.,* 1997); approximately 30% of the cells in these bronchioles produced a greater intensity of staining for the three ENaC subunits. It is unknown whether the cells expressing the greater level of ENaC are ciliated cells or nonciliated (Clara) cells. However, larger bronchioles, which lack Clara cells, expressed the ENaC subunits in equivalent amounts uniformly throughout the epithelium (Farman *et al.,* 1997), suggesting ciliated cells may be the site of ENaC expression. Equivalent expression of all three ENaC subunits at relatively high intensity was also observed in mouse small-to-medium airways (Talbot *et al.,* 1999). Intensity of expression in adult lung was greater in medium-sized bronchioles than in terminal bronchioles. The pattern of expression in the mouse was relatively uniform throughout the airway epithelium at all levels observed.

B. Proximal Lung and Nasal Epithelium ENaC Expression

Expression of ENaC subunit mRNA in upper airways produces a significantly different pattern from that found in bronchiolar epithelium. Similar to the alveolar epithelium, noncoordinate expression of the three ENaC subunits is typically observed. The alpha subunit is generally the most intensely expressed; however, the relative expression of βENaC or γENaC is species dependent.

1. Bronchial and Tracheal ENaC Expression

Human proximal airways surface epithelium produces a noncoordinate expression pattern for the ENaC subunits with αENaC the most strongly expressed, a low to moderate amount of βENaC, and very low γENaC expression. This pattern was observed using *in situ* hybridization techniques in bronchial sections (Fig. 4, see color insert), Northern analysis of tracheal surface epithelium, and primary cultures (Burch *et al.,* 1995). The Northern analysis of the human tracheal surface epithelium identified two bands, suggesting the presence of multiple messages for αENaC. In contrast to the pattern observed in human (*i.e.,* $\alpha >> \beta > \gamma$), rat trachea produced relative expression levels of the ENaC subunits of $\alpha > \gamma >> \beta$ (Farman *et al.,* 1997; see Fig. 5). Interestingly, these tracheal tissue sections also contained sections of the esophagus which expressed high and equivalent levels of all three ENaC subunits. Also in rats, Matsushita and colleagues (1996) noted that the intensity of expression of the three ENaC subunits decreased along the distal to proximal axis of the respiratory tract, with larger bronchi having a lower relative level of expression than the smaller bronchi and bronchioles. In that study, tracheal epithelium had the lowest level of expression,

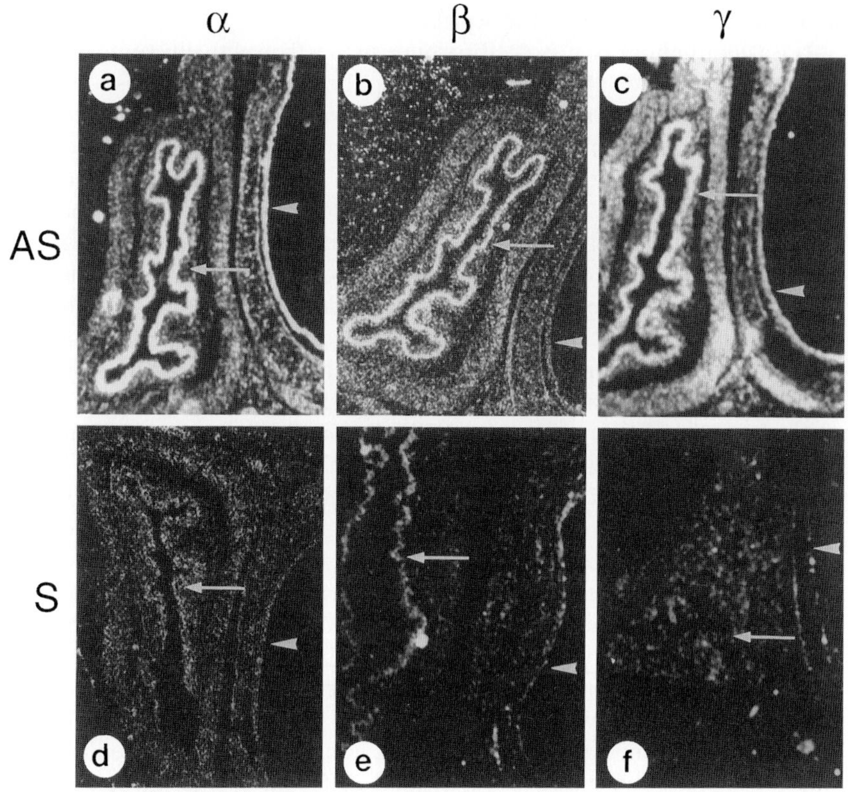

FIGURE 5 Dark field micrographs of transverse sections of trachea and esophagus after hybridization with antisense (top row) or sense (bottom row) riboprobes for α (A, D), β (B, E), or γENaC (C, F). Ventral side of neck is towards the right of each micrograph. Clear signal is evident over tracheal epithelium (arrowheads) for α and γENaC, whereas β signal is not different from the background. In contrast, the epithelium and muscularis mucosa of esophagus (arrows) express all three ENaC subunits at high, approximately equivalent levels. Magnification, X50. (Reprinted from Farman *et al.,* 1997, with permission).

with αENaC diffusely expressed in the trachea and β and γENaC not detected by *in situ* hybridization; however, all three subunits were clearly detected in tracheal epithelium by RT-PCR. A similar pattern of decreasing levels of expression can be seen along the distal to proximal axis of the mouse respiratory tract (C. R. Talbot and R. C. Boucher, unpublished; Rochelle *et al.,* 1998b). Figure 6 (see color insert) shows a longitudinal cross section of the mouse lung probed for the β and γENaC subunits and the intensity of expression shows a clear gradient along the major airway.

2. ENaC Expression in Nasal Epithelia

Nasal epithelium is a Na^+ absorbing tissue that actively removes Na^+ from the thin fluid layer covering the epithelium to produce iso-osmotic fluid absorption. This region of proximal airway shows many of the characteristics of Na^+ absorbing epithelia, including amiloride sensitivity, electrogenic Na^+ transport, and a serosal positive transepithelial potential difference (Willumsen and Boucher, 1991). Fluid absorption is likely produced by transcellular Na^+ transport in parallel with a passive paracellular Cl^- absorption. However, this tissue also shows some differences in its electrophysiological profile relative to classic Na^+ absorbing epithelia. It is interesting that these tissues (as above) show expression patterns of the ENaC subunits different from those observed in aldosterone-sensitive epithelia (*i.e.,* noncoordinate expression of the three subunits).

Noncoordinate expression of the ENaC subunits in nasal epithelium has been observed by a variety of techniques. RNase protection assays (RPA) and Northern blots of mRNA derived from human nasal epithelium produce an expression pattern of the ENaC subunits similar to that seen in human trachea and bronchus: $\alpha>\beta>>\gamma$ (Burch *et al.,* 1995). Human nasal surface epithelium also shows strong αENaC expression by *in situ* hybridization. Rat ciliated respiratory nasal epithelium also showed a predominate expression of the α subunit (Farman *et al.,* 1997), however, while the γ subunit was clearly present in this epithelium, it was expressed at levels less than the α subunit; rat βENaC showed no visible staining by *in situ* hybridization. Nonciliated squamous (*i.e.,* nonrespiratory) nasal epithelium did not express detectable levels of mRNA for any of the three subunits (Farman *et al.,* 1997). Mouse nasal epithelium potentially showed yet a different pattern (Fig. 7, see color insert) with strong, equivalent expression of the α and γ subunits (the β subunit was not done).

C. Expression in Submucosal Glands

A key advantage derived from using the technique of *in situ* hybridization is localization of mRNA expression to specific structures within a complex tissue. An example of this can be seen in the localization of ENaC expression to submucosal gland duct and acinar epithelium. In experiments on human proximal lung, Burch *et al.* (1995) found that in addition to expression of the three ENaC subunits in surface epithelium, there was also evidence of ENaC expression within bronchial gland duct and acinar epithelium (Fig. 4). Again, the alpha subunit was the most strongly expressed, in both ducts and acinar epithelia. The beta subunit was detected at a much lower signal intensity, especially in the acinar epithelium (where a very low, but significant, signal above background was observed); however, mRNA for the gamma subunit was not detected by *in situ* hybridization. Rat tracheal gland duct and acini also express ENaC (Farman *et al.,* 1997). In this species, α and γ ENaC are detected at reasonably high levels; however, there is little or no βENaC expression in either tracheal gland duct or acini (Fig. 8). Rat nasal gland acini have a pattern similar to the tracheal gland acini, with slightly less alpha expression; however, rat nasal submucosal gland ducts produce a different pattern than that observed in tracheal gland ducts — approximately equivalent expression of all three subunits. Mouse nasal submucosal gland ducts have strong expression of αENaC and little or no expression of γENaC; gland acini show no discernible expression of either of these two subunits (Fig. 7).

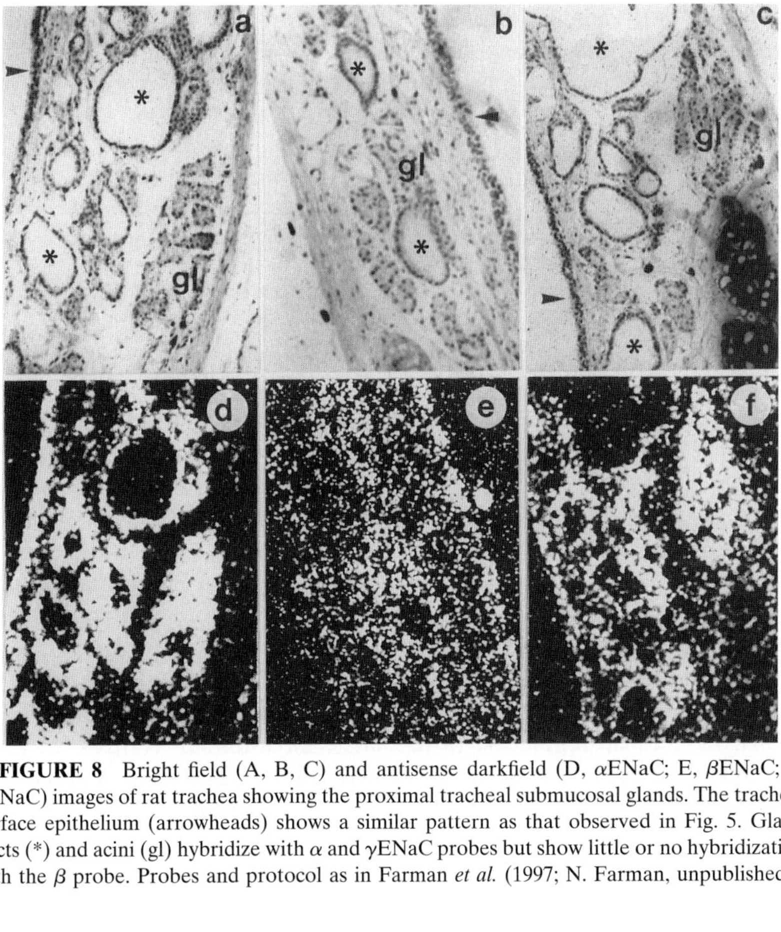

FIGURE 8 Bright field (A, B, C) and antisense darkfield (D, αENaC; E, βENaC; F, γENaC) images of rat trachea showing the proximal tracheal submucosal glands. The tracheal surface epithelium (arrowheads) shows a similar pattern as that observed in Fig. 5. Gland ducts (*) and acini (gl) hybridize with α and γENaC probes but show little or no hybridization with the β probe. Probes and protocol as in Farman *et al.* (1997; N. Farman, unpublished).

IV. DEVELOPMENTAL EXPRESSION OF ENaC IN THE LUNG

Fetal lung is a secretory organ. Liquid secretion results from the active transport of Cl^- from the interstitium into the lumen of the developing lung. This fluid must be rapidly cleared at birth to allow for aerial respiration (Strang, 1991). Fluid absorption at the time of birth has been shown to be mediated by amiloride-sensitive Na^+ transport (Olver *et al.,* 1986, O'Brodovich *et al.*, 1991) and specifically by ENaC. Mice that were deficient in αENaC were unable to clear liquid from their lungs at birth and died within 48 h (Hummler *et al.,* 1996). The switch from net fluid secretion to net fluid absorption occurs very rapidly (within minutes) at birth and is thought to be mediated by β-adrenergic stimulation (Brown *et al.,* 1983). Because of the rapid induction of fluid absorption observed at birth and because Na^+ absorption can be hormonally induced in a premature lung (Barker *et al.,* 1988), ENaC is likely present prior to birth but maintained in an inactive form. Studies have been performed examining the change in the expression of the ENaC subunits during late gestation on several species by both *in situ* hybridization and Northern analysis.

A. Fetal Expression in the Developing Rat Lung

The three ENaC subunits are differentially expressed in developing rat lung. In Northern analysis of whole lung, αENaC is first observed in the canalicular stage of lung development at approximately 18–19 days gestational age (GA, full term = 22 days) and the level of expression increases until birth (O'Brodovich *et al.,* 1993; Tchepichev *et al.,* 1995). The relative level of αENaC decreases immediately after birth and remains lower than that observed at 22 days GA at least through 2 weeks post natally; adult expression of αENaC is similar to that at 22 days GA (O'Brodovich *et al.,* 1993). The observed changes in αENaC expression pre- and postnatally correspond with endogenous fetal corticosteroid levels and the postnatal resistance to glucocorticoids observed in the neonate (Henning, 1978). Experimental application of the corticosteroid dexamethasone, either by maternal infusion (O'Brodovich *et al.,* 1993; Tchepichev *et al.,* 1995) or direct application to isolated fetal distal lung cultures (Champigny *et al.,* 1994), induces up-regulation of αENaC expression. In contrast to the α subunit, β and γENaC are not expressed until 20–21 days GA and do not greatly increase during the remaining time of gestation. These two subunits do show a postnatal increase in their relative expression levels; however, expression at 2 weeks of age was still lower than adult levels (Tchepichev *et al.,* 1995). Additionally, β and γENaC expression does not appear to be regulated by steroid hormones (Tchepichev *et al.,* 1995).

B. Fetal Expression in the Developing Mouse Lung

Fetal expression of ENaC in the developing mouse lung contrasts sharply to that observed in the rat (Talbot *et al.,* 1999). Low levels of all three subunits were observed by Northern analysis of whole, pooled fetal lung by 15 days GA (full term = 19 days). However, α and βENaC expression was not detected by *in situ* hybridization on individual fetal mouse lungs until 16 days gestational age and then, primarily in the central bronchi (Figs. 9 and 10, see color insert; Table II). In contast, expression of the γENaC is evident as early as 14 days GA. Very intense expression of this subunit was observed in the lobar airways and primitive distal tubules by 15 days GA (Fig. 11, see color insert; Table II). Both the α and γ subunits show strong expression throughout the developing airways and into the developing acinar structures by 18 days GA, with the gamma subunit producing the most intense signal throughout the developing lung. The signal for the β subunit is similar to that for α and γENaC in the developing

TABLE II

Ontogeny of Mouse ENaC Submit Expression by *in Situ* Hybridization[a]

	Fetal						Postnatal	
Gestational age	14d	15d	16d	17d	18d	19d	1 week	adult
αENaC								
Medium airway	−	−	++	+++	++	+++	++++	++++
Small airway	−	−	+	++	+	++	+++	+++
Alveolar acinus	NA	NA	NA	−	+	+	+	++[a]
βENaC								
Medium airway	−	−	+++	+++	++	+++	+++	++++
Small airway	−	−	+	−	+	++	++	+++
Alveolar acinus	NA	NA	NA	−	+/−	+	+	+
γENaC								
Medium airway	+/−	+++	+++	++++	++++	++++	++++	++++
Small airway	−	+	++	+++	+++	+++	++	+++
Alveolar acinus	NA	NA	NA	++	+++	+++	+++[b]	+++[b]

Note. NA, not applicable, since the acinus has not developed by these gestational ages.

[a] Reprinted with permission from Talbot *et al.* (1999). Quantitation and localization of ENaC subunit expression in fetal, newborn and adult mouse lung. *Am. J. Resp. Crit. Care Med.* **20,** 398–406. Official Journal of the American Thoracic Society. © American Lung Association.

[b] Patchy distribution in a pattern suggestive of alveolar type II localization.

airways, but βENaC is only faintly observed in the acini by the end of natural gestation. The equivalent, strong expression of all three subunits in the developing airways near the end of gestation suggests these regions are the major sites of Na^+ and fluid reabsorption at the time of birth. Indeed, these changes closely parallel those observed for the $\alpha1$ and $\beta1$ isoforms of the Na^+, K^+-ATPase, both in gestational timing and specific localization in the developing airways (Crump *et al.,* 1995).

V. EXPRESSION PATTERNS AND ENaC FUNCTION

The above studies show that the message encoding the amiloride-sensitive epithelial sodium channel, ENaC, is expressed throughout the respiratory tract in a region and cell-specific pattern, thus, ENaC is likely involved in maintaining the airway surface liquid layer. However, before conclusions are drawn regarding the physiological significance of variations in ENaC subunit mRNA expression, there are several caveats that may influence the interpretation of the data that must be examined. (1) The different expression patterns observed between subunits may be due to probe differences. While there are likely differences between the probes, this effect is ameliorated by the fact that the same probes have been used in a variety of tissues within each species studied and these probes have detected high levels of expression within some tissues (i.e., kidney tubules and colon; Duc *et al.,* 1994) and within the same tissue section where low expression was observed (i.e., esophagus versus tracheal expression of βrENaC; Farman *et al.,* 1997). (2) There may be cell-specific factors that limit the ability to detect subunit mRNA by *in situ* hybridization. This is suggested by Northern analyses and RT-PCR assays that detect approximately equivalent amounts of the three subunits while *in situ* hybridization experiments suggest differential expression (see Matsushita *et al.,* 1996). These types of experiments, however, must also be cautiously interpreted. In the case of distal lung samples, these may be primarily alveolar cells, but bronchiolar epithelium, which show approximately equivalent expression of the ENaC subunits, may be contaminating the sample. When mRNA from a specific cell type, human tracheal surface epithelium, was isolated, Burch *et al.* (1995) found comparable expression levels by Northern analysis and RPA relative to those observed by *in situ* hybridization with two different sets of riboprobes. (3) Low/absent expression of one of the subunits may be due to tissue or region specific expression of a subunit splice variant or novel isoform. There are splice variants known for the α and β subunits that may lead to functional differences in the expressed protein (Voilley *et al.,* 1994; Thomas *et al.,* 1998; Tucker *et al.,* 1998). Voilley and colleagues (1994) have identified a βhENaC splice variant in the lung that deletes a 463 bp fragment from the 3' region. Farman and associates (1997) tested this hypothesis and found no difference in relative signal for βrENaC using two different riboprobes that hybridized to either the 5' or 3' ends, respectively, of the beta subunit mRNA. It is, however, still possible that a unique isoform may exist in regions where one of the three subunits appears to be "limiting." This is the case for ENaC expression in the brain, pancreas, testis, and ovary where an analog of the alpha subunit, δENaC, substitutes to form an amiloride-sensitive Na^+ channel in these tissues (Waldmann *et al.,* 1995). (4) Factors such a post-translational modifications may influence the amount of functional protein at the apical cell membrane. This is likely to be a critical factor if one tries to correlate the level of expression with physiological function if inactive channels are stored in subapical vesicles.

Given the above caveats, relative expression levels and tissue specific patterns of expression may provide some information on the physiological function of Na^+ absorption in various epithelia. Aldosterone-sensitive tissues (i.e., distal renal tubules, distal colon, and salivary gland ducts) express mRNA encoding α, β, and γENaC in approximately equivalent amounts (Duc *et al.,* 1994); however, respiratory epithelia reveal distinct, region-specific, noncoordinated expression of the three subunits. This type of noncoordinated expression is common to all three mammalian species currently examined by *in situ* hybridization (Table I); however, the actual pattern is species specific. In adult tissue, in all three species and all regions examined, the alpha subunit was the most intensely expressed, at levels equal to or much greater than the other subunits within a given region. The relative levels of the β and γ subunits varied between species: human proximal airways showed relative expression levels of a $\alpha >> \beta > \gamma$, whereas rat and mouse lungs generally expressed the ENaC subunits as $\alpha \geq \gamma > \beta$. Some regions also appear not to express one of the subunits (*i.e.,* γENaC was not detected in human gland acini and βENac was nondetectable, or at very low levels, in rat nasal epithelia, trachea, gland acini, and alveolar epithelium). Similarly, noncoordinate expression was observed during gestational development of the lungs in rats and mice. Again, there were species specific differences. The rat follows the typical adult pattern of αENaC predominant (Tchepichev *et al.,* 1995), whereas in the mouse, γENaC was predominant throughout gestation to birth (Talbot *et al.,* 1999). The significance of the variation in ENaC subunit expression within the respiratory tract is difficult to interpret without a better understanding of how the ENaC subunits assemble to form a functional channel.

The original studies describing the cloning and function of ENaC using the *Xenopus* oocyte expression system (Canessa *et al.,* 1993, 1994) found that αENaC alone will form a Na^+ specific, amiloride-sensitive channel but that to get maximum current all three subunits must be expressed. Two stoichiometric models have currently been proposed for subunit assembly: 2α:1β:1γ (Firsov *et al.,* 1998; Kosari *et al.,* 1998) and 3α:3β:3γ (Snyder *et al.,* 1998). Both of these models again suggest that all three subunits must be present to produce a maximally functional channel. Bronchiolar epithelium of all three species does express approximately equivalent amounts of the three subunits and may be the primary site for Na^+ absorption along the airways. There is, however, evidence that the relative amounts of ENaC mRNA expressed within the cells along the airways decreases along a distal (maximal expression) to proximal axis in the airways (Fig. 6; Matsushita *et al.,* 1996; Rochelle *et al.,* 1998b) with the lowest level of expression occurring within the trachea. This supports the hypothesis that maximal Na^+ and fluid absorption may occur in the distal most airways (Fig. 1) to counteract excessive fluid accumulation in the alveoli.

Coordinate expression of the three subunits in respiratory epithelia is the exception rather than the rule. However, expression of α and βENaC or α and γENaC mRNA in oocytes does produce functional channels, but with a much lower current (Canessa *et al.,* 1994). Further studies have examined these $\alpha\beta$ENaC or $\alpha\gamma$ENaC channels to determine whether the three subunits of ENaC can associate in various combinations to form channels with distinct characteristics, as is common for other heteromultimeric channels (Nakanishi *et al.,* 1990; Bradley *et al.,* 1991; Chen *et al.,* 1993). McNicholas and Canessa (1997) found that $\alpha\beta$ENaC and $\alpha\gamma$ENaC channels have different functional properties that may correlate with the known physiological differences of amiloride-sensitive Na^+ channels that do not correspond to the classical ENaC properties. Firsov *et al.,* (1998) have suggested that in spite of the preferential assembly in *Xenopus* oocytes of a $\alpha_2\beta\gamma$ channel, native tissue that predominately express α and β or α and γENaC may do so in a manner that conserves the number of α subunits and the heterotetrameric structure and form $\alpha_2\beta_2$ or $\alpha_2\gamma_2$ channels. This corresponds with the 1:1 ratio of subunits McNicholas and Canessa (1997) suggested as the likely stoichiometry for the two subunit channels. Thus, the differential expression observed throughout the airways may have functional significance and may represent a mechanism to regulate Na^+ absorption in different tissues. Based on the differential expression of the ENaC subunit mRNA, other regions of the airway may be involved in Na^+ (and fluid) absorption as well, but with ENaC derived channels that exhibit different physiological and pharmacological characteristics. Thus, like other hetero-multimeric channels, ENaC may have the potential for complex regulation by differential subunit expression during development or in response to the specific neurohumoral environment within a given region within the respiratory tract.

References

Al Bazzaz, F. J. (1986). Regulation of salt and water transport across airway mucosa. *Clin. Chest Med.* **7,** 259–272.

Al Bazzaz, F. J. (1994). Regulation of Na and Cl transport in sheep distal airways. *Am. J. Physiol.* **267,** L193–L198.

Ballard, S. T. and Taylor, A. E. (1994). Bioelectric properties of proximal bronchiolar epithelium. *Am. J. Physiol.* **267,** L79–L84.

Ballard, S. T., Fountain, J. D., Inglis, S. K., Corboz, M. R., and Taylor, A. E. (1995). Chloride secretion across distal airway epithelium: Relationship to submucosal gland distribution. *Am. J. Physiol.* **268,** L526–L531.

Barker, P. M., Markiewicz, M., Parker, A., Walters, D. V., and Strang, L. B. (1988). Synergistic action of triiodothryonine and hydrocortisone on epinephrine-induced reabsorption of fetal lung liquid. *Pediatr. Res.* **27,** 588–591.

Boucher, R. C. (1994a). Human airway ion transport. Part 1. *Am. J. Respir. Crit. Care Med.* **150,** 271–281.

Boucher, R. C. (1994b). Human airway ion transport. Part 2. *Am. J. Respir. Crit. Care Med.* **150,** 581–593.

Bradley, J., Davidson, N., Lester, H. A., and Zinn, K. (1991). Heterotrimeric olfactory cyclic nucleotide-gated channels: A subunit that confers increased sensitivity to cAMP. *Proc. Natl. Acad. Sci. USA* **91,** 8890–8894.

Brown, M. J., Olver, R. E., Ramsden, C. A., Strang, L. B., and Walters, D. V. (1983). Effects of adrenaline and of spontaneous labour on the secretion and absorption of lung liquid in the fetal lamb. *J. Physiol. (London)* **344,** 137–152.

Burch, L. N., Talbot, C. R., Knowles, M. R., Canessa, C., Rossier, B., and Boucher, R. C. (1995). Relative expression of the human epithelial Na^+ channel (ENaC) subunits in normal and cystic fibrosis airways. *Am. J. Physiol.* **269,** C511–C518.

Canessa, C. M., Horisberger, J.-D., and Rossier, B. C. (1993). Epithelial sodium channel is related to proteins involved in neurodegeneration. *Nature (London)* **361,** 467–470.

Canessa, C. M., Schild, L., Buell, G., Thorens, B., Gautschi, I., Horisberger, J.-D., and Rossier, B. C. (1994). Amiloride-sensitive epithelial Na^+ channel is made of three homologous subunits. *Nature (London)* **367,** 463–467.

Champigny, G., Voilley, N., Lingueglia, E., Friend, V., Barbry, P., and Lazdunski, P. (1994). Regulation of expression of the lung amiloride-sensitive Na^+ channel by steroid hormones. *EMBO J.* **13,** 2177–2181.

Chen, T. Y., Peng, Y. W., Dahallan, R. S., Ahamed, B., Reed, R. R., and Yau, K. W. (1993). A new subunit of the cyclic nucleotide gated cation channel in retinal rods. *Nature (London)* **362,** 764–767.

Chinet, T. C., Fullton, J. M., Yankaskas, J. R., Boucher, R. C., and Stutts, M. J. (1993). Sodium-permeable channels in the apical membrane of human nasal epithelium. *Am. J. Physiol.* **265,** C1050–C1060.

Crump, R. G., Askew, G. R., Wert, S. E., Lingrel, J. B., and Joiner, C. H. (1995). In situ localization of sodium-potassium ATPase mRNA in developing mouse lung epithelium. *Am. J. Physiol.* **269,** L299–L308.

Diamond, J. M. (1979). Osmotic water flow in leaky epithelia. *J. Membr. Biol.* **51,** 195–216.

Duc, C., Farman, N., Canessa, C. M., Bonvalet, J., and Rossier, B. C. (1994). Cell specific expression of epithelial sodium channel alpha, beta and gamma subunits in aldosterone-responsive epithelia from the rat: Localization by in situ hybridization and immunocytochemistry. *J. Cell Biol.* **127,** 1907–1921.

Englehardt, J. F., Yankaskas, J. R., Ernst, S. A., Yang, Y., Marino, C. R., Boucher, R. C., and Wilson, J. M. (1992). Submucosal glands are the predominant site of CFTR expression in human bronchus. *Nat. Genet.* **2,** 240–247.

Farman, N., Talbot, C. R., Boucher, R., Fay, M., Canessa, C., Rossier, B., and Bonvalet, J. P. (1997). Noncoordinated expression of α-, β-, and γ-subunit mRNAs of epithelial Na^+ channel along rat respiratory tract. *Am. J. Physiol.* **272,** C131-C141.

Finkbeiner, W. E., Shen, B., and Widdicombe, J. H. (1994). Chloride secretion and function of serous and mucous cells of human airway glands. *Am. J. Physiol.* **267,** L206–L210.

Firsov, D., Gautschi, I., Merillat, A.-M., Rossier, B. C., and Schild, L. (1998). The heterotetrameric architecture of the epithelial sodium channel (ENaC). *EMBO J.* **17,** 344–352.

Henning, S. J. (1978). Plasma concentrations of total and free corticosterone during development in the rat. *Am. J. Physiol.* **235,** E451–E456.

Hummler, E., Barker, P. M., Gatzy, J. T., Beerman, F., Verdumo, C., Schmidt, A., Boucher, R. C., and Rossier, B. C. (1996). Early death due to defective neonatal lung liquid clearance in αENaC-deficient mice. *Nat. Genet.* **12,** 325–328.

Kalina, M., Mason, R. J., and Shannon, J. M. (1992). Surfactant protein C is expressed in alveolar type II cells but not in Clara cells of rat lung. *Am. J. Respir. Cell Mol. Biol.* **6,** 594–600.

Kilburn, K. H. (1968). A hypothesis for pulmonary clearance and its implications. *Am. Rev. Respir. Dis.* **98,** 449–463.

Kosari, F., Sheng, S., Li, J., Mak, D. O. D., Foskett, J. K., and Kleyman, T. R. (1998). Subunit stoichiometry of the epithelial sodium channel. *J. Biol. Chem.* **273,** 13469–13474.

Mason, R. J., Williams, M. C., Widdicombe, J. H., Sanders, M. J., Misfeldt, D. S., and Berry, L. C., Jr. (1982). Transepithelial transport by pulmonary alveolar type II cells in primary culture. *Proc. Natl. Acad. Sci. U.S.A.* **79,** 6033–6037.

Matalon, S., Kirk, K. L., Bubien, J. K., Oh, Y., Hu, P., Yue, G., Shoemaker, R., Cragoe, E. J., Jr., and Benos, D. J. (1992). Immunocytochemical and functional characterization of Na^+ conductance in adult alveolar pneumocytes. *Am. J. Physiol.* **262,** C1228–C1238.

Matsushita, K., McCray, P. B., Jr., Sigmund, R. D., Welsh, M. M., and Stokes, J. B. (1996). Localization of epithelial sodium channel subunit mRNAs in adult rat lung by in situ hybridization. *Am. J. Physiol.* **271,** L332–L339.

McNicholas, C. M., and Canessa, C. M. (1997). Diversity of channels generated by different combinations of epithelial sodium channel subunits. *J. Gen. Physiol.* **109,** 681–692.

Meyrick, B., Sturgess, J. M., and Reid, L. (1969). A reconstruction of the duct system and secretory tubules of the human bronchial submucosal gland. *Thorax* **24,** 729–736.

Nakanishi, N., Schneider, N. A., and Axel, R. (1990). A family of glutamate receptor genes — evidence for the formation of heteromultimeric receptors with distinct channel properties. *Neuron* **5,** 37–45.

Nielson, D. W. (1986). Electrolyte composition of pulmonary alveolar subphase in anesthetized rabbits. *J. Appl. Physiol.* **60,** 972–979.

O'Brodovich, H., Hannam, H., and Rafii, B. (1991). Sodium channel but neither Na^+-H^+ nor Na+-glucose symport inhibitors slow neonatal lung water clearance. *Am. J. Respir. Cell Mol. Biol.* **5,** 377–384.

O'Brodovich, H., Canessa, C., Ueda, J., Rafii, B., Rossier, B. C., and Edelson, J. (1993). Expression of the epithelial Na^+ channel in the developing rat lung. *Am. J. Physiol.* **265,** C491–C496.

Olver, R. E., Ramsden, C. A., Strang, L. B., and Walters, D. V. (1986). The role of amiloride-blockable sodium transport in adrenaline-induced liquid absorption in the foetal lamb. *J. Physiol.* (*London*) **176,** 321–340.

Palmer, L. G. (1992). Epithelial Na channels: Function and diversity. *Annu. Rev. Physiol.* **53,** 51–66.

Planès, C., Blot-Chabaud, M., Friedlander, G., Farman, N., and Clerici, C. (1997). Hypoxia downregulates expression and activity of epithelial sodium channels in rat alveolar epithelial cells. *Am. J. Respir. Cell Mol. Biol.* **17,** 508–518.

Rochelle, L. G., Li, D. C., Ye, H., Talbot, C., Yankaskas, J. R., and Boucher, R. C. (1998a). Coordinated expression of ENaC, CFTR, and Na^+, $K^+/2Cl^-$ cotransporter (cotrans) in human airways and lung. *Ped. Pulm.* **17S,** 232.

Rochelle, L. G., Li, D. C., Ye, H., Talbot, C., and Boucher, R. C. (1998b). ENaC mRNa expression suggests an absorptive capacity of mouse distal lung. *Ped. Pulm.* **17S,** 232.

Rossier, B. C., Canessa, C. M., Schild, L., and Horisberger, J.-D. (1994). Epithelial sodium channels. *Curr. Opin. in Nephrol. Hypertens.* **3,** 487–496.

Smith, P. R., and Benos, D. J. (1991). Epithelial Na^+ channels. *Annu. Rev. Physiol.* **53,** 509–530.

Snyder, P. M., Cheng, C., Prince, L. S., Rogers, J. C., and Welsh, M. J. (1998). Electophysiological and biochemical evidence that DEG/ENaC cation channels are composed of nine subunits. *J. Biol. Chem.* **273,** 681–681.

Strang, L. B. (1991). Fetal lung liquid: Secretion and absorption. *Physiol. Rev.* **71,** 991–1016.

Talbot, C. R., Bosworth, D. G., Briley, E. L., Fenstermacher, D. A., Gabriel, S. E., Boucher, R. C., and Barker, P. M. (1999). Quantitation and localization of ENaC subunit expression in fetal, newborn and adult mouse lung. *Am. J. Respir. Crit. Care Med.* **20,** 398–406.

Tchepichev, S., Ueda, J., Canessa, C., Rossier, B. C., and O'Brodovich, H. (1995). Lung epithelial Na channel subunits are differentially regulated during development and by steroids. *Am. J. Physiol.* **269,** C808–C812.

Thomas, C. P., Auerbach, S., Stokes, J. B., and Volk, K. A. (1998). 5' heterogeneity in epithelial sodium channel α-subunit mRNA leads to distinct NH_2-terminal variant proteins. *Am. J. Physiol.* **274,** C1312–C1323.

Tucker, J. K., Tamba, K., Lee, Y.-J., Shen, L.-L., Warnock, D. G., and Oh, Y. (1998). Cloning and functional studies of splice variants of the α-subunit of the amiloride-sensitive Na^+ channel. *Am. J. Physiol.* **274,** C1081–C1089.

Van Scott, M. R., Davis, C. W., and Boucher, R. C. (1989). Na^+ and Cl^- transport across rabbit nonciliated bronchiolar epithelial (Clara) cells. *Am. J. Physiol.* **256,** C893–C901.

Van Scott, M. R., Chinet, T., Burnette, A., and Paradiso, A. M. (1995). Purinergic regulation of ion transport across non-ciliated bronchiolar epithelial (Clara) cells. *Am. J. Physiol.* **269,** L30–L37.

Voilley, N., Bassilana, F., Mignon, C., Merscher, S., Mattei, G., Carle, F., Lazdunski, M., and Barbry, P. (1994). Cloning, chromosomal localization and physical linkage of the β and γ subunits (SCNN 1B and SCNN 1G) of the human epithelial amiloride-sensitive sodium channel. *Genomics* **28,** 36560–36565.

Waldmann, R., Champigny, G., Bassilana, F., and Lazdunski, M. (1995). Molecular cloning and functional expression of a novel amiloride-sensitive Na^+ channel. *J. Biol. Chem.* **270,** 27411–27414.

Widdicombe, J. H. (1994). Ion and fluid transport by airway epithelium. *In* "Airway Secretion: Physiological Basis for the Control of Mucous Hypersecretion" (T. Takishima and S. Shimura, eds.), pp. 399–431. Dekker, New York.

Willumsen, N. J., and Boucher, R. C. (1991). Sodium transport and intracellular sodium activity in cultured human nasal epithelium. *Am. J. Physiol.* **261,** C319–C331.

Yamaya, M., Finkbeiner, W. E., and Widdicombe, J. H. (1991). Ion transport by cultures of human tracheobronchial submucosal glands. *Am. J. Physiol.* **262,** L485–L490.

Yue, G., Russell, W. J., Benos, D. J., Jackson, R. M., Olman, M. A., and Matalon, S. (1995). Increased expression and activity of sodium channels in alveolar type II cells of hyperoxic rats. *Proc. Natl. Acad. Sci. U.S.A.* **92,** 8418–8422.

CHAPTER 13

Inhibition of Vectorial Na^+ Transport across Alveolar Epithelial Cells by Nitrogen–Oxygen Reactive Species

Sadis Matalon,*,†,‡ Ahmed Lazrak,* and Michael D. DuVall*
*Departments of Anesthesiology, †Physiology and Biophysics, and ‡Pediatrics, University of Alabama at Birmingham, Birmingham, Alabama 35233

I. INTRODUCTION

There have been many attempts to identify the cellular mechanisms responsible for the paucity of alveolar fluid in both normal and injured lungs. Originally, it was thought that the absence of any significant amount

1063-5823/99 $30.00

of fluid and proteins in the alveolar space was due to the low permeability of the alveolar epithelium to both electrolytes and plasma proteins (Matalon and Egan, 1981), and to the existence of passive Starling forces.

However, recent information indicates that active ion transport plays a role in the reabsorption of fluid across the alveolar epithelium. Several groups have demonstrated the reabsorption of intratracheally instilled isotonic fluid or plasma from the alveolar into the interstitial space both in anesthetized animals and isolated perfused lungs (Matthay *et al.*, 1996; Olivera *et al.*, 1994). Sodium transport inhibitors, amiloride or *N*-ethyl-*N*-isopropyl amiloride (EIPA), added to the alveolar space, and inhibition of the Na^+,K^+ ATPase with ouabain, injected into the vascular compartment, decreased the rate of fluid reabsorption by about 40–50 and 70–80%, respectively (Basset *et al.*, 1987b; Goodman *et al.*, 1987; Matthay *et al.*, 1985; Olivera *et al.*, 1995; Yue and Matalon, 1997). In addition, $^{22}Na^+$ uptake and bioelectric measurements across either freshly isolated or cultured ATII cells have confirmed the presence of channel-mediated transepithelial Na^+ transport. Na^+ in the alveolar lining fluid passively diffuses into ATII cells through channels in the apical membrane. The favorable electrochemical driving force for this is maintained by the ouabain-sensitive Na^+,K^+-ATPase localized to the basolateral membrane which ultimately transports Na^+ into the interstital space (Cheek *et al.*, 1989; Hu *et al.*, 1994; Russo *et al.*, 1992). K^+ which enter the ATII cell in exchange for Na^+, exits the cell down its electrochemical gradient through K^+ channels located in the basolateral membrane. It is the active Na^+ transport across the alveolar epithelium which creates an osmotic gradient that then drives fluid movement from the alveolar to the interstitial space.

Because of their location, alveolar epithelial cells are often exposed to increased intracellular and extracellular concentrations of reactive oxygen and nitrogen species generated by oxidant gases, cellular enzymatic generators such as xanthine oxidase and inflammatory cells. These agents may cause significant damage to alveolar epithelial cells resulting in functional and structural abnormalities. Interest has recently focused on the potential contribution of nitric oxide (•NO) to this injury. Pertinent sources of pulmonary •NO include activated macrophages (Assreuy *et al.*, 1994; Ischiropoulos *et al.*, 1992), endothelial cells (Kooy and Royall, 1994) and airway cells (Kobzik *et al.*, 1993; Robbins *et al.*, 1994), as well as ATII cells themselves (Kobzik *et al.*, 1993; Punjabi *et al.*, 1994). In addition, inhaled •NO is now administered to patients with inflammatory diseases to lower pulmonary artery pressure and improve ventilation-perfusion matching (Rossaint *et al.*, 1993). Consequently, in a number of pathological conditions, there may be significant levels of •NO present in the alveolar lining fluid and thus in close proximity to the alveolar epithelial cells. Herein, we will review what is known about the interaction of nitric oxide with other radicals and the effects of these products on active ion transport by alveolar epithelial cells.

II. INTERACTION OF •NO WITH BIOLOGICAL TARGETS: SIGNAL TRANSDUCER AND PATHOPHYSIOLOGICAL MEDIATOR

Most mammalian cells have the capacity to produce •NO from the oxidative deamination of L-arginine by either the Ca^{2+}-sensitive (cNOS) or the Ca^{2+}-insensitive (iNOS) forms of •NO synthase. This family of enzymes requires NADPH, heme, flavins, tetrahydrobiopterin, thiols and molecular oxygen as cofactors (Forstermann *et al.,* 1994). Although •NO may function as an endogenous signal transducer, the net biological effects depend on its concentration and the biochemical composition of the immediate environment. In recent years, a substantial body of work has demonstrated the importance of other radicals and their reactions with •NO in determining net biological effects.

A. Guanylate Cyclase

•NO binding to the heme group of soluble guanylate cyclase and the subsequent increase in cellular cGMP levels is probably the best characterized interaction of •NO with a biological target. Many of the physiological effects of •NO are mediated by cGMP through various isoforms of cGMP-dependent protein kinase (PKG) and protein phosphorylation (Ignarro, 1992; Kubes *et al.,* 1991; Varela *et al.,* 1992). However, this interaction may also contribute to sepsis-induced refractory hypotension and shock (Wei *et al.,* 1995).

There is convincing evidence that redox states of •NO modulate cation channel activity by increasing cGMP. Light *et al.* (1990) demonstrated the presence of a 28-pS cation channel in rat renal inner-medullary collecting duct cells, the activity of which was modulated both by cGMP *per se* and via PKG-induced phosphorylation. Subsequent studies on cultured collecting duct cells demonstrated that •NO released from endothelial cells specifically inhibited the apical membrane Na^+ conductance in permeabilized monolayers (Stoos *et al.,* 1994). These findings are consistent with a •NO-mediated inhibition of Na^+ reabsorption leading to an increased urinary Na^+ excretion (Light *et al.,* 1990; Stoos *et al.,* 1994). As discussed below, •NO was recently shown to decrease cation channel activity in isolated ATII cells through a cGMP-mediated pathway (Jain *et al.,* 1998). However, studies in intact monolayers indicate that the inhibitory effects of •NO on transepithelial Na^+ transport were cGMP independent (Guo *et al.*, 1998).

B. Superoxide (O_2•$^-$) and Other Reactive Species

•NO reacts with O_2•$^-$ at diffusion-limited rates (6.7×10^9 $M^{-1} \times s^{-1}$) to produce peroxynitrite ($ONOO^-$) as

$$O_2{\bullet}^- + {\bullet}NO \Rightarrow ONOO^- + H^+ \leftrightarrow ONOOH \Rightarrow \text{"}{\bullet}OH \cdots {\bullet}NO_2.\text{"}$$

This reaction reduces the steady state concentrations of both radicals, thus ameliorating the potentially harmful effects of each (Rubbo *et al.,* 1994; Wink *et al.,* 1993, 1994). However, $ONOO^-$ oxidizes thiols at rates at least 1000-fold greater than does H_2O_2 at pH 7 (Radi *et al.,* 1991), damages DNA and the mitochondria electron transport chain (Radi *et al.,* 1994), and causes iron-independent lipid peroxidation of human low density lipoproteins (Graham *et al.,* 1993). Furthermore, the protonated form of $ONOO^-$, peroxynitrous acid (ONOOH; pKa=6.8), forms an intermediate with a reactivity equivalent to nitrogen dioxide (•NO_2) and the hydroxyl radical (•OH) (Beckman *et al.,* 1990).

Peroxynitrite production has been demonstrated *in vitro* from human neutrophils (Carreras *et al.,* 1994), rat alveolar macrophages (Ischiropoulos *et al.,* 1992), and bovine aortic endothelial cells (Kooy and Royall, 1994). It can be argued that *in vivo* the alveolar epithelial lining fluid contains a number of antioxidant substances, such as superoxide dismutase (CuZnSOD), catalase, reduced glutathione, and urate, which will limit the steady-state concentrations of reactive oxygen and nitrogen species (Cantin *et al.,* 1987; Matalon *et al.,* 1990) and thus diminish the significance of the *in vitro* findings. However, because the rate of the reaction between •NO and O_2•$^-$ is much faster than the reactions of these radicals with antioxidant substances, and because of its high reactivity, $ONOO^-$ will attack biological targets even in the presence of antioxidant substances (van der Vliet *et al.,* 1994). Moreover, physiological concentrations of bicarbonate and carbon dioxide enhance the nitrating activity of $ONOO^-$ via the formation of the nitrosoperoxycarbonate anion ($O{=}N{-}OOCO_2^-$). Carbon dioxide increases the yield of nitration reactions (Denicola *et al.,* 1996; Gow *et al.,* 1996a; Uppu *et al.,* 1996). The detection of nitrotyrosine in the lungs of patients with acute respiratory distress syndrome (ARDS) (Haddad *et al.,* 1994b), the lungs of rats exposed to endotoxin (Wizemann *et al.,* 1994) or hyperoxia (Haddad *et al.,* 1994b), and atherosclerotic lesions of human coronary arteries (Beckman *et al.,* 1994), indicate that nitration reactions occur *in vivo* in both humans and animals. Although $ONOO^-$ is the most efficient nitrating agent, proteins may also become nitrated by nitryl chloride ($Cl\text{-}NO_2$), formed by the reaction of nitrite with hypochlorous acid (van der Vliet *et al.*, 1995), $\bullet NO_2$ present in cigarette smoke (Eiserich *et al.*, 1994), or nitronium (NO_2^+) ions, formed from $ONOO^-$ in the presence of CuZnSOD or other metal catalysts (Beckman *et al.*, 1992). It is very likely that injury to the alveolar epithelium, previously attributed to reactive oxygen species, is caused instead by protein nitration and/or oxidation induced by reactive oxygen-nitrogen intermediates including •NO, $ONOO^-$, and $O{=}N\text{-}OOCO_2^-$ (Gow *et al.*, 1996b; Haddad *et al.*, 1993, 1994a,b).

In a recent report, DuVall *et al.* (1998) demonstrated that 3-morpholino-sydnonimine (SIN-1), a generator of $ONOO^-$, profoundly inhibited the amiloride-sensitive whole-cell conductance in *Xenopus* oocytes expressing the three cloned subunits of the wild-type rat epithelial Na^+ channel (α, β, γ-rENaC) (Canessa *et al.*, 1993, 1994) (Fig. 1). Importantly, this effect was observed at very low $ONOO^-$ concentrations (~10 μM) suggesting that $ONOO^-$ may produce similar effects *in vivo* where concentrations have been estimated to occur at higher levels during inflammation (Ischiropoulos *et al.*, 1992). On the other hand, even supraphysiological concentrations of •NO, generated by a variety of •NO donors, had no effect on the amiloride-sensitive current. The inhibitory effects of SIN-1 were mimicked by tetranitromethane at pH 6, a condition that results in the oxidation of sulfhydryl groups (Haddad *et al.*, 1996). In contrast, at pH 7.4, tetranitromethane both

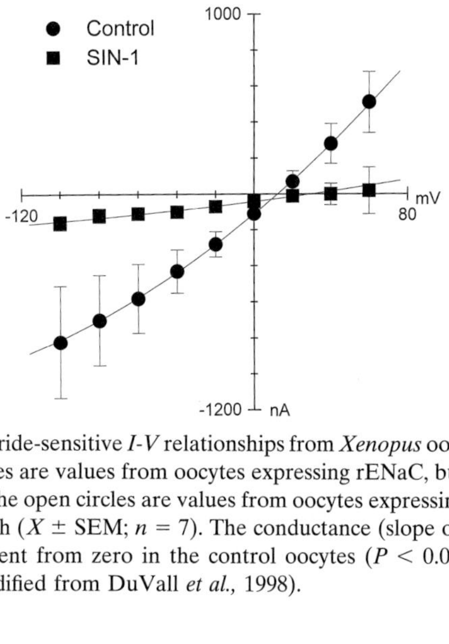

FIGURE 1 Amiloride-sensitive *I-V* relationships from *Xenopus* oocytes expressing α,β,γ-rENaC. The solid circles are values from oocytes expressing rENaC, but not exposed to SIN-1 ($X \pm$ SEM; $n = 6$). The open circles are values from oocytes expressing rENaC and exposed to SIN-1 (1 mM) for 2 h ($X \pm$ SEM; $n = 7$). The conductance (slope of the *I-V* relationship) was significantly different from zero in the control oocytes ($P < 0.01$), but not in oocytes exposed to SIN-1 (modified from DuVall *et al.*, 1998).

oxidizes sulfhydryl groups and nitrates tyrosine and tryptophan residues, but does not affect the amiloride-sensitive whole-cell current. These data suggested that oxidation of critical sulfhydryl groups within rENaC by $ONOO^-$ directly inhibits amiloride-sensitive Na^+ channel activity (DuVall *et al.*, 1998).

C. Thiols

In several studies, the biological effects of •NO on transport proteins have been associated with the formation of nitroso-thiols (RS-NO) (Bolotina *et al.*, 1994; Gupta *et al.*, 1995; Koiristo *et al.*, 1993; Lipton *et al.*, 1993). Although it should be stressed that the direct reaction of •NO with thiols is unfavorable, the presence of strong electron acceptors, such as Fe^{3+} and oxygen, in biological systems facilitate this reaction through the formation of the nitrosonium ion (NO^+) intermediate. Once formed, RS-NO adducts stabilize •NO and may decrease its cytotoxic potential while maintaining or promoting its bioactivity. This appears to be the instance in neurons where •NO donors generating NO^+, but not •NO *per se,* resulted in an S-nitrosylation of critical thiols at the NMDA receptor's redox modulatory site. This subsequently prevented excess Ca^{2+} entry into cells and reduced the neurotoxcity associated with •NO (Lipton *et al.*, 1993). In contrast, formation of S-nitrosoglutathione in the corpus cavernosum smooth muscle stabilized the bioactivity of •NO resulting in an upregulation of the Na^+,K^+-ATPase activity and subsequent muscle relaxation (Gupta *et al.*, 1995).

Micromolar concentrations of S-nitrosoglutathione have been detected in the airway fluid of normal subjects and substantially higher levels were observed in the lungs of patients with pneumonia or during inhalation of 80 ppm •NO (Gaston *et al.*, 1993). However, the effects of RS-NO on active ion transport and fluid clearance across the alveolar epithelium are poorly understood at this time.

III. MODULATION OF ION TRANSPORT ACROSS THE ADULT ALVEOLAR EPITHELIUM BY REDOX STATES OF •NO

A. ATII Cells Form Tight Monolayers Capable of Vectorial Na^+ Transport

The mammalian alveolar epithelium consists of two cell types: large squamous alveolar type I cells (ATI), which cover more than 97% of the alveolar surface area, but make up only 30% of the alveolar epithelial cells and the more numerous, cuboidal type II (ATII) cells, that make up 14%

of all parenchymal cells and 67% of the alveolar epithelial cells in the rat (Haies *et al.*, 1981).

Alveolar type II cells are located primarily in the alveolar corners and form only 3% of the alveolar surface area (Crapo *et al.*, 1980). The results of various *in vivo* and *ex vivo* studies indicate that fluid reabsorption across the alveolar space is, at least partly, due to active transport of Na^+ across these cells (Basset *et al.*, 1987a; Matthay *et al.*, 1982, 1996; Olivera *et al.*, 1994; Sznajder *et al.*, 1995). Moreover, immunocytochemical and Western blotting studies, utilizing polyclonal antibodies to Na^+ channels and Na^+,K^+-ATPase are consistent with the presence of proteins antigenically related to Na^+ channels in the apical membranes and Na^+,K^+-ATPase in the basolateral membranes of ATII cells (Matalon *et al.*, 1992; Nici *et al.*, 1991).

The mechanisms of active transepithelial Na^+ transport have been revealed from *in vitro* studies utilizing cultured ATII cells grown to confluence on permeable supports. Cheek *et al.* (1989) demonstrated that under these conditions, ATII cells form high transepithelial resistance ($R_t > 2000\ \Omega \cdot cm^2$) monolayers, capable of spontaneously generating a transepithelial potential (V_t). The short-circuit current (I_{sc}) across these monolayers corresponded closely to net Na^+ absorption as measured by the transepithelial movement of $^{22}Na^+$ (Cheek *et al.*, 1989). Amiloride (10 μM), added to the apical bathing solution, decreased the I_{sc} to 20% of its original value within 1 min of its application, while basolaterally applied ouabain (1 mM) abolished the I_{sc}. Substitution of Na^+ with equimolar concentrations of choline on the apical side of the monolayer also decreased the I_{sc} values to zero. These data indicate that the principal current carrying ion in cultured ATII cells is Na^+ and that absorption is dependent on an apical membrane Na^+ channel in addition to the basolateral Na^+,K^+-ATPase.

B. Nitric Oxide Inhibits Na^+ Absorption across Cultured Alveolar Type II Monolayers

We examined the mechanisms by which •NO decreased vectorial Na^+ transport across confluent monolayers of rat ATII cells grown on permeable supports (Guo *et al.*, 1998). When added to ATII cell monolayers in culture plates, the •NO donors spermine NONOate (200 μM) and papa NONOate (100 μM) generated a steady-state concentration of ~1.5 μM N•O. After 45 min, the equivalent current (I_{eq}; calculated from the Ohm's law under open-circuit conditions) was decreased by ~80% whereas the R_t increased by ~30%, consistent with the inhibition of a conductive pathway across the epithelium. Under these conditions, the effect of •NO on the I_{eq} was concentration dependent ($IC_{50} = 0.4\ \mu$M). When monolayers mounted in

Ussing chambers were treated with papa NONOate, the I_{sc} decreased ~60%. The residual I_{sc} was subsequently inhibited to ~10% of the basal level by amiloride added to the apical bath (Fig. 2A). In the presence of the •NO scavenger oxy-hemoglobin (50 μM), •NO donors did not increase •NO levels above baseline and the I_{eq} and I_{sc} were unaffected (Fig. 2B). Moreover, the I_{eq} recovered when monolayers were treated with oxy-hemoglobin 45 min after administration of the •NO donors. These findings demonstrated that •NO specifically inhibited transepithelial Na^+ transport across ATII monolayers.

•NO donors did not increase intracellular cGMP and the I_{sc} was not affected when monolayers were treated with 8-bromo-GMP (400 μM), suggesting that the effects of •NO on ATII cell Na^+ transport were not

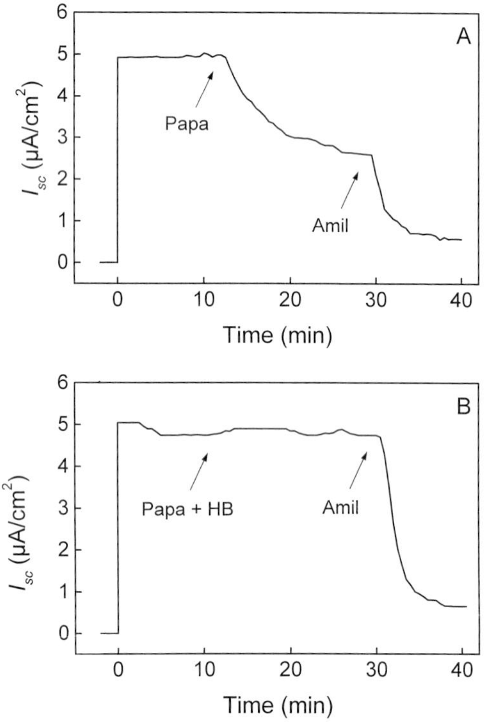

FIGURE 2 Effects of •NO on amiloride-sensitive I_{sc}. PAPA NONOate (100 μM), added to both bathing solutions, inhibited I_{sc} across ATII monolayers mounted to Ussing chamber (panel A; n = 4). Oxy-hemoglobin (HB; 50 μM) added to both bath solutions prior to PAPA NONOate completely blocked the inhibitory effects of PAPA NONOate (100 μM) on I_{sc} (panel B; n = 4). Oxy-hemoglobin alone had no effect on I_{sc} (from Guo *et al.*, 1988).

dependent on cGMP (Guo *et al.,* 1998). However, addition of the $ONOO^-$ generator SIN-1 also decreased the I_{sc}. This suggests that the decreased transport following •NO generation may be due, at least in part, to the reaction of •NO with $O_2•^-$ to form $ONOO^-$ (S. Matalon, Y. Guo, and M. D. DuVall, unpublished data). Moreover, this is consistent with our finding that $ONOO^-$ inhibits the amiloride-sensitive whole-cell conductance in oocytes that heterologously expressed α,β,γ-rENaC (DuVall *et al.,* 1998).

To obtain additional insights as to which transporters were affected by •NO, we performed a series of experiments in which we permeabilized either the apical or basolateral membranes of the monolayers with amphotericin B and measured the resulting currents. To examine the effects of •NO on the apical membrane Na^+ conductance, experiments were performed in the presence of a Na^+ concentration gradient ($[Na^+]_{mucosal}$:$[Na^+]_{serosal}$ = 145:25 mM) and the basolateral membrane was permeabilized. Under these conditions, in the absence of •NO, the amiloride-sensitive I_{sc} ($\Delta I_{sc}{}^{amil}$) was ~10 $\mu A/cm^2$. By contrast, the $\Delta I_{sc}{}^{amil}$ was decreased ~60% in monolayers pretreated with papa NONOate (Fig. 3). The effects of •NO on Na^+,K^+-ATPase were examined with symmetrical Na^+ concentrations (145 mM) and amiloride in the apical bath. Permeabilization of the apical membrane produced an ouabain-sensitive I_{sc} ($\Delta I_{sc}{}^{pump}$) which was inhibited by ~65% in the presence of papa NONOate (Fig. 4). In conjunction with these findings, we also found that •NO inhibited intracellular ATP, which might have accounted for the decreased Na^+,K^+-ATPase activity. However, 2-deoxy-D-glucose, which produced a comparable decrease in cellular ATP (~ 45% of control value), did not alter $\Delta I_{sc}{}^{pump}$. These data indicate that •NO, at noncytotoxic concentrations, decreased Na^+ absorption across cultured ATII monolayers by inhibiting both the amiloride-sensitive Na^+ channels and the Na^+,K^+-ATPase, through cGMP-independent mechanisms (Guo *et al.,* 1998).

An important question that needs to be addressed is whether the concentrations of •NO used in these experiments are likely to be encountered *in vivo.* As mentioned previously the IC_{50} for transport inhibition was about 0.4 μM. The concentration of •NO in the alveolar epithelial lining fluid has not been measured directly. However, various studies indicated that alveolar macrophages and alveolar epithelial cells release large amounts of •NO in the epithelial lining fluid in a variety of pathological situations (Punjabi *et al.,* 1994; Saleh *et al.,* 1997; Wizeman *et al.,* 1994). For example, Ischiropoulos *et al.* reported that activated rat alveolar macrophages may produce 0.1 nmol of •NO per min per 10^6 cells (Ischiropoulos *et al.,* 1992). which may generate micromolar concentrations in the epithelial lining fluid. In other studies, Malinski *et al.* (1993) measured 2–4 μM •NO in ischemic brain during cerebral ischemia and Gaston *et al.* (1993) reported the pres-

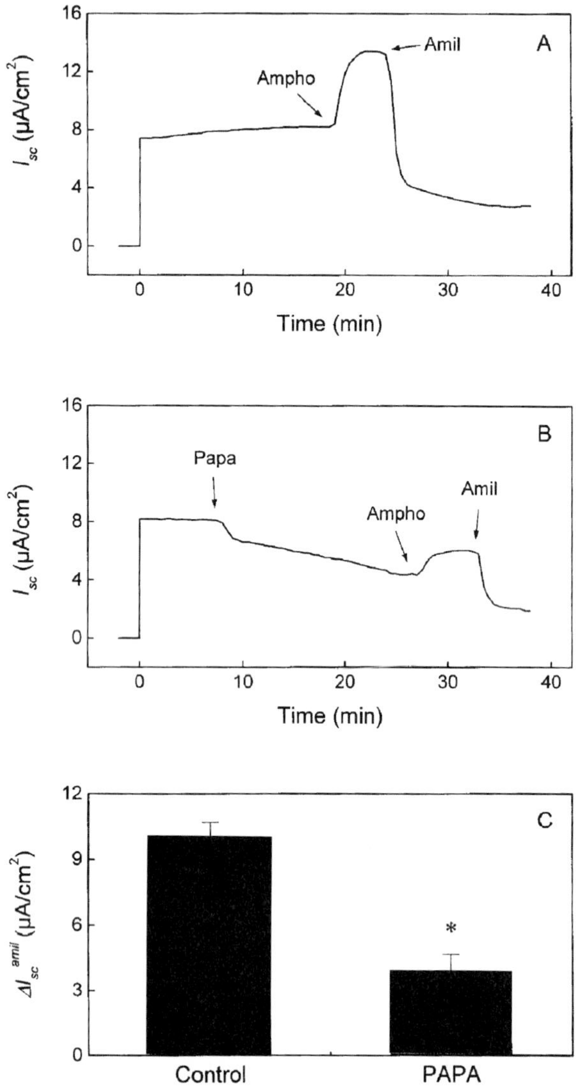

FIGURE 3 Effects of •NO on amiloride-sensitive apical membrane Na^+ channels. Amphotericin B (10 μM) increased the I_{sc} which was inhibited by amiloride (10 μM; panel A). Application of PAPA NONOate (100 mM) decreased the amiloride-sensitive current (I_{Na}; panel B). Mean values of amiloride-sensitive Na^+ currents across apical membranes in the presence or absence of •NO are shown in C (from Guo *et al.*, 1998). *, significantly different from control (student *t*-test; $n = 4$ for each group).

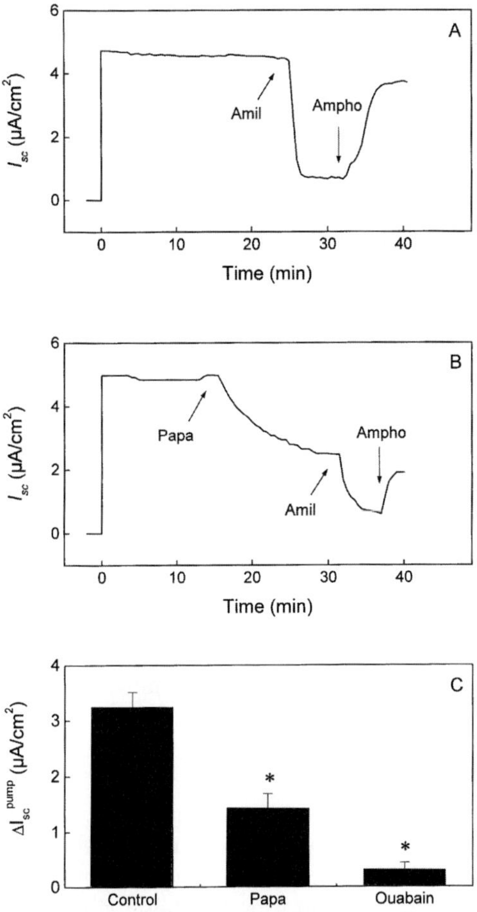

FIGURE 4 Effects of •NO on basolateral membrane Na^+,K^+-ATPase activity. Amiloride (10 μM) decreased the I_{sc} which was increased by amphotericin B (panel A). Addition of ouabain (1 mM) into the basolateral compartment totally abolished the amphotericin B-induced increase in I_{sc} (ΔI_{sc}^{pump}; panel C). Application of PAPA NONOate (100 μM) also decreased ΔI_{sc}^{pump}. The ΔI_{sc}^{pump} was completely inhibited by the addition of ouabain (1 mM) into the basolateral medium. Mean values of these variables are shown in panel C (from Guo *et al.*, 1998). *, significantly different from control (student *t*-test; $n \geq 4$ for each group).

ence of 4 μM nitrosothiols in the distal airway fluid of patients with pneumonia. Thus, it seems reasonable to assume that concentrations of •NO used in this study are likely to be encountered in the alveolar space in a number of pathologic situations.

IV. INHIBITION OF Na^+ TRANSPORT ACROSS FRESHLY ISOLATED ATII CELLS BY REDOX STATES OF •NO

A. Biophysical Properties of Na^+ Channels in Freshly Isolated ATII Cells

When ATII cells were patch-clamped in the whole-cell configuration in symmetrical Na^+ solutions (150 mM Na^+-glutamate; $[Ca^{2+}]_i$ <10 nM), both rabbit and rat ATII cells exhibited outwardly rectified Na^+ currents inhibited by amiloride, benzamil and EIPA (IC_{50} = 1 μM) (Matalon *et al.*, 1992; Yue *et al.*, 1993). Cell-attached and excised inside-out patch-clamp experiments on ATII cells revealed single channels which were Na^+ selective ($P_{Na}:P_K = 7:1$) and had unitary conductances (g) of 25–27 pS. Amiloride or EIPA (1 μM) in the pipette solution (extracellular side) blocked single-channel activity almost completely. It is interesting to note that another group of investigators have reported the presence of a Ca^{2+}-activated ($[Ca^{2+}]_i$> 100 μM), 20-pS cation channel, that were equally permeant to Na^+ and K^+ (Feng *et al.*, 1993) in rat ATII cells in primary culture. It is unknown whether this current is sensitive to EIPA. However, rat fetal distal lung epithelial (FDLE) cells are known to contain whole-cell currents across fetal distal lung epithelial cells which are known to contain Ca^{2+}-activated cation channels sensitive to both amiloride and EIPA (Wang *et al.*, 1993), raising the possibility that the adult ATII channel may also be inhibited by EIPA.

Amiloride-sensitive epithelial Na^+ channels are classified as either high (H-type) or low (L-type) affinity channels, depending on their binding affinities for amiloride and its structural analogues. H-type channels display the following structure/inhibitory pattern relationship: phenamil, benzamil > amiloride >>> EIPA. In contrast, L-type channels are inhibited to the same extent by EIPA and amiloride (Matalon *et al.*, 1996). In addition to the patch-clamp measurements mentioned above, the following data are consistent with the presence of EIPA-sensitive channels in freshly isolated ATII cells: (1) $^{22}Na^+$ fluxes across freshly isolated rabbit ATII cells were inhibited to the same extent by amiloride, benzamil, and EIPA (Matalon *et al.*, 1992). (2) Membrane vesicles prepared from freshly isolated rabbit ATII cells were inhibited to the same extent by amiloride and EIPA (IC_{50} = 10 μM) (Matalon *et al.*, 1991). (3) Reconstitution of a putative Na^+ channel protein purified from freshly isolated rabbit ATII into lipid bilayers exhibited Na^+ single-channel activity (g = 25 pS). Moreover, the channel open probability (P_o) was significantly increased by addition of the catalytic subunit of protein kinase A (PKA) and ATP to the presumed cytoplasmic side of the bilayer. However, the unitary conductance was unchanged. The P_o was inhibited by both amiloride and EIPA when added to the putative extracellular side of the bilayer (Senyk *et al.,* 1995). (4) Intratracheal instillation of EIPA inhibited Na^+ reabsorption across the rat alveolar epithelium *in vivo* to the same degree as amiloride (Yue and Matalon, 1997).

In contrast, the aforementioned data of Guo *et al.* (1998) indicate that amiloride added to the apical compartments of Ussing chambers containing monolayers of ATII cells was about two orders of magnitude more effective than EIPA in inhibiting short-circuit current across ATII cell monolayers. These results are consistent with the presence of high amiloride affinity (H-type) channels in cultured ATII cells. However, it should be kept in mind that with increasing time in culture, ATII cells undergo a number of important phenotypic changes, including decreased ability to secrete surfactant lipids (Dobbs *et al.,* 1985), decreased levels of both the amount and mRNA for surfactant apoproteins (Liley *et al.,* 1988), and internalization of Na^+ channel proteins (Yue *et al.,* 1993). It is unclear whether these alterations represent degenerative changes or transformation of ATII to ATI-type cells. Taken as a whole, these data coupled with our previous findings indicate that culture conditions will influence the expression of Na^+ channels in ATII cells.

B. Effects of Redox States of •NO on Whole-Cell Currents across ATII Cells

Addition of SIN-1 (1 mM), into the bath solutions of ATII cells, maintained in primary culture for 12 h, and patched in the whole-cell mode, completely inhibited Na^+ whole-cell currents (Fig. 5) within 3–5 min from its application. At that time, based on measurements of dihydrorhodamine oxidation, 1mM SIN-1 at 25°C produced about ~7 μM $ONOO^-$ at 25°C. Subsequent addition of amiloride (10 μM), produced no further inhibition, suggesting that $ONOO^-$ generated by SIN-1 completely inhibited the amiloride-sensitive whole-cell currents. Similar results were obtained following addition of SIN-1 (1 mM) plus superoxide dismutase (3000 U/ml). In the presence of SOD, SIN-1 released about 1 μM•NO, detected by an ISO-NO meter, but no peroxynitrite. These findings are in agreement with the aforementioned data on cultured cells and indicate that •NO and $ONOO^-$ down-regulate the activity of amiloride-sensitive Na^+ channels in ATII cells. In a recent study Jain *et al.* (1998) reported that various •NO donors (S-nitrosoglutathione and S-nitoso-N-acetylpenicillamine) decreased the open probability of a cation channel present in the apical membrane of cultured ATII cells. Since incubation of ATII cells with 8-BrcGMP also down-regulated the open probability of this channel, these authors conclude that the inhibitory effecs of •NO were due to an increase in intracellular cGMP. It should be noted that in other studies, an increase

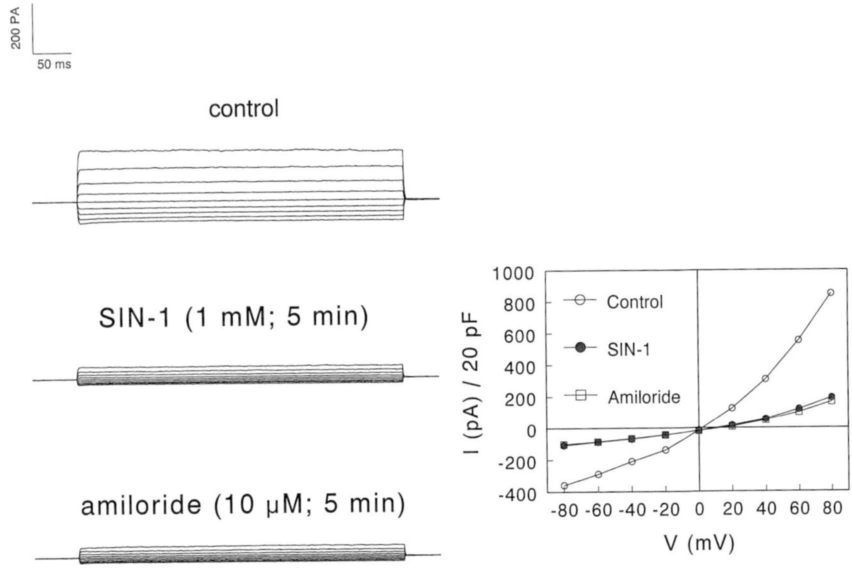

FIGURE 5 Whole cell recordings of rat ATII cells patched in the whole-cell mode. The standard compositions of the pipette solution were (in mM) 145 NaCl, 4 MgATP, 0.5 EGTA, and 15 HEPES; pH 7.4. The bath solution contained (im mM) 145 NaCl, 2 $MgCl_2$, 2 $CaCl_2$, and 15 HEPES; pH 7.4. The records show current values for voltage pulses from −80 to +80 mV in 20-mV increments with holding potential at 0 mV, under control conditions, and 5 min postaddition of SIN-1 (1 mM) or amiloride into the bath solution. The line graph shows average current-voltage relationships for these records ($n \geq 4$).

in intracellular cGMP failed to alter Na^+ trasnport across cultured rat ATII cells (Goodman *et al.,* 1984; Guo *et al.,* 1998).

Reactive oxygen–nitrogen species may interfere with Na^+ transport across epithelial cells by damaging important structural proteins, necessary for the proper function of these transporters. For example, Compeau *et al.* (1994) assessed changes in sodium transport across monolayers of rat distal fetal epithelial cells following incubation of these cells with macrophages, stimulated with endotoxin for 16 h. They reported a 75% decline in R_t and a selective 60% reduction in amiloride-sensitive I_{sc}. Single-channel patch-clamp analysis demonstrated a 60% decrease in the density of the 25-pS nonselective cation (NSC) channel present in the apical membrane of these cells. However, single-channel conductance and open probability were not affected. A concurrent reduction in epithelial F-actin content suggested a role for actin depolymerization in mediating this effect. Incubation of cocultures with the methylated L-arginine (Arg) derivative NG-monomethyl-L-arginine prevented the reduction in epithelial I_{sc}, as did substitution of L-Arg with D-Arg or incubation in L-Arg-free medium. These data indicate that reactive oxygen–nitrogen species may affect amiloride-sensitive cation channels by reducing F-actin or other cytoskeletal structures.

V. DOES •NO MODULATE ALVEOLAR EPITHELIAL FLUID TRANSPORT *IN VIVO?*

The toxicity of •NO *per se,* as well as the production of reactive oxygen–nitrogen intermediates formed by the interaction of •NO with superoxide, has raised concerns that increased levels of •NO or $ONOO^-$ may damage ion transporters and thus interfere with Na^+ reabsorption across the alveolar epithelium *in vivo.* In a recent study, Modelska *et al.* (1998) showed that reabsorption of isotonic fluid, secondary to Na^+ reabsorption across the alveolar space, was inhibited followed prolonged hemorrhagic shock. Moreover, instillation of aminoguanidine, an inhibitor of the inducible form of nitric oxide synthase, restored fluid reabsorption to normal levels. Thus, in agreement with the aforementioned *in vitro* studies across isolated type II cells, this study demonstrates that increased production of •NO by lung epithelial or inflammatory cells may damage Na^+ transporters leading to decreased sodium reabsorption.

VI. CONCLUSIONS

The results of various studies show that alveolar epithelial cells are capable of vectorial sodium transport both *in vitro* and *in vivo.* Sodium enters the apical membranes of alveolar cells, predominantly through amiloride-sensitive ion channels, and is transported across the basolateral membrane by the Na^+,K^+-ATPase. The vectorial movement of Na^+ ions creates an osmotic gradient resulting in the reabsorption of isosmotic fluid across the alveolar space *in vivo.* This fluid reabsorption helps to decrease alveolar fluid especially in situations when alveolar permeability to albumin is increased. Reactive nitrogen species, generated by the reaction of •NO with superoxide, damage both amiloride-sensitive ion channels and the Na^+,K^+-ATPase resulting in decreased vectorial sodium transport both *in vivo* and *in vitro.*

Acknowledgments

This project was supported by NIH Grants HL31197 and HL51173 and a grant from the Office of Naval Research (N00014-97-1-0309). Dr. DuVall is a Parker B. Francis Fellow.

References

Assreuy, J., Cunha, F. Q., Epperlein, M., Noronha-Dutra, A., O'Donnell, C. A., Liew, F. Y., and Moncada, S. (1994). Production of nitric oxide and superoxide by activated macrophages and killing of Leishmania major. *Eur. J. Immunol.* **24,** 672–676.

Basset, G., Crone, C., and Saumon, G. (1987a). Significance of active ion transport in transalveolar water absorption: A study on isolated rat lung. *J. Physiol. (London)* **384,** 311–324.

Basset, G., Crone, C., and Saumon, G. (1987b). Fluid absorption by rat lung in situ: Pathways for sodium entry in the luminal membrane of alveolar epithelium. *J. Physiol. (London)* **384,** 325–345.

Beckman, J. S., Beckman, T. W., Chen, J., Marshall, P. A., and Freeman, B. A. (1990). Apparent hydroxyl radical production by peroxynitrite: Implications for endothelial injury from nitric oxide and superoxide. *Proc. Natl. Acad. Sci. U.S.A* **87,** 1620–1624.

Beckman, J. S., Ischiropoulos, H., Zhu, L. van der Woerd, M., Smith, C., Chen, J., Harrison, J., Martin, J. C., and Tsai, M. (1992). Kinetics of superoxide dismutase- and iron-catalyzed nitration of phenolics by peroxynitrite. *Arch. Biochem. Biophys.* **298,** 438–445.

Beckman, J. S., Ye, Y. Z., Anderson, P. G., Chen, J., Accavitti, M. A., Tarpey, M. M., and White, C. R. (1994). Extensive nitration of protein tyrosines in human atherosclerosis detected by immunohistochemistry. *Biol. Chem. Hoppe-Seyler* **375,** 81–88.

Bolotina, V. M., Najibi, S., Palacino, J. J., Pagano, P. J., and Cohen, R. A. (1994). Nitric oxide directly activates calcium-dependent potassium channels in vascular smooth muscle. *Nature (London)* **368,** 850–853.

Canessa, C. M., Horisberger, J. D., and Rossier, B. C. (1993). Epithelial sodium channel related to proteins involved in neurodegeneration [see comments]. *Nature (London)* **361,** 467–470.

Canessa, C. M., Schild, L., Buell, G., Thorens, B., Gautschi, I., Horisberger, J. D., and Rossier, B. C. (1994). Amiloride-sensitive epithelial Na+ channel is made of three homologous subunits [see comments]. *Nature (London)* **367,** 463–467.

Cantin, A. M., North, S. L., Hubbard, R. C., and Crystal, R. G. (1987). Normal alveolar epithelial lining fluid contains high levels of glutathione. *J. Appl. Physiol.* **63,** 152–157.

Carreras, M. C., Pargament, G. A., Catz, S. D., Poderoso, J. J., and Boveris, A. (1994). Kinetics of nitric oxide and hydrogen peroxide production and formation of peroxynitrite during the respiratory burst of human neutrophils. *FEBS Lett.* **341,** 65–68.

Cheek, J. M., Kim, K. J., and Crandall, E. D. (1989). Tight monolayers of rat alveolar epithelial cells: Bioelectric properties and active sodium transport. *Am. J. Physiol.* **256,** C688–C693.

Compeau, C. G., Rotstein, O. D., Tohda, H., Marunaka, Y., Rafii, B., Slutsky, A. S., and O'Brodovich, H. (1994). Endotoxin-stimulated alveolar macrophages impair lung epithelial Na+ transport by an L-Arg-dependent mechanism. *Am. J. Physiol.* **266,** C1330–C1341.

Crapo, J. D., Barry, B. E., Foscue, H. A., and Shelburne, J. (1980). Structural and biochemical changes in rat lungs occurring during exposures to lethal and adaptive doses of oxygen. *Am. Rev. Respir. Dis.* **122,** 123–143.

Denicola, A., Freeman, B. A., Trujillo, M., and Radi. R. (1996). Peroxynitrite reaction with carbon dioxide/bicarbonate: Kinetics and influence on peroxynitrite-mediated oxidation. *Arch. Biochem. Biophys.* **333,** 49–58.

Dobbs, L. G., Williams, M. C., and Brandt, A. E. (1985). Changes in biochemical characteristics and pattern of lectin binding of alveolar type II cells with time in culture. *Biochim. Biophys. Acta* **846,** 155–166.

DuVall, M. D., Zhu, S., Fuller, C. M., and Matalon, S. (1998). Peroxynitrite inhibits amiloride-sensitve Na^+ currents in *Xenopus* oocytes expressing α,β,γrENaC. *Am. J. Physiol.* **274,** C1417-C1423.

Eiserich, J. P., Vossen, V., O'Neill, C. A., Halliwell, B., Cross, C. E., and van der Vliet, A. (1994). Molecular mechanisms of damage by excess nitrogen oxides: Nitration of tyrosine by gas-phase cigarette smoke. *FEBS Lett.* **353,** 53–56.

Feng, Z. P., Clark, R. B., and Berthiaume, Y. (1993). Identification of nonselective cation channels in cultured adult rat alveolar type II cells. *Am. J. Respir. Cell Mol. Biol.* **9,** 248–254.

Forstermann, U., Closs, E. I., Pollock, J. S., Nakane, M., Schwarz, P., Gath, I., and Kleinert, H. (1994). Nitric oxide synthase isozymes. Characterization, purification, molecular cloning, and functions. *Hypertension* **23,** 1121–1131.

Gaston, B., Reilly, J., Drazen, J. M., Fackler, J., Ramdev, P., Arnelle, D., Mullins, M. E., Sugarbaker, D. J., Chee, C., Singel, D. J., Loscalzo, J., and Stamler, J. S. (1993). Endogenous nitrogen oxides and bronchodilator S-nitrosothiols in human airways. *Proc. Natl. Acad. Sci.U.S.A* **90,** 10957–10961.

Goodman, B. E., Brown, S. E., and Crandall, E. D. (1984). Regulation of transport across pulmonary alveolar epithelial cell monolayers. *J. Appl. Physiol.* 57, 703–710.

Goodman, B. E., Kim, K. J., and Crandall, E. D. (1987). Evidence for active sodium transport across alveolar epithelium of isolated rat lung. *J. Appl. Physiol.* **62,** 2460–2466.

Gow, A., Duran, D., Thom, S. R., and Ischiropoulos, H. (1996a). Carbon dioxide enhancement of peroxynitrite-mediated protein tyrosine nitration. *Arch. Biochem. Biophys.* **333,** 42–48, 1996.

Gow, A. J., Duran, D., Malcolm, S., and Ischiropoulos, H. (1996b). Effects of peroxynitrite-induced protein modification on tyrosine phosphorylation and degradation. *FEBS Lett.* **385,** 63–66.

Graham, A., Hogg, N., Kalyanaraman, B., O'Leary, V., Darley-Usmar, V., and Moncada, S. (1993). Peroxynitrite modification of low-density lipoprotein leads to recognition by the macrophage scavenger receptor. *FEBS Lett.* **330,** 181–185.

Guo, Y., Duvall, M. D., Crow, J. P., and Matalon, S. (1998). Nitric oxide inhibits Na+ absorption across cultured alveolar type II monolayers. *Am. J. Physiol.* **274,** L369–L377.

Gupta, S., Moreland, R. B., Munarriz, R., Daley, J., Goldstein, I., and Saenz de Tejada, I. (1995). Possible role of Na(+)-K(+)-ATPase in the regulation of human corpus cavernosum smooth muscle contractility by nitric oxide. *Br. J. Pharmacol.* **116,** 2201–2206.

Haddad, I. Y., Ischiropoulos, H., Holm, B. A., Beckman, J. S., Baker, J. R., and Matalon, S. (1993). Mechanisms of peroxynitrite-induced injury to pulmonary surfactants. *Am. J. Physiol.* **265,** L555–L564.

Haddad, I. Y., Crow, J. P., Hu, P., Ye, Y., Beckman, J., and Matalon, S. (1994a). Concurrent generation of nitric oxide and superoxide damages surfactant protein A. *Am. J. Physiol.* **267,** L242–L249.

Haddad, I. Y., Pataki, G., Hu, P., Galliani, C., Beckman, J. S., and Matalon, S. (1994b). Quantitation of nitrotyrosine levels in lung sections of patients and animals with acute lung injury. *J. Clin. Invest.* **94,** 2407–2413.

Haddad, I. Y., Zhu, S., Ischiropoulos, H., and Matalon, S. (1996). Nitration of surfactant protein A results in decreased ability to aggregate lipids. *Am. J. Physiol.* **270,** L281–L288.

Haies, D. M., Gil, M., and Weibel, E. R. (1981). Morphometric study of rat lung cells. *Am. Rev. Respir. Dis.* **123,** 533–541.

Hu, P., Ischiropoulos, H., Beckman, J. S., and Matalon, S. (1994). Peroxynitrite inhibition of oxygen consumption and sodium transport in alveolar type II cells. *Am. J. Physiol.* **266,** L628–L634.

Ignarro, L. J. (1992). Haem-dependent activation of cytosolic guanylate cyclase by nitric oxide: A widespread signal transduction mechanism. *Biochem. Soc. Trans.* **20,** 465–469.

Ischiropoulos, H., Zhu, L., and Beckman, J. S. (1992). Peroxynitrite formation from macrophage-derived nitric oxide. *Arch. Biochem. Biophys.* **298,** 446–451.

Jain, L., Chen, X. J., Brown, L. A., and Eaton, D. C. (1998). Nitric oxide inhibits lung sodium transport through a cGMP-mediated inhibition of epithelial cation channels. *Am. J. Physiol.* **274,** L475–L484.

Kobzik, L., Bredt, D. S., Lowenstein, C. J., Drazen, J., Gaston, B., Sugarbaker, D., and Stamler, J. S. (1993). Nitric oxide synthase in human and rat lung: Immunocytochemical and histochemical localization. *Am. J. Respir. Cell Mol. Biol.* **9,** 371–377.

Koivisto, A., Siemen, D., and Nedergaard, J. (1993). Reversible blockade of the calcium-activated nonselective cation channel in brown fat cells by the sulfhydryl reagents mercury and thimerosal. *Pfluegers Arch.* **425,** 549–551.

Kooy, N. W., and Royall, J. A. (1994). Agonist-induced peroxynitrite production from endothelial cells. *Arch. Biochem. Biophys.* 310, 352–359.

Kubes, P., Suzuki, M., and Granger, D. N. (1991). Nitric oxide: An endogenous modulator of leukocyte adhesion. *Proc. Natl. Acad. Sci. U. S. A.* **88,** 4651–4655.

Light, D. B., Corbin, J. D., and Stanton, B. A. (1990). Dual ion-channel regulation by cyclic GMP and cyclic GMP- dependent protein kinase. *Nature* (*London*) **344,** 336–339.

Liley, H. G., Ertsey, R., Gonzales, L. W., Odom, M. W., Hawgood, S., Dobbs, L. G., and Ballard, P. L. (1988). Synthesis of surfactant components by cultured type II cells from human lung. *Biochim. Biophys. Acta* **961,** 86–95.

Lipton, S. A., Choi, Y. B., Pan, Z. H., Lei, S. Z., Chen, H. S., Sucher, N. J., Loscalzo, J., Singel, D. J., and Stamler, J. S. (1993). A redox-based mechanism for the neuroprotective and neurodestructive effects of nitric oxide and related nitroso- compounds [see comments]. *Nature* (*London*) **364,** 626–632.

Matalon, S., and Egan, E. A. (1981). Effects of 100% O_2 breathing on permeability of alveolar epithelium to solute. *J. Appl. Physiol.* **50,** 859–863.

Matalon, S., Holm, B. A., Baker, R. R., Whitfield, M. K., and Freeman, B. A. (1990). Characterization of antioxidant activities of pulmonary surfactant mixtures. *Biochim. Biophys. Acta* **1035,** 121–127.

Matalon, S., Bridges, R. J., and Benos, D. J. (1991). Amiloride-inhibitable Na+ conductive pathways in alveolar type II pneumocytes. *Am. J. Physiol.* **260,** L90–L96.

Matalon, S., Kirk, K. L., Bubien, J. K., Oh, Y., Hu, P., Yue, G., Shoemaker, R., Cragoe, E. J., Jr., and Benos, D. J. (1992). Immunocytochemical and functional characterization of Na+ conductance in adult alveolar pneumocytes. *Am. J. Physiol.* **262,** C1228–C1238.

Malinski, T., Bailey, F., Zhang, Z. G., and Chopp, M. (1993). Nitric oxide measured by a porphyrinic microsensor in rat brain after transient middle cerebral artery occlusion. *J. Cereb. Blood Flow Metab.* **13,** 355–358.

Matalon, S., Benos, D. J., and Jackson, R. M. (1996). Biophysical and molecular properties of amiloride-inhibitable Na+ channels in alveolar epithelial cells. *Am. J. Physiol.* **271,** L1–L22.

Matthay, M. A., Landolt, C. C., and Staub, N. C. (1982). Differential liquid and protein clearance from the alveoli of anesthetized sheep. *J. Appl. Physiol.* **53,** 96–104.

Matthay, M. A., Berthiaume, Y., and Staub, N. C. (1985). Long-term clearance of liquid and protein from the lungs of unanesthetized sheep. *J. Appl. Physiol.* **59,** 928–934.

Matthay, M. A., Folkesson, H. G., and Verkman, A. S. (1996). Salt and water transport across alveolar and distal airway epithelia in the adult lung. *Am. J. Physiol.* **270,** L487–L503.

Modelska, K., Matthay, M. A., and Pittet, J. F. (1998). Inhibition of inducible NO synthase activity (iNOS) after prolonged hemorrhagic shock attenuates oxidant-mediated decrease in alveolar epithelial fluid transport in rats. *FASEB J.* **12,** A39 (abstr.).

Nici, L., Dowin, R., Gilmore-Hebert, M., Jamieson, J. D., and Ingbar, D. H. Upregulation of rat lung Na-K-ATPase during hyperoxic injury. *Am. J. Physiol.* **261,** L307–L314.

Olivera, W., Ridge, K., Wood, L. D., and Sznajder, J. I. (1994). Active sodium transport and alveolar epithelial Na-K-ATPase increase during subacute hyperoxia in rats. *Am. J. Physiol.* **266,** L577–L584.

Olivera, W. G., Ridge, K. M., and Sznajder, J. I. (1995). Lung liquid clearance and Na,K-ATPase during acute hyperoxia and recovery in rats. *Am. J. Respir. Crit. Care Med.* **152,** 1229–1234.

Punjabi, C. J., Laskin, J. D., Pendino, K. J., Goller, N. L., Durham, S. K., and Laskin, D. L. (1994). Production of nitric oxide by rat type II pneumocytes: Increased expression of inducible nitric oxide synthase following inhalation of a pulmonary irritant. *Am. J. Respir. Cell Mol. Biol.* **11,** 165–172.

Radi, R., Beckman, J. S., Bush, K. M., and Freeman, B. A. (1991). Peroxynitrite oxidation of sulfhydryls. The cytotoxic potential of superoxide and nitric oxide. *J. Biol. Chem.* **266,** 4244–4250.

Radi, R., Rodriguez, M., Castro, L., and Telleri, R. (1994). Inhibition of mitochondrial electron transport by peroxynitrite. *Arch. Biochem. Biophys.* **308,** 89–95.

Robbins, R. A., Barnes, P. J., Springall, D. R., Warren, J. B., Kwon, O. J., Buttery, L. D., Wilson, A. J., Geller, D. A., and Polak, J. M. (1994). Expression of inducible nitric oxide in human lung epithelial cells. *Biochem. Biophys. Res. Commun.* **203,** 209–218.

Rossaint, R., Falke, K. J., Lopez, F., Slama, K., Pison, U., and Zapol, W. M. (1993). Inhaled nitric oxide for the adult respiratory distress syndrome [see comments]. *N. Engl. J. Med.* **328,** 399–405.

Rubbo, H., Radi, R., Trujillo, M., Telleri, R., Kalyanaraman, B., Barnes, S., Kirk, M., and Freeman, B. A. (1994). Nitric oxide regulation of superoxide and peroxynitrite-dependent lipid peroxidation. Formation of novel nitrogen-containing oxidized lipid derivatives. *J. Biol. Chem.* **269,** 26066–26075.

Russo, R. M., Lubman, R. L., and Crandall, E. D. (1992). Evidence for amiloride-sensitive sodium channels in alveolar epithelial cells. *Am. J. Physiol.* **262,** L405–L411.

Saleh, D., Barnes, P. J., and Giaid, A. (1997). Increased production of the potent oxidant peroxynitrite in the lungs of patients with idiopathic pulmonary fibrosis. *Am. J. Respir. Crit. Care Med.* **155,** 1763–1769.

Senyk, O., Ismailov, I., Bradford, A. L., Baker, R. R., Matalon, S., and Benos, D. J. (1995). Reconstitution of immunopurified alveolar type II cell Na+ channel protein into planar lipid bilayers. *Am. J. Physiol.* **268,** C1148–C1156.

Stoos, B. A., Carretero, O. A., and Garvin, J. L. (1994). Endothelial-derived nitric oxide inhibits sodium transport by affecting apical membrane channels in cultured collecting duct cells. *J. Am. Soc. Nephrol.* **4,** 1855–1860.

Sznajder, J. I., Olivera, W., Ridge, K. M., Rutschman, D. H., Olivera, W. G., and Ridge, K. M. (1995). Mechanisms of lung liquid clearance during hyperoxia in isolated rat lungs. *Am. J. Respir. Crit. Care Med.* **151,** 1519–1525.

Uppu, R. M., Squadrito, G. L., and Pryor, W. A. (1996). Accelaration of peroxynitrite oxidations by carbon dioxide. *Arch. Biochem. Biophys.* **327,** 335–343.

van der Vliet, A., Smith, D., O'Neill, C. A., Kaur, H., Darley-Usmar, V., Cross, C. E., and Halliwell, B. (1994). Interactions of peroxynitrite with human plasma and its constituents: Oxidative damage and antioxidant depletion. *Biochem. J.* **303,** 295–301.

van der Vliet, A., Eiserich, J. P., O'Neill, C. A., Halliwell, B., and Cross, C. E. (1995). Tyrosine modification by reactive nitrogen species: A closer look. *Arch. Biochem. Biophys.* **319,** 341–349.

Varela, A. F., Runge, A., Ignarro, L. J., and Chaudhuri, G. (1992). Nitric oxide and prostacyclin inhibit fetal platelet aggregation: A response similar to that observed in adults. *Am. J. Obstet. Gynecol.* **167,** 1599–1604.

Wang, X., Kleyman, T. R., Tohda, H., Marunaka, Y., and O'Brodovich, H. (1993). 5-(N-Ethyl-N-isopropyl)amiloride sensitive Na+ currents in intact fetal distal lung epithelial cells. *Can. J. Physiol. Pharmacol.* **71,** 58–62.

Wei, X. Q., Charles, I. G., Smith, A., Ure, J., Feng, G. J., Huang, F. P., Xu, D., Muller, W., Moncada, S., and Liew, F. Y. (1995). Altered immune responses in mice lacking inducible nitric oxide synthase. *Nature* (*London*) **375,** 408–411.

Wink, D. A., Hanbauer, I., Krishna, M. C., DeGraff, W., Gamson, J., and Mitchell, J. B. (1993). Nitric oxide protects against cellular damage and cytotoxicity from reactive oxygen species. *Proc. Natl. Acad. Sci. U.S.A.* **90,** 9813–9817.

Wink, D. A., Hanbauer, I., Laval, F., Cook, J. A., Krishna, M. C., and Mitchell, J. B. (1994). Nitric oxide protects against the cytotoxic effects of reactive oxygen species. *Ann. N.Y. Acad. Sci.* **738,** 265–278.

Wizemann, T. M., Gardner, C. R., Laskin, J. D., Quinones, S., Durham, S. K., Goller, N. L., Ohnishi, S. T., and Laskin, D. L. (1994). Production of nitric oxide and peroxynitrite in the lung during acute endotoxemia. *J. Leukocyte Biol.* **56,** 759–768.

Yue, G. and Matalon, S. (1997). Mechanisms and sequelae of increased alveolar fluid clearance in hyperoxic rats. *Am. J. Physiol.* **272,** L407–L412

Yue, G., Hu, P., Oh, Y., Jilling, T., Shoemaker, R. L., Benos, D. J., Cragoe, E. J., Jr., and Matalon, S. (1993). Culture-induced alterations in alveolar type II cell Na+ conductance. *Am. J. Physiol.* **265,** C630–C440.

CHAPTER 14

Induction of Epithelial Sodium Channel (ENaC) Expression and Sodium Transport in Distal Lung Epithelia by Oxygen

Bijan Rafii,* A. Keith Tanswell,*,† Olli Pitkänen,‡ and Hugh O'Brodovich*,†

*MRC Group in Lung Development, Program in Lung Biology Research, Hospital for Sick Children Research Institute and †Department of Pediatrics, University of Toronto, Toronto, Ontario, Canada M5G 1X8; and ‡Hospital for Children and Adolescents, University of Helsinki, Helsinki, Finland

I. INTRODUCTION

Prior to birth, the fetal lung is filled with liquid that has been secreted by its epithelia. This liquid secretion results from the active transport of Cl^- from the interstitium to the alveolar lumen (Adamson *et al.,* 1969; Olver and Strang, 1974), a phenomenon that is necessary for the normal development of the lung (Alcorn *et al.,* 1977). At birth, however, fluid secretion must cease and the lung liquid must be cleared before effective gas exchange can take place. This liquid clearance results from active epithe-

1063-5823/99 $30.00

lial Na^+ transport from the lumen to the interstitium (Olver *et al.*, 1986; O'Brodovich *et al.*, 1990b). If Na^+ transport is impaired at birth, respiratory distress and hypoxemia occurs (O'Brodovich *et al.*, 1990a, 1991). Abnormally low amounts of amiloride sensitive Na^+ transport have been documented in patients with transient tachypnea of the newborn (Gowen *et al.*, 1988) and hyaline membrane disease (Barker *et al.*, 1997). The functional cloning of an epithelial sodium channel (ENaC) (Canessa *et al.*, 1993, 1994; Lingueglia *et al.*, 1993) has allowed in-depth studies of the mechanisms and importance of Na^+ absorption in the lung. Indeed, a transgenic mouse deficient in the α subunit of ENaC fails to clear its fetal lung liquid and dies shortly after birth (Hummler *et al.*, 1996).

How the lung converts from a fluid-secreting to an absorbing organ is incompletely understood and has been the subject of many studies (reviewed in O'Brodovich, 1991, 1996)). β-receptor agonists can within minutes convert the mature *in utero* fetal lung from a fluid-secreting to a fluid-absorptive organ (Brown *et al.*, 1983). Similarly, instillation of membrane permeable cAMP analogues leads to reduced fluid secretion and increased fluid absorption (Walters *et al.*, 1990). Catecholamines are unlikely to be the only signal triggering Na^+ absorption at birth since the β-agonist effect is reversible. Additionally, only during the late stages of labor do β agonist levels become sufficiently elevated to induce Na^+ absorption, and airspace instillation of β adrenergic antagonists does not slow the lung clearance that is initiated by labor (Strang and Barker, 1993; McDonald *et al.*, 1986; Chapman *et al.*, 1990).

Arginine vasopressin (AVP) is another hormone that is postulated to increase Na^+ absorption at birth. Intravenous injection, but not air space instillation of AVP, at pharmacological doses, has been shown to reduce fluid secretion in goats and lambs (Cassin and Perks, 1993; Wallace *et al.*, 1990; Ross *et al.*, 1984; Perks and Cassin, 1982, 1989). However, AVP does not increase Na^+ transport in rabbit trachea (Boucher and Gatzy. 1983), adult type II alveolar epithelia (Cott *et al.*, 1986), or fetal distal lung epithelial cells (O'Brodovich *et al.*, 1992). Additionally, AVP's effect may be mediated by catecholamine release (reviewed in Strang and Barker, 1993).

The change in oxygen concentration from the fetal (~3%) to postnatal (21%) environment represents another potential signal for the conversion of the perinatal lung. Others have previously demonstrated that the ability of late gestation fetal distal lung explants to form fluid-filled cysts is dependent on oxygen concentration (Barker and Gatzy, 1993). In this chapter we will provide evidence that the increase in oxygen concentration, that normally occurs at birth, induces Na^+ absorption in fetal distal lung epithelial cells (FDLE), and we will examine possible mechanisms through which oxygen may induce conversion of the lung from a secreting mode to an absorptive mode thus preparing the newborn for its new environment.

II. PHYSIOLOGIC INCREASE IN OXYGEN AUGMENTS Na^+ TRANSPORT IN FDLE

FDLE cultured on permeable membranes, when switched from 3 to 21% oxygen for 18 h, have increased amiloride sensitive Na^+ transport (as reflected by amiloride-sensitive I_{sc} (Pitkänen *et al.,* 1996)). Although an 8-h incubation in 21% oxygen is insufficient for this response to take place, incubation of these cells for 48 h further augments their ability to transport Na^+ and also leads to increased expression of all three subunits of the ENaC.

This oxygen induction of Na transport is reversible since we have found that when oxygen-induced FDLE are later returned to 3% oxygen, amiloride-sensitive Na^+ transport returns to the same level as the FDLE which were continuously maintained at the 3% oxygen concentration (Fig. 1).

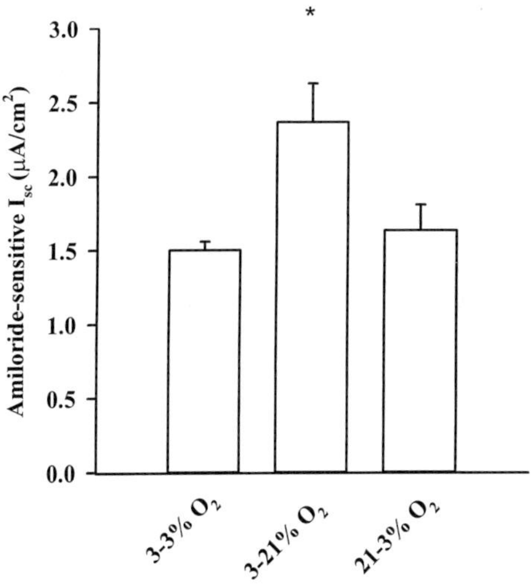

FIGURE 1 An increase in oxygen concentration reversibly induces amiloride-sensitive short-circuit current (amiloride-sensitive I_{sc}) in FDLE. FDLE were incubated under 3% O_2 for 96 h (3–3% O_2 group), or under 3% oxygen for 48 h followed by 21% oxygen for a further 48 h (3–21% O_2 group), or under 21% oxygen for 48 h followed by 3% oxygen for a further 48 h (21–3% O_2 group). The amiloride-sensitive I_{sc} was measured at the end of the 4 d experimental period. *, $P < 0.05$ relative to other groups, $n = 3$. Reproduced from Rafii *et al.* (1998).

Oxygen regulation of sodium transport has also been shown in adult lung epithelia. Other laboratories have shown that Na^{22} uptake, mRNA expression for all three ENaC subunits, protein expression of α ENaC, and activity of Na^+,K^+-ATPase all decline when either primary cultures of adult rat type II epithelial cells or a human lung carcinoma derived cell line are switched from 21% to hypoxic oxygen concentrations (Planes *et al.*, 1997; Mairbaurl *et al.*, 1997).

III. OXYGEN INDUCTION OF Na^+ TRANSPORT IS MEDIATED BY REACTIVE OXYGEN SPECIES

To elucidate the mechanism through which oxygen induces Na^+ transport in FDLE we conducted our 48-h oxygen induction protocol in the presence and absence of various inhibitors and antioxidants (see the legend of Table I for methods).

A. Role of Cyclo-oxygenase Derived Prostaglandins

An increased environmental oxygen concentration will increase intracellular reactive oxygen species (ROS) levels (Turrens *et al.*, 1982a,b; Boveris and Chance, 1973). Since an increased oxygen concentration, ROS, and hydroperoxides are known to stimulate prostaglandin synthesis and secretion (Chakraborti *et al.*, 1989; Herman *et al.*, 1995; Taylor *et al.*, 1983; Tanswell *et al.*, 1990a), and prostaglandins are known to alter Na^+ transport by otic epithelium (Herman *et al.*, 1995), we treated FDLE with ibuprofen prior to oxygen induction. Ibuprofen, at concentrations known to inhibit cyclo-oxygenase (Tanswell *et al.*, 1990a), did not prevent 21% oxygen induction of amiloride-sensitive I_{sc} (Table I), thus making cyclo-oxygenase derived prostaglandins unlikely mediators in induction of FDLE Na^+ transport by oxygen.

This phenomenon does not appear to be secondary to the secretion of a stable compound. Conditioned media derived from FDLE grown in 21% oxygen do not enhance Na^+ transport in FDLE cells which are incubated in 3% oxygen (Rafii *et al.*, 1998). Since cyclo-oxygenase derived prostaglandins are secreted molecules, these data also support the absence of a role for prostaglandins in oxygen induction of Na^+ transport in FDLE.

TABLE I

Effects of Enzyme Inhibitors and ROS Scavengers on Amiloride-Sensitive I_{sc}

		Amiloride-sensitive I_{sc} ($\mu A/cm^2$)[a]		
Treatment (*n*)	Dose	3–3% O_2	3–21% oxygen + vehicle	3–21% oxygen + treatment
TEMPO (6)	1 mM	1.2 ± 0.2	2.6 ± 0.1	1.4 ± 0.2[b]
EUK-8 (3)	30 μM	1.2 ± 0.2	2.1 ± 0.2	1.2 ± 0.2[b]
SOD (4)	500 U/ml	1.2 ± 0.2	2.5 ± 0.3	2.5 ± 0.3
SOD:liposome (3)	875 mU/ml	1.6 ± 0.2	3.1 ± 0.5	2.9 ± 0.7
Catalase:liposome (3)	150 mU/ml	1.4 ± 0.0	2.3 ± 0.2	2.1 ± 0.1
NAC (3)	20 mM	1.4 ± 0.1	2.5 ± 0.1	3.0 ± 0.3
U74389G (5)	30 μM	1.0 ± 0.2	1.7 ± 0.2	2.0 ± 0.4
Trolox (3)	100 μM	0.9 ± 0.1	2.3 ± 0.3	2.1 ± 0.2
NAME (3)	0.2 mM	1.2 ± 0.5	3.0 ± 0.3	2.8 ± 0.1
Ibuprofen (4)	50 μM	1.8 ± 0.2	3.2 ± 0.5	3.3 ± 0.4
Allopurionol (3)	200 μM	1.5 ± 0.1	2.8 ± 0.3	2.7 ± 0.2

Note. Effects of antioxidants and inhibitors on oxygen-induced amiloride-sensitive I_{sc}. FDLE cultures were prepared as previously reported (O'Brodovich *et al.*, 1990b) and seeded on Snapwell inserts at the concentration of 1×10^6 cells/cm² in Dulbecco's modified Eagle media (DMEM) + 10% fetal bovine serum (FBS) and kept overnight at 3% oxygen. After 24 h, the media was changed, antioxidants/inhibitors/vehicle added, and the 21% oxygen groups were transferred to a 21% oxygen incubator for a 48-h period. Amiloride-sensitive I_{sc} was measured by mounting the monolayers in an Ussing chamber (O'Brodovich *et al.*, 1990b) and recording the I_{sc} before and after apical addition of 1×10^{-4} M amiloride. U74389G was a gift from Upjohn Scientific (Kalamazoo, MI); TEMPO, ibuprofen, and trolox were purchased from Sigma-Aldrich; NAC from Roberts pharmaceutical (Mississauga, Ontario); NAME from Calbiochem (San Diego, CA); catalase from Cooper Biomedical (Malvern, PA); SOD from Boehringer Mannheim (Laval, Quebec); and EUK-8 was a gift from Eukarion (Bedford, MA). For catalase treatment, cationic liposomes (1:1 dioleoyldimethylammonium chloride:dioleoylphosphatidylethanolamine (DODAC:DOPE) INEX, Vancouver, BC) containing succinylated catalase were prepared as previously described (Tanswell *et al.*, 1990a). After the initial incubation in 3% O_2 for 20 h, FDLE were exposed to 150 mU/ml catalase encapsulated in cationic liposomes (10 nmoles/cm²) in DMEM + 2 % FBS for 2 h. This concentration was designed to increase cell catalase activity by 50%, which was confirmed by assay (Tanswell *et al.*,1990b). After this 2-h incubation, the monolayers were rinsed, the media was replaced with DMEM + 10% FBS, and the FDLE then remained at 3% O_2 for an additional 24 h prior to induction at 21% O_2. SOD without liposomes (500 U/ml) was added to media 2 h before induction and again after 24 h. Alternatively, SOD was encapsulated in pH sensitive liposomes (1,2-dioleoyl-sn-glycero-3-succinate: DOPE, 1:1) and added at a concentration of 875 mU/ml (5 nmoles/cm² liposome) to cell media at the time of O_2 induction. This concentration was based on studies using direct measurements of enzyme activity (Tanswell *et al.*, 1990b) to define the concentration required to achieve a 50% increase in enzyme activity. Preparation of pH sensitive liposomes has been described elsewhere (Briscoe *et al.*, 1995). As control in liposome-mediated studies, an equivalent mass of bovine serum albumin (Sigma-Aldrich, Mississauga, Ontario) was used instead of the antioxidant enzymes in the preparation of the liposomes. Data drawn from Rafii *et al.*, (1998). Amiloride-sensitive I_{sc}, amiloride-sensitive short-circuit current; TEMPO, tetramethyl piperidine-*N*-oxyl; NAME, N^G-nitro-L-arginine methyl ester; SOD, superoxide dismutase; NAC, *N*-acetylcysteine.

[a] Data presented are mean ± SEM. Each *n* represents 2–4 monolayers from a single primary culture of FILE.

[b] $P < 0.05$ from the 3–21% oxygen control group.

B. Role of NO

Nitric oxide (NO) is an essential mediator in many cellular processes and plays a role as both an oxidant and an antioxidant molecule (Darley-Usmar *et al.*, 1995). The mechanism by which NO acts varies from cell to cell but in general it acts through a guanosine 3′,5′-cyclic monophosphate (cGMP) pathway (Moncada *et al.*, 1991). Atrial naturetic peptide (ANP) acts via cGMP to regulate the 25-pS nonselective cation channel in kidney epithelium (Light *et al.*, 1989). NO, through both cGMP dependent (Jain *et al.*, 1998) and independent (Guo *et al.*, 1998) mechanisms, is known to modulate Na^+ transport in adult type II alveolar epithelial cells. Although studies from our laboratory have shown that a brief (less than 1 h) exposure to ANP is ineffective in regulating amiloride-sensitive I_{sc} in FDLE (O'Brodovich *et al.*, 1992) we have also observed that activated macrophages reduce amiloride-sensitive I_{sc} in FDLE by a NO-dependent mechanism (Compeau *et al.*, 1994).

Therefore, we thought it important to test the possible role NO may play in oxygen-induced Na^+ transport in FDLE by incubating the monolayers with an inhibitor of the inducible nitric oxide synthase, N^G-nitro-L- arginine methyl ester (NAME), for the duration of oxygen induction protocol. NAME was unable to block the oxygen induced increase in amiloride-sensitive I_{sc} (Table I) suggesting that the 21% oxygen induction of amiloride-sensitive I_{sc} in FDLE is not mediated by NO.

C. Role of Reactive Oxygen Species

There has been an increasing recognition of the role of ROS in signal transduction and gene regulation (reviewed in Suzuki *et al.*, 1997; Lander, 1997). Our results demonstrate that the superoxide scavenger tetramethyl piperidine-*N*-oxyl (TEMPO) inhibits the 21% oxygen-induced amiloride-sensitive Na^+ transport and ENaC mRNA expression (Table I, Fig. 2). We chose TEMPO for this study because of its high cellular permeability and its lack of hydroperoxide scavenging activity (Mitchell *et al.*, 1990; Samuni *et al.*, 1990; Caraceni *et al.*, 1995; B. Rafii, K. Tanswell, and H. O'Brodovich, unpublished observations). The superoxide dismutase (SOD)/catalase mimic EUK-8 (Doctrow *et al.*, 1997), at a concentration that was much lower than we used for TEMPO, also diminished the oxygen induced amiloride-sensitive I_{sc} to the same level as the cells kept at the 3% oxygen level. On the other hand, neither catalase encapsulated in cationic liposomes nor the cell permeable hydroperoxide scavengers U74389G and trolox were able to block oxygen induction of amiloride-sensitive I_{sc} (Table I). *N*-

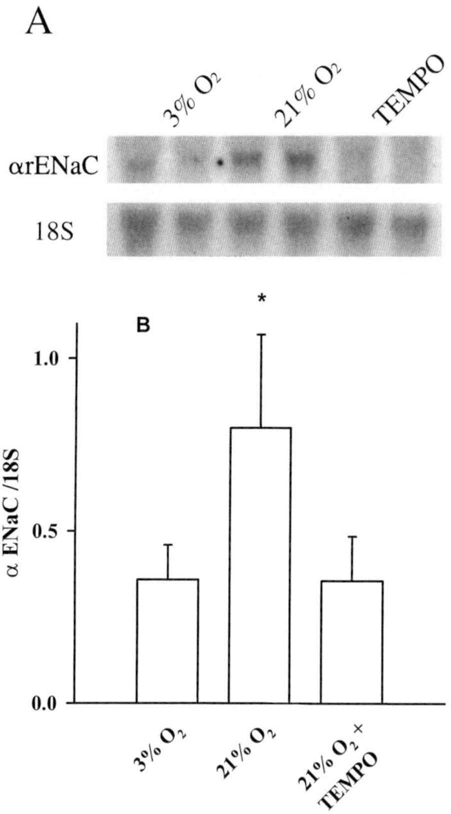

FIGURE 2 Effect of TEMPO on oxygen-induced α ENaC mRNA expression. The superoxide scavenger TEMPO blocks the oxygen-induced increase in α rENaC mRNA expression in FDLE. Cells were treated with 1 mM TEMPO at the time of oxygen induction. Control cells in 21% oxygen received vehicle (ethanol) only. (A) Representative Northern blot probed sequentially with cDNA to α rENaC and to 18S rRNA (two replicates per lane are shown). (B) Densitometric quantitation of α rENaC expression normalized to 18S rRNA from Northern blots. *, $P < 0.05$ relative to other groups, $n=5$. RNA was extracted from FDLE monolayers that had been grown on permeable supports using 4 ml of Trizol (Gibco/BRL) according to the manufacturer's instructions. The final pellet was dissolved in water treated with dimethyl pyrocarbonate (Sigma-Aldrich) and then 20 μg of total RNA was size fractioned on a 1% agarose / MOPS / 2% formaldehyde gel. The RNA was subsequently transferred to Hybond N^+ nylon membranes (Amersham, Oakville, Ontario). The blots were then UV cross-linked and hybridized with ^{32}P random primed α rENaC cDNA fragments (base pairs 74-403) (Canessa *et al.,* 1994) in Expresshyb solution (Clontech, Palo Alto, CA) using the manufacturer's instructions. After washing in 0.1 × SSC + 0.1% SDS at 50°C for 1 h, the blots were exposed to autoradiography film at −80°C. Autoradiographic bands were quantified using a LKB Ultrascan XL enhanced laser densitometer (Pharmacia, Montreal, Quebec). All mRNA expression was normalized to 18S ribosomal RNA content by hybridizing the blots with a full length mouse 18S ribosomal RNA ^{32}P random primed cDNA probe (American Type Culture Collection, Rockville, MD). Reproduced from Rafii *et al.* (1998).

Acetylcysteine (NAC), a substrate source for the synthesis of glutathione (co-factor for glutathione peroxidase which detoxifies hydroperoxides), was also ineffective (Table I).

SOD, either alone or when encapsulated with pH sensitive liposomes, similarly did not block oxygen-induced amiloride-sensitive I_{sc}. The failure of SOD to block oxygen-induced amiloride-sensitive I_{sc} may have been due to its lack of uptake by FDLE during the 48-h oxygen induction period. Unlike the small and diffusible TEMPO, large proteins such as SOD may also have stearic limitations in reaching relevant sites of superoxide production.

IV. INCREASE IN ENaC mRNA EXPRESSION IS ASSOCIATED WITH NF-κB ACTIVATION

Transcription factors NF-κB and AP-1 are known to be redox sensitive. NF-κB is activated in a pro-oxidant environment whereas AP-1 is activated by both oxidants and antioxidants (Amstad *et al.*, 1992; Muller *et al.*, 1997; Sen and Packer, 1996). Pyrrolidine dithiocabamate, *N*–acetylcysteine, or over expression of thioredoxin prevent activation of NF-κB but induce AP-1 (Schenk *et al.*, 1994; Meyer *et al.*, 1993). Both NF-κB and AP-1 are key transcription factors in the regulation of gene expression by the cellular redox state (Meyer *et al.*, 1994; Sen and Packer, 1996; Flohe *et al.*, 1997), and the promoter for α rENaC contains potential binding sites for both NF-κB (GGGGAGTTCC at −519 to −510) and tetradecanoylphorbol 13-acetate (TPA, CTAGTCA at −284 to −278) (Otulakowski *et al.*, 1999).

We tested nuclear extracts from FDLE grown under 3 or 21% oxygen by electrophoretic mobility shift assays using radiolabeled oligonucleotides corresponding to consensus binding sites for NF-κB and AP-1 transcription factors. Three hours after FDLE were switched from 3 to 21% oxygen, nuclear extracts possessed increased binding activity to an NF-κB consensus oligonucleotide (Fig. 3A). This activation was prevented when FDLE were incubated with 1 mM TEMPO during the oxygen induction period. In contrast, AP-1 binding activity was not induced under similar conditions (Fig. 3B). Our results are in agreement with another study which reported NF-κB is strongly activated when HeLa cells kept in hypoxic conditions were exposed to normal oxygen concentration (Rupec and Baeuerle, 1995).

NF-κB is most commonly composed of relA (p65) and p50 subunits, although other combinations of any two other subunits from the NF-κB/rel family (relB, c-rel, and P-52) are also known to dimerize in the cell (Flohe *et al.*, 1997). In the cytoplasm, the NF-κB dimer is bound to an inhibitory IkB family of proteins which prevents it from entering the nucleus

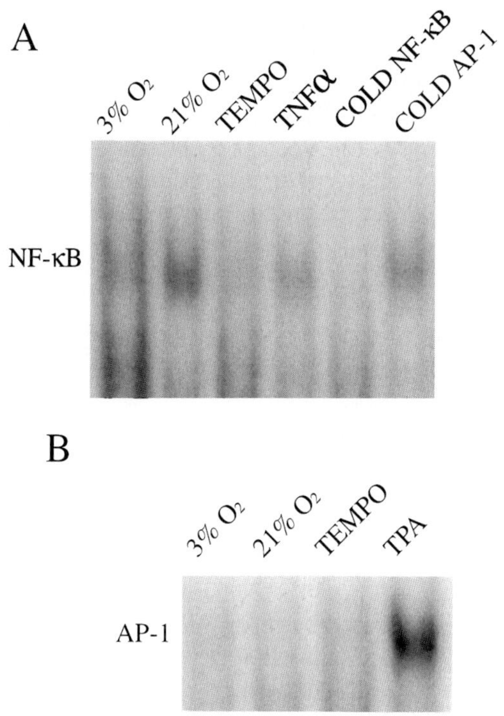

FIGURE 3 Postnatal oxygen activates NF-κB but not AP-1. (A) The 1 mM TEMPO prevents NF-κB activation induced by a switch from 3 to 21% oxygen concentration. Nuclear extracts from FDLE cultured on permeable supports under 21% oxygen (lane 2) show increased binding to an end-labeled NF-κB consensus oligo, compared to cells grown under 3% oxygen (lane 1). Specificity of binding is demonstrated by competition with a 200-fold excess of cold NF-κB oligo (lane 5) but not with an unrelated oligonucleotide (AP-1, lane 6). Induction of NF-κB binding activity by 21% oxygen can be blocked with 1 mM TEMPO (lane 3). Lane 4 represents a positive control in which NF-κB was induced with 1 ng/ml TNF-α for 4 h. (B) No activation of AP-1 occurs when FDLE are switched from 3 to 21% oxygen concentration (lanes 1 and 2). Lane 4 represents a positive control in which AP-1 was activated with 0.1 μM tetradecanoylphorbol 13-acetate (TPA) for 3 h. AP-1 activation was also absent in longer induction times of up to 5 h (data not shown). TEMPO did not activate AP-1 (lane 3). Each gel shift assay was repeated at least once. Nuclear extract preparations and binding reactions were performed according to the protocol of Kazmi *et al.* (1995) without further modification. Either 2 or 5 μg of nuclear extract and 10,000 cpm of ^{32}P end-labeled oligonucleotide were used for each binding reaction. The binding reaction mixture was run on a 4 or 6% nondenaturing acrylamide gel in 0.5x tris borate EDTA buffer (TBE), which after drying was exposed to autoradiography film at −80°C. The oligo containing the consensus sequence for NF-κB was 5′ AGC TTC AGA GGG GAC TTT CCG AGA GG 3′, and the one containing the AP-1 consensus sequence was AGC TTT CCA AAG AGT CAT CAG G. Reproduced from Rafii *et al.* (1998)

(Baeuerle and Henkel, 1994). However, after an appropriate stimulus (such as increased intracellular ROS levels) (Schreck *et al.*, 1991) IκB is phosphorylated and is consequently degraded by a ubiquitination mediated mechanism by the 26S proteosome (Traenckner *et al.*, 1995; Chen *et al.*, 1995). Removal of IκB uncovers the nuclear localization signal on the NF-κB subunits and thus allows it to enter the nucleus and initiate transcription by binding to NF-κB consensus binding sequences within the promoter of the target genes.

V. POSSIBLE SITES OF ROS GENERATION FOR OXYGEN INDUCTION OF Na^+ TRANSPORT IN FDLE

Molecular oxygen, as part of the oxidative phosphorylation phase of ATP generation, undergoes a four-electron reduction to form water. This reaction occurs at cytochrome *c* oxidase (complex IV) of the mitochondrial electron transport chain. Although ROS are transiently produced during this process, they remain tightly bound to the enzyme complex and are not released (Tyler, 1992). All but 1–2% of oxygen consumed by eukaryotes is completely reduced to water by this route (Boveris *et al.*, 1972; Forman and Kennedy, 1974; Boveris and Chance, 1973). "Leakage" of electrons from the electron transport chain occurs mainly at the level of NADH dehydrogenase (complex I) and coenzyme Q cytochrome *c* reductase (complex III). It is at these steps that oxygen molecules are partially reduced to form superoxide anions (Nohl and Jordan, 1986; Nohl and Hegner, 1978). About 80% of superoxide thus generated is dismutated to H_2O_2 by the mitochondrial SOD (Nohl and Jordan, 1986). Some of this H_2O_2 may also be converted to the highly reactive hydroxyl radical through reaction with superoxide in the presence of transition metals (Halliwell and Gutteridge, 1986; Halliwell, 1978).

Xanthine oxidase, a cytoplasmic enzyme involved in the catabolism of nucleic acids, is another potential source of ROS in oxygen induction of FDLE. Under reduced oxygen concentration xanthine dehydrogenase (the predominant form of this enzyme in normoxia in the cell) converts to xanthine oxidase by a sulfhydryl oxidation or limited proteolysis and then upon return of the cells to higher oxygen concentration becomes capable of utilizing xanthine and oxygen to produce superoxide (Corte and Stirpe, 1972; Battelli *et al.*, 1972; McCord, 1985). Indeed, a rise in xanthine/xanthine oxidase derived ROS has been implicated as a major cause of tissue injury in post hypoxic reoxygenation (e.g., Younes and Strubelt, 1988; McCord, 1985). However, others have found allopurinol, an inhibitor of xanthine oxidase, to be unable to provide any protection against this source of ROS (e.g., Metzger, *et al.,* 1988; Littauer and De Groot, 1992; Caraceni *et al.,* 1995). The reason for this discrepancy can be attributed to the fact that many studies which had shown protective effect of allopurinol in their reoxygenation studies had used allopurinol at concentrations so high that it can act as a nonspecific ROS scavenger (Moorhouse *et al.,* 1987).

Our experiment using allopurinol (Table I) argues against a role for xanthine/xanthine oxidase as a source of ROS in oxygen induced amiloride-sensitive I_{sc} in FDLE. Additionally, inability of SOD to prevent oxygen induction points to the prospect that the intracellular source of ROS is compartmentalized and therefore is inaccessible to large molecules such as SOD.

Beside mitochondria and xanthine/xanthine oxidase, a number of other potential ROS sources for ROS generation exist in the cell. These include a cytosolic NADH oxidoreductase with ROS generating capability that is responsive to changes in oxygen concentration over the physiological range (Mohazzab-H. and Wolin, 1994; Mohazzab-H. *et al.,* 1995), and the cytochrome P-450 family of enzymes in microsomes which catalyze a wide variety of biological oxidations which require molecular oxygen and NADPH (Guengerich and McDonald, 1990; Guengerich, 1991). However, more studies need to be conducted to determine the relative contributions of these sources to total intracellular ROS generated during oxygen induction of FDLE.

V. CONCLUSION

In conclusion, our studies demonstrate that the physiologic increase in oxygen concentration augments fetal distal lung epithelial Na^+ transport. Increase in Na^+ transport is a reversible phenomenon and is likely mediated by an increase in ROS leading to increased gene expression of ENaC. There are a number of possible sites for ROS generation during induction of FDLE Na^+ transport by oxygen and more studies need to be done to identify such sites. Since FDLE (O'Brodovich *et al.,* 1990b) and human distal lung epithelial cells (Barker *et al.,* 1995) have similar bioelectric properties, these findings may be relevant to the normal transition from fetal to postnatal life and in the recovery of patients with pulmonary edema.

Acknowledgments

This research was supported by the MRC Group in Lung Development. We thank Eukarion Inc. for providing EUK-8. A. K. Tanswell is the Hospital for Sick Children Women's Auxiliary and University of Toronto Chair in Neonatology, O. Pitkänen was supported by a grant from the Finnish Cultural Foundation, and H. O'Brodovich is the R. S. McLaughlin Foundation

Chair in Pediatrics at The Hospital for Sick Children. Results of this study were previously presented in Rafii *et al.* (1998).

References

Adamson, T. M., Boyd, R. D. H., Platt, H. S., and Strang, L.. (1969). Composition of alveolar liquid in the foetal lamb. *J. Physiol.* (*London*) **204,** 159–168.

Alcorn, D., Adamson, T. M., Lambert, T. F., Maloney, J. E., Ritchie, B. C., and Robinson, P. M. (1977). Morphological effects of chronic tracheal ligation and drainage in the fetal lamb lung. *J. Anat.* **123,** 649–660.

Amstad, P. A., Krupitza, G., and Cerutti, P. A. (1992). Mechanism of c-fos Induction by active oxygen. *Cancer Res.* **52,** 3952–3960.

Baeuerle, P. A., and Henkel, T. (1994). Function and activation of NF-kB in the immune system. *Annu. Rev. Immunol.* **12,** 141–179.

Barker, P. M., and Gatzy, J. T. (1993). Effect of gas composition on liquid secretion by explants of distal lung of fetal rat in submersion culture. *Am. J. Physiol.* **265,** L512–L517.

Barker, P. M., Boucher, R. C., and Yankaskas, J. R. (1995). Bioelectric properties of cultured monolayers from epithelium of distal human fetal lung. *Am. J. Physiol.* **268,** L270–L277.

Barker, P. M., Gowen, C. W., Lawson, E. E., and Knowles, M. (1997). Decreased sodium ion absorption across nasal epithelium of very premature infants with respiratory distress syndrome. *J. Pediatr.* **130,** 373–377.

Battelli, M. G., Corte, E. D., and Stirpe, F. (1972). Xanthine oxidase type D (dehydrogenase) in the intestine and other organs of the rat. *Biochem. J.* **126,** 747–749.

Boucher, R. C., and Gatzy, J.T. (1983). Characteristics of sodium transport by excised rabbit trachea. *J. Appl. Physiol.* **55,** 1877–1883.

Boveris, A., and Chance, B. (1973). The mitochondrial generation of hydrogen peroxide. General properties and effect of hyperbaric oxygen. *Biochem. J.* **134,** 707–716.

Boveris, A., Oshino, N., and Chance, B. (1972). The cellular production of hydrogen peroxide. *Biochem. J.* **128,** 617–630.

Briscoe, P., Caniggia, I., Graves, A., Benson, B., Huang, L., Tanswell, A. K., and Freeman, B. A. (1995). Delivery of superoxide dismutase to pulmonary epithelium via pH-sensitive liposomes. *Am. J. Physiol.* **268,** L374–L380.

Brown, M. J., Olver, R. E., Ramsden, C. A., Strang, L. B., and Walters, D. V. (1983). Effects of adrenaline and of spontaneous labour on the secretion and absorption of lung liquid in the fetal lamb. *J. Physiol.* (*London*) **344,** 137–152.

Canessa, C. M., Horisberger, J. D., and Rossier, B. C. (1993). Epithelial sodium channel related to proteins involved in neurodegeration. *Nature* (*London*) **361,** 467–470.

Canessa, C. M., Schild, L., Buell, G., Thorens, B., Gautschi, I., Horisberger, J.-D., and Rossier, B. C. (1994). Amiloride-sensitive epithelial Na^+ channel is made of three homologous subunits. *Nature* (*London*) **367,** 463–467.

Caraceni, P., Ryu, H. S., Van Thiel, D. H., and Borle, A. B. (1995). Source of oxygen free radicals produced by rat hepatocytes during postanoxic reoxygenation. *Biochim. Biophys. Acta* **1268,** 249–254.

Cassin, S., and Perks, A. M. (1993). Amiloride inhibits arginine vasopressin-induced decrease in fetal lung liquid secretion. *J. Appl. Physiol.* **75,** 1925–1929.

Chakraborti, S., Gurtner, G. H., and Michael, J. R. (1989). Oxidant-mediated activation of phospholipase A_2 in pulmonary endothelium. *Am. J. Physiol.* **257,** L430–L437.

Chapman, D. L., Carlton, D. P., Cummings, J. J., Poulain, F. R., and Bland, R. D. (1990). Propanolol does not prevent absorption of lung liquid in fetal lambs during labour. *FASEB J.* **4,** A411(abstr.).

Chen, Z., Hagler, J., Palombella, V. J., Melandri, F., Scherer, D., Ballard, D., and Maniatis, T. (1995). Signal induced site-specific phosphorylation targets IκBα to the uniquitin-proteosome pathway. *Genes Dev.* **9,** 1586–1597.

Compeau, C. G., Rotstein, O. D., Tohda, H., Marunaka, Y., Rafii, B., Slutsky, A. S., and O'Brodovich, H. M. (1994). Endotoxin-stimulated alveolar macrophages impair lung epithelial sodium transport by an L-arginine-dependent mechanism. *Am. J. Physiol.* **266,** C1330–C1341.

Corte, E. D., and Stirpe, F. (1972). The regulation of rat liver xanthine oxidase: Involvement of thiol groups in the conversion of of the enzyme activity from dehydrogenase (type D) into oxidase (type O) and purification of the enzyme. *Biochem. J.* **126,** 739–745.

Cott, G. R., Sugahara, K., and Mason, R. (1986). Stimulation of net active ion transport across alveolar type II cell monolayers. *Am. J. Physiol.* **250,** C222–C227.

Darley-Usmar, V., Wiseman, H., and Halliwell, B. (1995). Nitric oxide and oxygen radicals: A question of balance. *FEBS Lett.* **369,** 131–135.

Doctrow, S. R., Huffman, K., Marcus, C. B., Musleh, W., Bruce, A., Baudrey, M., and Malfroy, B. (1997). Salen-manganese complexes:combined superoxide dismutase/catalase mimics with broad pharmacological efficacy. *Adv. Pharmacol.* **38,** 247–269.

Flohe, L., Brigelius-Flohe, R., Saliou, C., Traber, M. G., and Packer, L. (1997). Redox regulation of NF-Kappa B activation. *Free Radicals Biol. Medi.* **22,** 1115–1126.

Forman, H. J., and Kennedy, J. A. (1974). Role of superoxide radical in mitochondrial dehydrogenase reactions. *Biochem. Biophys. Res. Commun.* **60,** 1044–1050.

Gowen, C. W., Lawson, E. E., Gingras, J., Boucher, R., Gatzy, J. T., and Knowles, M. (1988). Electrical potential difference and ion transport across nasal epithelium of term neonates: Correlation with mode of delivery, transient tachypnea of the newborn, and respiratory rate. *J. Pediatr.* **113,** 121–127.

Guengerich, F. P. (1991). Reactions and significance of cytochrome P-450 enzymes. *J. Biol. Chem.* **266,** 10019–10022.

Guengerich, F. P., and McDonald, T. L. (1990). Mechanisms of cytochrome P-450 catalysis. *FASEB J.* **4,** 2453–2459.

Guo, Y., DuVall, M. D., Crow, J. P., and Matalon, S. (1998). Nitric oxide inhibits Na^+ absorption across cultured alveolar type II monolayers. *Am. J. Physiol.* **274,** L369–L377.

Halliwell, B. (1978). Superoxide-dependant formation of hydroxyl radicals in the presence of iron salts. Its role in degredation of hyaluronic acid by a superoxide generating system. *FEBS Lett.* **96,** 238–242.

Halliwell, B., and Gutteridge, J. M. C. (1986). Oxygen free radicals and iron in relation to biology and medicine: Some problems and concepts. *Arch. Biochem. Biophys.* **246,** 501–514.

Herman, P., Tu, T. Y., Loiseau, A., Clerici, C., Cassingena, R., Grodet, A., Friedlander, G., Amiel, C., and Tran Ba Huy, P. (1995). Oxygen metabolites modulate sodium transport in gerbil middle ear epithelium: Involvement of PGE_2. *Am. J. Physiol.* **268,** L390–L398.

Hummler, E., Barker, P., Gatzy, J., Beermann, F., Verdumo, C., Schmidt, A., Boucher, R., and Rossier, B. C. (1996). Early death due to defective neonatal lung liquid clearance in $\alpha ENaC$-deficient mice. *Nat. Genet.* **12,** 325–328.

Jain, L., Chen, X.-J., Brown, L. A., and Eaton, D. C. (1998). Nitric oxide inhibits lung sodium transport through a cGMP-mediated inhibition of epithelial cation channels. *Am. J. Physiol.* **274,** L475–L484.

Kazmi, S. M. I., Plante, R. K., Visconti, V., Taylor, G. R., Zhou, L., and Lau, C. Y. (1995). Suppression of NFkB activation and NFkB-dependant gene expression by tepoxalin a dual inhibitor of cyclooxygenase and 5- lipoxigenase. *J. Cell. Biochem.* **57,** 299–310.

Lander, H. M. (1997). An essential role for free radicals and derived species in signal transduction. *FASEB J.* **11,** 118–124.

Light, D. B., Schwiebert, E. M., Karlson, K. H., and Stanton, B. A. (1989). Atrial natriuretic peptide inhibits cation channel in renal inner medullary collecting duct cells. *Science* **243,** 383–385.

Lingueglia, E., Voilley, N., Waldmann, R., Lazdunski, M., and Barbry, P. (1993). Expression cloning of an epithelial amiloride-sensitive Na^+ channel: A new channel type with homologies to *Caenorhabditis elegans* degenerins. *FEBS Lett.* **318,** 95–99.

Littauer, A., and De Groot, H. (1992). Release of reactive oxygen by hepatocytes on reoxygenation: Three phases and role of mitochondria. *Am. J. Physiol.* **262,** G1015–G1020.

Mairbaurl, H., Wodopia, R., Eckes, S., Schulz, S., and Bartsch, P. (1997). Impairment of cation transport in A549 cells and rat alveolar epithelial cells by hypoxia. *Am. J. Physiol.* **273,** L797–L806.

McCord, J. M. (1985). Oxygen-derived free radicals in postischemic tissue injury. *N. Engl. J. Med.* **312,** 159–163.

McDonald, J. V. J., Gonzales, L. W., Ballard, P. L., Pitha, J., and Roberts, J. M. (1986). Lung beta adrenoreceptor blockade affects perinatal surfactant release but not lung water. *J. Appl. Physiol.* **60,** 1727–1733.

Metzger, J., Dore, S. P., and Lauterburg, B. H. (1988). Oxidant stress during reperfusion of ischemic liver: No evidence for a role of xanthine oxidase. *Hepathology* **8,** 580–584.

Meyer, M., Schreck, R., and Baeuerle, P. A. (1993). H_2O_2 and antioxidants have opposite effects on activation of NF-kB and AP-1 in intact cells: AP-1 as secondary antioxidant-respose factor. *EMBO J.* **12,** 2005–2015.

Meyer, M., Pahl, H. L., and Baeuerle, P. A. (1994). Regulation of the transcription factors NF-kB and AP-1 by redox changes. *Chem.-Biol. Interact.* **91,** 91–100.

Mitchell, J. B., Samuni, A., Krishna, M. C., DeGraff, W. G., Ahn, M. S., Samuni, U., and Russo, A. (1990). Biologically active metal-independant superoxide mimics. *Biochemistry 29,*2802–2807.

Mohazzab-H., K. M., and Wolin, M. S. (1994). Properties of a superoxide anion generating microsomal NADH-oxidoreductase, a potential pulmonary artery PO_2 sensor. *Am. J. Physiol.* **267,** L823–L831.

Mohazzab-H., K. M., Fayngersh, R. P., Kaminski, P. M., and Wolin, M. S. (1995). Potential role of NADH oxidoreductase-derived reactive O_2 species in calf pulmonary arterial PO_2-elicited responses. *Am. J. Physiol.* **269,** L637–L644.

Moncada, S., Palmer, R. M. J., and Higgs, E. A. (1991). Nitric oxide: Physiology, pathophysiology, and pharmacology. *Pharmacol. Rev.* **43,** 109–142.

Moorhouse, P. C., Grootveld, M., Halliwell, B., Quinlan, J. G., and Gutteridge, J. M. C. (1987). Allopurinol and oxypurinol are hydroxyl radical scavengers. *FEBS Lett.* **213,** 23–28.

Muller, J. M., Cahill, M. A., Rupec, R. A., Baeuerle, P. A., and Nordheim, A. (1997). Antioxidants as well as oxidants activate c-fos via ras-dependant activation of extracellular-signal-regulated kinase 2 and elk-1. *Eur. J. Biochem.* **244,** 45–52.

Nohl, H., and Hegner, D. (1978). Do mitochondria produce oxygen radicals in vivo? *Eur. J. Biochem.* **82,** 563–567.

Nohl, H., and Jordan, W. (1986). The mitochondrial site of superoxide formation. *Biochem. Biophys. Res. Commun.* **138,** 533–539.

O'Brodovich, H. M. (1991). Epithelial ion transport in the fetal and perinatal lung. *Am. J. Physiol.* **261,** C555–C564.

O'Brodovich, H. M. (1996). Immature epithelial Na^+ channel expression is one of the pathogenic mechanisms leading to human neonatal respiratory distress syndrome. *Proc. Assoc. Am. Physicians* **108,** 345–355.

O'Brodovich, H. M., Hannam, V., Seear, M., and Mullen, J. B. M. (1990a). Amiloride impairs lung water clearance in newborn guinea pigs. *J. Appl. Physiol.* **68**(4), 1758–1762.
O'Brodovich, H. M., Rafii, B., and Post, M. (1990b). Bioelectric properties of fetal alveolar epithelial monolayers. *Am. J. Physiol.* **258,** L201–L206.
O'Brodovich, H. M., Hannam, V., and Rafii, B. (1991). Sodium channel but neither Na+-H+ nor Na-glucose symport inhibitors slow neonatal lung water clearance. *Am. J. Respir. Cell Mol. Biol.* **5,** 377–384.
O'Brodovich, H. M., Rafii, B., and Perlon, P. (1992). Arginine vasopressin and atrial natriuretic peptide do not alter ion transport by cultured fetal distal lung epithelium. *Pediatr. Res.* **31,** 318–322.
Olver, R. E., and Strang, L. B. (1974). Ion fluxes across the pulmonary epithelium and the secretion of lung liquid in the fetal lamb. *J. Physiol.* (*London*) **241,** 327–357.
Olver, R. E., Ramsden, C. A., Strang, L. B., and Walters, D. V. (1986). The role of amiloride-blockade sodium transport in adrenaline-induced lung liquid reabsorption in the fetal lamb. *J. Physiol.*(*London*) **376,** 321–340.
Otulakowski, G., Rafii, B., Bremner, H. R., and O'Brodovich, H. (1999). Structure and hormone responsiveness of the gene encoding the α-subunit of the rat amiloride-sensitive epithelial sodium channel. *Am. J. Resp. Cell Mol. Biol.* **20,** 1–13.
Perks, A. M., and Cassin, S. (1982). The effects of arginine vasopressin and other factors on the production of lung liquid in fetal goats. *Chest* **81,** 63S–65S.
Perks, A. M., and Cassin, S. (1989). The effects of arginine vasopressin and epinephrine on lung liquid production in fetal goats. *Can. J. Physiol. Pharmacol.* **67,** 491–498.
Pitkänen, O., Tanswell, A. K., Downey, G., and O'Brodovich, H. (1996). Increased PO_2 alters the bioelectric properties of fetal distal lung epithelium. *Am. J. Physiol.* **270,** L1060–L1066.
Planes, C., Escoubet, B., Blot-Chabaud, M., Friedlander, G., Farman, N., and Clerici, C. (1997). Hypoxia downregulates expression and activity of epithelial sodium channels in rat alveolar epithelial cells. *Am. J. Respir. Cell Mol. Biol.* **17,** 508–518.
Rafii, B., Tanswell, A.K., Otulakowski, G., Pitkänen, O., Belcastro-Taylor, R., and O'Brodovich, H. (1998). O_2-induced Na channel expression is associated with Nf-kB activation and blocked by superoxide scavenger. *Am. J. Physiol.* **275,** L764–L770.
Ross, M. G., Ervin, G., Leake, R. D., Fu, P., and Fisher, D. A. (1984). Fetal lung liquid regulation by neuropeptides. *Am. J. Obstet. Gynecol.* **150,** 421–425.
Rupec, R. A., and Baeuerle, P. A. (1995). The genomic response of tumor cells to hypoxia and reoxygenation: Differential activation of transcription factors AP-1 and NF-kB. *Eur. J. Biochem.* **234,** 632–640.
Samuni, A., Krishna, C. M., Mitchell, J. B., Collins, C. R., and Russo, A. (1990). Superoxide reactions with nitroxides. *Free Radical Res. Commun.* **9,** 241–249.
Schenk, H., Klein, M., Erdbrugger, W., Droge, W., and Schulze-Osthoff, K. (1994). Distinct effects of thioredoxin and other antioxidants on the activation of NF-kB and AP-1. *Proc. Nat. Acad. Sci. U.S.A.* **91,** 1672–1676.
Schreck, R., Rieber, P., and Baeuerle, P. A. (1991). Reactive oxygen intermediates as apparently widley used messangers in the activation of NF-kB transcription factor and HIV-1. *EMBO J.* **10,** 2247–2258.
Sen, C. K., and Packer, L. (1996). Antioxidant and redox regulation of gene transcription. *FASEB J.* **10,** 709–720.
Strang, L. B., and Barker, P. M. (1993). Invited editorial on "Amiloride inhibitis arginine vasopressin-induced decrease in fetal lung liquid secretion." *J. Appl. Physiol.* **75,** 1923–1924.
Suzuki, Y. J., Forman, H. J., and Sevanian, A. (1997). Oxidants as stimulators of signal transduction. *Free Radicals Biol. Med.* **22,** 269–285.

Tanswell, A. K., Olson, D. M., and Freeman, B. A. (1990a). Liposome-mediated augmentation of antioxidant defenses in fetal rat pneumocytes. *Am. J. Physiol.* **258,** L165–L172.

Tanswell, A. K., Olson, D. M., and Freeman, B. A. (1990b). Response of fetal rat lung fibroblasts to elevated oxygen concentrations after liposome-mediated augmentation of antioxidant enzymes. *Biochim. Biophys. Acta* **1044,** 269–274.

Taylor, L., Menconi, M. J., and Polgar, P. (1983). The participation of hydroperoxides and oxygen radicals in the control of prostaglandins synthesis. *J. Biol. Chem.* **258,** 6855–6857.

Traenckner, E. B.-M., Pahl, H. L., Henkel, T., Schmidt, K. N., Wilk, S., and Baeuerle, P. A. (1995). Phosphorylation of human IkB-α on serines 32 and 36 controls IkB-α proteolysis and NF-kB activation in response to diverse stimuli. *EMBO J.* **14,** 2876–2883.

Turrens, J. F., Freeman, B. A., Levitt, J. G., and Crapo, J. D. (1982a). The effect of hyperoxia on superoxide production by lung submitochondrial particles. *Arch. Biochem. Biophys.* **217,** 401–410.

Turrens, J. F., Freeman, B. A., and Crapo, J. D. (1982b). Hyperoxia increases H_2O_2 release by lung mitochondria and microsomes. *Arch. Biochem. Biophys.* **217,** 411–421.

Tyler, D. D. (1992). "The Mitochondrion in Health and Disease." VCH Publishers. New York.

Wallace, M. J., Hooper, S., and Harding, R. (1990). Regulation of lung liquid secretion by arginine vasopressin in fetal sheep. *Am. J. Physiol.* **258,** R104–R111.

Walters, D. V., Ramsden, C. A., and Olver, R. E. (1990). Dibutryrl cAMP induces a gestation-dependent absorption of fetal lung liquid. *J. Appl. Physiol.* **68,** 2054–2059.

Younes, M., and Strubelt, O. (1988). The involvement of reactive oxygen species in hypoxic injury to rat liver. *Res. Commun. Chem. Pathol. Pharmacol.* **59,** 369–381.

CHAPTER 15

Catecholamine Regulation of Amiloride-Sensitive Na^+ Transport in the Fetal Rat Alveolar Epithelium

Yoshinori Marunaka and Naomi Niisato
Lung Biology and MRC Group of Lung Development, The Hospital for Sick Children Research Institute, The University of Toronto Faculty of Medicine, Toronto, Ontario, M5G 1X8, Canada

1063-5823/99 $30.00

I. INTRODUCTION

Fetal lung fluid is secreted by the fetal lung epithelium via active Cl^- transport pathways and plays an important role in development, differentiation, and growth of the fetal lung (see Fig. 1) (Alcorn *et al.,* 1977; Olver and Strang, 1974; Brown *et al.,* 1983). However, this fluid is required to be rapidly absorbed as the alveolar cavity is utilized for gas exchange after birth (see Fig. 1) (Brown *et al.,* 1983). This absorption of lung fluid is activated by catecholamine, whose level in blood is increased at birth, via the β_2-adrenoceptor (O'Brodovich, 1991). This β_2 agonist effect has been reported to be mediated via cAMP (Barker *et al.,* 1988). Amiloride impairs this fluid clearance by blocking Na^+ channels but not Na^+/H^+ exchanger (O'Brodovich, 1991). Further, ouabain, an inhibitor of the Na^+,K^+-ATPase, applied to the basolateral membrane also blocks the fluid clearance (O'Brodovich, 1991). These observations suggest that active transepithelial Na^+ transport plays an important role in the fluid clearance via two steps: (1) a passive entry step of Na^+ across the apical membrane via amiloride-sensitive Na^+-permeable channels and (2) an active extrusion step of Na^+ across the basolateral membrane via the Na^+,K^+-ATPase (Fig. 2). In general, the entry step of Na^+ is thought to be the rate-limiting step of the transepithelial Na^+ transport (absorption) (O'Brodovich, 1991). Therefore, the study of amiloride-sensitive Na^+ (or Na^+-permeable) channels would

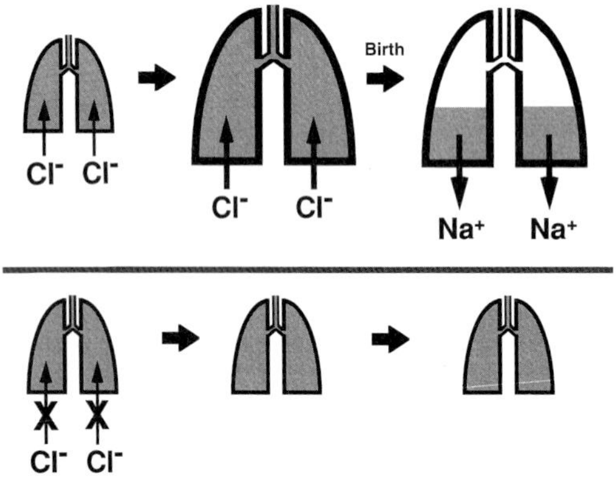

FIGURE 1 A model of fetal lung development. The fetal lung epithelium secretes Cl^- into the lung cavity, inducing the fetal lung fluid which produces a positive pressure from the fetal lung cavity to the fetal lung. The Cl^- secretion-induced positive pressure plays an essential role in development, differentiation, and growth of the fetal lung (Alcorn *et al.,* 1977). On the other hand, this fluid is required to be rapidly absorbed as the alveolar cavity is utilized for gas exchange after birth (Brown *et al.,* 1983). The absorption of the lung fluid is driven by amiloride-sensitive Na^+ transport which is activated by β_2-adrenergic agonists.

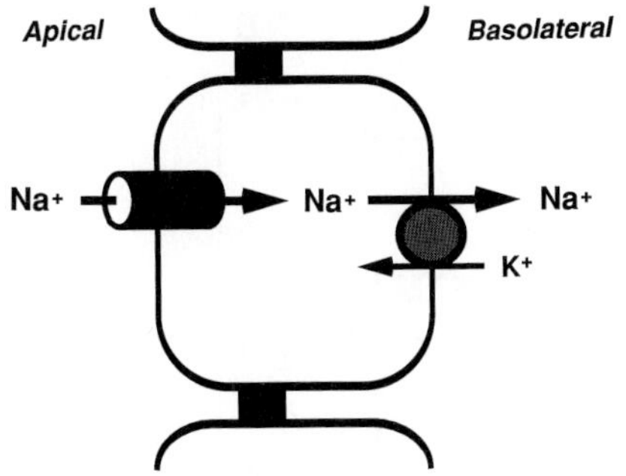

FIGURE 2 A model of the transepithelial Na^+ transport. There are, in general, two steps in the Na^+ transport, namely, (1) the entry step of Na^+ across the apical membrane via amiloride-sensitive Na^+-permeable channels from the lumen into the intracellular space, and (2) the extrusion step of Na^+ across the basolateral membrane via the Na^+,K^+-ATPase to the basolateral space from the intracellular space. It is generally thought that the Na^+ entry step through amiloride-sensitive Na^+-permeable channels is the rate-limiting step in the transepithelial Na^+ transport.

provide an important understanding for the mechanisms of the fluid clearance. An amiloride-sensitive Na^+-permeable 23-pS nonselective cation (NSC) channel has been found in primary cultures of rat fetal distal lung epithelium (FDLE) by Orser et al. (Orser *et al.,* 1991). We have also reported a 28-pS NSC channel in the FDLE (Marunaka *et al.,* 1992b; Tohda *et al.,* 1994; Marunaka, 1996), which may be the same as that reported by Orser *et al.* (1991), although the single-channel conductances are not identical. Further, we have found another type of amiloride-sensitive channel which has a high selectivity for Na^+ with a single channel conductance of 12 pS (Marunaka, 1996). In this chapter, we will describe the characteristics of two types of amiloride-sensitive Na^+-permeable channels, their roles in the catecholamine-induced Na^+ absorption, the regulatory mechanisms of these amiloride-sensitive channels, and the switching mechanisms of FDLE from the secretory to the absorptive epithelium.

II. CHARACTERISTICS OF AMILORIDE-SENSITIVE Na^+-PERMEABLE CHANNELS

A. Single-Channel Conductances and Current–Voltage (I-V) Relationships

Figure 3A shows actual traces of single-channel currents through two types of channels obtained from cell-attached patches formed on the apical membrane of FDLE. In the present study, we have used FDLE cells isolated

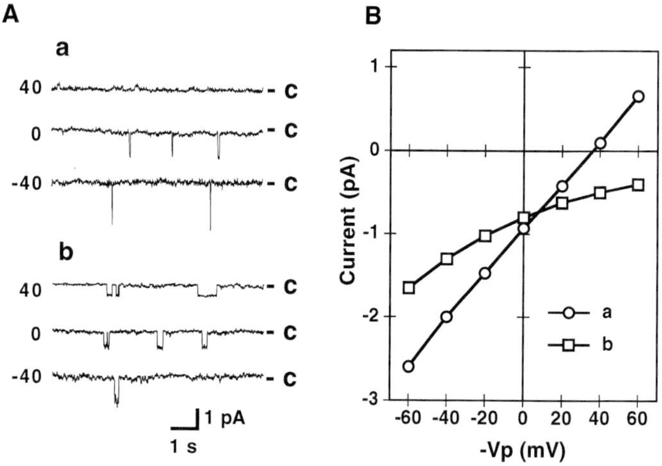

FIGURE 3 Two types of channels in rat fetal distal lung epithelium (FDLE) obtained from 20-day gestation in cell-attached patches. (A) Actual traces of single-channel currents through two types of channels. Voltages to the right of the traces are the displacement (in millivolts) of the intracellular potential. The 0 mV point means that the patch membrane is held at the resting membrane potential; 40 mV means that it is held at 40 mV more positive potential than the resting potential. Downward deflection indicates inward currents across the patch membrane. The horizontal bar and "C" beside each current trace indicate the closed level of the channel. (B) Current-voltage relationships of the two types of the channels shown in Fig. 3A. Circles and squares indicate current amplitudes through the channels shown in Figs. 3A-a and 3A-b, respectively. Modified from *Gen. Pharmacol.* **26,** Tohda and Marunaka (1995), Insulin-activated amiloride-blockable nonselective cation and Na^+ channels in fetal distal lung epithelium, pp. 755–763, Copyright (1995), with permission from Elsevier Science; and from *Pfluegers Arch.* **431,** Marunaka (1996), Amiloride-blockable Ca^{2+}-activated Na^+-permeant channels in the fetal distal lung epithelium, pp. 748–756, copyright notice of Springer-Verlag, with permission.

from the fetuses of pregnant Wistar rats at 20 days' gestation (term = 22 days) and have cultured these cells for two or three days as previously described (Tohda *et al.,* 1994; Nakahari and Marunaka, 1996). Figure 3B indicates the I-V relationships of two types of channels at cell-attached configuration, whose single channel conductances are 28 and 12 pS. The reversal potential of the 28-pS channel is about 40 mV more positive than the resting apical membrane potential. On the other hand, the reversal potential of the 12-pS channel is more positive by > 60 mV than the resting apical membrane potential, since only inward currents but no outward currents can be observed even at 60 mV more positive potential than the resting apical membrane potential.

B. Ion Selectivity

To determine the ion selectivity of both types of channels, we have formed inside-out patches and used the bathing solutions of various ionic compositions. Figure 4A indicates that the channel shown in Fig. 3A-a, which has a larger conductance (28 pS), has almost identical permeability to Na^+ and K^+ ($P_{Na}/P_K = 1.1 \pm 0.1$; $n = 6$) but much lower permeability to Cl^- ($P_{Cl}/P_{Na} < 0.01$, $n = 6$). Therefore, we classify it as a NSC channel. Figure 4B indicates that the channel shown in Figure 3A-b, which has a lower conductance (12 pS), is highly selective for Na^+ ($P_{Na}/P_K > 10$; $n = 3$). Therefore, we classify it as a Na^+ channel. The I-V relationship of the Na^+ channel is linear and the single channel conductance is 12 pS when a symmetrical solution containing 140 mM NaCl is used as pipette and cytosolic solutions in excised inside-out patches. This observation suggests that an inward rectifying I-V relationship observed in cell-attached patches (squares in Fig. 3B) would be caused by a low concentration of cytosolic Na^+ (Goldman type rectification).

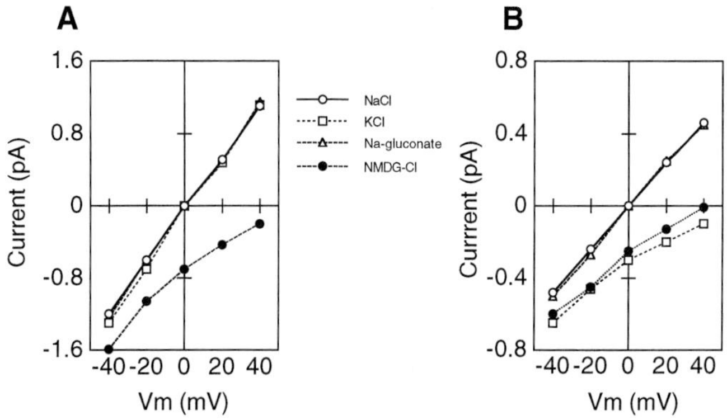

FIGURE 4 Ion selectivity of both types of the channels in excised inside-out patches (A, 28-pS channel; B, 12-pS channel). (A) The ion selectivity of the channel shown in Fig. 3A-a. The P_{Na}/P_K value is 1.1 ± 0.1 and $P_{Na}/P_{Cl} > 50$ ($n = 6$). Therefore, this channel is recognized as a nonselective cation (NSC) channel. (B) The ion selectivity of the channel shown in Fig. 3A-b. It is $P_{Na}/P_K > 10$ and $P_{Na}/P_{Cl} > 40$ ($n = 3$). Therefore, this channel is recognized as a Na^+ channel. In excised inside-out patches, Vm means the membrane potential referred to the extracellular side of the membrane (pipette). The pipette used for determination of ion selectivity contains the NaCl solution. Modified from *Gen. Pharmacol.* **26,** Tohda and Marunaka (1995), Insulin-activated amiloride-blockable nonselective cation and Na^+ channels in fetal distal lung epithelium, pp. 755–763, Copyright (1995), with permission from Elsevier Science; and from *Pfluegers Arch.* **431,** Marunaka (1996), Amiloride-blockable Ca^{2+}-activated Na^+-permeant channels in the fetal distal lung epithelium, pp. 748–756, copyright notice of Springer-Verlag, with permission.

C. Voltage Dependency

The open probability (P_o) of the 28- and 12-pS channels is increased by depolarization of the apical membrane (Fig. 5). However, the absolute values of the P_o of these channels are quite different; i.e., the P_o of the 28-pS channel is much smaller than that of the 12-pS channel (Fig. 5). The single-channel conductances and voltage dependencies of P_o of these channels are quite different from those of ENaC (*e*pithelial *Na* *c*hannel) cloned from rat distal colon by the functional expression in *Xenopus* oocytes (Canessa *et al.,* 1993, 1994). Namely, the single-channel conductance of rENaC is 4–5 pS and the P_o decreases as the membrane depolarizes.

D. Ca^{2+} Dependency

To study the effects of the cytosolic Ca^{2+} concentration ($[Ca^{2+}]_c$), we have varied the $[Ca^{2+}]_c$ of the bathing (cytosolic) solution in excised inside-out patches where the membrane potential is clamped at −40 mV. Figure

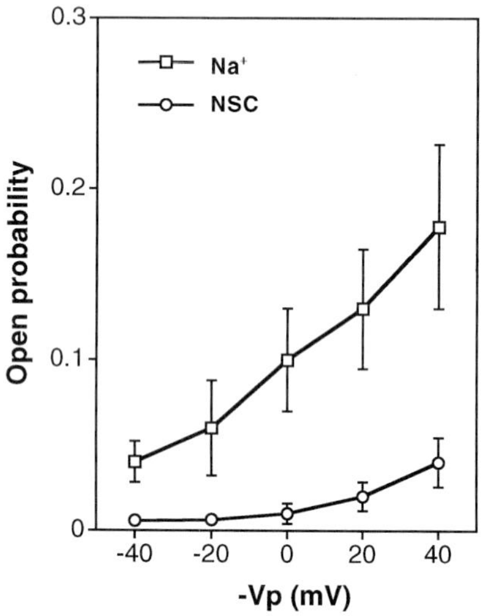

FIGURE 5 The dependency of the open probability (P_o) of the NSC (circles, $n = 5$) and Na^+ (squares, $n = 3$) channels on the membrane potential in cell-attached patches. Modified from *Gen. Pharmacol.* **26,** Tohda and Marunaka (1995), Insulin-activated amiloride-blockable nonselective cation and Na^+ channels in fetal distal lung epithelium, pp. 755–763, Copyright (1995), with permission from Elsevier Science; and from *Pfluegers Arch.* **431,** Marunaka (1996), Amiloride-blockable Ca^{2+}-activated Na^+-permeant channels in the fetal distal lung epithelium, pp. 748–756, copyright notice of Springer-Verlag, with permission.

6 shows the effects of $[Ca^{2+}]_c$ on the P_o of both channels. The P_o of both types of channels increases as $[Ca^{2+}]_c$ is raised. Although both types of channels show the dependency of the P_o on $[Ca^{2+}]_c$, the Na^+ channel is much more sensitive to cytosolic Ca^{2+} than the NSC channel. The P_o of the Na^+ channel almost reaches the maximum value (about 0.8) around 1 μM $[Ca^{2+}]_c$, while the NSC channel has little activity at 1 μM $[Ca^{2+}]_c$.

E. Sensitivity to Amiloride and Benzamil

Both channels are blocked by amiloride or benzamil (an analogue of amiloride but a more specific channel blocker (Kleyman and Cragoe, 1988)) applied from the extracellular side (i.e., in the patch pipette) in cell-attached patches at no applied potential (i.e., resting membrane potential) (Fig. 7). Figure 7A shows the effects of amiloride and benzamil on the P_o of the NSC channel. In the NSC channel, benzamil has larger blocking action than amiloride. Figure 7B shows the effects of amiloride and benzamil on the P_o of the Na^+ channel. In the Na^+ channel, benzamil shows blocking action similar to amiloride unlike in the NSC channel. The half-maximum

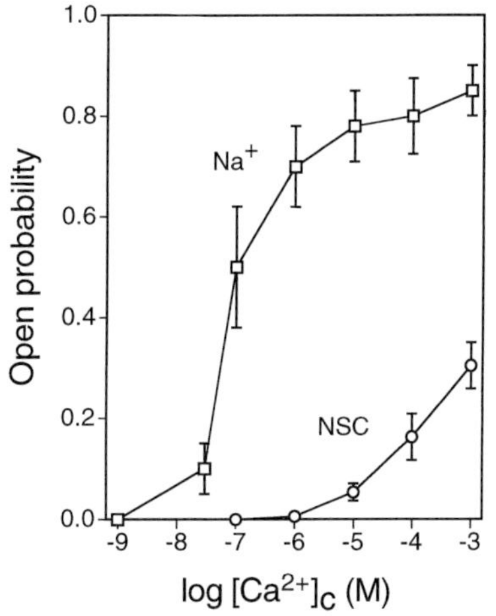

FIGURE 6 The dependency of the open probability (P_o) of the NSC (circles) and Na^+ (squares) channels on the cytosolic Ca^{2+} concentration ($[Ca^{2+}]_c$) in excised inside-out patches at the hold potential of −40 mV. The P_o of both types of channels is increased by elevation of $[Ca^{2+}]_c$. However, the sensitivity of the Na^+ channel to cytosolic Ca^{2+} is much higher than that of the NSC channel, $n = 3$. Modified from *Pfluegers Arch.* **431,** Marunaka (1996), Amiloride-blockable Ca^{2+}-activated Na^+-permeant channels in the fetal distal lung epithelium, pp. 748–756, copyright notice of Springer-Verlag, with permission.

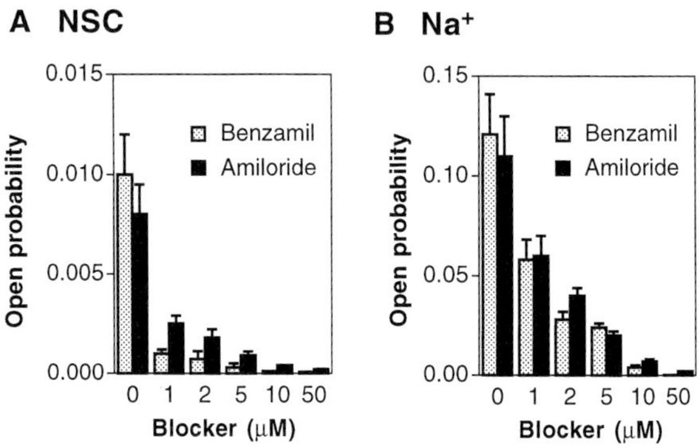

FIGURE 7 Effects of amiloride and benzamil applied into the pipette solution (the extracellular side) on the P_o of the NSC (A) and Na^+ (B) channels in cell-attached patches with no applied potential (the resting apical membrane potential), $n = 3$. Modified from *Gen. Pharmacol.* **26,** Tohda and Marunaka (1995), Insulin-activated amiloride-blockable nonselective cation and Na^+ channels in fetal distal lung epithelium, pp. 755–763, Copyright (1995), with permission from Elsevier Science; and from *Pfluegers Arch.* **431,** Marunaka (1996), Amiloride-blockable Ca^{2+}-activated Na^+-permeant channels in the fetal distal lung epithelium, pp. 748–756, copyright notice of Springer-Verlag, with permission.

inhibitory concentration (IC_{50}) of benzamil in the Na^+ channel is about 1 μM, while the IC_{50} of benzamil in the NSC channel is much less than 1 μM (Fig. 7); i.e., 1 μM benzamil almost completely blocks the NSC channel. This indicates that the NSC channel has a higher affinity site for benzamil than the Na^+ channel. Further, the IC_{50} of amiloride in the Na^+ channel is about 1 μM, while the IC_{50} of amiloride in the NSC channel is less than 1 μM (Fig. 7). This indicates that the NSC channel has a higher affinity site for amiloride than the Na^+ channel. These observations suggest that the NSC channel has higher affinity sites for benzamil and amiloride than the Na^+ channel.

There are at least two distinct categories of Na^+-permeable channels in epithelial cells, with varying sensitivity to amiloride or its analogues (Moran *et al.,* 1988; Barbry *et al.,* 1990; Smith and Benos, 1991). One type of channel has high affinity to benzamil and amiloride (referred to as H-type), and the other has low affinity to benzamil and amiloride (L-type) (Moran *et al.,* 1988). Recently, Matalon *et al.* (1993) have demonstrated that the fetal lung epithelium has two populations of amiloride-sensitive Na^+ channels by measuring $^{22}Na^+$ uptake into plasma membrane vesicles prepared from the epithelium. The plasma membrane of the fetal lung epithelium has two binding sites of benzamil (Matalon *et al.,* 1993); a high-affinity site (K_d = 19 nM) and a low-affinity site (K_d = 1.5 μM). Our present study has shown that 1 μM benzamil has almost completely blocked the NSC channel in cell-attached patches. On the other hand, the P_o of the Na^+ channel has decreased only about 50 % with 1 μM benzamil. These findings suggest that the NSC channel has a high-affinity benzamil binding site, while the Na^+ channel reported here has a relatively low-affinity benzamil binding site with an IC_{50} of about 1 μM, which is similar to that (1.5 μM) reported by Matalon *et al.* (1993). These observations suggest that the NSC channel would be an H-type and the Na^+ channel would be an L-type channel.

Other tissues have amiloride-sensitive Na^+-permeable channels. A 4-pS Na^+ channel in the distal nephron is highly sensitive to amiloride (Eaton and Marunaka, 1990; Marunaka *et al.*, 1992a; Marunaka, 1997; Garty and Palmer, 1997). Further, a Na^+-permeable NSC channel, which is activated by cAMP and stretch (Marunaka *et al.*, 1994, 1997), has been reported in a distal nephron cell line, A6 cell. This channel is sensitive to amiloride of high concentration ($\geq$ 10 μM) only when the channel is activated by stretch (Marunaka *et al.*, 1994).

III. CATECHOLAMINE ACTION

A. Regulation of the 28-pS Nonselective Cation (NSC) and 12-pS Na^+ Channels by Catecholamine

Figure 8 shows actual traces of single channel currents of the NSC and Na^+ channel obtained from cell-attached patches before and after terbutaline (a specific β_2 adrenergic agonist) application. The P_o of the NSC channel at cell-attached configuration is increased by terbutaline application (Fig. 8A). On the other hand, the P_o of the Na^+ channel is not increased by terbutaline (Fig. 8B). The effects of terbutaline on these channels are summarized in Fig. 9A. Further, although we describe the effect of terbutaline on the number of the NSC and Na^+ channels later in detail, we summarize the terbutaline action on the channel number in Fig. 9B; namely, terbutaline increases the number of the NSC channel, but not that of the Na^+ channel. These observations indicate that the regulation of Na^+ transport is mediated through the NSC channel but not through the Na^+ channel. Therefore, we focus our study on the terbutaline regulation of the NSC channel.

B. Catecholamine Action on $[Ca^{2+}]_c$

As described above, the NSC channel can be regulated by terbutaline through modification of P_o and channel number. First we have studied the

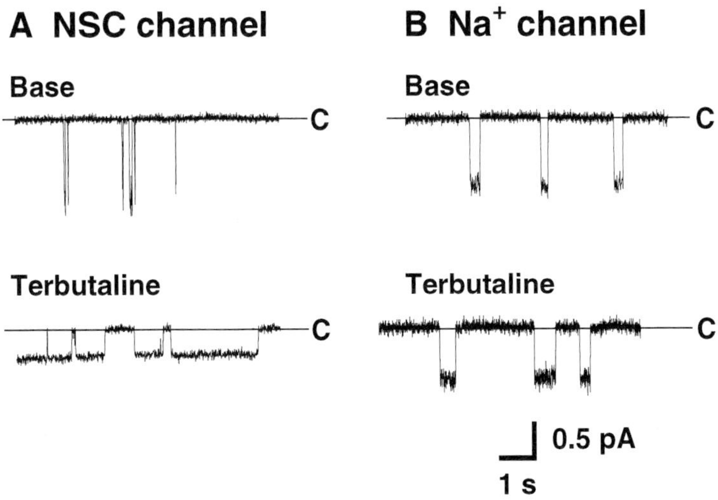

FIGURE 8 Actual traces of single-channel currents through the NSC (A) and Na^+ (B) channels before and after application of terbutaline (10 μM). The P_o of the NSC channel is increased by terbutaline application, but the P_o of the Na^+ channel is not increased. Terbutaline is applied to the basolateral bathing solution, while keeping the cell attached patch which has been made before application of terbutaline.

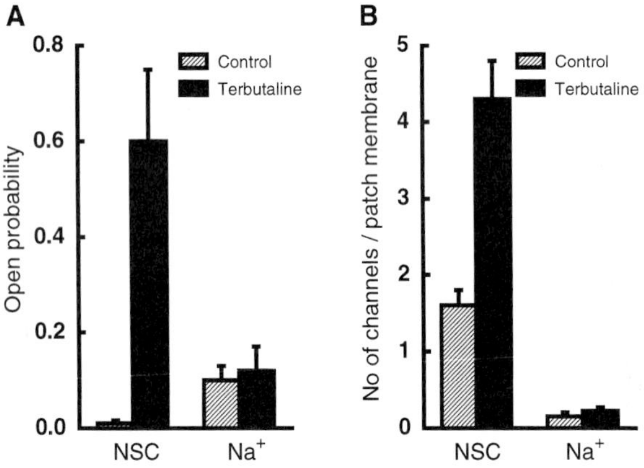

FIGURE 9 The summary of the effects of terbutaline on the P_o (A) and number (B) of the NSC and Na^+ channels. When the terbutaline action on the P_o is studied, terbutaline is applied, while keeping the cell attached patch which has been made before application of terbutaline. On the other hand, when the terbutaline action on the channel number is studied, terbutaline has been applied before the cell-attached patch is made.

regulatory mechanism of P_o of the NSC channel by terbutaline. Since the NSC channel is activated by cytosolic Ca^{2+} (Fig. 6), we have measured $[Ca^{2+}]_c$ (Niisato *et al.*, 1997). We have studied the effects of both terbutaline and DBcAMP (dibutyryl cAMP, a membrane-permeable analogue of cAMP, an intracellular second messenger of terbutaline) on $[Ca^{2+}]_c$. Terbutaline induces an increase in $[Ca^{2+}]_c$ in the presence of 1 mM extracellular Ca^{2+} (Fig. 10A-a). The terbutaline-induced increase in $[Ca^{2+}]_c$ is only tran-

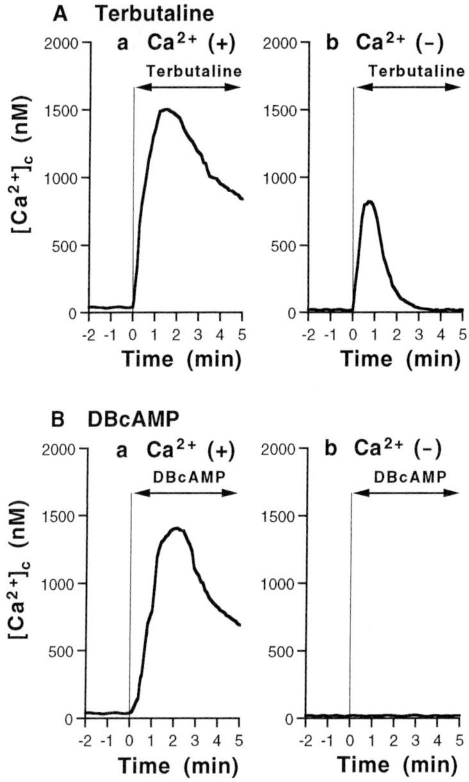

FIGURE 10 Effects of terbutaline (10 μM) and DBcAMP (1 mM) on the cytosolic Ca^{2+} concentration ($[Ca^{2+}]_c$) in the presence and absence of extracellular Ca^{2+}. Each trace shows a typical time course of $[Ca^{2+}]_c$. Similar observations have obtained from four or five more individual experiments. The basal $[Ca^{2+}]_c$ in the presence and absence of extracellular Ca^{2+} is 37 $\pm$ 5 nM (n = 5) and 20 $\pm$ 3 nM (n = 4), respectively. The terbutaline-induced transient increase in $[Ca^{2+}]_c$ in the absence of extracellular Ca^{2+} is blocked by pretreatment with 2 μM thapsigargin, an inhibitor of Ca^{2+}-ATPase which depletes the intracellular Ca^{2+} pool. From Niisato *et al.* (1997) with permission from National Research Council of Canada, Research Press.

siently observed in a Ca^{2+}-free solution (Fig. 10A-b). This transient increase in $[Ca^{2+}]_c$, in the absence of extracellular Ca^{2+}, is abolished by treatment with 2 μM thapsigargin (Niisato *et al.*, 1997), an inhibitor of Ca^{2+}-ATPase which depletes the intracellular Ca^{2+} pool. DBcAMP induces an increase in $[Ca^{2+}]_c$ in the presence of extracellular Ca^{2+} (Fig. 10B-a). In a Ca^{2+}-free solution, DBcAMP causes no significant change in $[Ca^{2+}]_c$ (Fig. 10B-b). These observations suggest that Ca^{2+} release from intracellular Ca^{2+} stores is stimulated by terbutaline but not DBcAMP, and that the stable increase in $[Ca^{2+}]_c$ caused by terbutaline or DBcAMP is due to elevation of Ca^{2+} influx from the extracellular space. Accordingly, these observations indicate that terbutaline, a β_2 agonist, increases a Ca^{2+} influx across the plasma membrane via a cAMP-dependent pathway and stimulates Ca^{2+} release from the intracellular Ca^{2+} pool via a cAMP-independent pathway. (Fig. 11).

C. Regulation of Cell Volume and Cytosolic Cl^- Concentration ($[Cl^-]_c$) by Catecholamine

We have demonstrated that terbutaline activates quinine-sensitive K^+ channels via an increase in $[Ca^{2+}]_c$, but does not activate Ba^{2+}-sensitive K^+ channels (Nakahari and Marunaka, 1996, 1997). On the other hand, the basal K^+ conductance is mainly due to Ba^{2+}-sensitive K^+ channels. Quinine-

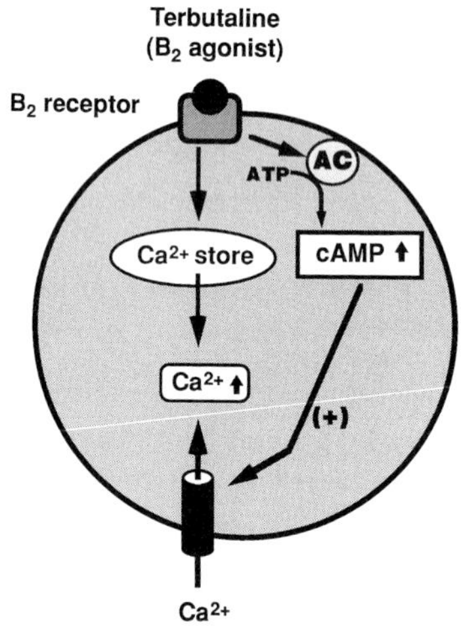

FIGURE 11 A model of the terbutaline- and cAMP-induced Ca^{2+} mobilization. From Niisato *et al.* (1997) with permission from National Research Council of Canada, Research Press.

sensitive K^+ channels also contribute to the basal K^+ conductance, however, the contribution is smaller than that of Ba^{2+}-sensitive K^+ channels (Nakahari and Marunaka, 1996, 1997). The terbutaline-induced activation of quinine-sensitive K^+ channels via an increase in $[Ca^{2+}]_c$ causes initial, rapid cell shrinkage (Fig. 12) (Nakahari and Marunaka, 1996, 1997). On the other hand, terbutaline also causes slow cell shrinkage following the initial, rapid shrinkage (Fig. 12). The slow cell shrinkage occurs through activation of NPPB (5-nitro-2-(3-phenylpropylamino)-benzoate) -sensitive Cl^- channels via a cAMP-dependent, Ca^{2+}-independent pathway (Nakahari and Marunaka, 1996, 1997). The activation of K^+ and Cl^- channels causes KCl release followed by water loss from the cell. The KCl and water loss from the cell induces cell shrinkage as described above and would be expected to cause a decrease in $[Cl^-]_c$ (Fig. 13). Indeed, $[Cl^-]_c$ of FDLE is decreased from 50 to 20 mM after application of terbutaline (Fig. 14) (Tohda *et al.,* 1994).

D. Regulation of Amiloride-Sensitive Nonselective Cation (NSC) Channel by Catecholamine-Induced Change in $[Cl^-]_c$

Next, we have studied the effect of $[Cl^-]_c$ on the P_o of the NSC channel, since terbutaline decreases $[Cl^-]_c$ of FDLE and some articles (Dinudom *et al.,* 1993; Nakahari and Marunaka, 1995a,b) suggest that cytosolic Cl^- regulates amiloride-sensitive Na^+ channels in various tissues. The decrease in $[Cl^-]_c$ drastically increases the P_o of the terbutaline-activated NSC channel but has only a little stimulatory effect on the P_o of the basal NSC channel (Fig. 15) (Tohda *et al.,* 1994; Marunaka *et al.,* 1999). This observation suggests that the cytosolic Cl^- has an inhibitory action on the NSC channel and the sensitivity of the channel to the cytosolic Cl^- is decreased by terbutaline, resulting in an increase of the NSC channel activity (P_o).

E. Relationship between the NSC Channel Activity and the Cl^- Secretion under the Catecholamine-Stimulated Condition

We have further studied the relationship between the regulation of the NSC channel activity by cytosolic Cl^- and the Cl^- secretion. Bumetanide, which inhibits the Cl^- secretion by blocking the $Na^+/K^+/2Cl^-$ cotransporter (Eaton *et al.,* 1992), would decrease the $[Cl^-]_c$. Therefore, we have studied the effect of bumetanide on the terbutaline-stimulated short-circuit current (I_{sc}), which has an amiloride-sensitive component (Hagiwara *et al.,* 1992). Bumetanide slightly decreases the basal I_{sc} (Ito *et al.,* 1997) due to a decrease in the Cl^- secretion, while bumetanide increases the terbutaline-stimulated

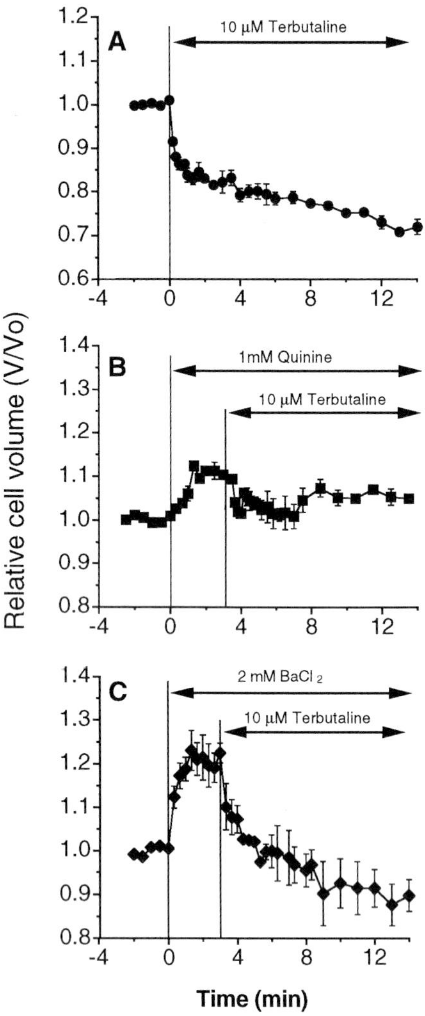

FIGURE 12 The effects of quinine and $BaCl_2$ on cell volume changes induced by 10 μM terbutaline. (A) terbutaline of 10 μM induces initial, rapid cell shrinkage followed by delayed slow cell shrinkage, n = 4. (B) Effects of quinine (1 mM) on the cell-volume changes induced by 10 μM terbutaline, n = 4. (C) Effects of $BaCl_2$ (2 mM) on cell-volume changes induced by 10 μM terbutaline, n = 5. Ba^{2+} has larger effects on the basal cell volume than quinine. On the other hand, Ba^{2+} has little effect on the terbutaline-induced cell-volume change, but quinine markedly diminishes the terbutaline-induced cell shrinkage. Modified from Nakahari and Marunaka (1997) with permission from the Physiological Society.

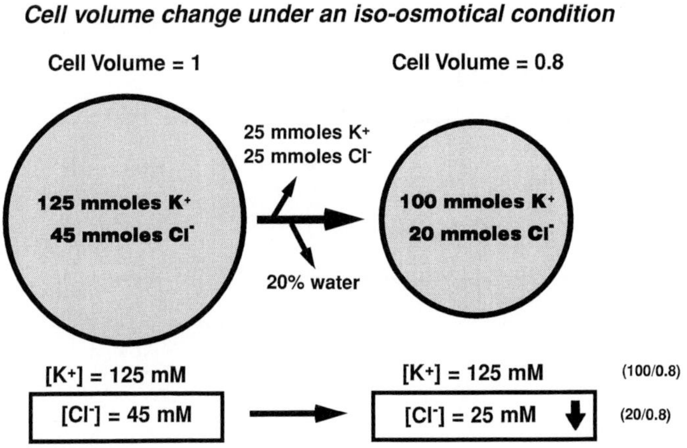

FIGURE 13 The relationship between cell volume (V) and cytosolic Cl^- concentration ($[Cl^-]_c$). When KCl loss occurs isosmotically, the $[Cl^-]_c$ decreases. The major membrane-permeable anion is Cl^-, and a large number of membrane-impermeable anions, such as proteins, are located in the cytosolic space. This situation causes the decrease in $[Cl^-]_c$ when cell shrinkage occurs isosmotically.

I_{sc} (Fig. 16A). The bumetanide-induced increase in I_{sc} is not observed in the presence of amiloride (Fig. 16B). Taken together, these observations indicate that bumetanide increases the terbutaline-stimulated amiloride-sensitive I_{sc} via a decrease in $[Cl^-]_c$ which activates the NSC channel.

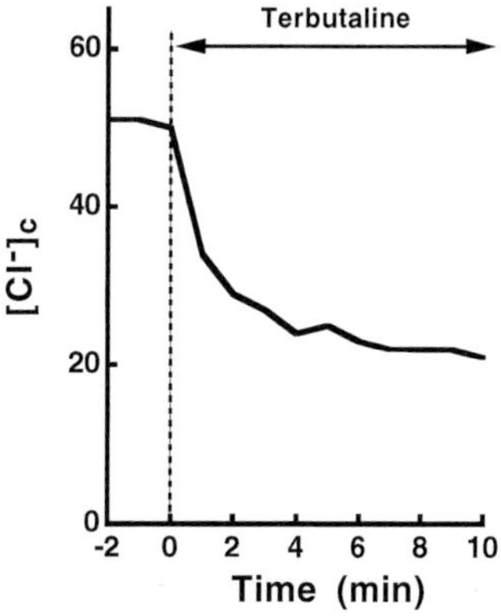

FIGURE 14 Terbutaline action on the cytosolic Cl^- concentration ($[Cl^-]_c$). Terbutaline decreases $[Cl^-]_c$.

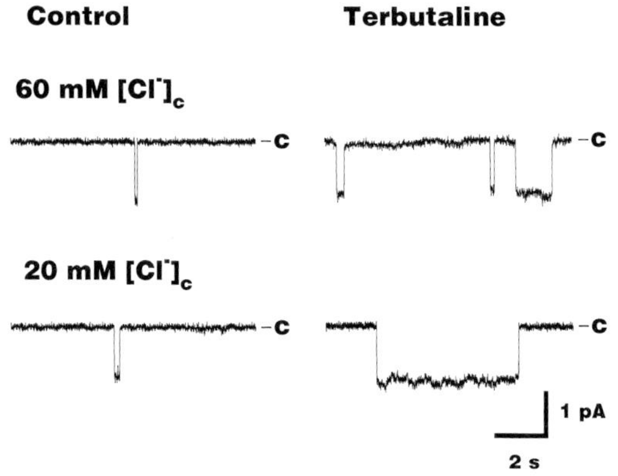

FIGURE 15 The effects of cytosolic Cl^- concentration ($[Cl^-]_c$) on the terbutaline-unstimulated and stimulated NSC channels. The P_o of the terbutaline-unstimulated NSC channel is not increased by lowering $[Cl^-]_c$. On the other hand, the P_o of the terbutaline-stimulated NSC channel is increased by lowering $[Cl^-]_c$.

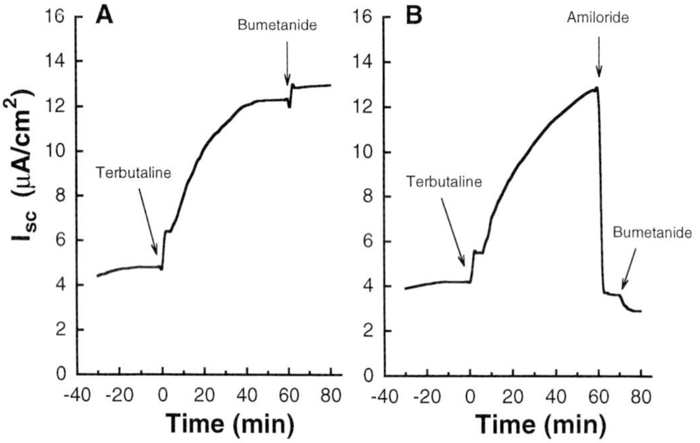

FIGURE 16 Effects of bumetanide on the terbutaline-stimulated I_{sc}. (A) Bumetanide (50μM, basolateral side) is added 60 min after addition of terbutaline. It causes a transient decrease in the I_{sc} followed by an increase. (B) Amiloride (10 μM, apical side) is added 60 min after addition of terbutaline. The application of amiloride immediately decreases the I_{sc}, which then maintains a steady state. Bumetanide (50 μM, basolateral side) is added 10 min after addition of amiloride (70 min after addition of terbutaline). In contrast to findings shown in Fig. 16A, it causes a decrease in the I_{sc} without a following increase. Reprinted from *Gen. Pharmacol.* **26,** Marunaka *et al.,* Bumetanide and bicarbonate increase short-circuit current in fetal lung epithelium, pp. 1513–1517, Copyright (1995), with permission from Elsevier Science.

F. Regulation of the NSC Channel Number by Catecholamine

As described above, terbutaline increases the number of amiloride-sensitive NSC channels at the apical membrane of FDLE. To investigate the mechanism of the terbutaline action on the channel number, we have applied brefeldin A (BFA, 1 μg/ml), an inhibitor of intracellular protein trafficking (Misumi *et al.,* 1986; Miller *et al.,* 1992). The stimulatory effect of terbutaline is completely abolished when BFA has been added into both the apical and basolateral solutions 30 min before terbutaline application (total application time of BFA = 80 min) (Figs. 17A and B). BFA also decreases the amiloride-sensitive I_{sc} of FDLE under the basal condition (Figs. 17A and B).

To directly evaluate the amiloride-sensitive NSC channel number, patch-clamp studies have been performed. Terbutaline treatment increases the NSC channel number per patch membrane by 3-fold ($n = 20$; $P < 0.0001$) (hatched bars in Fig. 17C). On the other hand, terbutaline does not increase the channel number in BFA-treated FDLE (solid bars in Fig. 17C).

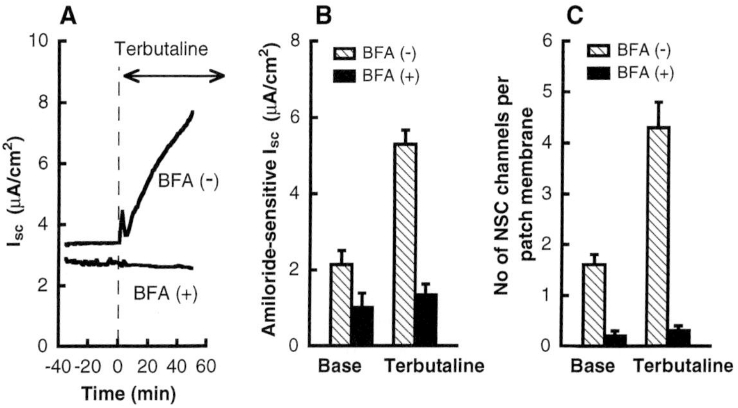

FIGURE 17 Effects of brefeldin A (BFA) on the basal and terbutaline-stimulated amiloride-sensitive I_{sc} and the number of amiloride-sensitive NSC channel. (A) The terbutaline action on the I_{sc} in the presence and absence of BFA. BFA treatment completely abolishes the stimulatory action on the I_{sc}. (B) The terbutaline action on the amiloride-sensitive I_{sc} in the presence and absence of BFA. In the presence of BFA, terbutaline cannot increase the amiloride-sensitive I_{sc}, while terbutaline increases the amiloride-sensitive I_{sc} in the absence of BFA. (C) The terbutaline action on the number of amiloride-sensitive NSC channels. Terbutaline increases the number of amiloride-sensitive NSC channels, and the terbutaline action on the channel number is completely blocked by BFA. Modified from *Pfluegers Arch.* **434,** Ito *et al.* (1997), The effect of brefeldin A on terbutalin-induced sodium absorption in fetal rat distal lung epithelium, pp. 492–494, copyright notice of Springer-Verlag, with permission.

These results indicate that terbutaline stimulates translocation of the NSC channel into the apical membrane from the intracellular store site. However, it is still unclear whether the terbutaline-stimulated translocation of the channel is mediated through a protein kinase A-dependent pathway. We have recently reported that the cAMP-stimulated translocation of Cl^- channels into the apical membrane from the intracellular store site is not mediated through PKA-dependent pathways (Shintani and Marunaka, 1996), suggesting a possibility that terbutaline also stimulates the translocation of the NSC channel via a PKA-independent pathway.

G. Regulation of the Na^+,K^+-ATPase by Catecholamine

As mentioned above (Fig. 2), the transepithelial Na^+ absorption is mediated through two steps: (1) Na^+ entry across the apical membrane, and (2) Na^+ extrusion across the basolateral membrane. Therefore, there are two possible sites which could be targets of the inhibitory effect of BFA on the amiloride-sensitive Na^+ absorption; i.e., BFA would decrease (1) the number of amiloride-sensitive NSC channels at the apical membrane or (2) the number (capacity) of the Na^+,K^+-ATPase at the basolateral membrane. As described above, BFA decreases the number of NSC channels at the apical membrane. Further, to evaluate the effect of BFA on the capacity of the Na^+,K^+-ATPase at the basolateral membrane, we have applied nystatin (50 μM) to the apical solution, which increases the cytosolic Na^+ concentration via elevation of the apical Na^+ conductance that has fully activated the Na^+,K^+-ATPase and can measure the Na^+,K^+-ATPase-generated I_{sc} as an ouabain-sensitive I_{sc}. Under this condition, the rate limiting step of the Na^+ absorption is not the Na^+ entry step through channels across the apical membrane but the Na^+ extrusion step by the Na^+,K^+-ATPase across the basolateral membrane. This nystatin application to the apical membrane has enabled us to study the effect of BFA on the Na^+,K^+-ATPase capacity without the effect of BFA on the apical Na^+ conductance (Na^+-permeable channels). BFA has no significant effects on the basal or terbutaline-stimulated Na^+,K^+-ATPase-generated I_{sc} (Fig. 18). This observation suggests that the Na^+,K^+-ATPase is not affected by BFA, but the inhibitory effect of BFA on the terbutaline-induced I_{sc} occurs by affecting the translocation of the amiloride-sensitive NSC channel into the apical membrane from the intracellular store site (Ito *et al.*, 1997).

Further, under the basal (unstimulated) condition, the maximum current generated by the Na^+,K^+-ATPase is larger than the amiloride-sensitive I_{sc} (Figs. 17B and 18), suggesting that the rate-limiting step of the transepithelial Na^+ transport is the entry step of Na^+ through amiloride-sensitive Na^+-

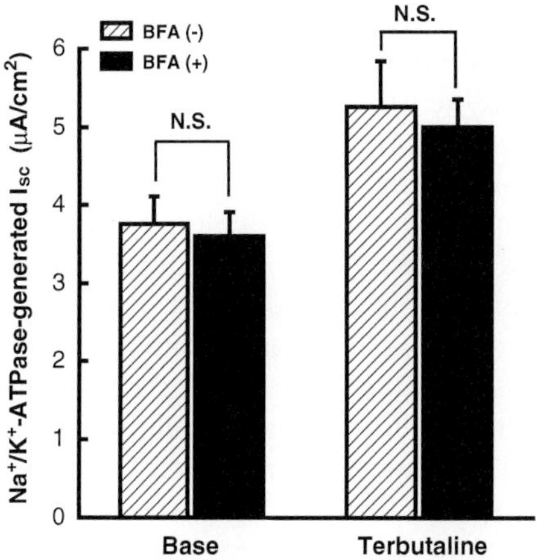

FIGURE 18 The effect of brefeldin A (BFA) on the basal and terbutaline-stimulated Na^+,K^+ATPase-generated I_{sc}. Irrespectively of terbutaline stimulation, BFA has no significant effects on the Na^+,K^+-ATPase-generated I_{sc}. Modified from *Pfluegers Arch.* **434,** Ito *et al.* (1997), The effect of brefeldin A on terbutalin-induced sodium absorption in fetal rat distal lung epithelium, pp. 492–494, copyright notice of Springer-Verlag, with permission.

permeable channels at the apical membrane. On the other hand, under the terbutaline-stimulated condition, the maximum current generated by the Na^+,K^+-ATPase does not exceed the amiloride-sensitive I_{sc} (Figs. 17B and 18). This indicates a possibility that the rate-limiting step of the transepithelial Na^+ transport is not the entry step of Na^+ through amiloride-sensitive Na^+-permeable channels at the apical membrane but the extrusion step of Na^+ by the Na^+,K^+ATPase at the basolateral membrane under the terbutaline-stimulated condition. Therefore, some regulators which increase the capacitance of the Na^+,K^+-ATPase could enhance the stimulatory action of the β_2-agonist on the amiloride-sensitive transepithelial Na^+ absorption via the stimulatory action on the Na^+,K^+-ATPase.

IV. SWITCHING MECHANISM OF FDLE TO THE ABSORPTIVE FROM THE SECRETORY TISSUE BY CATECHOLAMINE

To study the switching mechanism of FDLE from the secretory tissue to the absorptive tissue, we have measured the paracellular permeability under the terbutaline-unstimulated and stimulated conditions. To identify the paracellular permeability, we have tried to block the transcellular ion-transporting pathway by ion-channel blockers and ion-transporter inhibitors (amiloride, quinine, $BaCl_2$, DPC (a chloride channel blocker), ouabain, and bumetanide). When we measure the paracellular permeability under the terbutaline-stimulated condition, we have added these blockers after application of terbutaline. In the presence of these blockers, the transcellular potential difference becomes 0 mV from some negative value (usually −5 and −10 mV), suggesting that the transcellular ion-transporting pathways are blocked by these ion-channel blockers and ion-transporter inhibitors. In the presence of these blockers, we have changed the ionic compositions of the apical and basolateral solutions. Using these methods, we have determined the paracellular ion permeability. As shown in Fig. 19, Na^+ and Cl^- have almost identical paracellular permeability under both the terbutaline-unstimulated and stimulated conditions. These observations indicate that Na^+ and Cl^- can be transported between the air (apical) and blood (basolateral) spaces through the paracellular pathway according to the electrochemical gradient. Taken together, these observations suggest a model of the switching mechanism of FDLE from the secretory tissue to the absorptive tissue (Fig. 20). Under the terbutaline-unstimulated condition, FDLE secretes Cl^- into the future air space through transcellular pathways similar to other Cl^- secretory tissues; i.e., Cl^- is accumulated into the intracellular space across the basolateral membrane by the $Na^+/K^+/2Cl^-$ cotransporter and the Cl^-/HCO_3^- exchanger, and is released to the future air space through Cl^- channels at the apical membrane. Under

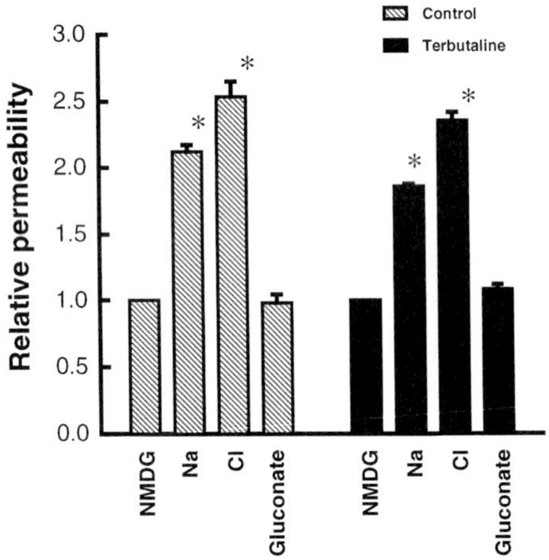

FIGURE 19 The relative values of the paracellular permeability to various ions of the terbutaline-unstimulated and stimulated FDLE in the presence of ion-channel blockers and ion-transporter inhibitors (amiloride, quinine, $BaCl_2$, DPC (a chloride channel blocker), ouabain, and bumetanide). The FDLE has almost identical permeability to Na^+ and Cl^-, irrespective of terbutaline treatment.

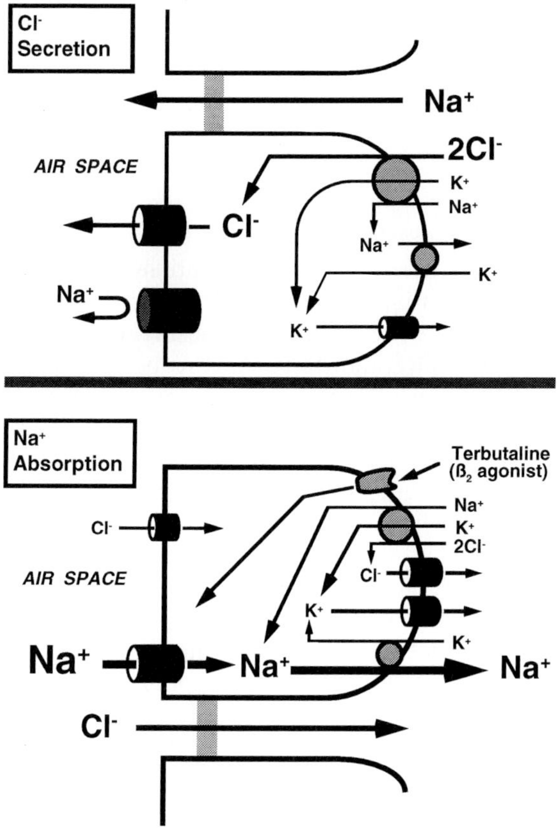

FIGURE 20 Models of ion movements in the FDLE unstimulated and stimulated by terbutaline. (A) Under the basal condition, where the Na^+ conductance is much smaller than the Cl^- conductance, Cl^- is transcellularly secreted followed by Na^+ secretion via the paracellular pathway according to the electrochemical potential generated by active Cl^- secretion. (B) Under the terbutaline-stimulated condition, where the Na^+ conductance is larger than the Cl^- conductance, Na^+ is transcellularly absorbed followed by Cl^- absorption via the paracellular pathway according to the electrochemical potential generated by active Na^+ absorption.

this condition, the Na^+ movement paracellularly occurs, following the Cl^- secretion. On the other hand, under the terbutaline-stimulated condition, the apical Na^+ conductance is increased by terbutaline. This increase in the apical Na^+ conductance depolarizes the apical membrane, decreasing the driving force for Cl^- release across the apical membrane from the intracellular space and producing the Cl^- movement to the basolateral space from the air space through the paracellular pathway. Thus, as a result from the terbutaline-induced increase in the apical Na^+ conductance, the transcellular Na^+ absorption is increased and the paracellular Cl^- absorption is induced (Fig. 20).

V. CONCLUSION

Active Cl^- secretion via the fetal lung epithelium producing the driving force for secretion of the fetal lung fluid essentially plays an important role in development, differentiation, and growth of the fetal lung. On the other hand, this fluid is required to be rapidly reabsorbed as the alveolar cavity is utilized for gas exchange after birth. This reabsorption of the fetal lung fluid is initiated by catecholamine, whose level in blood is increased at birth, via the following pathways. Catecholamine increases the apical Na^+ conductance via amiloride-sensitive Na^+-permeable nonselective cation channels due to the increases in activity and number of the channel. This catecholamine-induced increase in the apical Na^+ conductance essentially produces the active Na^+ reabsorption, by which catecholamine also decreases the driving force for Cl^- release across the apical membrane. This catecholamine-induced increase in the Na^+ absorption exceeding the Cl^- secretion provides the net NaCl absorption that generates osmotic gradient for water absorption, leading to fluid absorption. The catecholamine-induced increase in the apical Na^+ conductance is just a simple mechanism to switch the fetal lung from the secretory to the absorptive tissue.

Acknowledgments

This work was supported by grants from the Medical Research Council of Canada (Group Grant), the Kidney Foundation of Canada, the Ministry of Education, Science, Sports, and Culture of Japan (International Scientific Research Program), and the Ontario Thoracic Society (Block Term Grant) to Y. Marunaka.

References

Alcorn, D., Adamson, T. M., Lambert, T. F., Maloney, J. E., Ritchie, B. C., and Robinson, P. M. (1977). Morphological effects of chronic tracheal ligation and drainage in the fetal lamb lung. *J. Anat.* **123,** 649–660.

Barbry, P., Chassande, O., Marsault, R., Lazdunski, M., and Frelin, C. (1990). [^{3}H]Phenamil binding protein of the renal epithelium Na^+ channel. Purification, affinity labeling, and functional reconstitution. *Biochemistry* **29,** 1039–1045.

Barker, P. M., Brown, M. J., Ramsden, C. A., Strang, L. B., and Walters, D. V. (1988). The effect of thyroidectomy in the fetal sheep on lung liquid reabsorption induced by adrenaline or cyclic AMP. *J. Physiol.* (*London*) **407,** 373–383.

Brown, M. J., Olver, R. E., Ramsden, C. A., Strang, L. B., and Walters, D. V. (1983). Effects of adrenaline and of spontaneous labour on the secretion and absorption of lung liquid in the fetal lamb. *J. Physiol.* (*London*) **344,** 137–152.

Canessa, C. M., Horisberger, J.-D., and Rossier, B. C. (1993). Epithelial sodium channel related to proteins involved in neurodegeneration. *Nature (London)* **361,** 467–470.

Canessa, C. M., Schild, L., Buell, G., Thorens, B., Gautschi, I., Horisberger, J.-D., and Rossier, B. C. (1994). Amiloride-sensitive epithelial Na^+ channel is made of three homologous subunits. *Nature (London)* **367,** 463–467.

Dinudom, A., Young, J. A., and Cook, D. I. (1993). Na^+ and Cl^- conductances are controlled by cytosolic Cl^- concentration in the intralobular duct cells of mouse mandibular gland. *J. Membrane Biol.* **135,** 289–295.

Eaton, D. C. and Marunaka, Y. (1990). Ion channel fluctuation: "Noise" and single-channel measurements. *Curr. Top. Memb. Transp.* **37,** 61–114.

Eaton, D. C., Marunaka, Y., and Ling, B. N. (1992). Ion channels in epithelial tissue: Single channel properties. *In* "Membrane Transport in Biology" (J. A. Schafer, H. H. Ussing, P. Kristensen, and G. H. Giebisch, eds.), pp. 73–165. Springer-Verlag, Berlin.

Garty, H., and Palmer, L. G. (1997). Epithelial sodium channels: Function, structure, and regulation. *Physiol. Rev.* **77,** 359–396.

Hagiwara, N., Tohda, H., Doi, Y., O'Brodovich, H., and Marunaka, Y. (1992). Effects of insulin and tyrosine kinase inhibitor on ion transport in the alveolar cell of the fetal lung. *Biochem. Biophys. Res. Commun.* **187,** 802–808.

Ito, Y., Niisato, N., O'Brodovich, H., and Marunaka, Y. (1997). The effect of brefeldin A on terbutaline-induced sodium absorption in fetal rat distal lung epithelium. *Pfluegers Arch.—Eur. J. Physiol.* **434,** 492–494.

Kleyman, T. R., and Cragoe, E. J., Jr. (1988). Amiloride and its analogs as tools in the study of ion transport. *J. Membr. Biol.* **105,** 1–21.

Marunaka, Y. (1996). Amiloride-blockable Ca^{2+}-activated Na^+-permeant channels in the fetal distal lung epithelium. *Pfluegers Arch.—Eur. J. Physiol.* **431,** 748–756.

Marunaka, Y. (1997). Hormonal and osmotic regulation of NaCl transport in renal distal nephron epithelium. *Jpn. J. Physiol.* **47,** 499–511.

Marunaka, Y., Hagiwara, N., and Tohda, H. (1992a). Insulin activates single amiloride-blockable Na channels in a distal nephron cell line (A6). *Am. J. Physiol.* **263,** F392–F400.

Marunaka, Y., Tohda, H., Hagiwara, N., and O'Brodovich, H. (1992b). Cytosolic Ca^{2+}-induced modulation of ion selectivity and amiloride sensitivity of a cation channel and beta agonist action in fetal lung epithelium. *Biochem. Biophys. Res. Commun.* **187,** 648–656.

Marunaka, Y., Tohda, H., Hagiwara, N., and Nakahari, T. (1994). Antidiuretic hormone-responding non-selective cation channel in distal nephron epithelium (A6). *Am. J. Physiol.* **266,** C1513–C1522.

Marunaka, Y., Doi, Y., and Nakahari, T. (1995). Bumetanide and bicarbonate increase short-circuit current in fetal lung epithelium. *Gen. Pharmacol.* **26,** 1513–1517.

Marunaka, Y., Shintani, Y., Downey, G. P., and Niisato, N. (1997). Activation of Na^+-permeant cation channel by stretch and cAMP-dependent phosphorylation in renal epithelial A6 cells. *J. Gen. Physiol.* **110,** 327–336.

Marunaka, Y., Niisato, N., O'Brodovich, H., and Eaton, D. C. (1999). Regulation of an amiloride-sensitive Na^+-permeable channel by a β_2-adrenergic agonist, cytosolic Ca^{2+} and Cl^- in fetal rat alveolar epithelium. *J. Physiol. (London)* (in press).

Matalon, S., Bauer, M. L., Benos, D. J., Kleyman, T. R., Lin, C., Cragoe, E. J., Jr., and O'Brodovich, H. (1993). Fetal lung epithelial cells contain two populations of amiloride-sensitive Na^+ channels. *Am. J. Physiol.* **264,** L357–L364.

Miller, S. G., Carnell, L., and Moore, H. H. (1992). Post-Golgi membrane traffic: Brefeldin A inhibits export from distal Golgi compartments to the cell surface but not recycling. *J. Cell Biol.* **118,** 267–283.

Misumi, Y., Miki, K., Takatsuki, A., Tamura, G., and Ikehara, Y. (1986). Novel blockade by brefeldin A of intracellular transport of secretory proteins in cultured rat hepatocytes. *J. Biol. Chem.* **261,** 11398–11403.

Moran, A., Asher, C., Cragoe, E. J., Jr., and Garty, H. (1988). Conductive sodium pathway with low affinity to amiloride in LLC- PK1 cells and other epithelia. *J. Biol. Chem.* **263,** 19586–19591.

Nakahari, T. and Marunaka, Y. (1995a). ADH-evoked $[Cl^-]_i$-dependent transient in whole cell current of distal nephron cell line A6. *Am. J. Physiol.* **268,** F64–F72.

Nakahari, T., and Marunaka, Y. (1995b). Regulation of whole cell currents by cytosolic cAMP, Ca^{2+}, and Cl^- in rat fetal distal lung epithelium. *Am. J. Physiol.* **269,** C156–C162.

Nakahari, T., and Marunaka, Y. (1996). Regulation of cell volume by β_2-adrenergic stimulation in rat fetal distal lung epithelial cells. *J. Membr. Biol.* **151,** 91–100.

Nakahari, T., and Marunaka, Y. (1997). β-agonist-induced activation of Na^+ absorption and KCl release in rat fetal distal lung epithelium: A study of cell volume regulation. *Exp. Physiol.* **82,** 521–536.

Niisato, N., Nakahari, T., Tanswell, K., and Marunaka, Y. (1997). β_2-agonist regulation of cell volume in fetal distal lung epithelium by cAMP-independent Ca^{2+} release from intracellular stores. *Can. J. Physiol. Pharmacol.* **75,** 1030–1033.

O'Brodovich, H. (1991). Epithelial ion transport in the fetal and perinatal lung. *Am. J. Physiol.* **261,** C555–C564.

Olver, R. E., and Strang, L. B. (1974). Ion fluxes across the pulmonary epithelium and the secretion of lung liquid in the fetal lamb. *J. Physiol.* (*London*) **241,** 327–357.

Orser, B. A., Bertlik, M., Fedorko, L., and O'Brodovich, H. (1991). Cation channel in fetal alveolar type II epithelium. *Biochim. Biophys. Acta* **1094,** 19–26.

Shintani, Y., and Marunaka, Y. (1996). Regulation of chloride channel trafficking by cyclic AMP via protein kinase A-independent pathway in A6 renal epithelial cells. *Biochem. Biophys. Res. Commun.* **223,** 234–239.

Smith, P. R., and Benos, D. J. (1991). Epithelial Na^+ channels. *Annu. Rev. Physiol.* **53,** 509–530.

Tohda, H., and Marunaka, Y. (1995). Insulin-activated amiloride-blockable nonselective cation and Na^+ channels in fetal distal lung epithelium. *Gen. Pharmacol.* **26,** 755–763.

Tohda, H., Foskett, J. K., O'Brodovich, H., and Marunaka, Y. (1994). Cl^- regulation of a Ca^{2+}-activated nonselective cation channel in β agonist treated fetal lung alveolar epithelium. *Am. J. Physiol.* **266,** C104–C109.

CHAPTER 16

Cyclic Nucleotide-Gated Cation Channels Contribute to Sodium Absorption in Lung: Role of Nonselective Cation Channels

Sandra Guggino
Johns Hopkins University Medical School, Baltimore, Maryland 21205

I. NONSELECTIVE CATION CHANNELS IN EPITHELIA

When the single-channel patch-clamp technique was first used to investigate ENaC activity in various epithelial cells including airway cells, other types of sodium channels were also identified. These nonselective cation channels had a single-channel conductance between 10–30 pS and nearly equal selectivity for sodium and potassium (Duszyk *et al.*, 1989; Feng *et al.*, 1993; Joris *et al.*, 1989; Jorissen *et al.*, 1990; Matalon *et al.*, 1992), although some channels had a higher sodium to potassium ratio of about 7 : 1 (Chinet *et al.*, 1993; Yue *et al.*, 1994.). Some of these channels were activated by calcium (Chinet *et al.*, 1993; Jayr *et al.*, 1994; Joris *et al.*, 1989; Tohda *et al.*, 1994) consistent with being in the family of calcium-activated nonselective cation channels (Maruyama and Petersen, 1982). The amiloride sensitivity of these channels was also variable. Some channels were not sensitive to

1063-5823/99 $30.00

amiloride (Joris *et al.,* 1989) whereas others were blocked by amiloride (Light *et al.,* 1988, 1989, 1990). The sodium channel in alveolar type II pneumocytes was blocked by amiloride and ethylisopropylamiloride (Matalon *et al.,* 1993; Wang *et al.,* 1993), the latter an agent which does not inhibit ENaC (George *et al.,* 1989) except in mM concentrations.

Nonselective cation channels have been identified in fetal distal lung epithelium (Marunaka *et al.,* 1992; Matalon *et al.,* 1993; Tohda *et al.,* 1994), in alveolar type II cells (Feng *et al.,* 1993, MacGregor *et al.,* 1994, Yue *et al.,* 1994) and in human nasal epithelial cells (Chinet *et al.,* 1993, Duszyk *et al.,* 1989; Jorissen *et al.,* 1990). Other epithelial cells also have nonselective cation channels including sweat ducts cells (Joris *et al.,* 1989), sweat gland cells (Disser *et al.,* 1991, Krouse *et al.,* 1989), A6 cells (Hamilton and Eaton, 1985, Hamilton and Eaton 1986), cortical collecting duct cells (Ahmad *et al.,* 1992), and the apical membrane of cells from inner medullary collecting duct cells (Light *et al.,* 1988, 1989, 1990). Nonselective channels are quite abundant (21% of cell-attached patches) compared to ENaC (6% of cell-attached patches) in cultured human nasal polyp cells (Chinet *et al.,* 1993). Although nonselective cation channels are abundant in epithelial tissues, their function remains unexplained.

The growth of epithelial cells on collagen matrix increases the amiloride-sensitive $^{22}Na^+$ influx into A6 cells (Sariban-Sohraby *et al.,* 1983) and the amiloride-sensitive whole-cell sodium current in airway cells (Kunzelmann *et al.,* 1996). Addition of the differentiating agent butyrate further increases the amiloride-sensitive sodium current (Kunzelmann *et al.,* 1996). Since growth on collagen substrate results in better differentiation of the epithelium (Sariban-Sohraby *et al.,* 1983), it was originally thought that ENaC channels were associated only with completely differentiated epithelia and that the nonselective cation channels might be a tissue culture "artifact" the result of poor cellular differentiation. Dexamethasone, a well known tissue culture differentiating agent, enhances tissue development in cultured organ transplants of fetal lung as judged by earlier differentiation of lamellar bodies and increased incorporation of ^{3}[H]choline into phosphatidylcholine (Smith *et al.,* 1990; Tanswell *et al.,* 1983). Thus, a generalized epithelial differentiation may be a major part of the action of dexamethasone that increases ENaC message in cultured lung cells (Champigny *et al.,* 1994). The fact that the ENaC knockout mouse cannot clear liquid from the lung after birth (Hummler *et al.,* 1996) suggests that ENaC-mediated sodium absorption removes fluid from the lung during the transition from the fluid-filled fetal lung to the air breathing adult lung. Dexamethasone is given before a premature birth to encourage development of the human fetal lung and it is also thought that increased ENaC enhances clearance of lung fluid in newborn infants (O'Brodovich, 1996). Thus, in lung, like in A6 cells, the abundance of ENaC is increased in a differentiated epithelium, but no real quantitative information exists about the abundance of nonselective cation channels in relation to differentiation.

Another explanation that has been given for the high incidence of nonselective cation channels, when an investigation is performed using the patch-clamp technique, is that during patch excision the nonselective cation channels are spuriously activated perhaps because they are excised from intracellular organelles. The endoplasmic reticulum or endocytic vesicles that lie directly beneath the plasma membrane could contain different populations of channels from those expressed on the plasma membrane. For example, when nasal epithelial cells from cystic fibrosis patients or normal human controls were investigated using the cell-excised patch clamp mode compared to the cell-attached mode, there was a difference in the channels identified. In excised patches the channels appeared to be mainly nonselective cation channels. Many of these were calcium-activated nonselective cation channels because the incidence of these channels increased to 50% when the patches were bathed in 10^{-3} M calcium compared to an incidence of 7–15% when patches were exposed to 10^{-7} M calcium. The single-channel conductance of the excised channels was about 20 pS and most were insensitive to amiloride. However, in the cell-attached mode, although the channels still had a single-channel conductance of about 20 pS, they appeared to have a greater selectivity for sodium over potassium as judged by the reversal potential (Chinet *et al.,* 1993).

A final explanation for the presence of large nonselective cation channels is that some of these channels could represent ENaC in another conductance state. For example, when αbENaC was expressed in oocytes, then vesicles isolated from these oocytes were incorporated into bilayers and the αbENaC had a conductance of about 40 pS in 100 mM NaCl and often a lower selectivity (6:1) (Fuller *et al.,* 1997) than ENaC measured by the patch-clamp technique in native cortical collecting duct cells (Palmer and Frindt, *et al.,* 1986). Channels isolated from A6 cells and incorporated into bilayers had single-channel conductances of both 4 and 44 pS in 100 mM Na (Olans *et al.,* 1984). When $\alpha\beta\gamma$ rENaC was transfected into oocytes the single-channel conductance was 5 pS (Canessa *et al.,* 1994), but when the αrENaC subunit alone was transfected and subjected to stretch the single-channel conductance was 25 pS (Kinzer *et al.,* 1997). This may suggest that the stretching mechanics in a naked bilayer or expression of the αENaC subunit alone could account for some of the large conductance channels measured by the patch-clamp technique. Pertinent to this is the observation that the expression of the γ subunit is low in the alveoli (Matsushita *et al.,* 1996) where many nonselective channels have been reported.

However, if other sodium channels besides ENaC do exist on the apical membrane of epithelial cells, they could contribute to transepithelial transport because the rate-limiting step for transepithelial sodium transport is sodium entry at the apical membrane. These other nonselective cation channels must have adequate signals to increase their open probability in order to mediate sodium entry and transepithelial sodium transport, but it is still unclear what these signals are, although in the case of calcium-activated nonselective cation channels increased intracellular calcium should increase sodium entry. If all the patch-clamp information is taken into account, there may be multiple channels by which sodium entry into epithelial cells can be initiated.

II. ANALOGUES OF AMILORIDE BLOCK MULTIPLE TYPES OF CATION CHANNELS

The pyrazine diuretic, amiloride (Fig. 1), is best known as a reversible blocking agent of sodium absorption mediated by ENaC in the cortical collecting tubule of the kidney, but amiloride also blocks multiple sodium-mediated transport processes (Benos, 1982). The blocking affinity of amiloride for ENaC is generally about 1-5 $\times 10^{-7}$ M when measured by the patch-clamp technique on cultured kidncy A6 cells (Hamilton and Eaton, 1985, 1986) or in channels from kidney cortical collecting tubules (Palmer and Frindt, 1986). The cloned $\alpha\beta\gamma$ rENaC, when expressed in *Xenopus* oocytes, exhibits a K_i of 104 nM for amiloride and 11 nM for benzamil (Canessa *et al.,* 1993) which is similar to the K_i of the α subunit when expressed alone (Canessa *et al.,* 1993).

Some nonselective cation channels have comparable affinity for amiloride as ENaC. For example, the nonselective cation channel in the inner medullary collecting duct which is inhibited by cGMP and cGMP-dependent protein kinase is inhibited by amiloride with an affinity of 10^{-7} M (Light *et al.,* 1988, 1989, 1990). Thus, when amiloride is used as a specific pharmacological agent for blocking ENaC-mediated sodium conductance, it is difficult to know whether it is acting specifically on ENaC. This underscores the use of other blocking agents including other amiloride analogues like ethylisopropyl-amiloride (EIPA), benzamil, phenamil, and diclorobenzamil (Fig. 1) to give a pharmacological profile, including affinity and rank order potency that can pharmacologically identify ENaC-mediated transepithelial transport from that mediated via other epithelial sodium channels. There is progress in the evaluation of the pharmacology of the cloned αrENaC which has a greater affinity for benzamil than amiloride when conductance is evaluated and a greater affinity for phenamil than amiloride when binding

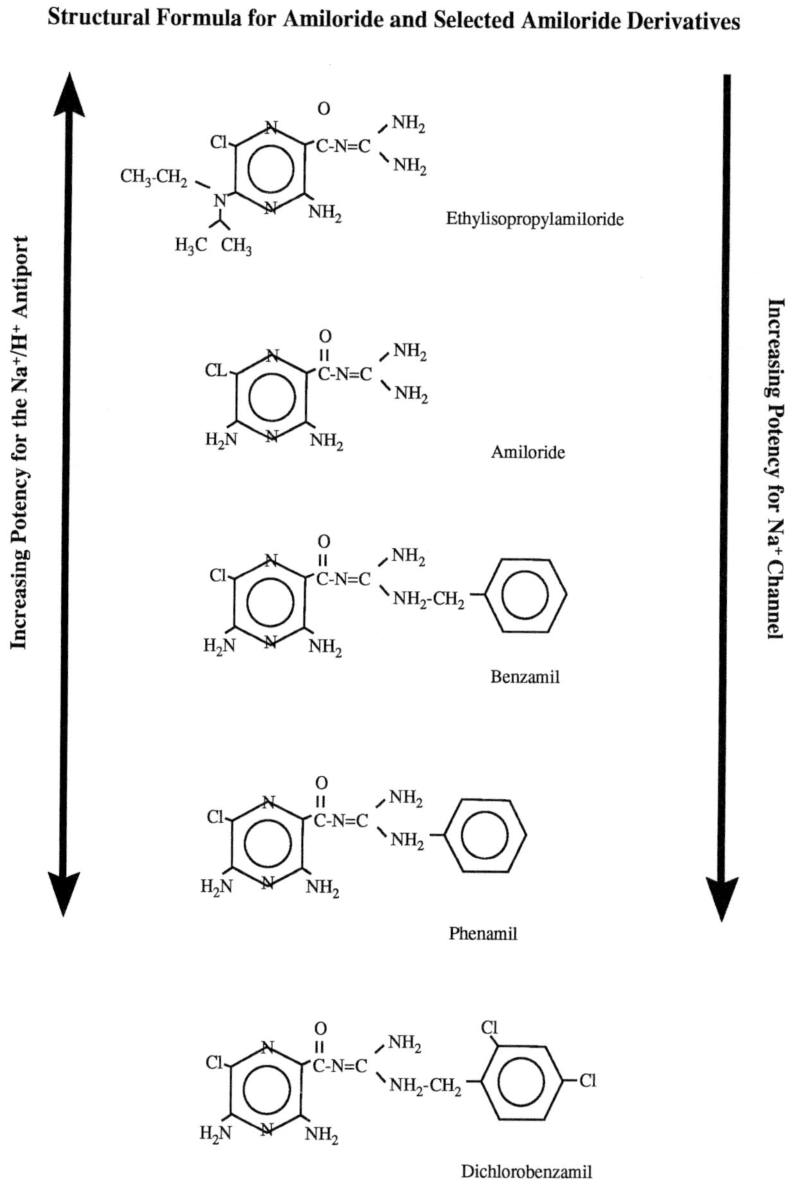

FIGURE 1 Amiloride and amiloride analogues. EIPA is best known as a blocker of the sodium proton exchanger but also blocks some channels in lung. Amiloride can block both sodium proton exchangers and the ENaC channel, but the affinity for ENaC is higher. Benzamil, phenamil, and dichlorobenzamil are high affinity blockers of ENaC, but do little to block sodium proton exchange. Dichlorobenzamil blocks CNC channels with 1 μM affinity but amiloride has a much lower affinity of over 50 μM at the intracellular face. Used with permission from Frelin *et al.* (1987), *Kidney International* **32,** 785–793.

is measured. MDCK cells transfected with $\alpha\beta\gamma$ rENaC show a rank order of potency of phenamil > benzamil > amiloride.

III. THE ROLE OF ENaC AND OTHER CHANNELS IN TRANSEPITHELIAL TRANSPORT IN LUNG

Amiloride was used in early transport (I_{sc} and $^{22}Na^+$ flux) experiments in airway cells to define the contribution of ENaC to sodium absorption and fluid absorption. Electrophysiological evidence suggests that amiloride-sensitive sodium conductance contributes to sodium absorption in the upper airway, especially nasal epithelia (Boucher *et al.*, 1986). Using $^{22}Na^+$ fluxes, amiloride-sensitive sodium absorption is increased in cells cultured from cystic fibrosis nasal polyps compared to normal controls (Boucher *et al.*, 1986; Willumsen and Boucher, 1991; Cotton *et al.*, 1987). ENaC mediates an enhanced sodium absorption in upper airways in cystic fibrosis (Boucher *et al.*, 1986) and this increase is regulated directly by CFTR (Stutts *et al.*, 1995). However, sodium transport in other portions of the lung may be mediated by other sodium conductances. For example, only a component of I_{Na} in bronchioles is decreased by amiloride (Stutts *et al.*, 1988, Van Scott *et al.*, 1989). Likewise not all sodium absorption or fluid absorption in the deep lung (alveoli) is blocked by amiloride. In fetal or young sheep, water absorption is coupled to sodium absorption measured by $^{22}Na^+$ net fluxes and both of these are blocked by amiloride. However, in the mature sheep lung, both volume absorption and net $^{22}Na^+$ flux become insensitive to amiloride (Ramsden *et al.*, 1992). Likewise in rabbit, rat, and human lung only 40–70% of the basal alveolar fluid absorption is blocked by amiloride (Jayr *et al.*, 1994; Smedira *et al.*, 1991; Sakuma *et al.*, 1994), whereas ouabain blocks over 90% of this absorption (Basset *et al.*, 1987).

Other amiloride analogues, like phenamil, have been useful in identifying transepithelial sodium absorptive pathways in epithelia that are not blocked by amiloride. For example, phenamil, but not 10^{-4} M amiloride, inhibits sodium-mediated short circuit current and net transepithelial $^{22}Na^+$ flux in rabbit ileum (Sellin *et al.*, 1989) and cecum (Sellin *et al.*, 1993). Hyperaldosteronism (Frizzell and Shultz, 1978; Will *et al.*, 1980) or dietary salt deprivation (Will *et al.*, 1985) greatly increases the ENaC-mediated sodium conductance in the distal colon and this sodium conductance is completely blocked by 10 μM amiloride. However, in high salt fed rats, the distal colon has a sodium-mediated short circuit current that is insensitive to 10^{-4} M amiloride (Will *et al.*, 1980). Similarly, in the proximal colon of rat, sodium absorption measured by short circuit current is not blocked by amiloride (Will *et al.*, 1980), suggesting that transepithelial sodium transport across this segment is not a result of sodium entry via ENaC and that another sodium channel may mediate sodium entry.

We recently characterized a cGMP-stimulated, sodium-mediated short circuit current in rat tracheal epithelial cells (Schwiebert *et al.* 1997), which in the presence of 10^{-5} M amiloride, to block currents through ENaC, is inhibited by *l-cis*-diltiazem or dichlorobenzamil with the same affinity of block as native cyclic nucleotide-gated nonselective cation channels (Kaupp, 1991). This suggests that other ion channels sensitive to other amiloride analogues like dichlorobenzamil, but not to amiloride itself, may mediate part of the sodium absorption in lung and other epithelial tissues.

IV. NUCLEOTIDE-GATED NONSELECTIVE CATION CHANNELS (αCNC1, αCNC2, αCNC3, βCNCab)

Nucleotide-gated nonselective cation channels (1) are not voltage dependent; (2) are equally selective for sodium and potassium; (3) exhibit a large conductance (25 pS) in the presence of monovalent cations, but have a smaller conductance of about 1 pS in the presence of 100 mM calcium; (4) are gated by cyclic nucleotides (either cAMP or cGMP or both); and (5) are blocked by *l-cis*-diltiazem (IC50 about 50 μM) (Nicol *et al.,* 1987), *d-cis*-diltiazem (IC50 of 200 μM) (Frings *et al.,* 1992) and dichlorobenzamil (IC50 of 1 μM) (Nicol *et al.,* 1987). The IC50 for amiloride is about 50 μM when measured at the intracellular face (Nicol *et al.,* 1987), and the IC50 for EIPA is 50 μM. Under physiological conditions these channels in epithelial cells could act to increase calcium and sodium influx and increase potassium efflux from a cell. One of the most interesting aspects of this channel is actually the fact that it can conduct both sodium and calcium, thereby allowing it to mediate transepithelial transport of both these ions if the channel is situated at the apical membrane.

The 69-kDa channel protein for CNC1 was isolated and the cDNA cloned from bovine rod outer segment cells (Kaupp *et al.,* 1989). The cDNA encoding the cGMP-gated cation channel from rat rod outer segment (CNC1) has 2148 nucleotides, with an open reading frame encoding 682 amino acids (Ding *et al.,* 1997). This channel is a member of a family of ligand-gated cation channels that have six hydrophobic segments (Fig. 2). The overall topology is similar to potassium channels of the Shaker family and the S4 domain has high sequence homology to the voltage gate in Shaker potassium channels. The cGMP-binding region, located on the first 80 amino acids from the C terminus, has a cGMP-binding domain that is homologous to the cGMP-binding site of cGMP-dependent protein kinase.

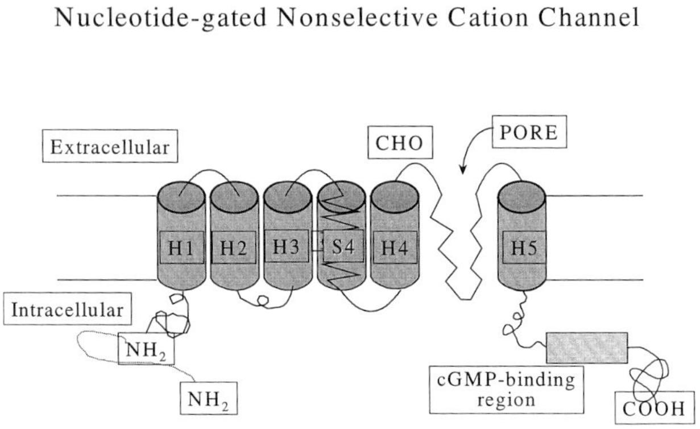

FIGURE 2 Cyclic nucleotide-gated nonselective cation channel structure. Cyclic nucleotide-gated channels have homology to the Shaker channel in the pore region and in their overall structure of six transmembrane domains and a voltage gate shown as S4. The carboxyl cytoplasmic tail region binds nucleotides.

The channels in rod are gated by cGMP which is 50% effective at opening the channel between 5–45 μM at the intracellular face (Biel *et al.,* 1994).

Cyclic nucleotide-gated nonselective cation channels (CNC1, CNC2) are known to play a critical role in sensory signal transduction, facilitating transmission of information in response to primary sensory stimuli (e.g., light and odor) (Fesenko *et al.,* 1985; Nakamura and Gold, 1987). In the dark, high intracellular cGMP levels cause nucleotide-gated cation channels (CNC1) in the retinal rod outer membrane to remain open, depolarizing the membrane potential and causing the "dark" current. In vertebrate photoreceptors light is absorbed by rhodopsin which catalyses the exchange of GTP for GDP bound to the G protein transducin. This causes the γ subunit of the G protein to dissociate, activating a cGMP-dependent phosphodiesterase which decreases the concentration of cGMP. Closure of nucleotide-gated cation channels results in a hyperpolarization of the plasma membrane potential thereby transducing the light signal (Kaupp, 1991). Odorant signal transduction shares many similarities with signal transduction in the visual system. Odorant signal transduction occurs in the cilia of the highly specialized olfactory receptor cells. These olfactory receptor cells generate a typical inside negative resting membrane potential until odorants increase cAMP causing the membrane potential to depolarize. The channel isoform in olfactory neuroepithelium, CNC2, has 60% amino acid homology to the retinal rod outer segment nucleotide-gated cation channel CNC1 (Dhallan *et al.,* 1990).

A third isoform of this family, CNC3, has been cloned from heart, kidney, testis and sperm (Biel *et al.,* 1994; Weyand *et al.,* 1994). But the contribution of this channel to control of membrane potential and the overall physiological function in these tissues remains speculative. Demonstration of mRNA transcripts in kidney cortical collecting duct cells (Ahmad *et al.,* 1992) that hybridized with partial sequences of the rod photoreceptor cGMP-gated channel (CNG1) provided the first evidence that cGMP-gated nonselective cation channels were present in kidney. Single-channel patch-clamp recordings from mouse kidney cortical collecting duct cells identified channels with the biophysical properties of nucleotide-gated cation channels (Ahmad *et al.,* 1992). The CNC3 isoform is thought to have a greater permeability to calcium than the CNC1 isoform.

Two β subunits of the CNC family have also been cloned. While these subunits do not conduct ions, they are known to increase the sensitivity of CNC1 to *l-cis*-diltiazem and increase the relative calcium permeability of the channel (Chen *et al.,* 1993). When the β subunit is coexpressed with CNC2 it increases the affinity to cAMP compared to cGMP (MacGregor *et al.,* 1994).

V. CYCLIC NUCLEOTIDE-GATED CHANNELS IN THE LUNG

Using ribonuclease protection assay (Fig. 3), *in situ* hybridization, and RT-PCR, the message of the CNC1 isoform of the cyclic nucleotide-gated cation channel was documented in adult rat lung and rat tracheal epithelial cells grown in primary culture (Ding *et al.,* 1997). Although CNC1 is present in the adult rat lung, very little message is present in fetal lung at day 18 gestation when αENaC message is known to be expressed (O'Brodovich *et al.,* 1993; Tchepichev *et al.,* 1995). Figure 4 shows a comparison of message for βENaC and CNC1 at 18 days. This suggests that the CNC channel may be expressed later in development and that this channel is not involved in removal of lung fluid at birth, which is a major function of ENaC (Hummler *et al.,* 1996).

Rat tracheal airway epithelial cells grown in monolayer culture exhibit a cGMP-stimulated, sodium-mediated short circuit current. These experiments were done in the presence of amiloride to block sodium movement through ENaC channels and zero chloride or in the presence of chloride channel blockers to block chloride short circuit currents (Figs. 5 and 6). The cGMP-stimulated current is inhibited by *l-cis*-diltiazem and dichlorobenzamil suggesting that cyclic nucleotide-gated cation channels contribute to transepithelial sodium transport (Figs. 5 and 6). In addition, because *l-cis*-diltiazem and dichlorobenzamil inhibit not only the cGMP-stimulated short circuit currents, but also the basal sodium current not stimulated by

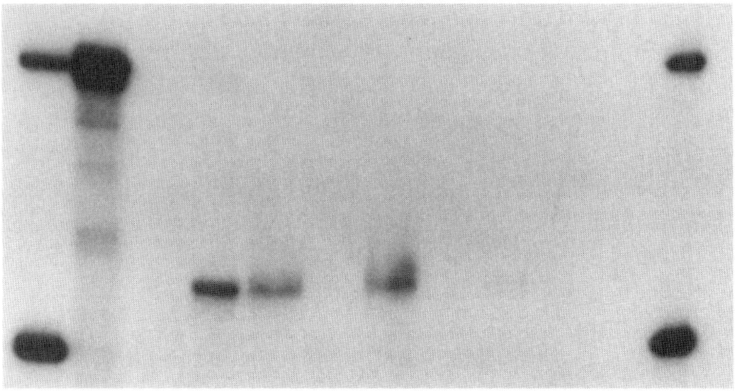

FIGURE 3 Ribonuclease protection assay performed on several rat tissues. Rat eye CNC1 cDNA™ II (1036-2148 bp) was digested with Cla I to generate a 227 bp (nt. 1921–2148)^{32}P-labeled riboprobe and was synthesized using the MAXI script transcript kit (Ambion). Aliquots of RNA were hybridized with 7.5x10^4 cpm probe and subjected to digestion with RNase T_1 (1/50 dilution) for 30 min at 37° using a commercial kit (RPA II, Ambion). Samples were electrophoresed through 5% polyacrylamide gel containing 8 M urea. The gel was exposed to autoradiographic film at −80°C for 4 days. Lane 1, molecular weight markers 300 bp (top) 200 bp (bottom). Lane 2, using yeast RNA control and full-length probe 308 bp (without RNase). Lane 3, 10 μg yeast RNA and probe digested with RNase. Lane 4, 10 μg rat eye total RNA. Lane 5–11 60 μg total rat RNA B (brain), H (heart), Lu (lung), Li (liver), Sp (spleen), K (kidney), T (testis). Lane 12, molecular weight markers. Rat eye was used as a positive control because this tissue has the highest expression of CNC1. Testis, kidney, and heart were used as negative control for expression of the CNC3 isoform which is highly expressed in these tissues. (Reprinted from Ding *et al.*, 1997, with permission.)

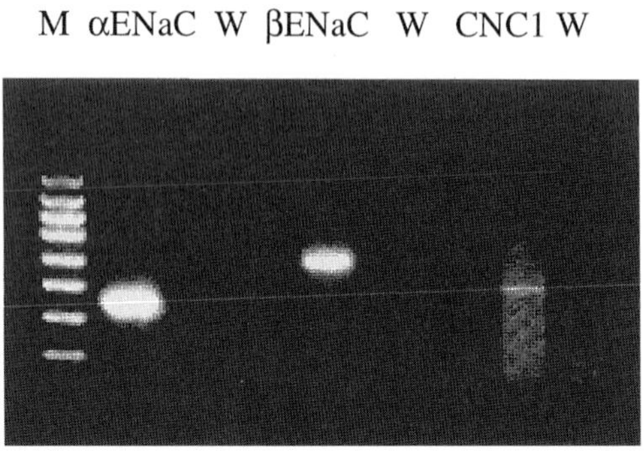

FIGURE 4 RT-PCR of ENaC and CNC1 subunits in 18-day gestation fetal rat lung. At 18 days the fetal rat expresses abundant message for the α and β ENaC subunits but very little message for CNC1 in whole fetal lung. M is the molecular weight marker. W is water blank control.

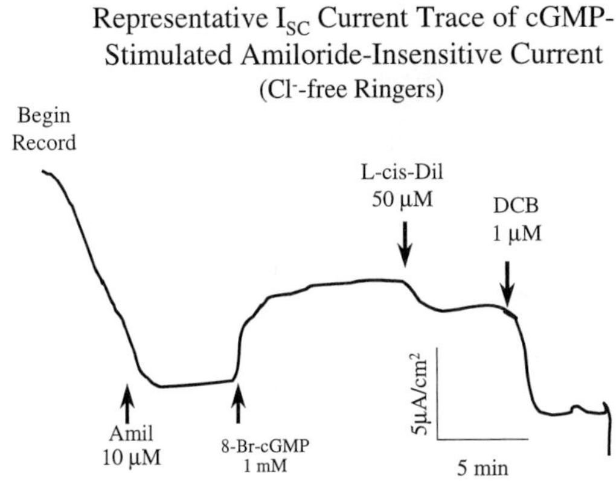

FIGURE 5 Representative trace of I_{sc} recordings from rat tracheal epithelial cell monolayers measured in the absence of chloride. I_{sc} was measured in chloride-free media to eliminate chloride conductance and in the continuous presence of 10 μM amiloride to eliminate currents from ENaC. The initial current of 9 $\mu A/cm^2$ represents a nonsteady state current measured immediately after monolayers are transferred from the culture medium to Ussing chambers containing a chloride free Ringer solution. In this recording, 10 μM amiloride causes a slight 1 μA reduction in the basal current. In the continued presence of 10 μM amiloride, addition of 1 mM 8-Br cGMP (a membrane permeable analogue of cGMP) increases the I_{sc} to 5 $\mu A/cm^2$. This current is inhibited about 35% with 50 μM l-*cis*-diltiazem (L-*cis*-Dil). Further addition of 1μM dichlorobenzamil, in the presence of all the other agents, inhibits all of the 8-Br-cGMP stimulated current and also reduces the current to below basal levels measured before addition of the drugs. (Reprinted from Schwiebert *et al.*, 1997, with permission.)

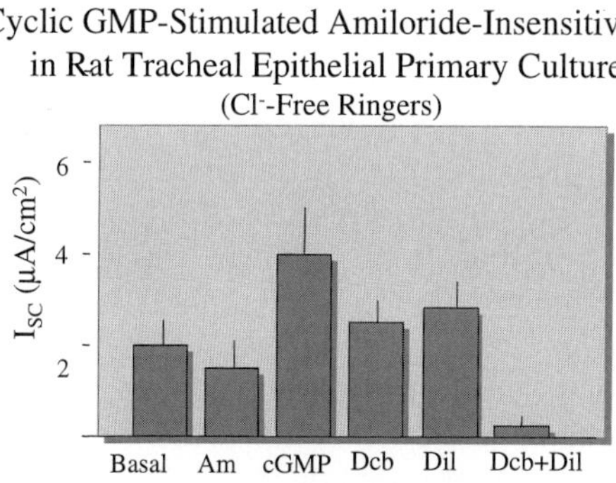

FIGURE 6 Summary of data for I_{sc} experiments measuring transepithelial sodium currents from rat tracheal epithelial monolayers in chloride-free medium. Steady-state (basal) current (bar 1) is not significantly inhibited by 10 μM amiloride used to block ENaC conductance (bar 2). In the continued presence of amiloride, addition of 1 mM 8-Br-cGMP stimulates I_{sc} 2.7-fold (bar 3). Addition of 0.1 μM DCB inhibits 60% of the 8-Br-cGMP stimulated current (bar 4). In paired experiments 60% of the current is also inhibited by 50 μM l-*cis*-diltiazem (bar 5). In the presence of both 0.1 μM DCB and 50 μM l-*cis*-diltiazem, all of the 8-Br-cGMP stimulated current and 96% of the basal current is inhibited (bar 6). Bars 4–6 represent paired data performed in different order of addition. (Schwiebert *et al.*, 1997).

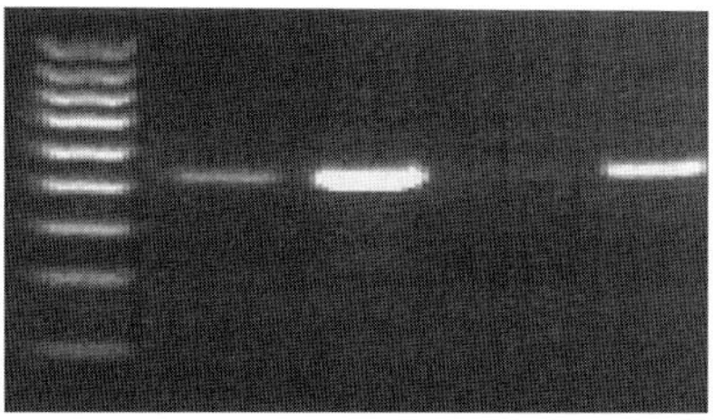

FIGURE 7 RT-PCR of the CNC2 subunit from adult rat lung. The rat lung (L) expresses some message for CNC2. The message is also highly expressed in olfactory cilia (O) as shown previously by Northern blot (Dhallan *et al.*, 1990). Some message for this isoform is also expressed in eye, but none in kidney (K) where the CNC3 isoform is most abundant (Biel *et al.*, 1994).

cGMP, it suggests that CNC channels may carry an underlying sodium current that is not blocked by amiloride.

We have now identified that message for CNC2 is also present in the adult rat lung. Performing RT-PCR from rat tracheal epithelial cells using degenerate primers, then sequencing of clones identified the presence of this isoform. Of ten clones sequenced, five were CNC1 and five were CNC2. RT-PCR from whole rat lung using primers specific for CNC2 also showed the presence of CNC2 (Fig. 7). The exact distribution of this isoform remains to be determined.

In conclusion, there is now evidence that cyclic nucleotide-gated cation channels may represent another sodium entry pathway in lung. While the message for this channel is low in fetal rat lung, suggesting that unlike ENaC this channel is not involved in removing fluid at birth, CNC1 message in the adult rat is localized in both alveolar type 2 and the more abundant alveolar type 1 cells. Because it is widely distributed, CNC1 is potentially a major pathway for sodium absorption in the lung. The hormonal signal transduction pathways that might activate cyclic nucleotide channels (cGMP or cAMP) in lung remain to be identified. Both soluble and particulate guanylate cyclase have been identified in type 1 alveolar cells (Secca *et al.*, 1991) suggesting that in this cell type sodium entry may be mostly a function of CNC1 channels. ENaC is not present in alveolar type 1 cells. The effect of CNC channel blockers on $^{22}Na^{+}$ flux and alveolar fluid clearance remains to be elucidated.

Acknowledgments

This work was funded by NIH DK 48977 and the Cystic Fibrosis Foundation.

References

Ahmad, I., Kormacher, C., Segal, A. S., Cheung, P., Boulpaep, E. L., and Barnstable, C. J. (1992). Mouse cortical collecting duct cells show nonselective cation channel activity and express a gene related to the cGMP-gated rod photoreceptor channel. *Proc. Natl. Acad. Sci. U.S.A.* **89,** 10262–10266.

Basset, G., Crone, C., and Saumon, G. (1987). Fluid absorption by rat lung in situ: Pathways for sodium entry in the luminal membrane of alveolar epithelium. *J. Physiol.* (*London*) **384,** 325–345.

Benos, D. J. (1982). Amiloride: A molecular probe of sodium transport in tissues and cells. *Am. J. Physiol.* **242** (Cell Physiol. 11), C131–145.

Biel, M., Zong, Z., Distler, M., Bosse, E., Klugbauer, N., Murakami, M., Flockerzi, V., and Hofmann, F. (1994). Another member of the cyclic nucleotide-gated channel family, expressed in testis, kidney, and heart. *Proc. Natl. Acad. Sci. U.S.A.* **91,** 3505–3509.

Boucher, R. C., Stutts, M. J. Knowles, M. R., Cantley, L., and Gatzy, J. T. (1986). Na^+ transport in cystic fibrosis respiratory epithelia. *J. Clin. Invest.* **78,** 1245–1252.

Canessa, C. M., Horisberger, J. D., and Rossier, B. C. (1993). Epithelial sodium channel related to proteins involved in neurodegeneration. *Nature* (*London*) **361,** 467–470.

Canessa, C. M., Schild, L., Buell, G., Thorens, B., Gautschi, I., Horisberger, J., and Rossier, B. C. (1994). Amiloride-sensitive epithelial Na channel is made of three homologous subunits. *Nature* (*London*) **367,** 463–467.

Champigny, G., Voilley, N., Lingueglia, E., Friend, V., Barbry, P., and Lazdunski, M. (1994). Regulation of expression of the lung amiloride-sensitive Na^+ channel by steroid hormones. *EMBO J.* **13**(9), 2177–2181.

Chen, T. Y., Peng, Y. W., Dhallan, R. S., Ahamedt, B., Reed, R. R., and Yau, K.-Y. (1993). A new subunit of the cyclic nucleotide-gated cation channel in retinal rods. *Nature* (*London*) **362,** 764–767.

Chinet, T. C., Fulton, J. M., Yankaskas, J. R., Boucher, R. C., and Stutts, M. J. (1993). Sodium-permeable channels in the apical membrane of human nasal epithelial cells. *Am. J. Physiol.* **265,** C1030–C1060.

Cotton, C. U., Stutts, M. J. Knowles, M. R., Gatzy, J. T., and Boucher, R. C. (1987). Abnormal apical cell membrane in cystic fibrosis respiratory epithelium. *J. Clin. Invest.* **79,** 80–85.

Dhallan, R. S., Yau, K.-Y., Schrader, K. W., and Reed, R. R. (1990). Primary structure and functional expression of a cyclic nucleotide-activated channel from olfactory neurons. *Nature* (*London*) **347,** 184–187.

Ding, C., Potter, E. D., Qiu, W., Coon, S. L., Levine, M. A., and Guggino, S. E. (1997). Cloning and widespread distribution of the rat rod-type cyclic nucleotide-gated cation channel. *Am. J. Physiol.* **272** (Cell Physiol. 41), C1335–C1344.

Disser, J., Hazama, A., and Fromter, E. (1991). Some properties of sodium and chloride channels in respiratory epithelia of CF and nonCF patients. *Adv. Exp. Med. Biol.* **290,** 133–144.

Duszyk, M., French, A. S., and Man, S. F. P. (1989). Cystic fibrosis affects chloride and sodium channels in human airway epithelia. *Can. J. Physiol. Pharmacol.* **67,** 1362–1365.

Feng, Z. P., Clark, R. B., and Berthiaume, Y. (1993). Identification of nonselective cation channels in cultured adult rat alveolar type II cells. *Am. J. Respir. Cell Mol. Biol.* **9**(3), 248–254.

Fesenko, E. E., Kolesnikov, S. S., and Lyubarsky, A. L. (1985). Induction of cGMP cationic conductance in plasma membrane of retinal rod outer segment. *Nature* (*London*) **313,** 310–313.

Frelin, C., Vigne, P., Barbry, P., and Lazdunski, M. (1987). Molecular properties of amiloride action and of its Na^+ transporting targets. *Kidney Int.* **32,** 785–793.

Frings, S., Lynch, J. W., and Lindemann, B. (1992). Properties of cyclic nucleotide-gated channels mediating olfactory transduction. *J. Gen. Physiol.* **100,** 45–65.

Frizzell, R. A., and Shultz, S. G. (1978). Effect of aldosterone on ion transport by rabbit colon in vitro. *J. Membr. Biol.* **39,** 1–26.

Fuller, C. M., Berdiev, B. K., Shlyonsky, V. G., Ismailov, I. I., and Benos, D. J. (1997). Point mutations in alpha bENaC regulate channel gating, ion selectivity, and sensitivity to amiloride. *Biophys. J.* **72**(4), 1622–1632.

George, A. L., Staub, O., Geering, K., Rossier, B. C., Kleyman, T. R., and Kraehenbuhl, J. R. (1989). Functional expression of the amiloride-sensitive sodium channel in Xenopus oocytes. *Proc. Natl. Acad. Sci. U.S.A.* **86,** 7295–7298.

Hamilton, K. L., and Eaton, D. C. (1985). Single-channel recordings from amiloride-sensitive sodium channel. *Am. J. Physiol.* **249,** C200–C207.

Hamilton, K. L., and Eaton, D. C. (1986). Single-channel recordings from two types of amiloride-sensitive epithelial Na channels. *Membr. Biochem.* **6**(2), 149–171.

Hummler, E., Barker, P., Gatzy, J., Beerman, F., Verdumo, C., Schmidt, A., Boucher, R., and Rossier, B. C. (1996). Early death due to defective neonatal lung liquid clearance in alpha-ENaC-deficient mice. *Nat. Genet.* **12**(3), 325–328.

Jayr, C., Garat, C., Meiganan, M., Pittet, F., Zelter, M., and Matthay, M. A. (1994). Alveolar liquid and protein clearance in anesthetized ventilated rats. *J. Appl. Physiol.* **76,** 2636–2642.

Joris, L., Krouse, M. E., Hagiwara, G., Bell, C. L., and Wine, J. J. (1989). Patch-clamp study of cultured human sweat duct cells: Amiloride-blockable Na^+ channel. *Pfluegers Arch.* **414,** 369–372.

Jorissen, M., Vereecke, J., Carmeliet, E., Van den Berghe, H., and Cassiman, J. (1990). Identification of a voltage and calcium-dependent non-selective cation channel in cultured adult and fetal human nasal epithelial cells. *Pfluegers Arch.* **415,** 616–623.

Kaupp, U. B. (1991). The cyclic nucleotide-gated channels of vertebrate photoreceptors and olfactory epithelium. *Trends Neurosci.* **14**(4), 149–151.

Kaupp, U. B., Niidome, T., Tanabe, T., Terada, S., Bonigk, W., Stuhmer, W., Cook, N. J., Kangawa, K., Matsuo, H., Hirose, T., Miyata, T., and Numa, S. (1989). Primary structure and functional expression from complementary DNA of the rod photoreceptor cyclic GMP-gated channel. *Nature* (*London*) **342,** 762–766.

Kinzer, N., Guo, X. L., and Hruska, K. (1997). Reconstitution of stretch-activated cation channels by expression of the alpha-subunit of the epithelial sodium channel cloned from osteoblasts. *Proc. Natl. Acad. Sci. U.S.A.* **94**(3), 1013–1018.

Krouse, M. E., Hagiwara, G., Chen, J., Lewiston, N. J., and Wine, J. J. (1989). Ion channels in normal human and cystic fibrosis sweat gland cells. *Am. J. Physiol.* **257,** C129–C140.

Kunzelmann, K., Kathofer, S., Hipper, A., Gruenert, D. C., and Greger, R. (1996). Culture-dependent expression of Na^+ conductances in airway epithelial cells. *Eur. J. Physiol.* **431,** 578–586.

Light, D. B., McCann, F. V., Keller, T. M., and Stanton, B. A. (1988). Amiloride-sensitive cation channel in apical membrane of inner medullary collecting duct. *Am. J. Physiol.* **255,** F278–F286.

Light, D. B., Ausiello, D. A., and Stanton, B. A. (1989). Guanine nucleotide-binding protein αi-3 directly activates a cation channel in renal inner medullary collecting duct cells. *J. Clin. Invest.* **84,** 352–356.

Light, D. B., Corbin, J. D., and Stanton, B. A. (1990). Dual ion-channel regulation by cyclic GMP and cyclicGMP–dependent protein kinase. *Nature* (*London*) **344,** 336–339.

MacGregor, G. G., Olver, R. E., and Kemp, P. J. (1994). Amiloride-sensitive Na^+ channels in fetal type II pneumocytes are regulated by G proteins. *Am. Physiol. Soc.* **11,** L1–L8.

Marunaka, Y., Tohda, H., Hagiwara, N., and O'Brodovich, H. (1992). Cytosolic calcium-induced modulation of ion selectivity and amiloride sensitivity of a cation channel and beta agonist function in fetal lung epithelium. *Biochem. Biophys. Res. Commun.* **187,** 648–656.

Maruyama, Y., and Petersen, O. H. (1982). Single-channel currents in isolated patches of plasma membrane from basal surface of pancreatic acini. *Nature* (*London*) **299,** 159–163.

Matalon, S., Kirk, K. L., Bubien, J. K., Oh, Y., Hu, P., Yue, G., Shoemaker, R., Cragoe, E. J., Jr., and Benos, D. J. (1992). Immunocytochemical and functional characterization of Na^+ conductance in adult alveolar pneumocytes. *Am. J. Physiol.* **262,** C1228–C1238.

Matalon, S., Bauer, M. L., Benos, D. J., Kleyman, T. R., Lin, C., Cragoe, E. J., Jr., and O'Brodovich, H. (1993). Fetal lung epithelial cells contain two populations of amiloride-sensitive Na^+ channels. *Am. J. Physiol.* **264,** L357–L364.

Matsushita, K., McCray, P. B., Jr., Sigmund, R. D., Welch, M. J., and Stokes, J. B. (1996). Localization of epithelial sodium channel subunit mRNAs in adult rat lung by in situ hybridization. *Am. J. Physiol.* **271** (Lung Cell. Mol. Physiol. 14), L332–L339.

Nakamura, T., and Gold, G. H. (1987). A cyclic nucleotide-gated conductance in olfactory receptor cilia. *Nature* (*London*) **325,** 442–444.

Nicol, G. D., Sachnetkamp, P. P. M., Saimi, Y., Cragoe, E. J., Jr., and Bownds, M. D.(1987). A derivative of amiloride blocks the light-regulated and cyclic GMP-regulated conductances in rod photoreceptors. *J. Gen. Physiol.* **90,** 651–669.

O'Brodovich, H. (1996). Immature epithelial Na^+ channel expression is one of the pathogenic mechanisms leading to neonatal respiratory distress syndrome. *Proc. Assoc. Am. Physicians* **108,**(5), 345–355.

O'Brodovich, H., Canessa, C., Ueda, J., Rafii, B., Rossier, B. C., and Edelson, J. (1993). Expression of the epithelial Na^+ channel in the developing rat lung. *Am. J. Physiol.* **34,** C491–C496.

Olans, L., Sariban-Sohraby, S., and Benos, D. J. (1984). Saturation behavior of single, amiloride-sensitive Na^+ channels in planar lipid bilayers. *Biophys. J.* **46,** 831–835.

Palmer, L. G., and Frindt, G. (1986). Amiloride-sensitive Na channels from the apical membrane of the rat cortical collecting tubule. *Proc. Natl. Acad. Sci U.S.A.* **83,** 2767–2770.

Ramsden, C. A., Markiewicz, M., Walters, D. V., Gabella, G., Parker, K. A., Barker, P. M., and Neil, H. L. (1992). Liquid flow across the epithelium of the artificially perfused lung of fetal and postnatal sheep. *J. Physiol.* (*London*) **448,** 579–597.

Sakuma, T., Okaniwa, G., Nagada, T., Nishimura, T., Fugimura, S., and Matthay, M. A. (1994). Alveolar clearances in the resected human lung. *Am. J. Respir. Crit. Care Med.* **150,** 305–310.

Sariban-Sohraby, S., Burg, M. B., and Turner, R. J. (1983). Apical sodium uptake in toad kidney epithelial cells. *Am. J. Physiol.* **245** (Cell Physiol. 14), C167–C171.

Schwiebert, E. M., Potter, E. D., Hwang, H., Woo, J. S., Ding, C., Qiu, W., Guggino, W. B., Levine, M. A., and Guggino, S. E. (1997). cGMP stimulates sodium and chloride currents in rat tracheal airway epithelia. *Am. J. Physiol.* **272** (Cell Physiol 41), C911–C922.

Secca, T., Vagnetti, D., Dolcini, B. M., and Di Rosa, I. (1991). Cytochemical and biochemical observations on alveolar guanylate cyclase of golden hamster lung. *Tissue Lung* **23,** 67–74.

Sellin, J. H., Oyarzabal, H. O., Cragoe, E. J., Jr., and Potter, G. D. (1989). Phenamil inhibits electrogenic sodium absorption in rabbit ileum. *Gastroenterology* **96,** 997–1003.

Sellin, J. H., Hall, A., Cragoe, E. J., Jr., and Dubinsky, W. P. (1993). Characterization of an apical sodium conductance in rabbit cecum. *Am. J. Physiol.* **264,** G13–G21.

Smedira, N., Gates, L., Hastings, R., Jayr, C., Sukama, T., Pittet J.-F., and Matthay, M. A. (1991). Alveolar and liquid lung clearance in anesthetized rabbits. *J. Appl. Physiol.* **70,** 1827–1835.

Smith, C. I., Searls, R. L., and Hilfer, S. R. (1990). Effects of hormones on functional differentiation of mouse respiratory epithelium. *Exp. Lung Res.* **16**(3), 191–209.

Stutts, M. J., Gatzy, J. T., and Boucher, R. C. (1988). Effects of metabolic inhibition of ion transport by dog bronchial epithelium. *J. Appl. Physiol.* **64**(1), 253–258.

Stutts, M. J., Canessa, C. M., Olsen, J. C., Hamrick, M., Cohn, J. A., Rossier, B. C., and Boucher, R. C. (1995). CFTR as a cAMP-dependent regulator of sodium channels. *Science* **269,** 847–850.

Tanswell, A. K., Joneja, M. G., Vreeken, E., and Lindsay, J. (1983). Differentiation-arrested fetal lung in primary monolayer cell culture. II. dexamethasone, triiodothyronine and insulin effects on different gestational age cultures. *Exp. Lung Res.* **5**(1), 49–60.

Tchepichev, S., Jueda, C., Canessa, M., Rossier, B. C., and O'Brodovich, H. (1995). Lung epithelial Na channels are differentially regulated during development and by steroids. *Am. J. Physiol.* **269**(3), C805–C812.

Tohda, H., Foskett, J. K., O'Brodovich, H., and Marunaka, Y. (1994). Chloride-regulation of a Ca^{2+}-activated nonselective cation channel in beta-agonist-treated fetal lung epithelium. *Am. J. Physiol.* **266,** C104–C109.

Van Scott, M. R., Davis, W. C., and Boucher, R. C. (1989). Na^+ and Chloride-transport across nonciliated bronchiolar epithelial (Clara) cells. *Am. J. Physiol.* **256,** C893–C901.

Wang, X., Kleyman, T. R., Tohda, H., Marunaka, Y., and O'Brodovich, H. (1993). 5-(N-ethyl-N-isopropyl) amiloride sensitive Na^+ currents in intact fetal distal lung epithelial cells. *Can. J. Physiol. Pharmacol.* **71**(1), 58–62.

Weyand, I., Godde M., Frings, S., Weiner, J., Muller, F., Altenhofen, W., Hatt, H., and Kaupp, U. B. (1994). Cloning and functional expression of a cyclic-nucleotide-gated channel from mammalian sperm. *Nature* (*London*) **368,** 859–863.

Will, P. C., Lebowitz, J. L., and Hopfer, U. (1980). Induction of amiloride-sensitive sodium transport in the rat colon by mineralocorticoids. *Am. J. Physiol.* **238**(7), F261–F268.

Will, P. C., Cortright, R. N., Grosecvlose, R. G., and Hopfer, U. (1985). Amiloride-sensitive salt and fluid absorption in small intestine of sodium-depleted rats. *Am. J. Physiol.* **248** (Gastrointest. Liver Physiol. 11), G133–G141.

Willumsen, N. J., and Boucher, R. C. (1991). Transcellular sodium transport in cultured cystic fibrosis human nasal epithelium. *Am. J. Physiol.* **261,** C332–C341.

Yue, G., Shoemaker, R. L., and Matalon, S. (1994). Regulation of low-amiloride-affinity sodium channels in alveolar type II cells. *Am. J. Physiol.* **267,** L94–L100.

PART IV

Sensory and Mechanical Transduction

CHAPTER 17

C. elegans Members of the DEG/ENaC Channel Superfamily: Form and Function

Heather A. Thieringer, Sukhvinder Sahota, Itzhak Mano, and Monica Driscoll
Department of Molecular Biology and Biochemistry, Center for Advanced Biotechnology and Medicine, Rutgers University, Piscataway, New Jersey 08855

I. INTRODUCTION

The first member of the DEG/ENaC superfamily of ion channel subunits, the *C. elegans deg-1* gene, was initially investigated because specific mutant alleles induced neuronal swelling and degeneration (Chalfie and Wolinsky, 1990). A second gene studied for similar genetic properties proved to encode the related protein, MEC-4 (Driscoll and Chalfie, 1991). The normal functions of the DEG-1 and MEC-4 proteins were initially uncertain, but their potential to mutate to toxic forms earned them the family name of "degenerins." With the identification of the first mammalian members of this family as subunits of the epithelial amiloride-sensitive sodium channel

1063-5823/99 $30.00

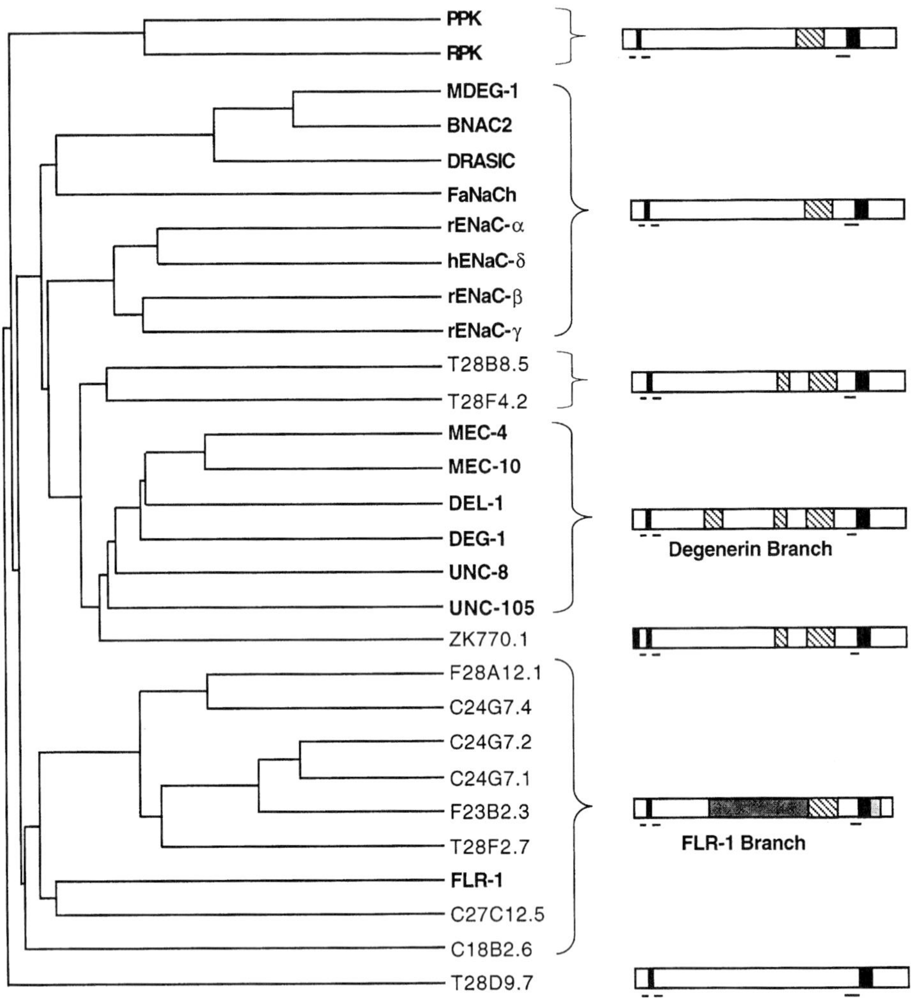

FIGURE 1 Dendogram showing the relationship between *C. elegans* ENaC-related proteins and some members of the ENaC family. Where protein names are not known, the open reading frames predicted by the *C. elegans* gene sequencing consortium are listed. Sequences were aligned using the Genetics Computer Group Sequence analysis software, version 7.2. Schematic diagrams show the conserved regions found in each of the subgroups within the ENaC superfamily. Black boxes show the predicted membrane spanning domains. Hatched boxes indicate Cys-rich regions. Underlines indicate short homologous regions common to all members of the DEG/ENaC superfamily. The grey box in the FLR-1 branch denotes the region of homolgy termed FEHD. The light grey box in this branch depicts the region of homolgy found just after the second membrane spanning domain. Protein sequences and predicted open reading frames which were lined up include the following sequences with GenBank accession numbers in parenthesis; PPK (Drosophila pickpocket, AF043263), RPK

(ENaC) (Canessa *et al.,* 1993, 1994; Lingueglia *et al.,* 1993), it became clear that this newly identified channel family was conserved throughout the animal kingdom. Since the identification of the first family members, the DEG/ENaC family has expanded considerably, with related genes identified from snails (Lingueglia *et al.,* 1995), flies (Adams *et al.,* 1998), and mammals (for example, Waldmann *et al.,* 1996, 1997; García-Añoveros *et al.,* 1997).

To date, the most extensive collection of identified family members from a single organism has been assembled from the nematode *C. elegans.* Here we review features of the *C. elegans* family members evident from a survey of genome sequence data. We also discuss genetic studies that implicate some *C. elegans* family members in mechanical signaling and that have extended models of channel complexes. Finally, we consider the potential of studies in the nematode to reveal the spectrum of activities executed by the entire family of channel subunits.

II. *C. elegans* PROTEINS RELATED TO ENaC CHANNELS DEFINE TWO SUBFAMILIES

At the time of this writing, essentially all of the *C. elegans* genomic sequence has been entered into accessible databases (The *C. elegans* Sequencing Consortium, 1998). Analysis of the *C. elegans* database thus provides a good overview of the composition of the entire ENaC-related gene family from this organism. In total, there are 19 nematode genes that exhibit extensive homology to the DEG/ENaC superfamily. Alignment algorithms reveal that, with a few exceptions, *C. elegans* family members fall into two distinct subfamilies (Fig. 1).

We refer to one subfamily as the degenerin subfamily. The degenerin subfamily includes the "classic" degenerins, DEG-1 (Chalfie and Wolinsky, 1990) and MEC-4 (Driscoll and Chalfie, 1991; Lai *et al.,* 1996), as well as other well-studied related proteins including MEC-10 (Huang and Chalfie,

---

(Drosophila ripped pocket, AF043264), MDEG (U50352), BNAC2 (U78181), DRASIC (AF01359), FaNaCh (X92113), rENaC-α (X70497), hENaC-δ (U38254), rENaC-β (X77932), rENaC-γ (X77933), *C. elegans* cosmid T28B8.5 (Z81133), *C. elegans* cosmid T28F4.2 (X72517), MEC-4 (X58982), MEC-10 (L25312), DEL-1 (U76403), DEG-1 (L34414), UNC-8 (U76402), UNC-105 (Z48045), *C. elegans* cosmid ZK770.1 (U97404), *C. elegans* cosmid F28A12.1 (U64851), *C. elegans* cosmid C24G7 (U88310), *C. elegans* cosmid T28F2.7 (AF000198), FLR-1 (Z67990), *C. elegans* cosmid C27C12.5 (Z69883), *C. elegans* cosmid C18B2.6 (U40413), and *C. elegans* cosmid T28D9.7 (U28738).

1994), UNC-8 (Tavernarakis *et al.,* 1997), and UNC-105 (Liu *et al.,* 1996). Closely related but more divergent proteins are predicted open reading frames T28B8.5, T28F4.2, and ZK770.1, which have been identified on cosmid clones. The genetics of these family members remain to be investigated.

We refer to the second subfamily as the FLR-1 family, because the family member identified as F02D10.5 corresponds to *flr-1* (I. Katsura, National Institute of Genetics, Mishima, Japan, personal communication), currently the only genetically characterized member of this group (Katsura *et al.,* 1994). At present, the FLR-1 subfamily includes 9 members.

A. Sequence Features of DEG/ENaC Superfamily Members

Certain sequence features are common to all members of the DEG/ENaC superfamily (summarized in Fig. 1). All members of this channel family include two hydrophobic domains. In addition, short regions of homology are situated before and after the first hydrophobic domain and before the second hydrophobic domain. We have referred to the two hydrophobic domains as membrane spanning domains I and II (MSDI and MSDII), since in several family members they have been experimentally defined to traverse the membrane (N- and C- termini are intracellular, one large loop is extracellular—see Fig. 2; Canessa *et al.,* 1994; Renard *et al.,* 1994; Snyder *et al.,* 1994; Lai *et al.,* 1996). In general, MSDI is not distinguished by striking sequence similarity except for the strict conservation of a Trp residue (corresponding to MEC-4(W111)) and the strong conservation of a Gln/Asn residue (corresponding to MEC-4(N125)). MSDII is more distinctive, exhibiting strong conservation of hydrophilic residues (consensus GLW_G_S__T__E) that has been implicated in pore function (Hong and Driscoll, 1994; Waldmann *et al.,* 1995). The short highly conserved region before the minimal transmembrane domain is thought to loop back into the membrane to contribute to the channel pore (Canessa *et al.,* 1994; Renard *et al.,* 1994; Schild *et al.,* 1997). The extended MSDII homology region (loop + transmembrane) can be considered a defining characteristic of superfamily members. A second characteristic of most superfamily members is the presence of a cysteine-rich domain in the extracellular loop that corresponds to the third Cys-rich domain in the *C. elegans* degenerins (CRDIII). The high degree of conservation of Cys residues suggests that the tertiary structure of this region is critical to the function of most channel subunits.

B. Sequence Features of the Degenerin Subfamily

The *C. elegans* degenerin subfamily is distinguished by the presence of three Cys-rich domains (CRDI, CRDII, and CRDIII) and an extracellular

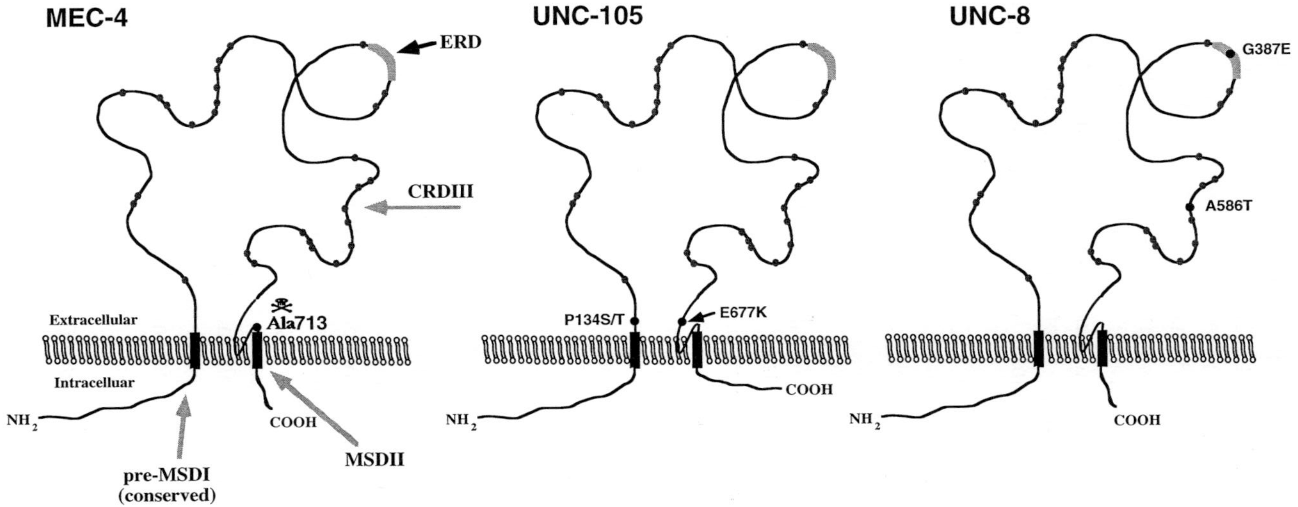

FIGURE 2 Structure/function summary on degenerin subfamily members. MEC-4, UNC-105, and UNC-8 are schematically diagrammed with transmembrane topology indicated. Channel subunits are depicted approximately, but not exactly, to scale. Dark bars indicate MSDI and MSDII; small dots represent Cys residues. Larger black dots mark sites of amino acid changes caused by channel activating gain-of-function mutations. For the MEC-4 subunit, sites of EMS-induced loss-of-function substitutions that inactivate the channel are indicated by grey arrows. Also noted are the extracellular regulatory domain (ERD) and conserved Cys-rich domain III.

regulatory domain (ERD) that has been implicated in channel closing (García-Añoveros *et al.,* 1995; Tavernarakis *et al.,* 1997) (Figs. 1, 2). In general, degenerin subfamily members have a large extracellular loop and a short intracellular C-terminal domain approximately 30–50 amino acids in length. C-termini are not conserved within the degenerin subfamily. Several degenerins (MEC-4 (Driscoll and Chalfie, 1991), MEC-10 (Huang and Chalfie, 1994), UNC-105 (Liu *et al.,* 1996), UNC-8, and DEL-1 (Tavernarakis *et al.,* 1997)) have been implicated in mechanotransduction (see below), although their exact cellular functions remain to be experimentally demonstrated.

C. Sequence Features of the FLR-1 Subfamily

FLR-1 subfamily members include 9 of 10 conserved Cys residues common to CRDIII and exhibit similarity within MSDII, but they lack clear regions of homology to degenerin CRDI. FLR-1 subfamily members are better described as sharing a distinct and extended region of similarity in the predicted extracellular domain that we refer to as FEHD (*F*LR-1 subfamily *e*xtracellular *h*omology *d*omain, corresponding to amino acids 106 to 282 of FLR-1; I. Katsura, personal communication). Included in FEHD is a somewhat modified version of the CRDII which is found in the degenerin subfamily. In general, the extracellular domains of the FLR-1 subfamily are shorter than the degenerin subfamily extracellular loop and the predicted C-terminal intracellular domains are longer than degenerin C-termini. In addition, FLR-1 subfamily members include a block of conserved sequence in the C-terminal domain (situated close to MSDII) that is not found in the degenerin subfamily.

Little can be predicted about possible functions of FLR-1 subfamily channels since genetic defects have been reported in only one family member, *flr-1. flr-1* loss-of-function mutations confer resistance to fluoride toxicity (Katsura, 1993), show slow growth, and cause frequent defecation (Iwasaki *et al.,* 1995). Unfortunately, the mechanism of fluoride poisoning is unknown, so it is difficult to build models of FLR-1 activity. One possibility is that the flouride ion may activate the channel via a signal transduction mechanism (I. Katsura, personal communication; see also Iwasaki *et al.,* 1995).

D. Additional Observations on the Genome

Genes encoding most of the degenerin/FLR-1 family members are dispersed throughout the genome. However, there is one cluster of three independent genes on chromosome I. Predicted genes C24G7.1, C24G7.2, and C24G7.4 are all members of the FLR-1 subfamily and are approximately 40–60% identical to one another. Thus it appears that relatively recent gene duplication events generated this cluster. The other predicted gene, T28D9.7, is similar to other family members within and adjacent to the membrane spanning domains, but it lacks certain Cys-rich domains (see Fig. 1). This more distant gene is not categorized into either the degenerin or the FLR-1 branches.

1. Atypical and Truncated Degenerin/FLR-1-like Genes

Genome sequence analysis has identified some peculiarities within the family. For example, one member (ZK770.1) appears to include an additional hydrophobic domain at its N-terminus. CRDI is also highly divergent in this family member.

A more frequent finding is the presence of genes that show similarity to only a subset of degenerin/FLR-1 domains. For example, the gene *del-2* exhibits sequence similarity only to the intracellular N-terminus of the *C. elegans* degenerins (Harbinder *et al.,* 1997). No *del-2* mutations have yet been generated, and thus the biological function of the encoded protein remains unknown. However, it is highly intriguing that *del-2* appears to be expressed only in cells that mediate nose touch sensation (Kaplan and Horvitz, 1993; Harbinder *et al.,* 1997), a pattern consistent with the possibility that the protein could be involved in the mechanosensory functions of these neurons. Moreover, given that the expression of N-terminal MEC-4 fragments has been shown to interfere with touch receptor channel function (Hong, *et al.,* 1999) and that expression of the N-terminus of γhENaC can inhibit channel function (Adams *et al.,* 1997), it is tempting to speculate that proteins with homology to individual degenerin domains might act as modulators that regulate channel activities *in vivo.*

Other truncated degenerins predicted by the *C. elegans* genome sequencing project are open reading frames F55G1.12 (which exhibits CRDIII and MSDII homology, but does not include similarities to more N-terminal domains) and C46A5.2 (which includes homology within the conserved FEHD domain and within more C-terminal domains of the FLR-1 subfamily). Searches for sequences similar to N-terminal degenerin/FLR-1 domains in the vicinities of these loci have failed to reveal additional exons. These "truncated degenerins" could associate with channels to influence function. Alternatively, these genes may be defective pseudogenes. Their existence underscores the importance of accurate predictions of gene structures and the need to test for expression of identified genes to evaluate their biological significance.

III. *C. elegans* DEGENERINS HAVE BEEN IMPLICATED IN MECHANOTRANSDUCTION

A. Touch Sensitivity

In *C. elegans* six neurons sense gentle touch delivered over the body (Chalfie and Sulston, 1981). A saturating screen for touch-insensitive mutant worms led to the identification of a number of genes needed for the mechanosensory function of these neurons (termed *mec* genes because when they are defective, mutants are *mec*hanosensory abnormal (Chalfie and Sulston, 1981; Chalfie and Au, 1989). Two genes required for mechanosensation are members of the DEG/ENaC superfamily, *mec-4* (Driscoll and Chalfie, 1991) and *mec-10* (Huang and Chalfie, 1994). The MEC-4 and MEC-10 subunits have been shown to interact with each other *in vitro* through coimmunoprecipitation studies (Lai, 1995). Genetic data also support that MEC-4 and MEC-10 subunits associate in the touch neurons and that more than one subunit of each is included within the channel complex (Hong and Driscoll, 1994; Huang and Chalfie, 1994; Gu *et al.*, 1996). Thus it appears that the touch receptor ion channel is multimeric and the channel complex includes at least two MEC-4 and two MEC-10 subunits. Genetic data from *C. elegans* cannot be used to predict more precise subunit stoichiometry. To date, no other degenerin family members coexpressed with *mec-4* and *mec-10* in the touch receptor neurons have been identified among several degenerin family members tested (N. Tavernarakis, S. Wang, and M. Driscoll, unpublished observations).

Genetic and molecular analyses of the *mec* genes have resulted in the formulation of a working model for a touch-transducing complex in the touch neurons (Chalfie, 1997; Tavernarakis and Driscoll, 1997). Central to this model is the heteromeric degenerin channel, composed of MEC-4 and MEC-10 subunits. Subunits are oriented such that their amino- and carboxy-termini project into the cytoplasm and their Cys-rich regions extend in a single loop outside the touch cell (Lai *et al.*, 1996). Localized tension (which is expected to be required for regulated mechanical opening and closing of the channel) is postulated to be administered by tethering the extracellular channel domains to the specialized extracellular matrix and anchoring intracellular domains to the microtubule cytoskeleton. On the extracellular side, channel subunits may interact with MEC-5 collagen and/or MEC-9 (a protein that features multiple protein interaction domains such as EGF repeats, Kunitz protease inhibitor motifs, and snake venom toxin motifs) in the touch receptor mantle (Du *et al.*, 1996). Inside the cell, channel subunits may interact with large diameter microtubules that are unique to the touch receptor neurons (Chalfie and Thomson, 1979). These specialized microtubules have been proposed to contact the channel at their distal ends via MEC-2, a linker protein that may interact both with the MEC-12 α-tubulin and with intracellular channel domains (Fukushige *et al.*, 1999; Huang *et al.*, 1995; Gu *et al.*, 1996). A touch stimulus may deform the microtubule network, which could tug the channel open from the intracellular side. Alternatively, a touch stimulus could perturb the mantle connections and pull the channel open from the extracellular side. While direct assay of MEC-4/MEC-10 channel activity has yet to be published, studies in which nematode and mammalian chimeras were expressed in *C. elegans* (Hong and Driscoll, 1994; Waldmann *et al.*, 1995) or in which *C. elegans unc-105* was expressed in oocytes (García-Añoveros *et al.*, 1998) suggest that MEC-4 and MEC-10 are likely to form channels.

1. Structure/Function Studies of MEC-4 Highlight Domains That Are Critical for Subunit Activity

Genetic screens in *C. elegans* have provided a wealth of material that has yielded insight into the structure/function relationships of degenerins and of superfamily members in general (key findings summarized in Fig. 2). For example, more than 50 recessive *mec-4* loss-of-function mutations have been identified in searches for touch-insensitive mutants (Chalfie and Sulston, 1981; Chalfie and Au, 1989). Sequence analysis of these *mec-4* loss-of-function alleles has revealed that single amino acid substitutions that inactivate MEC-4 are clustered, highlighting domains essential for *in vivo* channel function (Hong *et al.*, 1999). More specifically, inactivating substitutions are concentrated in three regions: (1) a conserved N-terminal intracellular region defined by amino acids 91–95 that has been implicated in gating by studies of the human pseudohypoaldosteronism type I G37S mutation (Grunder *et al.*, 1997); (2) an area near and within the conserved extracellular CRDIII that might mediate interactions with components of the extracellular mantle; and (3) MSDII, an amphipathic transmembrane helix that features conserved hydrophilic residues on one face (Hong and Driscoll, 1994). Based on genetic manipulations that indicated conserved charged and polar amino acids were essential for normal and toxic *mec-4* activities (see below), it was predicted that conserved MSDII residues contribute to the channel pore (Hong and Driscoll, 1994), a hypothesis supported by electrophysiological studies on some mammalian family members. For example, substitution for some of the polar residues S589 and S593 in MDEG does influence channel conductance properties (Waldmann *et al.*, 1995).

Toxic degenerin alleles also provide insight into *in vivo* channel function. Three dominant gain-of-function *mec-4* alleles induce swelling and degeneration of the touch receptor neurons (Chalfie and Sulston, 1981; Driscoll and Chalfie, 1991). Dominant death-inducing substitutions all map to the same site—a conserved Ala (position 713) adjacent to MSDII (Driscoll and Chalfie, 1991; Lai *et al.,* 1996). Because large side chain amino acids at this site are toxic but small side chain amino acids are not, we hypothesized that steric hindrance resulting from such substitutions "locks" the channel in the open conformation, resulting in constitutive cation uptake and consequent death. Electrophysiological study of superfamily members engineered to encode the cognate toxic amino acid substitution supports this model (Waldmann *et al.,* 1996; Adams *et al.,* 1998).

Degenerins can mutate in a second way to cause toxicity. Studies of recessive *deg-1(u506)* established that this allele encodes substitution A393T situated within a 22 amino acid sequence in the predicted extracellular regulatory domain (ERD) that is conserved among the *C. elegans* degenerin subfamily, but is missing from mammalian family members (García-Añoveros *et al.,* 1995). Introduction of the same amino acid change in MEC-4 (or a deletion of this region) also creates a toxic allele. Furthermore, dominant *unc-8* alleles that induce neuronal swelling also map to this domain (Tavernarakis *et al.,* 1997). Elevated ion influx has been implicated in toxicity (García-Añoveros *et al.,* 1995) and thus models that explain the genetic properties postulate that this region may correspond to an extracellular channel gate or may be needed for conformational changes that close or stabilize a closed state of the channel.

B. A Stretch-Activated Ion Channel in Muscle?

Many *C. elegans* mutations cause abnormal animal motility. Nematodes that harbor semidominant mutations in *unc-105* exhibit muscle disorganization and hypercontraction that result in severe paralysis of the animal (Park and Horvitz, 1986a,b). Molecular studies have revealed that *unc-105* encodes a member of the degenerin subfamily (Liu *et al.,* 1996). The hypercontracted phenotype is hypothesized to be the consequence of inappropriately increased channel activity. However, the *unc-105(sd)* mutations do not encode substitutions that affect the residue corresponding to the MEC-4(Ala713) site or to the ERD region (see Fig. 2). Instead, *unc-105(sd)* mutations alter two different residues in the predicted extracellular loop (substitutions are P134T and P134S in the conserved region after MSDI and E677K in the conserved region preceding MSDII (Liu *et al.,* 1996)). Interestingly, the severe paralysis induced by *unc-105(sd)* can be genetically reversed by specific mutations in a second gene, *let-2,* which encodes a type IV collagen (Liu *et al.,* 1996). This genetic interaction suggests that the UNC-105 channel associates with collagen in the extracellular matrix of the muscle cell. It is striking that a MEC-5 collagen-touch receptor channel interaction has also been suggested by genetic data (Du *et al.,* 1996; Gu *et al.,* 1996). Perhaps all *C. elegans* degenerins will prove to associate with specific extracellular collagens. Such an association could be a key contact required for maintenance of channel gating tension.

C. A Channel That Functions in Nematode Proprioception?

Another recently cloned member of the DEG/ENaC family is *unc-8.* This gene is expressed in about 45 neurons including the nose-touch sensory neurons, interneurons, and several motor neuron classes (Tavernarakis *et al.,* 1997). Dominant gain-of-function *unc-8* mutations cause severe uncoordination such that the worm, which normally moves forward and backward in a sinusoidal trajectory, is unable to backup and instead coils (Park and Horvitz, 1986a). Some *unc-8(d)* alleles cause neurons to swell, although cells do not degenerate as they do with *mec-4(d)* or *deg-1(d)* mutations (Shreffler *et al.,* 1995).

In contrast to the dramatic paralysis conferred by dominant mutations, *unc-8* loss-of-function mutations confer a very subtle phenotype. The *unc-8(lf)* mutants move nearly normally except that the amplitude and the wavelength of the tracks that they leave as they travel over an *E. coli* lawn are markedly reduced as compared to wild-type worms (Tavernarakis *et al.,* 1997). What might the contribution of the UNC-8 channel to normal locomotion be? A clue may come from neuronal anatomy: some of the *unc-8*-expressing motor neurons have long synapse-free processes that resemble sensory processes (White *et al.,* 1986). In fact, such motor neurons have been postulated to serve as stretch sensors in *C. elegans* and have been shown to be stretch sensors in the highly analogous nervous system of *Ascaris suum* (Davis and Stretton, 1996). By analogy to the MEC-4/MEC-10 channel, we have proposed that the UNC-8 channel may be mechanically gated and specialized to respond to the localized body stretch that occurs during locomotion (model summarized in Fig. 3). In this model the UNC-8 channel, which may be preferentially localized to the stretch sensory domain of the neuronal process, responds to displacement bending of the process during locomotion. Activation of the UNC-8 channel would potentiate signaling at the distant neuromuscular junction. In other words, the stretch signal would enhance motorneuron excitation of muscle, increasing the strength and duration of the pending muscle contraction and directing a deep body bend. In this way the UNC-8 channel can be thought of as acting in nematode proprioception. Although this model links neuroanatomy and behavioral phenotype, the actual contribution of the UNC-8 chan-

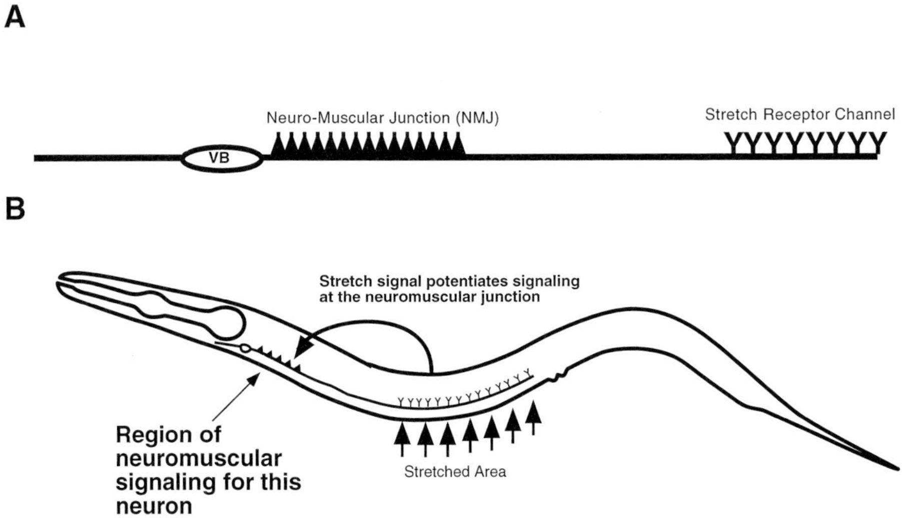

FIGURE 3 A model for the UNC-8 channel in nematode proprioception. (A) Features of the VB motorneuron. A typical VB motorneuron makes synapses to muscle and to other neurons near to the cell body; a long nerve terminal is synapse-free and thought to be sensory. UNC-8 stretch receptor channels are postulated to be situated in the sensory domain of this process. (B) Depiction of a model for stretch activation of body bends. The worm normally moves in a sinusoidal motion. Motorneurons should be maximally stretched at their termini at the time when the motor neuron activates an anterior localized muscle contraction. We postulate that the stretch signal acts on an UNC-8 channel to potentiate signaling at the neuromuscular junction, increasing the strength and duration of the distant anterior muscle contraction and consequent bend.

nel to locomotion and stretch-reception remains to be experimentally verified.

Like other members of the DEG/ENaC superfamily, UNC-8 is likely to participate in a heteromeric channel complex. At least one other degenerin gene, named *del-1* (for degenerin-like), is coexpressed with *unc-8* in subsets of motorneurons (Tavernarakis *et al.,* 1997). In light of working hypotheses on mechanosensory activity of degenerins, it is interesting that *del-1* is expressed exclusively in the FLP harsh-touch sensory neurons and in the VA and VB motor neurons, motor neurons that are the strongest candidate stretch sensors. Characterization of other degenerins and determination of their expression patterns is underway. It appears likely that different heteromeric channels will be assembled in different cells so that distinct UNC-8-containing channels with distinct properties contribute to regulated locomotion.

1. Candidate Genes That Affect UNC-8 Channel Function

As discussed above for the MEC-4/MEC-10 touch receptor channel, working models for mechanotransducing channels must include proteins that interact with degenerin subunits both inside and outside the cell to confer gating tension on the channel. In addition, it is likely that proteins that modify the activities of channel subunits function in various cells. One of the great advantages of the *C. elegans* system is that such proteins can be identified using genetic approaches. For example, the strikingly coiled phenotype of *unc-8(d)* alleles enabled exhaustive genetic screens for rare extragenic suppressor mutations that restore normal locomotion to be conducted. Screens of approximately 600,000 mutagenized chromosomes identified extragenic mutations in 5 genes that could reverse the severely Unc phenotype (Shreffler *et al.*, 1995; Shreffler and Wolinsky, 1997). One class of mutations identified as extragenic *unc-8(d)* suppressors are loss-of-function alleles of *mec-6*. *mec-6* also has a role in the function of the MEC-4/MEC-10 mechanotransducing complex (Chalfie and Wolinsky, 1990; Harbinder *et al.*, 1997).

Mutations in the four other genes identified as suppressors of *unc-8(d)* alleles, designated *sup-40–sup-43* (for *sup*presors), were isolated at a frequency considerably lower than that expected for a simple knockout of gene activity. This suggests that the existing alleles are not null (complete loss-of-function) mutations. Rather, they most likely encode single amino acid changes that reduce the activities of SUP proteins or alter an interaction with the UNC-8 channel in a highly specific manner. Features of these alleles are outlined in Table I and discussed in more detail below.

TABLE I

Summary of Genetic Suppressors of the *unc-8(d)* Hyperactive Channel

unc-8(d) suppressor	Mode of suppression	Phenotype independent of *unc-8(d)*	Comments
sup-40(lb130)	Dominant	Recessive sterility, hypodermal swelling, NDG resistance	Interacts with *mec-4(d)* channel as well
sup-41(lb125)	Dominant	?	Interacts with *deg-1(d)* channel as well
sup-42(lb88)	Recessive	None obvious	Suppresses all *unc-8(d)* alleles
sup-43(lb141)	Recessive	Uncoordinated	Allele-specific suppressor of *unc-8(e49)*, which affects extracellular domain CRDIII

In considering the possible activities of the suppressor mutations, it is useful to reiterate that there are two types of *unc-8(d)* alleles. *e15* and *n491* both encode substitution G387E, situated in the extracellular regulatory domain; these alleles induce neuronal swelling. Allele *e49* encodes substitution A586T, which is situated within conserved CRDIII. Some *unc-8* suppressors affect all alleles, suggesting they may act generally to alter or circumvent UNC-8 channel activity. In contrast, *sup-43(lb141)* is an allele-specific suppressor that exclusively affects the mutant variant UNC-8(A586T) channel encoded by *e49*. Allele-specificity is genetically consistent with models of direct protein–protein interactions.

The *sup-40(lb130)* mutation is a dominant suppressor of all *unc-8(d)* mutations (Shreffler *et al.*, 1995). In addition, *sup-40(lb130)* can suppress neurodegeneration induced by *mec-4(d)*. This suggests that *sup-40(lb130)* can act generally to counteract the toxic effects of hyperactive degenerin channels. *sup-40(lb130)* homozygotes exhibit additional phenotypes. *sup-40* animals are sterile, apparently because their oocytes swell and burst. Hypodermal nuclei also swell in *sup-40(lb130)* homozygotes. Another feature of *sup-40* mutants is their resistance to nordihydroguaieretic acid (NDG, a lipoxygenase inhibitor that interferes with the synthesis of the second messenger hepoxilin; Belardetti *et al.*, 1989; Schacher *et al.*, 1993). Taken together, these data suggest that SUP-40 may normally contribute to the regulation of degenerin channel function and/or may play a role in cellular osmotic balance.

Dominant suppressor *sup-41(lb125)* suppresses both *unc-8(d)* and *deg-1(d)* mutations. Thus *sup-41* appears to encode a protein that can interact with multiple degenerin channel complexes. The *sup-41(lb125)* homozygotes do not have any other readily apparent phenotypes. *sup-42(lb88)* has a different character. The *sup-42(lb88)* allele is a recessive suppressor of all *unc-8(d)* mutations. However, it does not interact with *deg-1(d)* channels suggesting that the protein encoded by *sup-42* may interact specifically with the UNC-8 channel.

A particularly interesting mutation has been identified in the *sup-43* gene. *sup-43(lb141)* is a recessive allele-specific suppressor of *unc-8(e49)*. As noted above, allele specificity suggests possible direct protein–protein interaction between UNC-8 and SUP-43. Since *unc-8(e49)* encodes a substitution in the extracellular CRDIII, one possibility would be that *sup-43* could encode a component of the extracellular matrix that associates with CRDIII. Since CRDIII is highly conserved in superfamily members, identification of this protein may shed light on the biology of the channel class as a whole. Finally, one enhancer of the uncoordinated phenotype of *unc-8(e49)*, *enu-2*, has been identifed (Shreffler and Wolinsky, 1997). *enu-2* may encode another protein that interacts in a direct way with the UNC-8 channel.

IV. FUTURE DIRECTIONS

The identification of nematode and mammalian DEG/ENaC family members that are expressed in neurons and muscle prompts the important question of what the nonepithelial DEG/ENaC channels contribute to cell function. As we have discussed, genetic studies in *C. elegans* implicate neuronal and muscle degenerin channels in mechanotransduction, suggesting that the MEC-4/MEC-10 channel may be the key mediator of touch transduction, the UNC-105 channel may respond to muscle stretch, and the UNC-8 channel (or a subset of UNC-8-containing channels) may contribute to proprioception. Although mutant phenotypes are consistent with models that degenerin subfamily members are mechanically gated channels that play the central roles in mechanotransduction in touch and locomotory systems, it should be emphasized that the channel gating mechanism remains to be tested. To date it has proved difficult to assay *C. elegans* channel function *in vivo* using electrophysiological approaches since the neurons of this nematode have prohibitively small cell bodies (approximately 1 μ in diameter). An exciting recent technological development is that methods for recording from *C. elegans* neurons have been reported (Goodman *et al.*, 1998). Although the technique, which involves extrusion of neurons from the nerve ring, may not be immediately applicable to the touch receptor neurons which most likely have to be embedded in the hypodermis and attached to the cuticle to function properly, modifications may soon render it feasible to record directly from touch neurons. Such a breakthrough would enable investigators to record from the wealth of existing touch-insensitive mutants to decipher channel structure/function relationships *in vivo*. An alternative approach is to assay degenerin subunits in expression systems. Such studies will clearly elaborate some channel properties. However, it should be cautioned that assays of mechanically gated channels may require reconstitution of protein interactions that transfer tension to the channels to be fully informative.

Another challenge that remains for the future is the determination of the composition of degenerin complexes. Again, while existing models of complexes are consistent with available genetic data, biochemical evidence of proposed protein associations has not been reported. Moreover, not all proteins that interact with degenerin complexes are expected to be identified by genetic approaches. Any genes that encode redundant products, any genes that encode products essential for viability or development, or any genes that encode products needed globally for coordinated locomotion would have been missed in screens for touch-insensitive mutants or for *unc-8(d)* suppressors. Thus, future biochemical studies will be needed before full understanding of the *C. elegans* channel complexes is accomplished.

Finally, it is particularly exciting that the complete genomic sequence of *C. elegans* is expected to be reported in the near future, making it possible to identify all family members from one organism. Given sequence information and standard molecular biology techniques, it will be straightforward to determine which cells express which degenerin. Moreover, recent development of techniques to isolate deletion alleles of genes by reverse genetics (Plasterk, 1996; Jansen *et al.*, 1997) and techniques to knock out expression (Fire *et al.*, 1998) should enable us to determine the contributions of these channels in cell function. Thus, the combination of genetic, biochemical, molecular and physiological techniques applicable in *C. elegans* should enable us to derive a detailed understanding of the biology of the entire channel family. Both structure/function relationships and behavioral studies should provide insight into the workings of the many related mammalian ENaC-like channels that are expected to be identified during analysis of the human genome.

References

Adams, C. M., Snyder, P. M., and Welsh, M. J. (1997). Interactions between subunits of the human epithelial sodium channel. *J. Biol. Chem.* **272,** 27295–27300.

Adams, C. M., Anderson, M. G., Motto, D. G., Johnson, W. A., and Welsh, M. J. (1998). Ripped pocket and pickpocket, novel Drosophila DEG/ENaC subunits expressed in early development and in mechanosensory neurons. *J. Cell Biol.* **140,** 143–152.

Belardetti, F., Campbell, W. B., Falck, J. R., Demontis, G., and Roslowsky, M. (1989). Products of hem-catalyzed transformation of the arachidonate derivative 12-HPETE open S-type K+ channels in Aplysia. *Neuron* **3,** 497–505.

Canessa, C. M., Horisberger, J. D., and Rossier, B. C. (1993). Epithelial sodium channel related to proteins involved in neurodegeneration. *Nature (London)* **361,** 467–470.

Canessa, C. M., Schild, L., Buell, G., Thorens, B., Gautschi, I., Horisberger, J. D., and Rossier, B. C. (1994). Amiloride-sensitive epithelial Na^+ channel is made of three homologous subunits. *Nature (London)* **367,** 412–413.

Chalfie, M. (1997). A molecular model for mechanosensation in *Caenorhabditis elegans. Biol. Bull. (Woods Hole, Mass.)* **192,** 125.

Chalfie, M., and Au, M. (1989). Genetic control of differentiation of the *Caenorhabditis elegans* touch receptor neurons. *Science* **243,** 1027–1033.

Chalfie, M., and Sulston, J. (1981). Developmental genetics of the mechanosensory neurons of *Caenorhabditis elegans. Dev. Biol.* **82,** 358–370.

Chalfie, M., and Thomson, J. N. (1979). Organization of neuronal microtubules in the nematode *Caenorhabditis elegans. J. Cell. Biol.* **82,** 278–289.

Chalfie, M., and Wolinsky, E. (1990). The identification and suppression of inherited neurodegeneration in *Caenorhabditis elegans. Nature (London)* **345,** 410–416.

Davis, R. E., and Stretton, A. O. (1996). The motornervous system of Ascaris: Electrophysiology and anatomy of the neurons and their control by neuromodulators. *Parasitology* **113,** Suppl., S97–S117.

Driscoll, M., and Chalfie, M. (1991). The *mec-4* gene is a member of a family of *Caenorhabditis elegans* genes that can mutate to induce neuronal degeneration. *Nature (London)* **349,** 588–593.

Du, H., Gu, G., Williams, C., and Chalfie, M. (1996). Extracellular proteins needed for *C. elegans* mechanosensation. *Neuron* **16,** 183–194.

Fire, A., Xu, S., M.K., M., Kostas, S. A., Driver, S. E., and Mello, C. C. (1998). Potent and specific genetic interference by double-stranded RNA in *Caenorhabditis elegans. Nature* (*London*) **391,** 806–811.

Fukushige, T., Siddiqui, Z. K., Chou, M., Culotti, J. G., Gogonea, C. B., Siddiqui, S. S., and Hamelin, M. (1999). MEC-12, an α-tubulin required for touch sensitivity in *C. elegans. J. Cell. Sci.* **112,** 395–403.

García-Añoveros, J., Ma, C., and Chalfie, M. (1995). Regulation of *Caenorhabditis elegans* degenerin proteins by a putative extracellular domain. *Curr Biol.* **5,** 441–448, errata: *Ibid.,* p. 686.

García-Añoveros, J., Derfler, B., Neville-Golden, J., Hyman, B. T., and Corey, D. P. (1997). BNaC1 and BNaC2 constitute a new family of human neuronal sodium channels related to degenerins and epithelial sodium channels. *Proc. Natl. Acad. Sci. U.S.A.* **94,** 1459–1464.

García-Añoveros, J., Garcia, J. A., Liu, J. D., and Corey, D. P. (1998). The nematode degenerin UNC-105 forms ion channels that are activated by degeneration- or hypercontraction-causing mutations. *Neuron* **20,** 1231–1241.

Goodman, M. B., Hall, D. H., Avery, L., and Lockery, S. R. (1998). Active currents regulate sensitivity and dynamic range in *C. elegans* neurons. *Neuron* **20,** 763–772.

Grunder, S., Firsov, D., Chang, S. S., Jaeger, N. F., Gautschi, I., Schild, L., Lifton, R. P., and Rossier, B. C. (1997). A mutation causing pseudohypoaldosteronism type 1 identifies a conserved glycine that is involved in the gating of the epithelial sodium channel. *EMBO J.* **16,** 899–907.

Gu, G., Caldwell, G. A., and Chalfie, M. (1996). Genetic interactions affecting touch sensitivity in *Caenorhabditis elegans. Proc. Natl. Acad. Sci. U.S.A.* **93,** 6577–6582.

Harbinder, S., Tavernarakis, N., Herndon, L., Kinnell, M., Xu, S. Q., Fire, A., and Driscoll, M. (1997). Genetically targeted cell disruption in *Caenorhabditis elegans. Proc. Natl. Acad. Sci. U.S.A.* **94,** 13128–13133.

Hong, K., and Driscoll, M. (1994). A transmembrane domain of the putative channel subunit MEC-4 influences mechanotransduction and neurodegeneration in *C. elegans. Nature* (*London*) **367,** 470–473.

Hong, K., Mano, I., and Driscoll, M. (1999). In preparation.

Huang, M., and Chalfie, M. (1994). Gene interactions affecting mechanosensory transduction in *Caenorhabditis elegans. Nature* (*London*) **367,** 467–470.

Huang, M., Gu, G., Ferguson, E. L., and Chalfie, M. (1995). A stomatin-like protein necessary for mechanosensation in *C. elegans. Nature* (*London*) **378,** 292–295.

Iwasaki, K., Liu, D. W., and Thomas, J. H. (1995). Genes that control a temperature-compensated ultadian clock in *Caenorhabditis elegans. Proc. Natl. Acad. Sci. U.S.A.* **92,** 10317–10321.

Jansen, G., Hazendonk, E., Thijssen, K. L., and Plasterk, R. H. A. (1997). Reverse genetics by chemical mutagenesis in *Caenorhabditis elegans. Nat. Genet.* **17,** 119–121.

Kaplan, J. M., and Horvitz, H. R. (1993). A dual mechanosensory and chemosensory neuron in *Caenorhabditis elegans. Proc. Natl. Acad. Sci. U.S.A.* **90,** 2227–2231.

Katsura, I. (1993). In search of new mutants in cell-signaling systems of the nematode *Caenorhabditis elegans. Genetica* **88,** 137–146.

Katsura, I., Kondo, K., Amano, T., Ishihara, T., and Kawakami, M. (1994). Isolation, characterization and epistasis of fluoride-resistant mutants of *Caenorhabditis elegans. Genetics* **136,** 145–154.

Lai, C. C., Hong, K., Kinnell, M., Chalfie, M., and Driscoll, M. (1996). Sequence and transmembrane topology of MEC-4, an ion channel subunit required for mechanotransduction in *Caenorhabditis elegans. J. Cell Biol.* **133,** 1071–1081.

Lai, C. C. (1995). PhD Thesis, Rutgers University. Part I: The mechanism of interaction between CDC25 and RAS2 in *Saccharomyces cerevisiae.* Part II: Structure and topology studies of MEC-4, a mechanoreceptor subunit of *Caenorhabditis elegans.*

Lingueglia, E., Voilley, N., Waldmann, R., Lazdunski, M., and Barbry, P. (1993). Expression cloning of an epithelial amiloride-sensitive Na+ channel. A new channel type with homologies to *Caenorhabditis elegans* degenerins. *FEBS Lett.* **318,** 95–99.

Lingueglia, E., Champigny, G., Lazdunski, M., and Barbry, P. (1995). Cloning of the amiloride-sensitive FMRFamide peptide-gated sodium channel. *Nature* (*London*) **378,** 730–733.

Liu, J., Schrank, B., and Waterston, R. H. (1996). Interaction between a putative mechanosensory membrane channel and a collagen. *Science* **273,** 361–364.

Park, E.-C., and Horvitz, H. R. (1986a). Mutations with dominant effects on the behavior and morphology of the nematode *C. elegans. Genetics* **113,** 821–852.

Park, E.-C., and Horvitz, H. R. (1986b). *C. elegans* unc-105 mutations affect muscle and are suppressed by other mutations that affect muscle. *Genetics* **113,** 853–867.

Plasterk, R. H. A. (1996). Postsequence genetics of *Caenorhabditis elegans. Genome Research* **6,** 169–175.

Renard, S., Lingueglia, E., Voilley, N., Lazdunski, M., and Barbry, P. (1994). Biochemical analysis of the membrane topology of the amiloride-sensitive Na+ channel. *J Biol. Chem.* **269,** 12981–12986.

Schacher, S., Kandel, E. R., and Montarolo, P. (1993). cAMP and arachidonic acid simulate long-term structural and functional changes produced by neurotransmitters in *Aplysia* sensory neurons. *Neuron* **10,** 1079–1088.

Schild, L., Schneeberger, E., Gautschi, I., and Firsov, D. (1997). Identification of amino acid residues in the alpha, beta, and gamma subunits of the epithelial sodium channel (ENaC) involved in amiloride block and ion permeation. *J. Gen. Physiol.* **109,** 15–26.

Shreffler, W., and Wolinsky, E. (1997). Genes controlling ion permeability in both motorneurons and muscle. *Behav. Genet.* **27,** 211–221.

Shreffler, W., Magardino, T., Shekdar, K., and Wolinsky, E. (1995). The *unc-8* and *sup-40* genes regulate ion channel function in *Caenorhadbitis elegans* motorneurons. *Genetics* **139,** 1261–1272.

Snyder, P. M., McDonald, F. J., Stokes, J. B., and Welsh, M. J. (1994). Membrane topology of the amiloride sensitive epithelial sodium channel. *J. Biol. Chem.* **269,** 24379–24383.

Tavernarakis, N., and Driscoll, M. (1997). Molecular modeling of mechanotransduction in the nematode *Caenorhabditis elegans. Annu. Rev. Physiol.* **59,** 659–689.

Tavernarakis, N., Shreffler, W., Wang, S., and Driscoll, M. (1997). unc-8, a DEG/ENaC family member, encodes a subunit of a candidate mechanically gated channel that modulates *C. elegans* locomotion. *Neuron* **18,** 107–119.

The *C. elegans* Sequencing Consortium. (1998). Genome sequence of the nematode *C. elegans:* A platform for investigating biology. *Science* **282,** 2012–2018.

Waldmann, R., Champigny, G., and Lazdunski, M. (1995). Functional degenerin-containing chimeras identify residues essential for amiloride-sensitive Na^+ channel function. *J. Biol. Chem.* **270,** 11735–11737.

Waldmann, R., Champigny, G., Voilley, N., Lauritzen, I., and Lazdunski, M. (1996). The mammalian degenerin MDEG, an amiloride-sensitive cation channel activated by mutations causing neurodegeneration in *Caenorhabditis elegans. J. Biol. Chem.* **271,** 10433–10436.

Waldmann, R., Bassilana, F., de Weille, J., Champigny, G., Heurteaux, C., and Lazdunski, M. (1997). Molecular cloning of a non-inactivating proton-gated Na+ channel specific for sensory neurons. *J. Biol. Chem.* **272,** 20975–20978.

Waterston, R., and Sulston, J. (1995). The genome of *Caenorhabditis elegans. Proc. Natl. Acad. Sci. U.S.A.* **92,** 10836–10840.

White, J. G., Southgate, E., Thomson, J. N., and Brenner, S. (1986). The structure of the nervous system of *Caenorhabditis elegans. Philos. Trans. R. Soc. London, Ser. B* **314,** 1–340.

CHAPTER 18

Amiloride-Sensitive Sodium Channels in Taste

Bernd Lindemann,* Timothy A. Gilbertson,† and Sue C. Kinnamon‡
*Department of Physiology, Saar University, D-66421 Homburg, Germany, †Pennington Biomedical Research Center, Louisiana State University, Baton Rouge, Louisiana 70808–4124, and ‡Department of Anatomy and Neurobiology, Colorado State University, Fort Collins, Colorado 80523

I. INTRODUCTION AND HISTORICAL BACKGROUND

The concept that salt taste is mediated by amiloride-blockable Na^+ channels located in the apical membranes of taste cells, introduced by John DeSimone and colleagues, is an attractive one (DeSimone *et al.,* 1984). It suggests economy of design, in that a typical transport protein of NaCl-retrieving epithelia, the apical Na^+ channel, is also used by taste receptor cells. These, after all, are derived from surrounding epithelial cells by differentiation. Due to its apical location, the Na^+ channel would permit current to flow into the receptor cell and the strength of this current would mirror the Na^+ concentration in the oral compartment. Furthermore, the

1063-5823/99 $30.00

current would necessarily pass outward through the basolateral membrane which, in consequence, would depolarize and release transmitter. A sodium salt taste receptor cell would thus be possible based on principles already familiar from the polarization model of Na^+-retrieving epithelia (Figs. 1A and B). This much-quoted model was proposed in 1958 (Koefoed-Johnsen and Ussing, 1958).

The original experiments showed that high concentrations of mucosal NaCl caused current to flow into the lingual epithelium of the dog or rat tongue. The current could be blocked by amiloride at a concentration of 100 μM (DeSimone *et al.,* Heck 1981; *et al.,* 1984). In retrospect, these seminal observations are not without surprises because (a) rodents but not carnivores like the dog have a Na^+ specific salt taste; (b) the current flowing into the lingual *epithelium* need not reflect the current flowing into taste cells as the latter occupy a minute fraction of the lingual surface only; and (c) the amiloride concentration required was much higher than the concentration expected to block the epithelial sodium channel, which would be submicromolar, and a preincubation of minutes was required before amiloride blocked the current. However, recordings from the taste nerve showed that mucosal amiloride did indeed block the taste response to mucosal NaCl (Heck *et al.,* 1984). This was confirmed in subsequent experi-

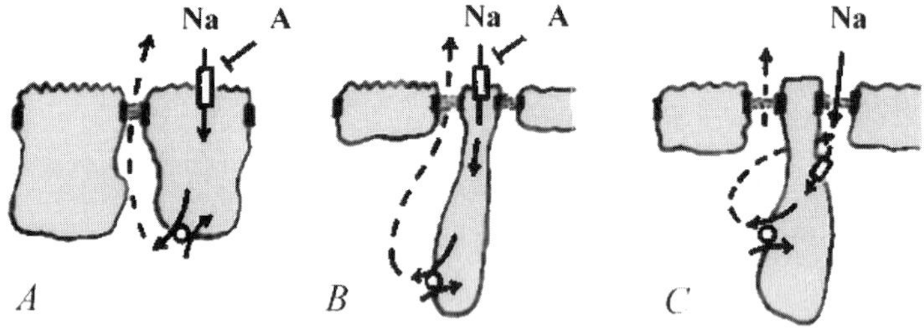

FIGURE 1 Polarization models for epithelia and taste cells. (A) Na^+ uptake through "distal" epithelia like frog skin and toad urinary bladder (Koefoed-Johnsen and Ussing, 1958). The Na^+ and Li^+ selective channels contained in the apical membrane are sensitive to mucosal amiloride (A). The current flows inward through these channels and outward through the Na^+ pump (circle). The current loop is closed (dashed arrow) through a tight junction permeable for small ions. (B) In elongated taste cells, Na^+ channels contained in the apical membrane are accessible to mucosal amiloride. The current loop is identical with that of panel A. (C) Tentative model for frog taste cells. Here the inward Na^+ current has first to overcome the tight junction and Na^+ channels located below the tight junction are not accessible to mucosal amiloride. Two current loops are shown: one for flow through the tight junction, the other for flow through the basolateral membrane. Current through the apical membrane may also contribute to the loops. Not shown are some characteristics of elongated taste cells like a variety of ion channels, including voltage-gated channels, and the synapses with sensory nerve fibers.

ments with rodents like rat and hamster, which do have a Na^+ (and Li^+) specific salt taste arising from the anterior tongue. This taste is blocked by mucosal amiloride in low micromolar concentrations, as evidenced by recordings from the taste nerve, from the taste pore, and by psychophysical experiments (Avenet and Lindemann, 1991; Brand, *et al.,* 1985; DeSimone and Ferrell, 1985; Gilbertson *et al.,* 1992; Spector *et al.,* 1996; Yoshii, *et al.,* 1986).

Thus, at least in the anterior tongue of rodents, the model of Na^+ taste transduction outlined above may apply. It predicts an apical location of the epithelial Na^+ channel which, since its successful cloning, is known as ENaC (Canessa *et al.,* 1993, 1994). Below we shall review results which support this model and others, which contradict it or necessitate modifications.

II. THE FROG Na^+ TASTE CHANNEL

At the cellular level, the first demonstration of amiloride-blockable inward currents was achieved with taste receptor cells (TRCs) isolated from taste buds of the frog. The cells, which are of slender, bipolar shape, were patch-clamped and the whole-cell current and the cell potential recorded (Avenet and Lindemann, 1987). A subset of such cells (more than 50% of those successfully patched) was found steadily depolarized in Na^+ Ringers but assumed resting potentials near −70 mV in the presence of 30 μM amiloride. In these cells amiloride blocked part of a Na^+-dependent sustained inward current. The K_M was near 0.3 μM (Avenet and Lindemann, 1988). In addition to the amiloride-blockable inward Na^+ currents, the cells showed TTX sensitive, voltage-gated Na^+ currents and voltage-gated K^+ currents, in full support of the original observation of Steve Roper, that taste receptor cells fire action potentials (Roper, 1983).

Outside-out membrane patches, excised from the basolateral membrane of frog TRCs (Avenet and Lindemann, 1989), had stationary inward currents which were partially blocked by amiloride (K_M near 0.3 μM). Power spectra of this current, recorded in the presence of 0.3 μM amiloride, were evaluated in terms of the rate constants of channel blockage by amiloride. Surprisingly, the rate constants turned out to be more than 10 times higher than those known from the high-affinity epithelial Na^+ channel present, for instance, in amphibian skin and bladder (Avenet and Lindemann, 1990). The single-channel conductance, estimated from noise data, was 1–2 pS when the patch was exposed to 110 mM Na^+ (Avenet and Lindemann, 1990). It was thus significantly smaller than the conductance of the high-affinity epithelial Na^+ channel (6–10 pS) of amphibia (Lindemann, 1984;

Lindemann and Van Driessche, 1977; Palmer and Frindt, 1986; Van Driessche and Lindemann, 1979). Furthermore, in the frog TRC Na^+ channel, the sequence of blocking potencies of amiloride analogs was different from, and the Na^+/K^+ selectivity smaller than that of the high-affinity epithelial Na^+ channel (Avenet and Lindemann, 1989). Thus, the biophysical data available on the amiloride-blockable taste channel of frogs showed clear differences to the high-affinity epithelial Na^+ channel, even to the well-known Na^+ channel of frog epidermis, and did not indicate one of the Na^+ channel varieties listed by Palmer (1992). Meanwhile the amphibian Na^+ channel (xENaC) has also been cloned and splice variants were detected (Puoti *et al.*, 1995, 1997). Now it would be important to look for molecular variants in the taste buds of frogs, where channel properties appear to differ remarkably from those found in the Na^+ retrieving amphibian epithelia (Lindemann, 1980, 1984).

Taken together, the current recordings from isolated taste cells of the frog indicated that Na^+ ions can enter a subset of TRCs through channels which are blockable by amiloride in submicromolar concentrations. In addition, there were other pathways which also supported Na^+ inward current but which were not blocked by amiloride. These experiments with isolated cells were not designed to decide whether the amiloride-blockable channels are located in the apical membrane and whether they are the "Na^+ receptors of salt taste." In fact, recent recordings from intact tongues revealed that in the frog the taste response to mucosal NaCl is *not* sensitive to mucosal amiloride (Kitada and Mitoh, 1997). In addition, recordings from the taste nerve of frogs also failed to show the expected amiloride sensitivity (Stevens and Lindemann, unpublished observations).

How is it possible, then, that whole-cell recordings with isolated TRCs indicate strong amiloride-sensitive currents while nerve recordings from intact taste buds show no sensitivity to mucosal amiloride? The likely explanation is that the cells have amiloride-blockable Na^+ channels which are, however, not located in the apical but in the basolateral membrane. Accordingly, membrane patches containing 20–50 of such channels could be excised from the somal membrane of frog TRCs (Avenet and Lindemann, 1989, 1990). These observations provided the first indication that the epithelial Na^+ channel, believed to be always apically located, may be present in basolateral membranes of some types of epithelial cells.

Do we already understand how salt taste works in the frog? No, presently we can only offer a speculation. Part of the inward Na^+ current may flow not through channels in the apical membrane but through the tight junctions. It would then enter the taste cells via amiloride-blockable channels contained in the basolateral membrane close to the tight junction (Fig. 1C). Supposing that Na^+ ions but not amiloride can diffuse through the tight junction, the lack of sensitivity to mucosal amiloride would be explained. [A related model was first put up by others to explain the anion paradox of salt taste in mammals (Ye *et al.,* 1991, 1993a)]. The advantage of such a design is obvious: The Na^+ sensors are located in a *protected space* where they are difficult to reach by xenobiotics (like amiloride), chemicals which might be present in the nutriments and able to interfere with the taste process.

III. AMILORIDE-SENSITIVE Na^+ CHANNELS IN RODENT TONGUE

A. Physiological Properties and Distribution

Numerous studies in rodents have implicated the involvement of amiloride-sensitive sodium channels in the transduction of sodium salts in the peripheral gustatory system. Electrophysiological experiments at the level of the taste receptor cell, lingual epithelium, afferent nerve fiber, and taste nuclei in the central nervous system all consistently implicate amiloride-sensitive pathways as mediating, at least in part, the gustatory response to sodium salts (Avenet, 1992; Lindemann, 1997). Recent studies utilizing patch clamp recording on isolated taste receptor cells have begun to explore the physiological properties of the amiloride-sensitive Na^+ channels in rodent tongue. Interestingly, there appear to be significant differences among rodents in terms of the contribution of amiloride-sensitive pathways to overall sodium salt taste transduction. This ranges from little or no amiloride-sensitivity of NaCl responses in certain mouse strains (Tonosaki and Funakoshi, 1989) to a majority of the NaCl responsiveness being due to amiloride-sensitive pathways in hamsters and rats (Gilbertson and Gilbertson, 1994). Moreover, there appears to be distinct regionalization of amiloride-sensitive channels within the tongues of certain rodents. It is clear that having a basic understanding of the physiology of amiloride-sensitive Na^+ channels in taste tissue and their distribution will lead to a significant advancement in elucidating the mechanisms of salt taste transduction.

Despite their importance for salt taste transduction, there have been surprisingly few studies that have attempted to directly characterize the function of amiloride-sensitive Na^+ channels in taste tissue. Those that have reveal that amiloride-sensitive Na^+ channels in taste tissue appear remarkably functionally similar to the ENaC expressed in a variety of transporting epithelia, including kidney, lung, and colon (Garty and Palmer, 1997). That is, they are apparently highly selective for Na^+ over K^+, exhibit saturation, respond to a variety of hormones involved in the regulation of salt and water balance, and they are inhibited by amiloride and its analogs with inhibition constants in the submicromolar range (for review, see Lindemann, 1996, 1997; Gilbertson, 1998).

Distribution of Amiloride-Sensitive Na^+ Channels in the Oral Cavity. In hamsters and rats, the two species that have been examined in detail, there appears to be significant differences in the regional localization of amiloride-sensitive Na^+ channels in the oral cavity. In the rat, electrophysiological studies have demonstrated that functional amiloride-sensitive Na^+ channels are found only in the anterior tongue (Doolin and Gilbertson, 1996) and soft palate (Gilbertson and Zhang, 1998b; Gilbertson and Fontenot, 1998). That is, approximately two-thirds of taste cells from the fungiform papillae and one-third from the foliate papillae contain amiloride-sensitive Na^+ channels as assayed by patch-clamp recording. The taste cells from the vallate papillae had no functional amiloride-sensitive Na^+ channels. This distribution is in contrast with immunocytochemical and *in situ* hybridization studies that reveal amiloride-sensitive Na^+ channel protein and mRNA in the vallate taste tissue (Simon *et al.,* 1993; Li *et al.,* 1994; Li and Snyder, 1994), but in agreement with afferent nerve recordings showing that sodium salt responses recorded in the glossopharyngeal nerve are amiloride-insensitive (Formaker and Hill, 1991). Approximately 40% of taste cells in the soft palate and 67% in the geschmacksstreifen in the rat contain functional amiloride-sensitive Na^+ channels (Gilbertson and Fontenot, 1998) consistent with afferent nerve recordings from the greater superficial petrosal nerve (VIIth nerve) that innervates the palate and with transepithelial sodium current recordings (Gilbertson and Zhang, 1998b). In rat it appears that expression of functional amiloride-sensitive Na^+ channels may be linked with innervation by branches of the VIIth nerve (chorda tympani or greater superficial petrosal). Figure 2 summarizes the distribution of amiloride-sensitive Na^+ channels in the rat.

An alternative explanation that may serve to rectify the apparent conflict between the presence of amiloride-sensitive Na^+ channel protein and mRNA in vallate papillae with the aforementioned electrophysiological studies is that there may be alternatively spliced forms of amiloride-sensitive Na^+ channels or unique subunit arrangements of amiloride-sensitive Na^+ channels in this area. That is, unique forms of amiloride-sensitive Na^+ channels may be present in the rat vallate papillae that, while functional in terms of sodium permeability, may lack measurable amiloride sensitivity. Clearly, such ENaCs have been reported that possess this phenotype (see Benos *et al.,* 1997). Support for this theory must come from a combination of molecular and physiological studies designed to directly examine this issue.

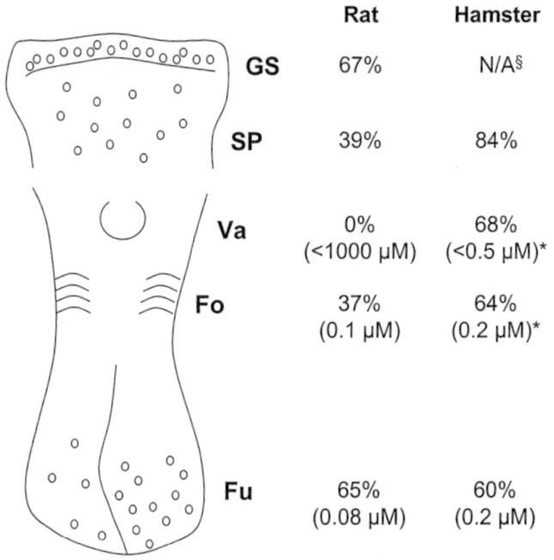

FIGURE 2 Distribution of amiloride-sensitive sodium channels in rat and hamster oral cavity determined by patch-clamp recording. Numbers show percentage of taste cells responsive to amiloride (30 μM) or benzamil (10 μM) in the fungiform (Fu), foliate (Fo), vallate (Va), soft palate (SP), and geschmacksstreifen (GS). Numbers in parentheses represent the approximate EC_{50} for amiloride, where determined. *, T. A. Gilbertson and D. T. Fontenot, unpublished observations. §, hamsters do not have a geschmacksstreifen.

This lack of functional amiloride-sensitive Na^+ channels in glossopharyngeally innervated taste buds is apparently not true, however, in all mammalian species. The hamster has apparently a roughly equal distribution of amiloride-sensitive Na^+ channels in all areas of the oral cavity. In the three classes of lingual taste buds and in the palate, greater than 60% of all taste cells examined contained functional amiloride-sensitive Na^+ channels (Fig. 2; Gilbertson and Fontenot, 1998). The presence of apical amiloride-sensitive Na^+ channels in the vallate has been confirmed by sodium transport measurements in Ussing chambers (Gilbertson and Zhang, 1998b). Interestingly, in the hamster epithelia containing vallate taste buds (and a subset of foliate taste buds), potassium transport was also inhibited by mucosal amiloride (Gilbertson and Zhang, 1998b) suggesting that there may be a class of amiloride-sensitive Na^+ channel in this region that has a significantly higher potassium permeability. Obviously, this will require direct confirmation at the cellular level, yet it is intriguing given that in the kidney, the relative potassium (to sodium) permeability in ENaC may be enhanced by phosphorylation and ADP ribosylation (see Benos *et al.,* 1997).

This fundamental difference between the rat and hamster in terms of distribution of this transduction pathway is interesting in light of reports that suggest that the glossopharyngeal nerve may be more responsive to aversive stimuli than is the chorda tympani (Ninomiya *et al.,* 1994). Behaviorally, the hamster finds most sodium salt concentrations aversive (Hettinger and Frank, 1990; Gilbertson and Gilbertson, 1994) and rats find them appetitive (Breslin *et al.,* 1993) and this may suggest a link between the physiology of the amiloride-sensitive pathways and, ultimately, behavior. It will be of considerable interest to explore this potential relationship further. Though substantially less work has been done in the mouse, it appears that certain mouse strains may lack functional amiloride-sensitive Na^+ channels altogether (Tonosaki and Funakoshi, 1989). Clearly, there appears to be a significant degree of variability even within mammalian species.

Cellular Distribution of Amiloride-Sensitive Na^+ Channels. Generally, the majority of studies are consistent with the functional amiloride-sensitive Na^+ channels in taste cells being located on the apical membrane, though this is far from unequivocal. This is the typical case for most, if not all, transporting epithelia. The presence of amiloride-sensitive Na^+ channels on the basolateral membrane has been speculated (Simon *et al.,* 1993), where they may play a role in taste cell response following the paracellular movement of Na^+ ions. Mierson and colleagues (1996), using transepithelial recording, have demonstrated the presence of a class of basolateral amiloride-sensitive Na^+ channels, with a significantly higher inhibition constant for amiloride ($\sim$50 μM). However, subsequent studies in other strains of rats and hamsters have failed to replicate this finding using similar techniques (Gilbertson and Zhang, 1998a,b). This may be reflective of species or methodological differences. Patch clamp studies on isolated taste buds, however, are generally more consistent with the latter finding as no amiloride-sensitive Na^+ channels have been identified with an inhibition constant greater than $\sim$0.5 μM (Fig. 2; Avenet and Lindemann, 1991; Gilbertson *et al.,* 1993; Doolin and Gilbertson, 1996).

B. Role of Amiloride-Sensitive Na^+ Channels in Sodium and Acid Taste

Due to their significant Na^+ permeability and distribution, at least partially, in the apical membranes of taste cells, it is not surprising that these channels provide a transduction channel that contributes to the perception of the prototypical "salty" stimulus, NaCl. The involvement of amiloride-sensitive Na^+ channels in mammalian taste responses to NaCl was first demonstrated by showing the amiloride-sensitivity of sodium transport in dog tongue by Heck *et al.,* (1984). Since that time, reports in a variety of species and using a wide range of techniques have shown that the diuretic amiloride and its analogs inhibit gustatory responses to NaCl applied to the tongue (for review, see Avenet, 1992; Lindemann, 1997).

The permeability of amiloride-sensitive Na^+ channels to other ions may contribute to their perception as "salty" and may participate in the transduction of other classes of taste stimuli. In mammalian taste cells, amiloride-sensitive Na^+ channels are apparently significantly permeant to both Li^+ and H^+. Like NaCl, LiCl is perceived generally as salty and somewhat sour (Ossebaard *et al.,* 1997) since it apparently acts through this amiloride-sensitive Na^+ transduction pathway. Amiloride-sensitive Na^+ channels in taste cells have also been shown to have a high H^+ permeability (Gilbertson *et al.,* 1992; 1993), similar to that reported for amiloride-sensitive Na^+ channels in other tissues (Palmer, 1987). Protons, which are generally perceived as "sour," permeate amiloride-sensitive Na^+ channels to depolarize hamster taste cells (Gilbertson *et al.,* 1993) in a manner analogous to Na^+ influx. Further, amiloride inhibits the acid responses from N-best fibers in recordings from the hamster nucleus tractus solitarius (Boughter and Smith, 1997) and reduces the aversiveness to acids in preference tests in the same species (Gilbertson and Gilbertson, 1994). The contribution of this mechanism to acid transduction in other species is presently unclear. Because of the interaction between sodium ions and protons (Gilbertson *et al.,* 1992), proton permeability may be significantly more important to acid taste in species, like hamster, that have relatively low salivary sodium levels ($\sim$ 6 mM; Rehnberg *et al.,* 1992). While proton permeability of amiloride-sensitive Na^+ channels may contribute to acid taste transduction, it also may explain the human psychophysical data that shows an interaction between acid and salt taste (Pangborn and Trabue, 1967; Helleman, 1992). Consistent with this interpretation, in humans amiloride blocks the total intensity of NaCl (Tennison, 1992; McCutcheon, 1992) which may in part be due to the inhibition of the "sour" side taste of NaCl (Ossebaard and Smith, 1995; Ossebaard *et al.,* 1997).

C. Regulation of Amiloride-Sensitive Na^+ Channels in Taste Tissue

The similarity between ENaC found in transporting epithelia and the amiloride-sensitive Na^+ channels in mammalian taste tissue extends to their ability to be regulated by a variety of extrinsic signals. In recordings from afferent nerves, lingual epithelia or isolated taste cells, sodium salt responsiveness has been shown, for example, to be regulated by various hormones involved in the balance of salt and water. Consistent with their effects in other systems (for review, see Garty and Palmer, 1997), aldosterone (Herness, 1992) and arginine^8-vasopressin (AVP; Gilbertson *et al.,* 1993) lead to increases in Na^+ transport in taste tissue. AVP apparently works via V_2-type vasopressin receptors since the effect on amiloride-sensitive Na^+ transport in isolated taste cells is mimicked by cAMP (Gilbertson *et al.*, 1993). The hormones, atrial natriuretic peptide (ANP), and oxytocin act to decrease Na^+ transport in lingual epithelia (Gilbertson, 1998) demonstrating a dual regulation common to a variety of transporting epithelia.

Other processes have been suggested to be involved in the regulation of amiloride-sensitive Na^+ channels in taste tissue. Recently, it has been demonstrated that sodium ions influence the activity of amiloride-sensitive Na^+ channels in isolated taste buds and lingual tissue. Sodium ions have been shown to act at ENaCs to decrease sodium permeability in a concentration-dependent fashion (Lindemann, 1984). One theory to explain the apparent saturation of ENaC has been suggested to be a result of increases in intracellular sodium (feedback inhibition; Ling and Eaton, 1989; Komwatana *et al.,* 1996a,b). The feedback inhibition of ENaCs typically takes minutes and may be direct, due to a decrease in driving force resulting from the increase in intracellular sodium, or indirect, involving activation of G proteins (Komwatana *et al.,* 1996a,b), protein kinase C and presumably, phosphorylation of the ENaCs. A second, more rapid mechanism for the sodium-dependent decrease in sodium permeability has been attributed to inhibition by extracellular sodium ions (self-inhibition; Fuchs *et al.,* 1977; Palmer *et al.,* 1998). Self-inhibition occurs by sodium ions binding to a sulfhdryl reagent-sensitive site and occurs within seconds to decrease sodium permeability. The most rapid of potential causes of ENaC would be expected to occur within milliseconds and be due to actual saturation of the binding sites for sodium ions within the pore of the channel (Hille, 1992).

Gilbertson and Zhang (1998a) have demonstrated that the sodium-dependent saturation of amiloride-sensitive Na^+ channels in rat taste tissue is due, at least in part, to the self-inhibition phenomenon. Both patch clamp and epithelial transport studies reveal the saturation (i.e., permeability decrease) of amiloride-sensitive Na^+ channels with increasing extracellular (or mucosal) sodium concentrations. This saturation, which occurs within 10–15 s, may be ameliorated by *p*-hydroxymercuribenzoate (*p*-HMB), a sulfhydryl compound that may act at cysteine residues in the channel (Luger and Turnheim, 1981). Self-inhibition is found only in those taste cells that show amiloride sensitivity, linking this phenomenon to the presence of functional amiloride-sensitive Na^+ channels. Interestingly, self-inhibition is most evident at ~70 mM sodium, due to the balance between the increases in driving force for sodium and in self-inhibition with increasing extracellular sodium, which is close to the levels of sodium found in rat saliva (Rehnberg *et al.,* 1992). These experiments do not, however, rule out that feedback inhibition or channel saturation may also be contributing to the change in sodium permeability. Self-inhibition has been suggested to potentially play roles in taste cell adaptation to sodium, which follows a similar time course (Avenet and Lindemann, 1991; Matsuo and Yamamoto, 1992; Nakamura and Norgren, 1991), the conservation of cellular resources during chronic sodium exposure and in the gustatory response to water (Gilbertson and Zhang, 1998a).

It is clear from these studies that both hormones and sodium ions may regulate the activity of amiloride-sensitive Na^+ channels in taste cells and lingual epithelia (Fig. 3), consistent with their ability to modulate the function of the more completely described ENaCs. Presently, however, little is known about how the regulation of amiloride-sensitive Na^+ channels in taste cells by either hormones or sodium ions may lead to changes, if any, in the gustatory responses to sodium salts. The further exploration of the regulation of amiloride-sensitive Na^+ channels will continue to be of interest since it provides a unique insight into the flexibility of the peripheral taste system to respond to systemic changes in the organism.

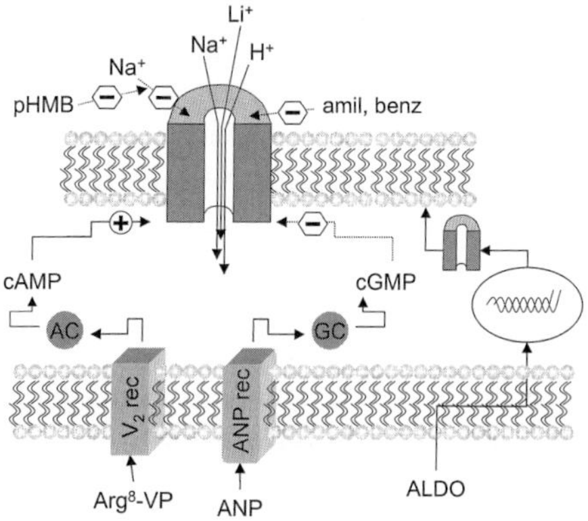

FIGURE 3 Regulatory mechanisms of amiloride-sensitive Na^+ channels reported in taste tissue. Amiloride-sensitive Na^+ channels, which in taste cells are permeable to Na^+, Li^+, and H^+, are inhibited by amiloride and benzamil in the sub- to low-micromolar range. These channels also demonstrate sodium self-inhibition at an extracellular site sensitive to sulfhydryl compounds like *p*-hydroxymercuribenzoate (pHMB). Hormones, like arginine-vasopressin (Arg^8-VP), atrial natriuretic peptide (ANP), and aldosterone (ALDO) also regulate amiloride-sensitive pathways in taste cells. Arg^8-VP leads to increases in channel activity via the second messenger cAMP, while ANP has the opposite effect, acting via cGMP. ALDO apparently may lead to increases in amiloride-sensitive sodium channel activity, presumably via its regulation of gene transcription, as reported in other transporting epithelia (see Garty and Palmer, 1997). See text for additional details.

IV. ENaC EXPRESSION IN THE LINGUAL EPITHELIUM (LE)

Taste cells originate from the surrounding lingual epithelium by differentiation. It is interesting, therefore, to view them in the context of general epithelial physiology. In those epithelia which develop from the ectoderm and which retrieve sodium ions from luminal compartments, ENaC is found in the apical membrane (Fig. 1A), as in colon, distal nephron, respiratory tract, and sweat and salivary glands (Duc *et al.,* 1994; Farman *et al.,* 1997; Kleyman *et al.,* 1994; Renard *et al.,* 1995; Tousson *et al.,* 1989). For the epidermis, which consists of several layers of connected epithelial cells, including outermost cornified layers, one would by analogy expect that ENaC is restricted to the membrane areas "outside of" the tight junction, i.e., to the apical membranes of the outermost living cell layer. Such a location was indicated by functional data obtained with the skin of frogs and toads (Fuchs *et al.,* 1977; Koefoed-Johnsen and Ussing, 1958; Lindemann and Voûte, 1976). In the epidermis of the rat, the distribution of channel protein between apical and basolateral membranes has not yet been investigated, even though ENaC mRNA was found (Roudier-Pujol *et al.,* 1996). In sweat-secreting epidermal areas of terrestrial mammals the channels may serve for sodium retrieval from excreted fluid.

Like the epidermis, the lingual epithelium (LE) of mammals is a multilayered cornifying epithelium. In the LE of rat and dog, the short circuit current increased following a step increase of the mucosal Na^+ concentration. The time course required minutes, however; i.e., it was surprisingly slow when compared to the response of other epithelia expressing ENaC (e.g. Fuchs *et al.,* 1977). The current was only in part blockable by *mucosal* amiloride and the effect of amiloride was not instantaneous but also took minutes (Mierson *et al.,* 1996; Simon and Garvin, 1985).

Thus, the response of the LE to mucosal NaCl and mucosal amiloride was slow, indicating a diffusion hindrance between mucosal solution and sodium channels. The cornified cell layer may be this hindrance. In contrast, the response of *in situ* taste buds to mucosal NaCl and amiloride is much faster (Avenet and Lindemann, 1991). This difference in speed suggests that the contribution of currents flowing through taste cells to the total lingual short circuit current must be negligible, as there is no discernible fast current component in the response of the LE short circuit current to mucosal NaCl.

One study indicated that amiloride applied from the *interstitial* side had a fully inhibitory effect in the rat LE. The inhibition exerted by interstitial amiloride was faster than the partial inhibition exerted by mucosal amiloride (Mierson *et al.,* 1996). The observation suggests that basolateral sodium uptake somehow participates in transepithelial sodium retrieval. This is consistent with the immunohistochemical occurrence of ENaC in the basolateral membranes of the LE (Simon *et al.,* 1993). In another study of the isolated rat LE, however, interstitial amiloride had no effect (Gilbertson and Zhang, 1998b). If the effect of interstitial amiloride can be confirmed, the possibility may be considered that mucosal sodium ions overcome the diffusion resistance of the tight junction and enter the cellular compartment of the LE through basolateral amiloride-sensitive pathways, as discussed above in the context of Fig.1C.

V. ENaC EXPRESSION IN TASTE BUDS

As detailed in Sections II and III, amiloride-sensitive membrane currents were recorded in subsets of taste cells. While whole-cell patch-clamp experiments did not discriminate between apical and basolateral currents (Avenet and Lindemann, 1988; Doolin and Gilbertson, 1996; Gilbertson and Zhang, 1998b), recordings from *in situ* taste buds did show the apical currents and their inhibition by mucosal amiloride at least in the anterior taste buds of rodents (Avenet and Lindemann, 1991; Gilbertson *et al.,* 1992). Furthermore, nerve recordings and psychophysical experiments also showed the inhibitory effect of mucosal amiloride in the anterior tongue of rodents (e.g., Heck *et al.,* 1984; Spector *et al.,* 1996).

Based on these results one would predict that ENaC or ENaC-like channels are contained in the apical membranes of taste cells which mediate sodium sensory signals. The expectation was, therefore, that immunohistochemistry would find ENaC on apical membranes of some taste cells of the rat anterior tongue but not, for instance, in taste cells of the rat vallate papilla, because the latter does not mediate amiloride-sensitive salt taste (Formaker and Hill, 1991). Furthermore, RT-PCR would be expected to show mRNA of ENaC subunits in anterior buds of the rat but not in those from the vallate papilla.

The experimental outcome was somewhat different from these expectations. RT-PCR showed ENaC subunits in the lingual epithelium (Li *et al.,* 1994) but also in isolated taste buds of the rat vallate papilla (Lindemann *et al.,* 1998). Immunohistochemistry showed ENaC-like reactivity in taste cells of the vallate papilla of dog and rat (Lin *et al.,* 1999; Lindemann *et al.,* 1998; Simon *et al.,* 1993; Kretz *et al.,* 1999) as well as in anterior taste buds of the rat (Lin *et al.,* 1999; Stewart *et al.,* 1995; Kretz *et al.,* 1999). Curiously, the immunoreactivity was not restricted to the apical membrane but was to a large extent intracellular, sometimes emphasizing basolateral cellular borders. While the intracellular ENaC-like reactivity may represent channel protein in transit to the apical membrane, the finding of ENaC

RNA in the rat vallate papilla taste buds and immunoreactivity in the same location is unexpected and functionally unexplained.

VI. DEVELOPMENT AND PLASTICITY OF THE CHANNEL

Taste buds are derived from local epithelium and begin to appear around birth in rodents (for recent reviews, see Mistretta and Hill, 1995; Stewart *et al.,* 1997; Northcutt and Barlow, 1998). Afferent nerve fibers appear at approximately the same time, and taste cells are innervated as early as postnatal day 2. Taste buds, as well as the number of taste cells per bud increase dramatically during the first two weeks of postnatal development and responses to taste stimuli can be observed as early as postnatal day 2 (Ganchrow *et al.,* 1986). Overlying these changes during development, taste cells undergo constant turnover within the taste bud throughout the life of the animal. Old taste cells are continuously replaced by new cells arising from basal cell progenitors at the base of the taste bud. Typically, a taste cell has a life span of only 10 days to 2 weeks in mammals. It has long been known that taste buds are dependent on afferent innervation for their survival. If the afferent nerve is severed, taste buds degenerate within a few days. Upon regeneration of the nerve, taste buds reappear and become fully functional in about 2 months. Clearly, the taste bud is a dynamic structure. Numerous studies have focused on the nerve-dependent disappearance/reappearance of taste buds and on the developmental regulation of their amiloride-sensitive Na^+ channel. These will be reviewed below. The time course for development and regeneration of taste buds is summarized in Fig. 4.

Na^+ Taste Development. Electrophysiological studies in rats (Ferrell *et al.,* 1981), sheep (Mistretta and Bradley, 1983), and mice (Ninomiya *et al.,* 1991) have shown that Na^+ taste develops rather late compared with that of other salts. Chorda tympani responses to NaCl, expressed relative to NH_4Cl responses, are low during the first 2 postnatal weeks. Responses then increase dramatically, reaching adult levels in about 3 months. Hill and Bour (1985) showed that the increase in the response to NaCl occurs concomitantly with an increase in amiloride sensitivity. They hypothesized the increase in Na^+ sensitivity occurred as a result of the addition of functional amiloride-sensitive Na^+ channels to the apical membrane of taste cells. Similar conclusions were reached by Settles and Mierson (1993), who showed that amiloride-sensitive transepithelial Na^+ currents developed with approximately the same time course.

Surprisingly, immunohistochemical studies using polyclonal antibodies raised against amiloride-sensitive Na^+ channels showed that the channel

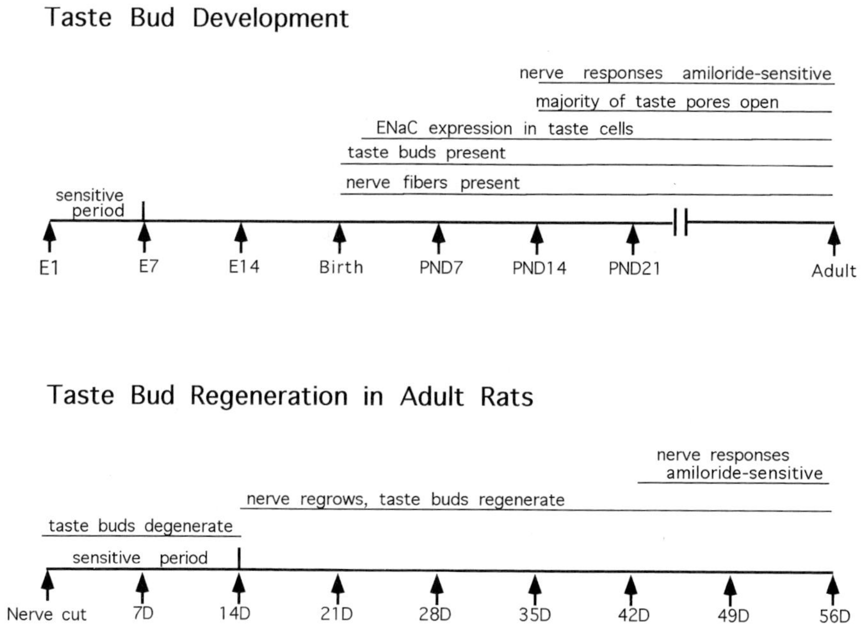

FIGURE 4 Timetable of taste bud development in anterior tongue of neonatal rat pups and taste bud regeneration following nerve section in adult rats. Taste buds begin to appear around birth and some taste buds are innervated as early as postnatal day 2. Although ENaC is expressed in young neonates, responses to NaCl are small and amiloride-insensitive until the majority of taste pores open during the second postnatal week. If pregnant rats are Na^+ restricted during the first embryonic week of development, the rat pups do not develop amiloride-sensitive Na^+ responses until the rats are exposed to NaCl in their diet. When the chorda tympani nerve is sectioned in adult rats, taste buds degenerate in 1–2 weeks. Upon regrowth of the sectioned nerve, taste buds regenerate, with amiloride-sensitive Na^+ responses nearly complete by 4 weeks after nerve section. In a manner similar to that observed during development of taste buds, if the rat is Na^+ deprived during the process of degeneration, the newly regenerated taste buds do not develop amiloride-sensitivity. The uncut contralateral nerve also shows changes when the rat is Na^+ deprived; but these changes result in a supersensitivity to Na^+ and the increase in Na^+ sensitivity is amiloride sensitive.

protein is expressed in neonatal rats as early as postnatal day 1, long before the Na^+ taste responses in the chorda tympani nerve show any amiloride sensitivity (Stewart *et al.,* 1995). These data suggest that the channel protein may be expressed in a quiescent state before becoming functional, or that the channels in young neonates may not be selectively localized in the apical membrane. Recent electrophysiological studies support the latter hypothesis. Whole-cell patch-clamp recordings were obtained from taste cells isolated from neonatal rat pups of varying ages (Kossel *et al.,* 1997).

Amiloride-sensitive Na^+ currents were observed as early as postnatal day 2, in agreement with the histochemical studies. Both the magnitude of the amiloride-sensitive current and the percentage of amiloride-sensitive taste cells remained the same throughout all age groups. Furthermore, the sensitivity of the current to amiloride block was similar in young neonates ($K_i = 0.1\mu M$) and adults ($K_i = 0.2\mu M$), suggesting that the channel expressed in young neonates is the same channel that is expressed in adult taste cells. Since taste pores and tight junctions are not present in the taste cells of early neonates, it is apparent that fully functional channels can be expressed in taste cell membranes prior to apical/basolateral differentiation. It was hypothesized that the lack of amiloride-sensitivity observed in chorda tympani recordings may reflect the lack of a functional taste pore, or the absence of channels clustered in the apical membrane.

Studies by David Hill and colleagues have shown that the development of amiloride-sensitive Na^+ channels in rodent taste cells is sensitive to dietary manipulation. If rats are deprived of dietary NaCl during pregnancy, and the rat pups are maintained on a low Na^+ diet, chorda tympani nerve responses to NaCl are reduced relative to control rats, and the responses are amiloride-insensitive. The effect is reversible; after early deprived rats are fed a Na^+ replete diet, responses to Na^+ salts fully recover within 15 days and the response becomes amiloride-sensitive (Hill *et al.*, 1986; Hill, 1987). Further studies showed that maternal dietary Na^+ must be restricted on or before embryonic day 8 to produce functional changes in the offspring (Hill and Przekop, 1988). Since taste buds do not form in rats prior to birth, these data suggest that the effects of Na^+ deprivation may be mediated by hormonal influences, rather than a direct effect of the low Na^+ on the taste cells. In addition, the recovery does not appear to require a direct interaction of Na^+ with the taste cells, since a single ingestion of isotonic NaCl is sufficient to restore normal Na^+ sensitivity and the sensitivity is maintained throughout many generations of taste cells (Przekop *et al.*, 1990; Stewart and Hill, 1996). Since the recovery process takes approximately 15 days, most taste cells that were directly exposed to the saline will have been have been replaced by new cells. A recent study by Phillips *et al.* (1995) suggests that the milk of Na^+ deprived rats contains factors that modulate the expression of the amiloride-sensitive Na^+ channel. Control rat pups that were fostered by Na^+ deprived mothers exhibited a supersensitivity to NaCl compared to normal pups fostered by Na^+ replete mothers. It will be of interest to determine the identity of these factors so that the mechanisms involved can be elucidated.

An electrophysiological study utilizing *in vivo* perturbation of the lingual voltage field suggested that the offspring of Na^+ restricted rats lack functional amiloride-sensitive Na^+ channels in the apical membrane (Ye *et al.*, 1993b).

Histochemical studies showed, however, that the channel protein is present (Stewart *et al.,* 1995). New studies should focus on whether these channels are functional, and if the taste pores are influenced by dietary manipulation.

Plasticity of Na^+ Channels in Regenerating Taste Buds. Since new taste buds can be induced in adult animals by cutting the chorda tympani nerve and subsequently allowing it to regenerate, it is of interest to determine if the plasticity observed in developing taste buds is also present in regenerating taste buds. Hill and Phillips (1994) showed that when unilateral chorda tympani nerve sectioning was combined with a Na^+ restricted diet in adult rats, the regenerated taste buds showed reductions in NaCl sensitivity and a lack of amiloride sensitivity similar to that observed with developing rats. Surprisingly, responses on the uncut contralateral nerve were also effected, but in a different manner. Immediately following transection of the contralateral nerve, responses to NaCl in the uncut nerve were significantly reduced compared to uncut controls. The responses then increased monotonically and eventually became supersensitive. Recent studies suggest that the immune system may play a role in this response. When the immune system was stimulated with a systemic injection of lipopolysaccharide (LPS), normal Na^+ responses were restored in the uncut contralateral nerve (Phillips and Hill, 1996). These data are provocative and suggest that cytokines or other immune factors may regulate functional expression of the amiloride-sensitive Na^+ channel. It will be of interest to determine if immune system regulation is observed with amiloride-sensitive Na^+ channels in other epithelial tissues.

VII. OUTLOOK

While convincing evidence shows that ENaC is the Na^+ salt receptor in the anterior tongue of rodents, several important questions remain to be elucidated. One perplexing problem relates to the observation that ENaC appears to be expressed in taste buds that have no apparent amiloride sensitivity. Do these channels fail to reach the membrane; do they reach the membrane but are nonconducting; or are they simply not sensitive to amiloride? Immunohistochemistry suggests that at least two ENaC subunits are present, but possibly the "silent" channels contain alternatively spliced ENaC transcripts that lack an amiloride binding site. Thus, the relationship between ENaC protein expression and the presence of functional amiloride-sensitive Na^+ currents in taste cells needs to be addressed experimentally. A biophysical analysis of taste ENaC channels in terms of single-channel electrical properties remains to be worked out, especially for mammals. Salt taste is highly adaptive, but the cellular mechanisms which cause this adaptation have remained unexplored. The molecular basis of the frog taste channel, which has properties different from the rodent taste channel, needs to be worked out. The hormonal basis of the diet-induced plasticity of the rodent channel constitutes a challenging problem. Finally, it will be important to identify the Na^+ taste channel in humans, who exhibit only weak amiloride-sensitivity of the NaCl response. In this regard it would be of interest to determine if Liddle's patients have altered Na^+ taste perception, since this would provide clues to a role of ENaC in human taste perception.

Note added in proof. Recent studies suggest that while α ENaC is expressed in both fungiform and vallate taste cells, expression of β and γ ENaC is significantly reduced in vallate compared to fungiform taste cells (Kretz *et al.,* 1999; Lin *et al.,* 1999). These findings may partially explain the lack of functional amiloride-sensitive Na^+ currents in vallate taste cells of rats in Na^+ balance. In contrast, rats that have been subjected to Na^+ deprivation or aldosterone treatment show an increased expression of β and γ ENaC in both vallate and fungiform taste cells, concomitant with an increase in the expression of functional amiloride-sensitive Na^+ currents (Lin *et al.,* 1999). Thus, the functional expression of β and γ ENaC in taste cells appears to be regulated by diet and aldosterone, as has been shown for other tissues. Another member of the ENaC family of channels, MDEG1, was recently cloned from rat vallate papillae (Ugawa *et al.,* 1998). This cation channel, also known as BNaC1, is gated by protons and mediates acid transduction in other tissues. Further studies will be required to determine the role of MDEG1 in sour taste transduction.

References

Avenet, P. (1992). Role of amiloride-sensitive sodium channels in taste. *In* "Sensory Transduction" (D. P. Corey and S. D. Roper, eds.), pp. 271–280. Rockefeller University Press, New York.

Avenet, P., and Lindemann, B. (1987). Patch-clamp study of isolated taste receptor cells of the frog. *J. Membr. Biol.* **97;** 223–240.

Avenet, P., and Lindemann, B. (1988). Amiloride-blockable sodium currents in isolated taste receptor cells. *J. Membr. Biol.* **105;** 245–255.

Avenet, P., and Lindemann, B. (1989). Chemoreception of salt taste: The blockage of stationary sodium currents by amiloride in isolated receptor cells and excised membrane patches. *In* "Chemical Senses: Molecular Aspects of Taste and Odor Reception" (J. G. Brand, J. H. Teeter, R. H. Cagan, and M. R. Kare, eds.) pp. 171–182. Dekker, New York.

Avenet, P., and Lindemann, B. (1990). Fluctuation analysis of amiloride-blockable currents in membrane patches excised from salt-taste receptor cells. *J. Basic Clin. Physiol. Pharmacol.* **1.** 383–391.

Avenet, P., and Lindemann, B. (1991). Noninvasive recording of receptor cell action potentials and sustained currents from single taste buds maintained in the tongue: The response to mucosal NaCL and amiloride. *J. Membr. Biol.* **124,** 33–41.

Benos, D. J., Fuller, C. M., Shlyonsky, V. G., Berdiev, B. K., and Ismailov, I. I. (1997). Amiloride-sensitive Na^+ channels: Insights and outlooks. *News Physiol. Sci.* **12,** 55–61.

Boughter, J. D., Jr., and Smith, D. V. (1997). Amiloride suppresses the responses to acids in NaCl-best neurons of the hamster solitary nucleus. *Chem. Senses* **22,** 648 (abstr.).

Brand, J. G., Teeter, J. H., and Silver, W. L. (1985). Inhibition by amiloride of chorda tympani responses evoked by monovalent salts. *Brain Res.* **334,** 207–214.

Breslin, P. A. S., Kaplan, J. M., Spector, A. C., Zambito, C. M., and Grill, H. J. (1993). Lick-rate analysis of sodium taste-state combinations. *Am. J. Physiol.* **264,** R312–R318.

Canessa, C. M., Horisberger, J.-D. and Rossier, B. C. (1993). Epithelial sodium channel related to proteins involved in neurodegeneration. *Nature* (*London*) **361,** 467–470.

Canessa, C. M., Schild, L., Buell, G., Thorens, B., Gautschi, I., Horisberger, J.-D., and Rossier, B. C. (1994). Amiloride-sensitive epithelial Na^+ channel is made of three homologous subunits. *Nature* (*London*) **367,** 463–467.

DeSimone, J. A., and Ferrell, F. (1985). Analysis of amiloride Inhibition of chorda tympani taste response of Rat to NaCl. *Am. J. Physiol.* **249,** R52–R61.

DeSimone, J. A., Heck, G. L., and DeSimone, S. K. (1981). Active ion transport in dog tongue: A possible role in taste. *Science* **214,** 1039–1041.

DeSimone, J. A., Heck, G. L., Mierson, S., and DeSimone, S. K. (1984). The active ion transport properties of canine lingual epithelia in vitro. *J. Gen. Physiol.* **83,** 633–656.

Doolin, R. E., and Gilbertson, T. A. (1996). Distribution and characterization of functional amiloride-sensitive sodium channels in rat tongue. *J. Gen. Physiol.* **107,** 545–554.

Duc, C., Farman, N., Canessa, C. M., Bonvalet, J.-P., and Rossier, B. C. (1994). Cell-specific expression of epithelial sodium channel α, β, and gamma subunits in aldosterone-responsive epithelia from the rat: Localization by in situ hybridization and immunocytochemistry. *J. Cell Biol.* **127,** 1907–1921.

Farman, N., Talbot, R. R., Boucher, R., Fay, M., Canessa, C., Rossier, B., and Bonvalet, J. P. (1997). Noncoordinated expression of α-, β-, and γ-subunit mRNAs of epithelial Na^+ channel along rat respiratory tract. *Am. J. Physiol.* **272,** C131–C141.

Ferrell, M. F., Mistretta, C. M., and Bradley, R. M. (1981). Development of chorda tympani taste responses in rat. *J. Comp. Neurol.* **198,** 37–44.

Formaker, B. K., and Hill, D. L. (1991). Lack of amiloride sensitivity in SHR and WKY glossopharyngeal taste responses to NaCl. *Physiol. Behav.* **50,** 765–769.

Fuchs, W., Larsen, E. H., and Lindemann, B. (1977). Current-voltage curve of sodium channels and concentration dependence of sodium permeability in frog skin. *J. Physiol.* (*London*) **267,** 137–166.

Ganchrow, J. R., Steiner, J. E. and Canetto, S. (1986). Behavioral displays to gustatory stimuli in newborn rat pups. *Dev. Psychobiol.***19,** 163–174.

Garty, H., and Palmer, L. G. (1997). Epithelial sodium channels: Function, structure and regulation. *Physiol. Rev.* **77,** 359–396.

Gilbertson, D. M., and Gilbertson T. A. (1994). Amiloride reduces the aversiveness of acids in preference tests. *Physiol. Behav.* **56,** 649–654.

Gilbertson, T. A. (1998). Peripheral mechanisms of taste. *Front. Oral Biol.* **9,** (in press).

Gilbertson, T. A., and Fontenot, D. T. (1998). Distribution of amiloride-sensitive sodium channels in the oral cavity of the hamster. *Chem. Senses* **23,** 495–499.

Gilbertson, T. A., and Zhang, H. (1998a). Self-inhibition in amiloride-sensitive sodium channels in taste receptor cells. *J. Gen. Physiol.* **111,** 667–677.

Gilbertson, T. A., and Zhang, H. (1998b). Characterization of sodium transport in gustatory epithelia from the hamster and rat. *Chem. Senses* **23,** 283–293.

Gilbertson, T. A., Avenet, P., Kinnamon, S. C., and Roper, S. D. (1992). Proton currents through Amiloride-sensitive Na channels in hamster taste cells: Role in acid transduction. *J. Gen. Physiol.* **100,** 803–824.

Gilbertson, T. A., Roper, S. D., and Kinnamon, S. C. (1993). Proton currents through amiloride-sensitive Na channels in isolated hamster taste cells: Enhancement by vasopressin and cAMP. *Neuron* **10,** 931–942.

Heck, G. L., Mierson, S., and DeSimone, J. A. (1984). Salt taste transduction occurs through an amiloride- sensitive sodium transport pathway. *Science* **223,** 403–405.

Helleman, U. (1992). Perceived taste of NaCl and acid mixtures in water and bread. *Int. J. Food Sci. Technol.* **27,** 201–211.

Herness, M. S. (1992). Aldosterone increases the amiloride-sensitivity of the rat gustatory neural response to NaCl. *Comp. Biochem. Physiol. A* **103,** 269–273.
Hettinger, T. P., and Frank, M.E. (1990). Specificity of amiloride inhibition of hamster taste responses. *Brain Res.* **513,** 24–34.
Hill, D. L. (1987). Susceptibility of the developing rat gustatory system to the physiological effects of dietary sodium deprivation. *J. Physiol. (London)* **393,** 413–424.
Hill, D. L., and Bour, T. C. (1985). Addition of functional amiloride-sensitive components to the receptor membrane: A possible mechanism for altered taste responses during development. *Dev. Brain Res.* **20,** 310–313.
Hill, D. L., and Phillips, L. M. (1994). Functional plasticity of regenerated and intact taste receptors in adult rats unmasked by dietary sodium restriction. *J. Neurosci.* **14,** 2904–2910.
Hill, D. L., and Przekop, P. R. J. (1988). Influences of dietary sodium on functional taste receptor development: A sensitive period. *Science* **241,** 1826–1828.
Hill, D. L., Mistretta, C. M., and Bradley, R. M. (1986). Effects of dietary NaCl deprivation during early development of behavioral and neurophysiological taste responses. *Behav. Neurosci.* **100,** 390–398.
Hille, B. (1992). "Ionic Channels in Excitable Membranes," 2nd ed. Sinauer Assoc., Sunderland, MA.
Kitada, Y., and Mitoh, Y. (1997). Amiloride does not affect the taste responses of the frog glossopharyngeal nerve to NaCl. *Chem. Senses* **22,** 720–721.
Kleyman, T. R., Smith, P. R., and Benos, D. J. (1994). Characterization and localization of epithelial Na^+ channels in toad urinary bladder. *Am. J. Physiol.* **266,** C1105–C1111.
Koefoed-Johnsen, V., and Ussing, H. H. (1958). The nature of the frog skin potential. *Acta Physiol. Scand.* **42,** 298–308.
Komwatana, P., Dinudom, A., Young, J. A., and Cook, D. I. (1996a). Control of the amiloride-sensitive Na^+ channel in salivary duct by extracellular sodium. *J. Membr. Biol.* **150,** 133–141.
Komwatana, P., Dinudom, A., Young, J. A., and Cook, D. I. (1996b). Cytosolic Na^+ controls an epithelial Na^+ channel via the GO guanosine nucleotide-binding regulatory protein. *Proc. Natl. Acad. Sci. U.S.A.* **93,** 8107–8111.
Kossel, A. H., McPheeters, M., Lin, W., and Kinnamon, S. C. (1997). Development of membrane properties in taste cells of fungiform papillae: Functional evidence for early presence of amiloride-sensitive sodium channels. *J. Neurosci.* **17,** 9634–9641.
Kretz, O., Barbry, P., Bock, R., and Lindemann, B. (1999). Differential expression of RNA and protein of the three pore-forming subunits of the amiloride-sensitive epithelial sodium channel in taste buds of the rat. *J. Histochem. Cytochem.* **47,** 51–64.
Li, X. J., and Snyder, S. H. (1994). Heterologous expression of amiloride-sensitive sodium channel subunits and an alternatively spliced form in taste tissues. *Soc. Neurosci. Abstr.* **20,** 1472.
Li, X. J., Blackshaw, S., and Snyder, S. H. (1994). Expression and localization of amiloride-sensitive sodium channels indicate a role for non-taste cells in taste perception. *Proc. Natl. Acad. Sci. U.S.A.*. **91,** 1814–1818.
Lin, W., Finger, T. E., Rossier, B. C., and Kinnamon, S. C. (1999). Epithelial Na^+ channel subunits in rat taste cells: localization and regulation by aldosterone. *J. Comp. Neurol.* **405,** 406–420.
Lindemann, B. (1980). The beginning of fluctuation analysis of epithelial ion transport. Topical review. *J. Membr. Biol.* **54,** 1–11.
Lindemann, B. (1984). Fluctuation analysis of sodium channels in epithelia. *Annu. Rev. Physiol.* **46,** 497–515.
Lindemann, B. (1996). Taste reception. *Physiol. Rev.* **76,** 719–766.
Lindemann, B. (1997). Sodium taste. *Curr. Opin. Nephrol. Hypertens.* **6,** 425–429.

Lindemann, B., and Van Driessche, W. (1977). Sodium-specific membrane channels of frog skin are pores: current fluctuations reveal high turnover. *Science* **195,** 292–294.
Lindemann, B., and Voûte, C. (1976). Structure and function of the epidermis. *In* "Frog Neurobiology" (R. Llinas and W. Precht, eds.) pp. 169–210. Springer-Verlag, Berlin and Heidelberg, New York.
Lindemann, B., Barbry, P., Kretz, O., and Bock, R. (1998). Occurrence of ENaC subunit mRNA and immunocytochemistry of the channel subunits in taste buds of the rat vallate papilla. *Ann. N.Y. Acad. Sci.* **855,** 116–127.
Ling, B. N., and Eaton, D. C. (1989). Effects of luminal Na^+ on single Na^+ channels in A6 cells, a regulatory role for protein kinase C. *Am. J. Physiol.***256,** F1094–F1103.
Luger, A., and Turnheim, K. (1981). Modification of cation permeability of rabbit descending colon by sulfhydryl reagents. *J. Physiol.* (*London*) **317,** 49–66.
Matsuo, R., and Yamamoto, T. (1992). Effects of inorganic constituents of saliva on taste responses of the rat chorda tympani. *Brain Res.* **583,** 71–80.
McCutcheon, N. B. (1992). Human psychophysical studies of saltiness suppression by amiloride. *Physiol. Behav.* **51,** 1069–1074.
Mierson, S., Olson, M. M., and Tietz, A. E. (1996). Basolateral amiloride-sensitive Na^+-Transport pathway in rat tongue epithelium. *J. Neurophysiol.* **76,** 1297–1309.
Mistretta, C. M., and Bradley, R. M. (1983). Neural basis of developing salt taste sensation: Response changes in fetal, postnatal, and adult sheep. *J. Comp. Neurol.* **215,** 199–210.
Mistretta, C. M., and Hill, D. L. (1995). Development of the taste system. *In* "Handbook of Olfaction and Gustation" (R. L. Doty, ed.) pp. 635–668. Dekker, New York.
Nakamura, K., and Norgren, R. (1991). Gustatory responses of neurons in the nucleus of the solitary tract of behaving animals. *J. Neurophysiol.* **66,** 1232–1248.
Ninomiya, Y., Tanimukai, T., Yoshida, S., and Funakoshi, M. (1991). Gustatory neural responses in preweanling mice. *Physiol. Behav.* **49,** 913–918.
Ninomiya, Y., Kajiura, H., Naito, Y., Mochizuki, K., and Katsukawa, H. (1994). Glossopharyngeal denervation alters responses to nutrients and toxic substances. *Physiol. Behav.* **56,** 1179–1184.
Northcutt, R. G., and Barlow, L. A. (1998). Amphibians provide new insights into taste-bud development. *Trends Neurosci.* **21,** 38–43.
Ossebaard, C. A., and Smith, D. V. (1995). Effect of amiloride on the taste of NaCl, Na-gluconate and KCl in humans: Implications for Na^+ receptor mechanisms. *Chem. Senses* **20,** 37–46.
Ossebaard, C. A., Polet, I. A., and Smith, D. V. (1997). Amiloride effects on taste quality: Comparison of single and multiple response category procedures. *Chem. Senses* **22,** 267–275.
Palmer, L. G. (1987). Ion selectivity of epithelial Na channels. *J. Membr. Biol.* **96,** 97–106.
Palmer, L. G. (1992). Epithelial Na channels: Function and diversity. *Annu. Rev. Physiol.* **54,** 51–66.
Palmer, L. G., and Frindt, G. (1986). Amiloride-sensitive Na channels from the apical membrane of the rat cortical collecting tubule. *Proc. Natl. Acad. Sci. U.S.A.* **83,** 2767–2770.
Palmer, L. G., Sackin, H., and Frindt, G. (1998). Regulation of Na^+ channels by luminal Na^+ in rat cortical collecting tubule. *J. Physiol.* **15,** 151–162.
Pangborn, R. M., and Trabue, I. M. (1967). Detection and apparent taste intensity of salt-acid mixtures in two media. *Percept. Psychophysiol.* **2,** 503–509.
Phillips, L. M., and Hill, D. L. (1996). Novel regulation of peripheral gustatory function by the immune system. *Am. J. Physiol.* **271,** R857–R862.
Phillips, L. M., Steward, R. E., and Hill, D. L. (1995). Cross fostering between normal and sodium-restricted rats: Effects on peripheral gustatory function. *Am. J. Physiol.* **269,** R603–R607.
Przekop, P. R., Mook, D. G., and Hill, D. L. (1990). Functional recovery of the gustatory system after sodium deprivation during development: How much sodium and where. *J. Physiol.* (*London*) **259,** R786.

Puoti, A., May, A., Canessa, C. M., Horisberger, J.-D., Schild, L., and Rossier, B. C. (1995). The highly selective low-conductance epithelial Na channel of *Xenopus laevis* A6 kidney cells. *Am. J. Physiol.* **269,** C188–C197.

Puoti, A., May, A., Rossier, B. C., and Horisberger, J. D. (1997). Novel isoforms of the beta and gamma subunits of the Xenopus epithelial Na channel provide information about the amiloride binding site and extracellular sodium sensing. *Proc. Natl. Acad Sci. U.S.A.* **94,** 5949–5954.

Rehnberg, B. G., Hettinger, T. P., and Frank, M. E. (1992). Salivary ions and neural taste responses in the hamster. *Chem. Senses* **17,** 179–190.

Renard, S., Voilley, N., Bassilana, F., Lazdunski, M., and Barbry, P. (1995). Localization and regulation by steroids of the α, β and gamma subunits of the amiloride-sensitive Na^+ channel in colon, lung and kidney. *Pfluegers Arch.* **430,** 299–307.

Roper, S. D. (1983). Regenerative impulses in taste cells. *Science* **220,** 1311–1312.

Roudier-Pujol, C., Rochat, A., Escoubet, B., Eugene, E., Barrandon, Y., Bonvalet, J.-P., and Farman, N. (1996). Differential expression of epithelial sodium channel subunit mRNAs in rat skin. *J. Cell Sci.* **109,** 379–385.

Settles, A. M., and Mierson, S. (1993). Ion transport in rat tongue epithelium in vitro: A developmental study. *Pharmacol., Biochem. Behav.* **46,** 83–88.

Simon, S. A., and Garvin, J. L. (1985). Salt and acid studies on canine lingual epithelium. *Am. J. Physiol.* **249,** C398–C408.

Simon, S. A., Holland, V. F., Benos, D. J., and Zamphighi, G. A. (1993). Transcellular and paracellular pathways in lingual epithelia and their influence in taste transduction. *Microsc. Res. Tech.* **26,** 196–208.

Spector, A. C., Guagliardo, N. A., and St. John, S. J. (1996). Amiloride disrupts NaCL vs. KCI discrimination performance: Implications for salt taste coding in rats. *J. Neurosci.* **16,** 8115–8122.

Stewart, R. E., and Hill, D. L. (1996). Time course of saline-induced recovery of the gustatory system in sodium-restricted rats. *Am. J. Physiol.* **270,** R704–R712.

Stewart, R. E., Lasiter, P. S., Benos, D. J., and Hill, D. L. (1995). Immunohistochemical correlates of peripheral gustatory sensitivity to sodium and amiloride. *Acta Anat.* **153,** 310–319.

Stewart, R. E., DeSimone, J. A., and Hill, D. L. (1997). New perspectives in gustatoryphysiology: transduction, development, and plasticity. *Am. J. Physiol.* **272,** C1-C26.

Tennison, A. M. (1992). Amiloride reduces intensity responses of human fungiform papillae. *Physiol. Behav.* **51,** 1061–1068.

Tonosaki, K., and Funakoshi, M. (1989). Amiloride does not block taste transduction in the mouse. *Comp. Biochem. Physiol. A* **94,** 659–661.

Tousson, A., Alley, C. D., Sorscher, E. J., Brinkley, B. R., and Benos, D. J. (1989). Immunochemical localization of amiloride-sensitive sodium channels in sodium-transporting epithelia. *J. Cell Sci.* **93,** 349–362.

Ugawa, S., Minami, Y., Guo, W., Saishin, Y., Takatsuji, K., Yamamoto, T., Tohyama, M., and Shimada, S. (1998). Receptor that leaves a sour taste in the mouth. *Nature* **395,** 555–556.

Van Driessche, W., and Lindemann, B. (1979). Concentration-dependence of currents through single sodium-selective pores in frog skin. *Nature* (*London*) **282,** 519–520.

Ye, Q., Heck, G. L. and DeSimone, J. D. (1991). The anion paradox in sodium taste reception: Resolution by voltage-clamp studies. *Science* **254,** 724–726.

Ye, Q., Heck, G. L., and DeSimone, J. A. (1993a). Voltage dependence of the rat chorda tympani response to Na^+ salts: Implications for the functional organization of taste receptor cells. *J. Neurophysiol.* **70,** 167–178.

Ye, Q., Stewart, R. E., Heck, G. L., Hill, D. L., and DeSimone, J. A. (1993b). Dietary Na-restriction prevents development of functional Na channels in taste cell apical membranes: Proof by in vivo membrane voltage perturbation. *J. Neurophysiol.* **70,** 1713–1716.

Yoshii, K., Kiyomoto, Y., and Kurihara, K. (1986). Taste receptor mechanism of salts in frog and rat. *Comp. Biochem. Physiol. A* **85A,** 501–507.

PART V

Clinical Relevance

CHAPTER 19

The Involvement of Amiloride-Sensitive Na^+ Channels in Human Genetic Hypertension: Liddle's Syndrome

Dale J. Benos
Department of Physiology and Biophysics, University of Alabama at Birmingham, Birmingham, Alabama 35294-0005

I. INTRODUCTION

Over 60 million Americans suffer from hypertension. This disease not only produces severe cardiovascular problems, but also can affect the kidneys. Although antihypertensive therapy may delay the onset of end-stage renal disease (ESRD) due to hypertension, hypertension-induced ESRD nonetheless accounts for over 25% of all ESRD in the United States (National Institutes of Health, 1994). The vast majority of hypertensive cases is of unknown etiology. However, in recent years, great strides have been made in understanding the causes of human genetic hypertension because of advances in molecular genetics. Specifically, the pathophysiology underlying glucocorticoid remedial aldosteronism (Lifton *et al.,* 1992; Rich *et al.,* 1992; Pascoe *et al.,* 1995; Yiu *et al.,* 1997), Liddle's syndrome (Liddle *et al.,* 1963; Shimkets *et al.,* 1994; Hansson *et al.,* 1995a,b; Jeunemaitre *et al.,* 1997), the apparent mineralocorticoid excess syndrome (Wilson *et al.,* 1995), and pseudohypoaldosteronism type I (Chang *et al.,* 1996; Grunder *et al.,* 1997)

1063-5823/99 $30.00

have now been elucidated. In each of these diseases, inappropriate renal sodium retention with subsequent salt and water reabsorption and volume expansion occurs. In these patients, blood pressures in excess of 200/110 mm Hg can be recorded. In each of these syndromes, the functional effects of the genetic mutations have been confirmed by expression of the relevant proteins in various systems. In the case of Liddle's syndrome, otherwise known as pseudohyperaldosteronism, mutations that constituently activate an amiloride-sensitive sodium channel are the culprits (Botero-Velez *et al.,* 1994). In this form of hypertension, plasma ionic concentrations are similar to those produced by a mineralocorticoid excess, even though plasma aldosterone levels are normal or below normal, and renin levels are usually below normal values (Warnock, 1998).

Hypertension is a multifactorial disease. Both genetic and nongenetic factors play a significant role in its development. Among the nongenetic factors, excesses in body weight, smoking, contraceptive agents, lack of exercise, and excessive salt intake have been proposed as risk factors. As indicated above, understanding genetic factors, specifically as related to amiloride-sensitive Na^+ channels, has greatly increased our understanding and potential treatments of at least a subset of the hypertensive population. Although the original proband described by Liddle was Caucasian (Liddle *et al.,* 1963), Japanese (Ohno *et al.,* 1975; Takeuchi *et al.,* 1989; Tamura *et al.,* 1996), Hispanic (Rodriquez *et al.,* 1981), and African-American (Gadallah *et al.,* 1995; Su *et al.,* 1996) pedigrees have been identified. In fact, the frequency of a Liddle-like hypertensive disease may be much greater than initially appreciated (Warnock and Bubien, 1994, 1995; Lifton, 1996).

Recently, cDNAs encoding polypeptides that display amiloride-sensitive Na^+ channel activity when expressed in oocytes or other cells have been cloned from low salt or dexamethasone-treated rat distal colon (Lingueglia *et al.,* 1993; Canessa *et al.,* 1993, 1994b; Barbry and Hofman, 1997; Garty and Palmer, 1997). This channel is a heterooligomeric complex consisting of at least three homologous subunits (α, β, and γ) that have predicted nonglycosylated molecular masses of 79, 72, and 75 kDa, respectively. The subunits of this channel termed ENaC (for epithelial Na^+ channel) share a high degree of structural homology with each other and with the stretch-activated channels of *C. elegans,* even though the amino acid identity can be low (<30%). It has been postulated that epithelial Na^+ channels and these stretch-activated cation channels belong to the same gene family (Canessa *et al.,* 1993, 1994b; Rossier, 1997). When expressed in oocytes, this channel has a single-channel conductance of 5–10 pS and exhibits ion selectivity, gating kinetics, and an amiloride pharmacological profile similar to the low conductance, highly selective Na^+ channel expressed in native renal Na^+ reabsorbing epithelia. At least 30 different homologs of ENaC have been cloned and studied (see Preface).

The general membrane topography for ENaC and related members are similar (Canessa *et al.,* 1994a; Renard *et al.,* 1994; Snyder *et al.,* 1994). There are two short transmembrane domains with a large extracellular domain and two short amino and carboxy termini located within the cytoplasm. Several papers have appeared with conflicting results concerning the exact number of subunits required for channel function. The group of Snyder *et al.* (1998) argues that three alphas, three betas, and three gammas comprise a functional channel, whereas the groups of Rossier (Firsov *et al.,* 1998) and Kleyman (Kosari *et al.,* 1998) argue that the stoichiometry is 2α:1β:1γ. Berdiev *et al.* (1998) propose that four α subunits comprise the core conduction element of a functional ENaC. Thus, there is still no consensus as to how many subunits, or in what relative proportion, comprise a single functional ENaC.

The role of each subunit in determining channel activity is also a subject of controversy. Alpha subunits alone can form functional amiloride-sensitive Na^+ channels. The role of the β and γ subunits are unclear, but they can influence trafficking to the surface of the membrane (Canessa *et al.,* 1994b; Rotin *et al.,* 1994; Hansson *et al.,* 1995a,b; Snyder *et al.,* 1996; Schild *et al.,* 1996; Staub *et al.,* 1996; Shimkets *et al.,* 1997), as well as contribute to the conduction/ion selectivity pathway and amiloride binding properties of the channel (Schild *et al.,* 1997).

Liddle's syndrome is a specific example of an autosomal dominant, low renin, volume expanded, salt-sensitive hypertension. The basic clinical features of this syndrome were first described in an Alabama family in 1963 by Grant Liddle and colleagues (1963). The original patient developed renal failure in 1989 and underwent successful renal transplantation at the University of Alabama at Birmingham's hospital. The subsequent resolution of this patient's hypertension (Botero-Velez *et al.,* 1994) supported Liddle's original contention that this disorder results from a greatly enhanced ability of the kidneys to reabsorb salt. Further support for this hypothesis came from clinical observations that the hypokalemia and hypertension associated with Liddle's syndrome could be alleviated with triamterene or amiloride therapy (Gadallah *et al.,* 1995), two drugs that specifically block ENaCs. The nephrology group in Alabama extended the early studies of a Liddle's syndrome patient's red blood cell Na^+ fluxes (Gardner *et al.,* 1971) and showed, in affected members of this pedigree, that there was a constituent activation of amiloride-sensitive Na^+ channels in lymphocytes (Bubien and Warnock, 1993; Bubien *et al.,* 1996). Genetic studies identified the causal mutation of this hypertensive disease to exist in the β subunit of ENaC (Warnock, 1996; Oh and Warnock, 1997). Subsequently, mutations in the γ subunit of this channel have also been identified; these mutations also produce a constitutive activation of these channels (Schild *et al.,* 1995) and indicate genetic heterogeneity of this disease.

There are two classes of mutations that produce constitutive activation of ENaC and resultant Liddle's disease (see Fig. 1, color insert). The first class involves introductions of premature stop codons or frameship mutations in the segments encoding the cytoplasmically located C termini of either β- or γ-ENaC, in effect deleting the last 44–75 amino acids. The second class involves point mutations, most notably P616L, Y618H, and T594M in the β-hENaC subunits. Expression of mutated β and/or γ subunits in *Xenopus* oocytes with wild-type α-ENaC produced a significantly enhanced (3- to 9-fold) amiloride-sensitive Na^+ current, as compared to oocytes injected with wild-type $\alpha\beta\gamma$-rENaC or hENaC (see Warnock, 1998, for references). Subsequent mutagenesis studies altering the proline-rich (PPP) region in the cytoplasmic tails of either $\alpha\beta\gamma$-ENaC likewise produced channel activation, as assessed in macroscopic oocyte expression studies (Hansson *et al.,* 1995b; Shimkets *et al.,* 1997). Because this region binds proteins bearing a ww domain like Nedd4, several groups hypothesize an increase in functional channel surface density as a mechanism responsible for the enhanced Na^+ currents (Snyder *et al.,* 1996; Staub *et al.,* 1997). It is interesting to note, however, that the T594M pedigree described by Su *et al.* (1996) does not occur within this proline-rich motif, nor does it eliminate it from the C-terminal β-hENaC tail as does the truncation mutations. I will present information below contrasting this view of enhanced macroscopic Na^+ current with those of two other groups that suggest that this increased macroscopic current results in addition from an increase in single-channel open probability (Firsov *et al.,* 1996; Ismailov *et al.,* 1996). In any case, these initial experiments demonstrate that alterations in the primary structure of the $\beta$and γ subunits of ENaC produce constitutive activation of amiloride-sensitive Na^+ channels. These results clearly demonstrate that the β and γ subunits of ENaC are critical for the modulation of ENaC activity, and that Liddle's syndrome is a disease of abnormally regulated amiloride-sensitive Na^+ channels.

II. MOLECULAR MECHANISM OF ENaC-GATED FUNCTION: INCREASED SURFACE DENSITY OR INCREASED SINGLE-CHANNEL OPEN PROBABILITY, OR BOTH

Several groups have reported that the mechanism by which Liddle's syndrome mutations constitutively activate macroscopic currents is to change the surface density of functional ENaC. Surface expression was detected immunologically, either by indirect immunofluorescence or by tagging an ENaC subunit with a FLAG epitope in its extracellular domain (Schild *et al.*, 1995; Firsov *et al.*, 1996; Staub *et al.*, 1996; Snyder *et al.*, 1996). These experiments were conducted in either *Xenopus* oocytes or in transfected mammalian cells such as MDCK cells. These studies were coupled with patch measurements of single channels showing that the conductance properties of the channels were not altered by these mutations. The authors also state that single-channel open probability was not altered. This same conclusion was reached by Jeunemaitre *et al.* (1997) in their studies of the β599del32 mutant. However, in all of the single-channel, patch-clamp studies that have been reported for these Liddle's mutations, no detailed analysis of single-channel open probability was presented. In other words, only very short stretches of single-channel records devoid of kinetic analyses were presented. In the absence of detailed kinetic analyses and exact knowledge concerning the number of functional channels in the membrane, actual single-channel open probability measurements are very difficult to determine and, thus, these conclusions about no change in ENaC open probability in Liddle's mutants versus wild-type channels should be viewed with caution.

The identification of specific Nedd4 and α spectrin binding domains in the cytoplasmic region of the ENaC subunits has posited the hypothesis that interactions with cytoskeletal elements may control the surface expression of ENaC complexes (Staub *et al.*, 1997). Previous work has described interactions of ENaC with actin and other cytoskeletal elements (Smith *et al.*, 1991; Cantiello, 1995; Berdiev *et al.*, 1996). Thus, several groups have hypothesized that alterations in membrane insertions and retrieval of ENaC may underly the increased macroscopic channel activity seen in Liddle's syndrome. Of note is the observation that many of the mutations described in Liddle's syndrome either truncate the $\beta\gamma$-subunits before the critical PPXY domain or cause point mutations within this domain. As indicated above, none of these studies has definitively distinguished between constitutive ENaC activation due to changes in channel open probability or surface expression. Firsov *et al.* (1996) made a comprehensive biochemical and electrophysiological examination of this issue in the oocyte expression system using ENaC subunits to which a defined FLAG epitope was added to the external loop of each subunit. They reported that the insertion of this epitope into any one of the ENaC subunits did not influence channel activity. The presence of this epitope within ENaC permits the use of high affinity, monoclonal antibodies that can be labeled to high-specific activity with ^{125}I and, thus, surface expression of these subunits can be quantitatively assessed.

In their experiments, the authors reported a 5-fold increase in amiloride-sensitive sodium current when ENaC was expressed in oocytes with the originally described β-truncation mutation of the last 75 amino acids. The authors reported a concurrent 1.9-fold increase in surface binding, compared to wild-type. However, there was a much larger, i.e., >5-fold increase, in the macroscopic conductance. The authors therefore concluded that the major effect of the originally described truncated mutation was to increase the open probability of the ENaC complex, although an increase in the number of channels in the plasma membrane does contribute as well. These essential observations were confirmed in *Xenopus* oocytes using an entirely different approach (Awayda *et al.,* 1997). Unfortunately, comparisons at the single-channel level were not made in either study.

Additional insights into the regulation of amiloride-sensitive Na^+ activity has also been obtained in reconstitution experiments in planar lipid bilayers. Amiloride-sensitive Na^+ channel activity can be recorded in bilayers with protein immunopurified from lymphocytes obtained from either normal or Liddle's individuals (Ismailov *et al.,* 1996). Notably, the Na^+ channel complex immunopurified from the lymphocytes from the Liddle's patients showed constitutive activation, i.e., a single channel P_o of 0.8, compared to 0.14 in affected cells, was seen. Unlike the channel from normal individuals, the addition of protein kinase A + ATP to the presumptive cytoplasmic surface resulted in no further increase in P_o. Comparable results were found in whole-cell, patch-clamp studies of the lymphocytes. Moreover, 10–30 amino-acid-long peptides constricted from the terminal portions of either the β- or γ-COOH-tails could block these constitutively activated in both the intact lymphocytes and the planar lipid bilayer reconstitution system (see Fig. 2). These observations suggest that the C-termini of either or both the β- and γ-ENaC subunits are important in the gating properties of the channel.

Thus, it is likely that the two mechanisms proposed for constitutive activation of Na^+ channel activity seen in Liddle's disease, namely, membrane insertion retrieval versus increase in single-channel P_o, can be simultaneously operative. It is likely that one or the other mechanism dominates in a cell-specific manner. It is also likely that the predominance of one mechanism versus another may be determined in large measure by the nature of the mutation. Nonsense mutations resulting in a single amino acid change in or in proximity to the proline-rich region in the cytoplasmic β- and γ-ENaC tail may entail a predominant effect on the membrane insertion/retrieval regulatory pathways of the Na^+ channel. On the other hand, the truncation mutations may produce both a change in surface density and single-channel open probability. A direct test of this hypothesis awaits experimentation.

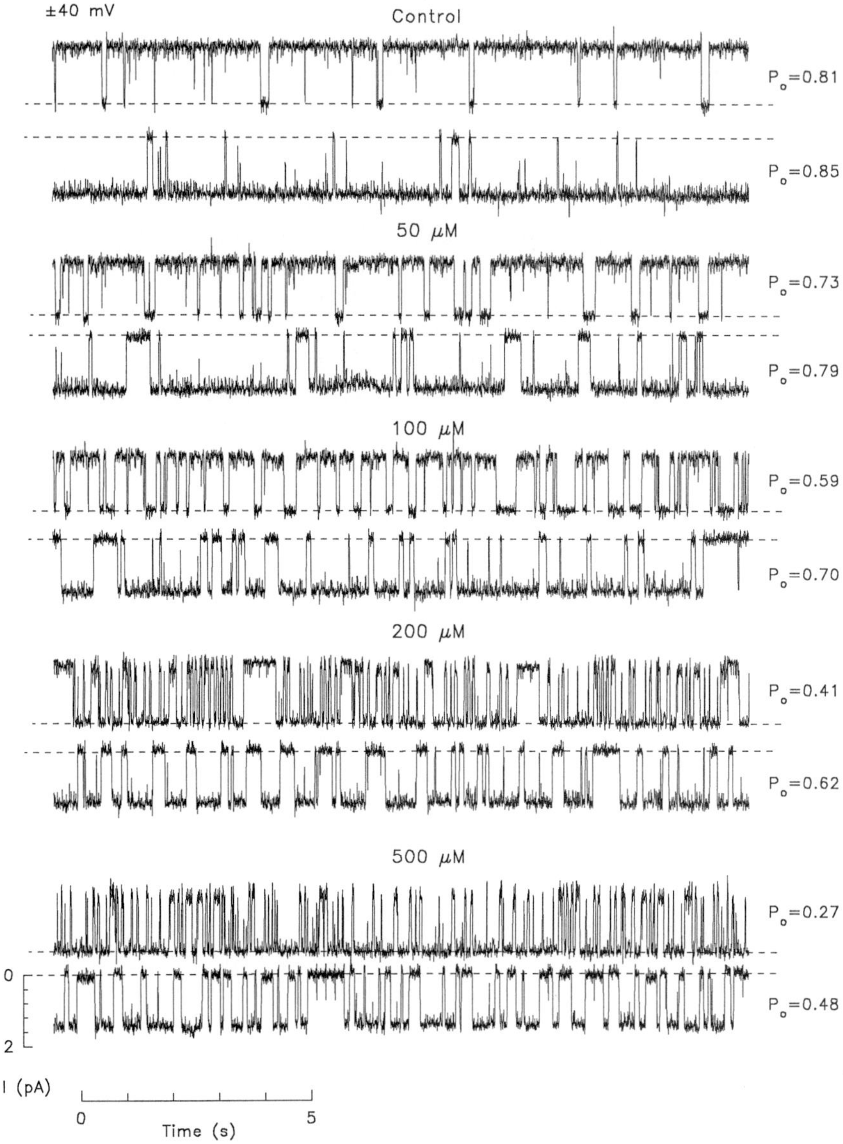

FIGURE 2 Block of constitutively activated Na^+ channels, immunopurified by synthetic peptide βrC10, from lymphocytes of a patient with Liddle's syndrome. Records are shown for ± 40 mV. The peptide was comprised of last 10 carboxy-terminal amino acids of β-subunit of rat epithelial Na^+ channel (ENaC). Its sequence (in single-letter code) was NH_2-MESDSEVEAI-COOH. The peptide was only effective in the presence of 100 μM GTPγS and worked only from the *cis* (or presumed cytoplasmic) side. Reprinted with permission from Ismailov *et al.* (1996) *Am. J. Physiol.* **270,** C214–C223.

III. SUMMARY

Liddle's disease is an autosomal-dominant form of genetic hypertension in which the precise loci of the defect within the cloned epithelial Na^+ channel was identified first by Lifton and co-workers (Shimkets *et al.,* 1994). These gain-of-function mutations produce constitutive activation of ENaC. Effective treatment of Liddle's syndrome can be realized through the use of the potassium-sparing diuretics, amiloride or triamterene, coupled with dietary salt restriction. Experimental approaches to understand the molecular physiology of the disease have provided new insights not only into how amiloride-sensitive Na^+ channels operate and are processed, but also of how alterations in protein–protein interaction and channel regulatory pathways may contribute to both the manifestations and treatments of overt hypertensive disease.

Acknowledgments

I thank Mrs. Cathleen Guy and Ms. Lisa Shea Thompson for superb secretarial assistance. This work was supported by N.I.H. Grant DK 37206.

References

Awayda, M. S., Tousson, A., and Benos, D. J. (1997). Regulation of a cloned epithelial Na^+ channel by its β and γ subunits. *Am. J. Physiol.* **273,** C1889–C1899.

Barbry, P., and Hofman, P. (1997). Molecular biology of Na^+ absorption. *Am. J. Physiol.* **273,** G571–G585.

Berdiev, B. K., Prat, A. G., Cantiello, H. F., Ausiello, D. A., Fuller, C. M., Jovov, B., Benos, D. J., and Ismailov, I. I. (1996). Regulation of epithelial sodium channels by short actin filaments. *J. Biol. Chem.* **273,** 17704–17710.

Berdiev, B. K., Karlson, K. H., Jovov, B., Ripoll, P.-J., Morris, R., Loffing-Cueni, D., Halpin, P., Stanton, B. A., Kleyman, T. R., and Ismailov, I. I. (1998). Subunit stoichiometry of a core conduction element in a cloned epithelial amiloride-sensitive Na^+ channel. *Biophys. J.,* **75,** 2292–2301.

Botero-Velez, M., Curtis, J. J., and Warnock, D. G. (1994). Brief report: Liddle's syndrome revisited. *N. Engl. J. Med.* **330,** 178–181.

Bubien, J. K., and Warnock, D. G. (1993). Amiloride-sensitive sodium conductance in human B lymphoid cells. *Am. J. Physiol.* **265,** C1175–C1183.

Bubien, J. K., Ismailov, I. I., Berdiev, B. K., Cornwell, T., Lifton, R. P., Fuller, C.M., Achard, J. M., Benos, D. J., and Warnock, D. G. (1996). Liddle's disease: Abnormal regulation of amiloride-sensitive Na^+ channels by β-subunit mutation. *Am. J. Physiol.* **270,** C208-C213.

Canessa, C. M., Horisberger, J.-D., and Rossier, B. C. (1993). Epithelial sodium channel related to proteins involved in neurodegeneration. *Nature* (*London*) **361,** 467-470.

Canessa, C. M., Meriallat A. M., and Rossier, B. C. (1994a). Membrane topology of the epithelial sodium channel in intact cell. *Am. J. Physiol.* **267,** C1682–1690.

Canessa, C. M., Schild, L., Buell, G., Thorens, B., Gautschil, Y., Horisberger, J.-D., and Rossier, B. C. (1994b). The amiloride-sensitive epithelial sodium channel is made of three homologous subunits. *Nature* (*London*) **367,** 463–467.

Cantiello, H. F. (1995). Role of actin cytoskeleton on epithelial Na^+ channel regulation. *Kidney Int.* **48,** 970–984.

Chang, S. S., Grunder, S., Hanukoglu, A., Rosler, A., Mathew, P. M., Hanukoglu, I., Schild, L., Lu, Y., Shimkets, R. A., Nelson-Williams, C., Rossier, B. C., and Lifton, R. P. (1996). *Nature Genet.* **12,** 248–253.

Firsov, D., Schild, L., Gautschi, I., Merillat, A. M., Schneeberger, E., and Rossier, B. C. (1996). Cell surface expression of the epithelial Na^+ channel and a mutant causing Liddle syndrome: A quantitative approach. *Proc. Natl. Acad. Sci. U.S.A.* **93,** 15370–15375.

Firsov, D., Gautschi, I., Merillat, A. M., Rossier, B. C., and Schild, L. (1998). The heterotetrameric architecture of the epithelial sodium channel (ENaC). *EMBO J.* **17,** 344–352.

Gadallah, M.F., Abreo, K., and Work, J. (1995). Liddle's syndrome: An underrecognized entity: A report of four cases including the first report in black individuals. *Am. J. Kidney Dis.* **25,** 924–927.

Gardner, J. D., Lapey, A., Simopoulous, A. P., and Bravo, E. L. (1971). Abnormal membrane sodium transport in Liddle's syndrome. *J. Clin. Invest.* **50,** 2253–2258.

Garty, H., and Palmer, L. G. (1997). Epithelial sodium channels: Function, structure, and regulation. *Physiol. Rev.* **77,** 359–396.

Gründer, S., Firsov, D., Chang, S. S., Jaeger, N. F., Gautshi, I., Schild, L., Lifton, R. P., and Rossier, B. C. (1997). A mutation causing pseudohypoaldosteronism type I identifies a conserved glycine that is involved in the gating of the epithelial sodium channel. *EMBO J.* **16,** 899–907.

Hansson, J. H., Nelson-Williams, C., Suzuki, H., Schild, L., Shimkets, R., Lu, Y., Canessa, C., Iwasaki, T., Rossier, B., and Lifton R. (1995a). Hypertension caused by a truncated epithelial sodium channel γ subunit: Genetic heterogeneity of Liddle syndrome. *Nat. Genet.* **11,** 76–82.

Hansson, J. H., Schild, L., Lu, Y., Wilson, T. A., Gautschi, I., Shimkets, R., Nelson-Williams, C., Rossier, B., and Lifton, R. (1995b). A de novo missense mutation of the β subunit of the epithelial sodium channel causes hypertension and Liddle syndrome, identifying a proline-rich segment critical for regulation of channel activity. *Proc. Natl. Acad. Sci. U.S.A.* **92,** 11495–11499.

Ismailov, I. I., Berdiev, B. K., Fuller, C. M., Bradford, A. L., Lifton, R. P., Warnock, D. G., Bubien, J. K., and Benos, D. J. (1996). Peptide block of constitutively activated Na^+ channels in Liddle's disease. *Am. J. Physiol.* **270,** C214–C223.

Jeunemaitre, X., Bassilana, F., Persu, A., Dumont, C., Champigny, G., Lazdunski, M., Corvol, P., and Barbry, P. (1997). Genotype-phenotype analysis of a newly discovered family with Liddle's syndrome. *J. Hypertens.* **15,** 1091–1100.

Kosari, F., Sheng, S., Li, J., Mak, D. O. D., Foskett, K., and Kleyman, T. R. (1998). Subunit stoichiometry of the epithelial sodium channel. *J. Biol. Chem.* **273,** 13469–13474.

Liddle, G. W., Bledsoe, T., and Coppage, W. S. (1963). A familial renal disorder simulating primary aldosteronism but with negligible aldosterone secretion. *Trans. Assoc. Am. Physicians* **76,** 199–213.

Lifton, R. P. (1996). Molecular genetics of human blood pressure variation. *Science* **272,** 676–680.

Lifton, R. P., Diuhy, R. G., Powers, M., Rich, G., Cook, S., Ulick, S., and Lalouel, J. M. (1992). A chimaeric 11 β-bydroxylase/aldosterone synthase gene causes glucocorticoid-remediable aldosteronism and human hypertension. *Nature (London)* **355,** 262–265.

Lingueglia, E., Voilley, N., Waldmann, R., Lazdunski, M., and Barbry, P. (1993). Expression cloning of an epithelial amiloride-sensitive Na^+ channel. A new channel type with homologies to *Caenorhabditis elegans* degenerins. *FEBS Lett.* **318,** 95–99.

National Institutes of Health. (1994) "U.S. Renal Data System: USRDS Annual Report" pp. 43–54. N.I.H., Bethesda, MD.

Oh, Y., and Warnock, D. G. (1997). Expression of the amiloride-sensitive sodium channel β subunit gene in human B lymphocytes. *J. Am. Soc. Nephrol.* **8,** 126–129.

Ohno, F., Harada, H., Komatsu, K., Saijo, K., and Miyoshi, K. (1975). Two cases of pseudoaldosteronism (Liddle's syndrome) in siblings. *Endocrinol. Jpn.* **22,** 163–167.

Pascoe, L., Jeunemaitre, X., Levrethon, M. C., Curnow, K. M., Gomez–Sanchez, C. E., Gasc, J. M., Saez, J. M., and Corvol, P. (1995). Glucocorticoid-suppressible hyperaldosteronism and adrenal tumors occurring in a single French pedigree. *J. Clin. Invest.* **96,** 2236–2246.

Renard, S., Lingueglia, E., Voilley N., Lazdunski, M., and Barbry, P. (1994). Biochemical analysis of the membrane topology of the amiloride–sensitive Na^+ channel. *J. Biol. Chem.* **269,** 12981–12986.

Rich, G., Ulick, S., Cook, S., Wang, J., Lifton, R., and Dluhy, R. (1992). Glucocorticoid-remediable aldosteronism in a large kindred: Clinical spectrum and diagnosis using a characteristic biochemical phenotype. *Ann. Intern. Med.* **116,** 813–820.

Rodriquez, J. A., Biglieri, E. G., and Schambelan, M. (1981). Pseudohyperaldosteronism with renal tubular resistance to mineralocorticoid horomones. *Trans. Assoc. Am. Physicians* **94,** 172-182.

Rossier, B. C. (1997). Cum grano salis: The epithelial sodium channel and the control of blood pressure. *J. Am. Soc. Nephrol.* **8,** 980–992.

Rotin, D., Bar-Sagi, D., O'Brodovich, H., Merlainen, J., Lehto, V. P., Caness, C. M., Rossier, B. C., and Downey, G. P. (1994). An SH3 binding region in the epithelial Na^+ channel (α rENaC) mediates its localization at the apical membrane. *EMBO J.* **13,** 4440–4450.

Schild, L., Canessa, C. M., Shimkets, R. A., Gautschi, I., Lifton, R. P., and Rossier, B. C. (1995). A mutation in the epithelial sodium channel causing Liddle disease increases channel activity in the *Xenopus laevis* oocyte expression system. *Proc. Natl. Acad. Sci. U.S.A.* **92,** 5699–5703.

Schild, L., Lu, Y., Gautschi, I., Schneeberger, E., Lifton, R. P., and Rossier, B. C. (1996). Identification of a PY motif in the epithelial Na^+ channel subunits as a target sequence for mutations causing channel activation found in Liddle's syndrome. *EMBO J.* **15,** 2381–2387.

Schild, L., Schneeberger, E., Gautschi, I., and Firsov, D. (1997). Identification of amino acid residues in the alpha, beta, and gamma subunits of the epithelial sodium channel (ENaC) involved in amiloride block and ion permeation. *J. Gen. Physiol.* **109,** 15–26.

Shimkets, R. A., Warnock, D. G., Bositis, C. M., Nelson-Williams, C., Hansson, J. H., Schambelan, M., Gill, J. R. J., Ulick, S., Milora, R. V., Findling, J. W., Canessa, C. M., Rossier, B. C., and Lifton, R. P. (1994). Liddle's syndrome: Heritable human hypertension caused by mutations in the β subunit of the epithelial sodium channel. *Cell (Cambridge, Mass.)* **79,** 407–414.

Shimkets, R. A., Lifton, R. P., and Canessa, C. M. (1997). The activity of the epithelial sodium channel is regulated by clathrin-mediated endocytosis. *J. Biol. Chem.* **272,** 25537–25541.

Smith, P. R., Saccomani, G., Joe, E.-H., Angelides, K. J., and Benos, D. J. (1991). Amiloride-sensitive sodium channel is linked to the cytoskeleton in renal epithelial cells. *Proc. Natl. Acad. Sci. U.S.A.* **88,** 6971–6975.

Snyder, P. M., McDonald, F. J., Stokes, J. B., and Welsh, M. J. (1994). Membrane topology of the amiloride-sensitive epithelial sodium channel. *J. Biol. Chem.* **269,** 24379–24383.

Snyder, P. M., Price, M. P., McDonald, F. J., Adams, C. M., Volk, K. A., Zelher, B. G., Stokes, J. B., and Welsh, M. J. (1996). Mechanism by which Liddle's syndrome mutations increase activity of a human epithelial Na^+ channel. *Cell (Cambridge, Mass.)* **83,** 969–978.

Snyder, P. M., Cheng, C., Prince, L. S., Rogers, J. C., and Welsh, M. J. (1998). Electrophysiological and biochemical evidence that DEG/ENaC cation channels are composed of nine subunits. *J. Biol. Chem.* **273,** 681–684.

Staub, O., Dho, S., Henry, P. C., Correa, J., Ishikawa, T., McGlade, J., and Rotin, D. (1996). WW domains of Nedd4 bind to the proline-rich PY motifs in the epithelial Na^+ channel deleted in Liddle's syndrome. *EMBO J.* **15,** 2371–2380.

Staub, O., Yeger, H., Plant, P. J., Kim, H., Ernst, S. A., and Rotin D. (1997). Immunolocalizaton of the ubiquitin-protein ligase Nedd4 in tissues expressing the epithelial Na^+ channel (ENaC). *Am. J. Physiol.* **272,** C1871–C1880.

Su, Y. R., Rutkowski, M. P., Klanke, C. A., Wu, X., Cui, Y., Pun, R. Y. K., Carter, V., Reif, M., and Menon, A. G. (1996). A novel variant of the β-subunit of the amiloride-sensitive sodium channel in African Americans. *J. Am. Soc. Nephrol.* **7,** 2543–2549.

Takeuchi, K., Abe, K., Sato, M., Yasujima, M., Omata, K., Murakami, O., and Yoshinaga, K. (1989). Plasma aldosterone level in a female case of pseudohyperaldosteronism (Liddle's syndrome). *Endocrinol. Jpn.* **36,** 167–173.

Tamura, H., Schild, L., Enomoto, N., Matsui, N., Marumo, F., Rossier, B. C., and Sasaki, S. (1996). Liddle disease caused by a missense mutation of β subunit of the epithelial sodium channel gene. *J. Clin. Invest.* **97,** 1780–1784.

Warnock, D. G. (1996). Polymorphism in the β subunit and Na^+ transport. *J. Am. Soc. Nephrol.* **7,** 2490–2494.

Warnock, D. G. (1998). Liddle syndrome: An autosomal dominant form of human hypertension. *Kidney Int.* **53,** 18–24.

Warnock, D. G., and Bubien, J. K. (1994). Liddle syndrome: Clinical and cellular abnormalities. *Hosp. Pract.* **29,** 95-105.

Warnock, D. G., and Bubien, J. K. (1995). Liddle's syndrome: A public health menace? *Am. J. Kidney Dis.* **25,** 924–927.

Wilson, R. C., Krozowski, Z. S., Li, K., Obeyesekere, V. R., Razzaghy-Azar, M., Harbison, M. D., Wei, J. Q., Shackleton, C. H., Funder, J. W., and New, M. I. (1995). A mutation in the HSD11B2 gene in a family with apparent mineralocorticoid excess. *J. Clin. Endocrinol. Metab.* **80,** 2263–2266.

Yiu, V. W. Y., Dluhy, R. P., Lifton, R. P., and Guay-Woodford, L. M. (1997). Low peripheral plasma renin activity (PRA) as a critical screening indicator of glucocorticoid remediable hypertension. *Pediatr. Nephrol.* **11,** 343–346.

CHAPTER 20

Epithelial Sodium Channels in Cystic Fibrosis

Bakhrom K. Berdiev and Iskander I. Ismailov
Department of Physiology and Biophysics, University of Alabama at Birmingham, Birmingham, Alabama 35294

I. INTRODUCTION

Cystic fibrosis (CF) is an autosomal recessive disease that results because of severely disordered transport function of several epithelia (Tsui and Buchwald, 1991; Collins, 1992; Mearns, 1993). This physiological defect results in production of viscid secretions of low water content (which gave the disease another name, *mucoviscidosis*), with the most dire consequences when the disease affects airways, and the patients suffer from recurrent and chronic bacterial lung infections. The nature of these phenomena has been a subject of extensive study over the last two decades. It was found that, while fluid and chloride secretion is inhibited in airways of CF patients due to mutational defects in the cystic fibrosis transmembrane conductance regulator (CFTR), fluid absorption across this tissue is concomitantly enhanced due to an increase in the rate of Na^+ transport (Quinton, 1990; Boucher, 1994a,b). The CFTR molecule was described to subserve many

1063-5823/99 $30.00

functions such as acting as a cAMP-dependent Cl^- channel, and as a protein regulator of other transporters (Boucher *et al.*, 1987a; Higgins, 1992; Fuller and Benos, 1992; Riordan, 1993; Welsh and Smith, 1993; Biwersi and Verkman, 1994; Hanrahan *et al.*, 1995; Frizzell, 1995; Jilling and Kirk, 1997). Numerous electrophysiological and flux studies of Na^+ absorbing epithelium lining the airways demonstrated that Na^+ movement in this tissue is consistent with the classic model of Koefoed-Johnson and Ussing (1958), and is rate-limited by amiloride-sensitive facilitated diffusion down an electrochemical gradient of Na^+ at the apical membrane (Boucher, 1994a,b). Further, the experimental evidence emerging since the cloning of the cDNAs encoding three polypeptides (named α-, β-, and γ-ENaC, for *e*pithelial Na^+ *c*hannel, Canessa *et al.*, 1993, 1994) that, when expressed in *Xenopus* oocytes, display characteristics similar to the amiloride-sensitive channel expressed in renal Na^+ reabsorbing epithelia (see Garty and Palmer, 1997, for review), suggested that CFTR may act as a negative regulator of these $\alpha\beta\gamma$-ENaC channels, that "balances the rates of Na^+ absorption and Cl^- secretion to properly hydrate the airways" (Stutts *et al.*, 1995). Furthermore, the inhibitory effects of CFTR on ENaCs in MDCK cells and 3T3 fibroblasts reported by Stutts *et al.* (1995) were subsequently generalized to other cells. Such evidence has been provided in: (i) A6 cells which are derived from amphibian kidneys and express endogenous CFTR and amiloride-sensitive Na^+ channels (Ling *et al.*, 1997); (ii) the *Xenopus* oocyte expression system (Mall *et al.*, 1996; Kunzelmann *et al.*, 1997; Briel *et al.*, 1998); and (iii) the cell-free planar lipid bilayer system (Ismailov *et al.*, 1996b, 1997). The present article aims to discuss the mechanisms that could account for the observations made in these experimental systems, and their relevance to transport of Na^+ in normal and the cystic fibrosis airway epithelia.

II. Na^+ Hyperabsorption in CF

The first, and essentially the most unequivocal, evidence for involvement of amiloride-sensitive Na^+ channels in the pathogenesis of mucoviscidosis came from direct measurements of transepithelial electric potential difference (PD) in both upper and lower airways, and in the microperfused ducts of sweat glands of the patients affected with this disease (Knowles *et al.*, 1981a,b; Quinton and Bijman, 1983). These *in vivo* bioelectric measurements permit an estimation of the rates of electrogenic ion transport across epithelia and have demonstrated a 2- to 3-fold greater activity of the amiloride-sensitive entry along the entire respiratory tract of CF patients, as compared to nonaffected individuals (Table I). Moreover, following the first reports by the Chapel Hill research group, this technique has proven

TABLE I

Transepithelial Potential Differences (mV) in Airways of Non-CF and CF Affected Individuals

Epithelia		Non-CF	CF	References
		−20–30	−50–60	Knowles *et al.* (1981a,b, 1982)
		−17	−36	Hay and Geddes (1985)
	In vivo	−10–30	−30–70	Sauder *et al.* (1987)
		−5–15	−15–50	Middleton *et al.* (1994)
Nasal		−15	−43	Ho *et al.* (1997)
	Freshly	−4–5	−10–12	Yankaskas *et al.* (1987)
	Excised	−0.5	−0.9	Mall *et al.* (1998)
	Specimen			
Tracheal	*In vivo*	−20–30	−40–50	
Main stem bronchus	*In vivo*	−20	−30–40	
Lobar bronchus	*In vivo*	−20	−30–40	Knowles *et al.* (1981a,b, 1982)
Segmental bronchus	*In vivo*	−10–20	−35–40	

its specificity for the differential diagnosis between cystic fibrosis and other acute and chronic illnesses of the respiratory system (Knowles *et al.*, 1981a,b, 1982, 1983a,b; Hay and Geddes, 1985; Gowen *et al.*, 1986; Sauder *et al.*, 1987; Alton *et al.*, 1987, 1990; Ho *et al.*, 1997; Delmarco *et al.*, 1997; Barker *et al.*, 1997), and its usefullness for assessing the efficacy of treatment of CF (Middleton *et al.*, 1994; Knowles *et al.*, 1995a; Hay *et al.*, 1995). Furthermore, based on these findings, amiloride delivery via aerosol to the lumen of CF airways was introduced into the therapy of CF patients and has been shown to promote improved biorheology and clearance of airway epithelia (Kohler *et al.*, 1986; Knowles *et al.*, 1990a,b, 1992, 1995b; App *et al.*, 1990; Gallo, 1990; Middleton *et al.*, 1993; Tomkiewicz *et al.*, 1993).

Much has been learned from these *in vivo* measurements of transepithelial PD in human airways, and when studying the effects of disruption of the CFTR gene in mice. In particular, the experiments of Grubb *et al.* (1994) revealed that the homozygous CFTR (−/−) "knockout" animals, like CF-affected humans, have increased basal and amiloride-sensitive transepithelial PD, as compared to heterozygous CFTR (−/+) or homozygous CFTR (+/+) controls. However, the utility of this technique for systematic physiological studies of the airway ion transport mechanisms is hampered, mostly due to limited accessibility of distal airways and to the difficulties associated with exposure of the tissue to drugs and solutions. These could be circumvented when the freshly excised specimens were mounted in Ussing flux chambers; and bioelectric properties, solute permeability and ion transport could be measured by conventional techniques (Boucher *et al.,* 1980, 1981; Widdicombe and Welsh, 1980; Cotton *et al.,* 1983, 1988; Knowles *et al.,* 1984; Ballard *et al.,* 1992). These studies not only provided the essential knowledge of Na^+ movement in airway epithelia and demonstrated its consistency with the model developed by Koefoed-Johnson and Ussing in 1958 for the frog skin (see reviews by Boucher, 1994a,b), but confirmed an approximately twofold greater amiloride-sensitive Na^+ current in freshly excised nasal epithelia of the CF patients (compared to non-CF individuals, Cotton *et al.,* 1987) and from the CFTR (−/−) "knockout" mice (compared to the littermate heterozygous (+/−) or homozygous (+/+) control animals, Grubb *et al.,* 1994).

Two other approaches to the investigation of ion transport in airways involve isolation of epithelial cells from this tissue and measurements of macroscopic and microscopic currents in cell culture, which helps to solve the problem of tissue quantity. While highly dependent on the cell culture conditions, these elaborate experiments faithfully reproduce the reduced Cl^- and increased Na^+ permeability in cystic fibrosis epithelia, as compared to normal both at the macroscopic and the microscopic level (Yankaskas *et al.,* 1985a,b, 1987; Boucher *et al.,* 1987b; Duszyk *et al.,* 1989; Jefferson *et al.,* 1990; Willumsen and Boucher, 1991a,b; Chinet *et al.,* 1994; Kunzelmann *et al.,* 1995; Rückes *et al.,* 1997; see Table II).

III. ENaCs AND CFTR

In 1993 the DNAs of encoding polypeptides capable of inducing a finite amiloride-sensitive current in *Xenopus* oocytes were cloned from the distal colon of dexamethasone-treated rats (α-rENaC; Canessa *et al.,* 1993; Lingueglia *et al.,* 1993) and from the human lung (Voilley *et al.,* 1994). Subsequently, α-ENaC was shown to induce single channels with conductance of ~5 pS and ion selectivity, gating kinetics, and an amiloride inhibition similar to the amiloride-sensitive channel expressed in renal Na^+ reabsorbing epithelia (see Garty and Palmer, 1997, for review), when coexpressed in oocytes together with the two additional homologous polypeptides (β- and γ-ENaCs, also cloned from the rat colon; Canessa *et al.,* 1994). These studies were taken further by Stutts *et al.* (1995), who found that the Madin-Darby canine kidney (MDCK) cells or 3T3 fibroblasts expressing the three ENaC subunits in the presence of CFTR exhibited reduced macroscopic amiloride-sensitive Na^+ current compared with that exhibited by control cells expressing $\alpha\beta\gamma$-ENaC in the absence of CFTR. Entirely consistent with the aforementioned idea that airways in CF display Na^+ hyper-

TABLE II

Macroscopic and Microscopic Electrical Properties of Cultured Airway Epithelial Cells from CF and Non-CF Individuals

Epithelia	Parameter	Normal	CF	References
	ΔV_T (mV)	−8–12[1,2,4]	−30–35[1,2,3]	
		−0.1–1.1[3]	−5.7–6.1[3]	
		−3–4[7,8]	−5–7[7,8]	Yankaskas *et al.* (1985a)[1]
	I_{sc} (μA/cm^2)	28[2]	80[2]	Boucher *et al.* (1988)[2]
		0.8–1.8[3]	5–6.1[3]	Jefferson *et al.* (1990)[3]
		40–50[4]	90–100[5]	Willumsen and Boucher (1991a)[4]
		13–15[7,8]	17[7,8]	
Nasal	R_T($\Omega\cdot$cm^2)	430[2]	470[2]	Willumsen and Boucher (1991b)[5]
		200–250[4]	400–450[5]	Chinet *et al.* (1994)[6]
		260–300[7,8]	290–350[7,8]	Rückes *et al.* (1997)[7]
	I_{amil} (%)	~50–60[1,2,4,8]	~85–95[1,2,5,8]	Blank *et al.* (1997)[8]
		10–40[3]	55–70[3]	
	Single-channel conductance, *i* (pS)	21.4 ± 1.5[6]	17.2 ± 2.8[6]	
	Open probability	0.15 ± 0.04[6]	0.30 ± 0.03[6]	
	φ_{Rev} (mV)	−54[6]	−66[6]	
Tracheal	I_{sc} (μA/cm^2)	58.0 ± 10.6	69 ± 18	Yamaya *et al.* (1992)
	I_{amil} (%)	~34	~70	

Note. ΔV_T, transepithelial potential difference (mV); I_{sc}, short-circuit current (μA/cm^2); R_T, transepithelial resistance ($\Omega\cdot$cm^2); I_{amil}/[Amil], amiloride sensitive fraction of I_{sc} (%); φ_{Rev}, reversal potential determined in cell-attached patches with 140 mM NaCl in the pipette (mV). Superscript numbers next to data correspond to the references for respective reports.

absorption, these initial findings did not determine its precise molecular mechanism, but suggested that CFTR may act as a negative regulator of these $\alpha\beta\gamma$-ENaC channels.

Several issues regarding the relationship of ENaCs to Na^+ absorption in airway epithelia *per se* should be discussed. First, the conductance, ion selectivity, and the inhibition of single α- or $\alpha\beta\gamma$-ENaC with amiloride (in either cell-free or cell expression systems) are rather different from those of the channels found in airways (Table III). Although gating of these channels was not analyzed in detail in the vast majority of these studies, simple visual examination of the representative recordings reveals that both open and closed times of native channels also vary significantly and are not even close to the range of seconds which is characteristic for ENaCs in *Xenopus* oocytes and other cell expression systems (Canessa *et al.*, 1994; Stutts *et al.*, 1997). It is not clear whether these differences result from the

TABLE III

Biophysical Properties of Single ENaCs and Amiloride-Sensitive Channels from Airway Epithelia

	G (pS)	P_{Na}/P_K	K_i^{Amil} (μM)	Recording conditions	Comments	Reference
α-ENaC	ND	>20[a]	0.1[a]	Whole-cell	Expressed in *Xenopus* oocytes	Canessa *et al.* (1993)
	ND	>20[a]	0.08[a]	Whole-cell		Voilley *et al.* (1994)
	13 (6[b])	10	0.18	BLM	*In vitro* translated proteins/oocytes vesicles	Ismailov *et al.* (1996a) Berdiev *et al.* (1996)
$\alpha\beta\gamma$-ENaC	4.6	>20[a]	0.1[a]	Cell-attached	Expressed in *Xenopus* oocytes	Canessa *et al.* (1994)
	4–5	>20	0.3	Inside-out	Expressed in 3T3 cells	Stutts *et al.* (1997)
	13 (6[b])	10	0.18	BLM	*In vitro* translated proteins/oocytes vesicles	Ismailov *et al.* (1996a) Berdiev *et al.* (1996)
Human nasal epithelia	22	>6	Sensitive[c]	Cell-attached	Primary culture	Chinet *et al.* (1993)[d]
	21	~10	ND	Cell-attached		Chinet *et al.* (1994)
	5–30	ND	Sensitive	Inside-out		Duszyk *et al.* (1989)

Bovine trachea	6	>35	0.1	BLM	Amiloride affinity purified protein	Cherksey (1988)
Guinea pig fetal distal lung	11	1.8	0.4–4	Inside-out	Freshly isolated cells	MacGregor *et al.* (1994)
Rat fetal distal lung	12	ND	ND	Cell-attached		Orser *et al.* (1991)
	23	0.9	<1	Inside-out		
	27	1.1	<1	Inside-out; cell-attached	Primary culture	Tohda and Marunaka (1995)
	12	>10	1–2	Inside-out; cell-attached		
	4	≥10	0.09	Outside-out		Voilley *et al.* (1994)
Rabbit adult lung	25	7	8	BLM	Protein immunopurified from ATII cells	Senyka *et al.* (1995)
Rat adult lung	27	~10		Cell-attached		Yue *et al.* (1994)
	25	7	<1	Inside-out	ATII cells primary culture	
	20	1.15	1	Inside-out		Feng *et al.* (1993)

[a] Determined from macroscopic current.

[b] Determined in the presence of actin (Berdiev *et al.*, 1996).

[c] In the presence of 10^{-4} M amiloride in the pipette, the incidence of Na permeable channels was 6% (8 of 134 patches).

[d] Upon excision, gating (but not conductance ~22 pS) of the channel changed dramatically: the channel lost the ability to discriminate between Na^+ and K^+ and became amiloride-insensitive. Properties similar to the nonselective amiloride-insensitive channel described by Jorissen *et al.* (1990) in excised patches of human nasal epithelial cells.

fact that the properties of native channels were assessed in isolated and cultured cells, and, therefore, are altered (Yue *et al.*, 1993; Kunzelmann *et al.*, 1996), or the properties of ENaCs determined in the artificial environment of planar lipid bilayer or in any of these cell (yet model) expression systems do not reflect accurately properties of ENaCs in airways. On the other hand, although the possibility that amiloride-sensitive channels in airways are distinct from the ENaCs (Rückes *et al.*, 1997) cannot be excluded, the presence of all three ENaC subunits ($\alpha > \beta > \gamma$) was detected along entire respiratory tract in both non-CF and CF human tissue by Northern blot, *in situ* hybridization and ribonuclease protection assays (Burch *et al.*, 1995). All three ENaCs were also found in the airways of rats (Renard *et al.*, 1995; Matsushita *et al.*, 1996). Interestingly, in rat airways, this distribution of the three ENaC subunits is different compared to humans: ENaCs are proportionally expressed ($\alpha \cong \beta \cong \gamma$) in bronchiolar epithelium and in nasal gland ducts, but in unequal proportions ($\alpha \geq \gamma >> \beta$) in trachea, in a subpopulation of alveolar cells, and in nasal and tracheal gland acini (Farman *et al.*, 1997). However, in contrast to the aforementioned electrophysiological studies that demonstrate Na^+ hyperabsorption in CF airways, the relative abundance of the RNA for all three subunits appears to be similar in CF and in non-CF tissue (Burch *et al.*, 1995). Considering the fact that α-ENaC by itself (but not β- or γ-) is sufficient to form functional amiloride-sensitive Na^+ channels in lipid bilayers and in *Xenopus* oocytes (Canessa *et al.*, 1993, Ismailov *et al.*, 1996a), and that β- and γ- appear to substitute 2 α-s in an otherwise tetrameric complex (Snyder *et al.*, 1998; Firsov *et al.*, 1998; Kosari *et al.*, 1998; Berdiev *et al.*, 1998), this variability of ENaC expression along the airways may form an alternative explanation for the differences in single-channel properties of ENaCs and native Na^+ channels in airways. Moreover, consistent with the most critical role of α-subunit in formation of the channel (Canessa *et al.*, 1993, Lingueglia *et al.*, 1993; Voilley *et al.*, 1994; Ismailov *et al.*, 1996a), the α-ENaC ($-/-$) "knockout" mice neonates developed respiratory distress and died within 40 h after the birth from a failure to clear their lungs of liquid (Hummler *et al.*, 1996). Measurements of transepithelial potential difference in the tracheal explants from these mice demonstrated an abolished electrogenic amiloride sensitive Na^+ transport. Taken together, these findings support the hypothesis that ENaCs play an important role in fluid and electrolyte transport in airways.

Subsequent to initial observations of Stutts *et al.* (1995) made in MDCK cells and 3T3 fibroblasts, the inhibitory effects of CFTR on ENaCs were generalized to other cells. Ling *et al.* (1997) demonstrated that amphibian kidney epithelial cell line, A6, expresses endogenous CFTR and amiloride-sensitive Na^+ channels and exhibits a 1.2-fold higher amiloride-sensitive short circuit current, when treated with CFTR-antisense oligonucleotides and pretreated with forskolin. Recordings in cell-attached patches in these cells reveal a 2.0-fold increase in open probability of Na^+ channels in response to treatment with CFTR-antisense oligonucleotides + forskolin. Further, in contrast to the generally accepted notion, elevations of intracellular cAMP did not activate, but inhibited highly selective low-conductance amiloride-sensitive Na^+ channels endogenously present in mouse cortical collecting duct cell line, M1 (Letz and Korbmacher, 1997). Based on the finding that the extent of this inhibition correlated with the extent of activation of an endogenous CFTR-like channel in these cells these authors attributed their findings to inhibitory interaction between ENaC and CFTR. A more direct evidence was presented by Kunzelmann and coauthors in a series of papers reporting significantly reduced amiloride-sensitive currents in *Xenopus* oocytes coexpressing $\alpha\beta\gamma$-ENaCs and CFTR, than in the oocytes expressing ENaCs alone (Mall *et al.,* 1996; Kunzelmann *et al.,* 1997; Briel *et al.,* 1998). All these findings, together with the reduced open probability of ENaCs co-reconstituted with CFTR in the cell-free planar lipid bilayer system (Ismailov *et al.,* 1996b, 1997), support the idea that CFTR can inhibit ENaCs in epithelia.

IV. CFTR-INDUCED INHIBITION OF ENaCs AND CELL REGULATORY MACHINERY

A summary of findings of different investigators directly studying the CFTR-induced inhibition of ENaCs in different cell expression systems is shown in Tables IV and V. One major difference that can be noted when comparing all these observations is related to the role of the cAMP dependent regulatory mechanism in this inhibition. For example, in the experiments of Stutts *et al.* (1995, 1997) in MDCK and 3T3 cells, CFTR reversed the effects of a permeant analog of cAMP (cpt-cAMP) on $\alpha\beta\gamma$-rENaC currents. In particular, cpt-cAMP stimulated the inward Na current in the $\alpha\beta\gamma$-rENaC-only expressing cells and inhibited these currents in the cells coexpressing $\alpha\beta\gamma$-rENaC subunits and CFTR. Therefore, the authors attributed the effect of CFTR to modulating the PKA-induced phosphorylation of ENaCs. In contrast, Mall *et al.* (1996) and Briel *et al.* (1998) found no effect of cAMP (elevated by the inhibitor of phosphodiesterase, 3-isobutyl-1-methylxanthine, IBMX) on $\alpha\beta\gamma$-currents in oocytes, unless CFTR was coexpressed. These authors explain their findings by the hypothesis in which CFTR has to be phosphorylated before it can down-regulate ENaCs, via direct protein–protein interaction(s).

TABLE IV

cAMP-Dependent Effect of CFTR on $\alpha\beta\gamma$-ENaC in Different Cell Expression Systems

Expression system	Parameter	ENaC		ENaC + CFTR		Stimulation agent	Reference
		Basal	Stimulated	Basal	Stimulated		
MDCK cells	I_{sc}	60–70	↑ 25–35	40–50	↓ 4–5	Forskolin	
	I_{inward}	20–30	↑ 3–6	20–30	↓ 4–5		Stutts *et al.* (1995)
	I_{amil}	16.9 ± 2.8	↑ 3–5	10.7 ± 2.0	↓ 3–4	cpt-cAMP	
3T3 cells		0.3–0.5	0.65–0.75	0.3–0.4	0.1–0.15	PKA-cs + ATP	
	P_o	0.2–0.4	0.6–0.8	0.2–0.4	0.05–0.15	cpt-cAMP/forskolin	Stutts *et al.* (1997)

Note. Underlined data included in the table represent Mean ± SEM presented in the respective report (where available); other numeric data were estimated from the bar-graphs presented in the respective report. The ↑ and ↓ represent increase and decrease of the parameter as compared with the respective basal condition. I_{sc} is a short circuit current measured in monolayers of cells in Cl-free KBR ($\mu A/cm^2$); I_{inward} is a whole-cell inward current measured in Cl^- free (both bath and pipette) conditions (pA/pF); P_o represents open probability of ENaCs determined in inside-out patches.

TABLE V

Distribution of Whole-Cell Conductances (G_m) in *Xenopus* Oocytes Coexpressing $\alpha\beta\gamma$-ENaC and WT or Mutant CFTR

G_m, μS	ENaC alone		ENaC + WT CFTR		ENaC + ΔF508 CFTR		ENaC + G551D CFTR	
	Basal	+IBMX	Basal	+IBMX	Basal	+IBMX	Basal	+IBMX
Total	34 ± 6.2[1a]	~35[1†]	16.3 ± 2.5[1a] 17.4 ± 2.7[1c] 20.3 ± 2.9[1c]	46.4 ± 9.2[1b] 38.7 ± 6.7[1d] 37.3 ± 5.1[1f]	21.9 ± 6.6[a] 21.0 ± 6.4[c]	25.4 ± 7.6[b] 21.9 ± 7.2[d]	—	—
Anion selective*	—	—	10.3 ± 2.1[1c]	35.5 ± 6.3[1d]	14.8 ± 5.1[c]	14.8 ± 5.3[d]	~9[†]	~8[†]
Cation selective**	~22[†]	24[2†]	14.8 ± 2.3[1a]	26.4 ± 2.3[1b]	21.0 ± 6.5[a]	21.6 ± 7.5[b]	—	—
Amiloride-insensitive[#]	7.6 ± 2.0[1a]	~8[†]	7.2 ± 1.2[1c]	33.3 ± 4.8[1f]	—	—	—	—
Amiloride-sensitive[##]	25 ± 4.3[3]	26 ± 4.9[3]	13.3 ± 4.3[2]	6.2 ± 0.4[2]	7.9 ± 1.8	9.2 ± 2.2	~13[†]	~12[†]
Reference	Mall *et al.* (1996)[1] Kunzelmann *et al.* (1997)[2] Briel *et al.* (1998)[3]		Mall *et al.* (1996)[1] Briel *et al.* (1998)[2]		Mall *et al.* (1996)		Kunzelmann *et al.* (1997)	

Note. Anion selective* and Cation selective** conductances were determined by substituting chloride and sodium in bath solution with gluconate and *N*-methyl-D-glucamine, respectively [not corrected for the background conductance determined in water injected oocytes as 2.5 ± 0.4 (−IBMX) and 2.6 ± 0.4 (+IBMX); Mall *et al.*, 1996]. Amiloride-insensitive[#] and amiloride-sensitive[##] conductances were determined by adding amiloride to bath solution to reach final concentration of 10 μM. Superscript letters next to data refer to paired measurements within respective reports (superscript numbers). Data included in the table represent mean ± SEM presented in the respective report, or estimated from the bar graphs if numeric data were not available.([†]).

The details of regulation of airway Na^+ reabsorption via the cAMP second messenger system have been under extensive investigation. The findings of Stutts *et al.* (1995, 1997) are consistent with results of the earlier studies of the same group where beta adrenergic agonist induced elevations of intracellular cAMP activate Na^+ absorption in CF, but not in normal airway epithelia (Boucher *et al.*, 1986). PKA-phosphorylation of β- and γ-rENaCs, but not of α-rENaC, was confirmed recently in MDCK cells stimulated by IBMX/forskolin (Shimkets *et al.*, 1998), but not *in vitro* (Awayda *et al.*, 1996), or in any other cell expression system. It is not clear at this point which experimental system reflects regulation of $\alpha\beta\gamma$-ENaCs by PKA-induced phosphorylation (and the role of this regulation in functional interaction of ENaCs with CFTR) more accurately, but the phenomenon clearly appears to be cell specific. On the other hand, the fact that CFTR-induced inhibition of ENaCs has been observed in several different types of cells, including the cells as distant from epithelia as *Xenopus* oocytes, suggests that this effect of CFTR must be general, and independent of cell specificity.

We employed the planar lipid bilayer as an integrative system that affords the opportunity of functionally addressing this hypothesis in a well-controlled and well-defined environment. The $\alpha\beta\gamma$-rENaCs were expressed in *Xenopus* oocytes either alone, or in the presence of CFTR; membrane vesicles were made from these oocytes and were fused to bilayers. For the simultaneous studies of two different (i.e., sodium and chloride) channels, the bilayers were bathed initially with symmetrical 100 mM NaMOPS solutions to assess the single-channel activity of $\alpha\beta\gamma$-rENaC not confounded by any Cl^- current (both rENaC and CFTR are impermeable to $MOPS^-$), followed by addition of CsCl to the bathing solution, thus supplying Cl^- to reveal CFTR current in the presence of PKA and ATP. In these experiments, we found that the open probability (P_o) of single $\alpha\beta\gamma$-ENaC was lower in the presence of CFTR than in its absence (Fig. 1). A possibility that the observed effects are produced by contaminants applies to all *in vitro* experiments, including experiments in planar lipid bilayers. However, the decrease in ENaC activity occurred in ~ ½ of co-incorporations of CFTR and ENaC and these occurred only when these proteins oriented in the membrane in the same direction (as determined by the sidedness of PKA+ATP activation of CFTR and of amiloride inhibition of ENaC), which speaks to the specificity of this effect. Nonetheless, for the experiments testing possible effects of the cytoskeletal protein actin on the interaction between CFTR and single Na^+ channels, we employed *in vitro* translated $\alpha\beta\gamma$-rENaCs. The attempt of using the *in vitro* translated protein to produce CFTR channels in planar lipid bilayers was unsuccessful, thus limiting our choice to immunopurified bovine tracheal CFTR (Jovov *et al.*,

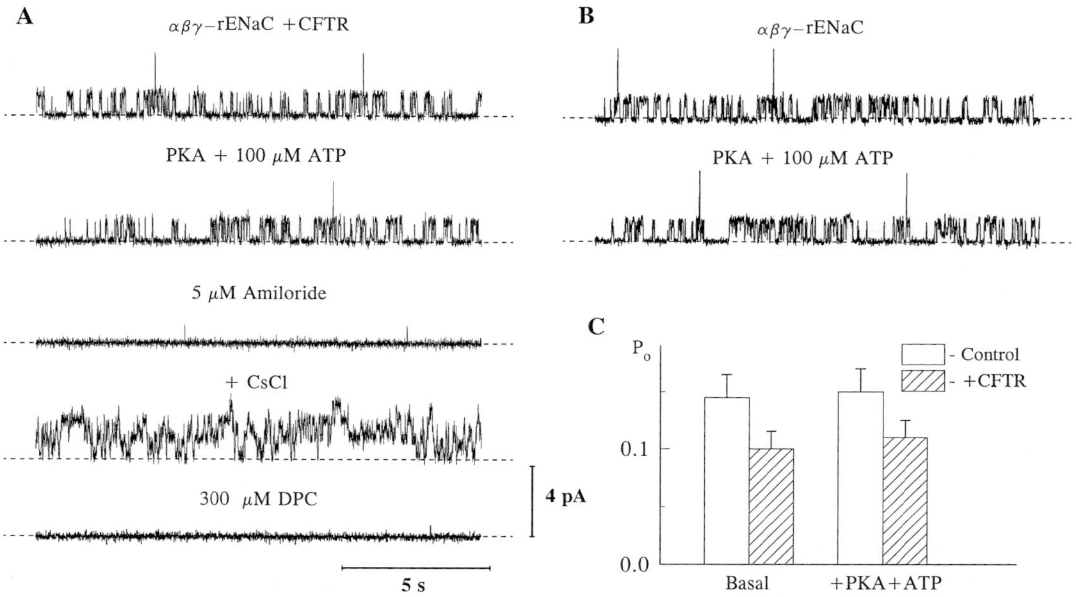

FIGURE 1 αβγ-rENaC and CFTR in planar lipid bilayers. Single-channel recordings in the presence (A, modified from Ismailov *et al.*, 1996b), and in the absence (B, modified from Awayda *et al.*, 1996, reproduced from *The Journal of General Physiology* by copyright permission of the Rockefeller University Press) of CFTR. Bathing solutions contained symmetrical 100 mM NaCl, 10 mM MOPS-Tris (pH 7.4). Additions were made sequentially as shown in the figure. Catalytic subunits of protein kinase A (PKA, 1.85 ng/ml), ATP (100 μM), dry CsCl (100 mM), and DPC were added to both bathing compartments; amiloride was added only to the *trans* bathing solution. The holding potential was + 100 mV. (C) Single αβγ-rENaC open probability (P_o) in the absence (clear bars) and presence (dark bars) of CFTR before and after addition of PKA + ATP (modified from Ismailov *et al.*, 1996b).

1995). Biophysical properties of these *in vitro* translated $\alpha\beta\gamma$-rENaC and immunopurified bovine tracheal CFTR were essentially identical to those of $\alpha\beta\gamma$-rENaC and CFTR reconstituted into bilayers from the oocyte membranes. Based on the hypothesis of Cantiello and colleagues of involvement of this protein in cAMP-dependent regulation of amiloride-sensitive Na^+ channels in amphibian kidney cell line, A6 (Cantiello *et al.,* 1991; Prat *et al.,* 1993; Cantiello, 1995), actin was proposed as a candidate for a role of the protein conferring PKA-sensitivity to $\alpha\beta\gamma$-rENaC (Berdiev *et al.,* 1996). The studies testing this hypothesis have not only shown its viability, but demonstrated that in the presence of actin ENaCs in bilayers gained the hallmark features of ENaCs in oocytes, namely, low conductance, long open/closed time (Fig. 2A). Furthermore, in the presence of actin and immunopurified bovine tracheal CFTR, basal open probability of $\alpha\beta\gamma$-rENaC was significantly lower than in the absence of CFTR (Fig. 2B), and, although ENaC could be transiently activated by polymerizing actin (induced by the catalytic subunit of protein kinase A and ATP, or by ATP alone) even in the presence of CFTR, the extent of this activation was lower than in its absence (Figs. 2C and D).

Extrapolated to macroscopic currents, our findings in planar lipid bilayers were consistent with the hypothesis that CFTR can exert tonic inhibition of ENaCs, as was suggested by Stutts *et al.* (1995). This conclusion requires additional discussion. By definition, macroscopic net current ($\bar{\mathbf{I}}$) is a product of the number of channels (N) at the cell surface and their ability to conduct ions by means of probability to be open (P_o) and single channel current (I):

$$\bar{\mathbf{I}} = N \cdot P_o \cdot I \; . \qquad (1)$$

Although subsequent experiments of Stutts *et al.* (1997) in 3T3 fibroblasts also attributed CFTR-induced inhibition of ENaCs to changes in the channel open probability (P_o), similar to that described above in bilayers, this evidence is not sufficient to exclude totally the possibility that presence/absence of CFTR could affect the number of ENaCs at the cell surface. Consideration of such a possibility is especially important given the fact that: (1) CFTR was demonstrated to have a significant impact on cellular membrane trafficking and recycling (Bradbury *et al.,* 1992; Biwersi *et al.,* 1996); (2) expression of the three ENaC subunits at the cell surface is complex by itself, as was demonstrated in recent cell surface labeling experiments (Awayda *et al.,* 1997; Firsov *et al.,* 1998). In these experiments (Firsov *et al.,* 1998), the magnitude of an amiloride sensitive current (I_{amil}) in oocytes correlated linearly with the specific binding of [^{125}I]anti-FLAG antibody to any two epitope tag-labeled (F) ENaC subunits ($\alpha^F + \beta^F$, $\alpha^F + \gamma^F$, or $\beta^F + \gamma^F$) out of three expressed. On the other hand, increasing

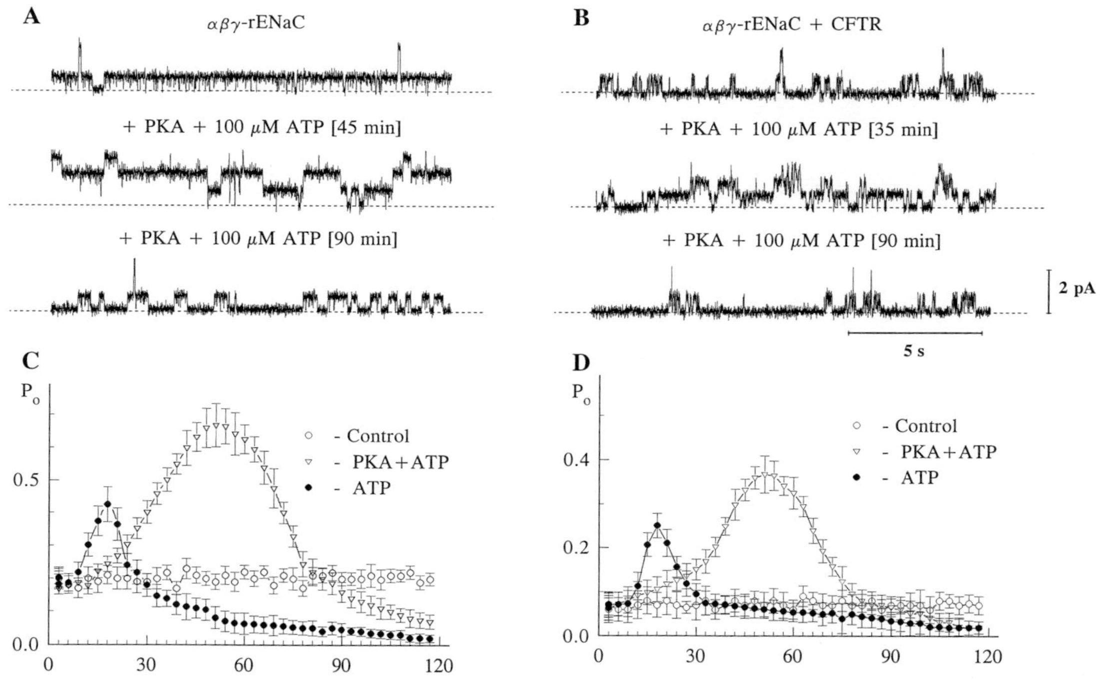

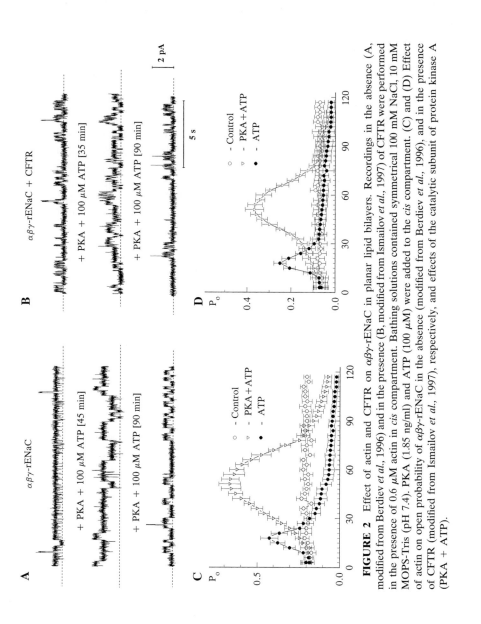

FIGURE 2 Effect of actin and CFTR on $\alpha\beta\gamma$-rENaC in planar lipid bilayers. Recordings in the absence (A, modified from Berdiev *et al.*, 1996) and in the presence (B, modified from Ismailov *et al.*, 1997) of CFTR were performed in the presence of 0.6 μM actin in *cis* compartment. Bathing solutions contained symmetrical 100 mM NaCl, 10 mM MOPS-Tris (pH 7.4). PKA (1.85 ng/ml) and ATP (100 μM) were added to the *cis* compartment. (C) and (D) Effect of actin on open probability of $\alpha\beta\gamma$-rENaC in the absence (modified from Berdiev *et al.*, 1996), and in the presence of CFTR (modified from Ismailov *et al.*, 1997), respectively, and effects of the catalytic subunit of protein kinase A (PKA + ATP).

the amount of RNA for the third (nontagged) subunit (while keeping the amount of RNA for two tagged subunits constant) decreased both the surface expression of channels and the I_{amil}. This decrease was linear in the case of α- and β-subunits, and bell-shaped in the case of the γ-subunit (with a maximum of both I_{amil} and channel expression at a three- to tenfold excess of γ-subunit cRNA over β- and α-, $1\alpha:1\beta:3\text{-}10\gamma$). Furthermore, the specific binding of a monoclonal antibody to α- and γ- subunits in $\alpha^F\beta\gamma^F$ was an order of magnitude lower than that in $\alpha^F\beta^F\gamma$ and $\alpha\beta^F\gamma^F$. All these findings are in contrast with the relative abundance of α subunit over β- and γ-rENaC ($\alpha>\beta>\gamma$) as concluded in these experiments when only one out of three subunits was FLAG-epitope tagged, and they provide enough room for the possibility that the changes in macroscopic amiloride sensitive current in cells expressing $\alpha\beta\gamma$-ENaC and CFTR observed by Stutts *et al.* (1995) could be accounted for by the changes in the number of channels at the cell surface.

Another issue important in understanding the mechanism of physiological regulation of ENaCs by CFTR is related to the question: If these proteins interact physically in a relatively large two-dimensional artificial bilayer membrane then what could be the range of such direct protein–protein interaction? Or, alternatively, does the lipid matrix transduce a signal(s) from one protein to another over a distance exceeding the length of physical interaction? An estimate of the likelihood of channel collision in the bilayer (which gives an independent measure for the probability of direct interaction between proteins) refers to a problem of two-dimensional diffusion (Adam and Delbrück, 1967; Saffman and Delbrück, 1975; Fahey and Webb, 1978; Cherry, 1979; Peters and Cherry, 1982; Edidin, 1987). The mean time for an encounter to occur on a two-dimensional surface (assuming one of the molecules to be fixed) depends on the ratio of the radius of diffusion area to the radius of the target area (Edidin, 1987). Assuming a globular radius of 10 Å for a heterotrimeric ENaC, that one functional channel complex exists in the bilayer membrane of ~75 μm in radius, and that this complex is immobile (Smith *et al.*, 1995), whereas a single (for simplification purposes) CFTR molecule has a diffusion coefficient of 10^{-8} cm$^2\cdot$s^{-1}, then the time for one collision to occur (τ_c) can be determined following Berg and Purcell (1977) as

$$\tau_c = (1.1a^2/ND)\ln(1.2a^2/Ns^2) \ , \tag{2}$$

where N is the number of CFTR molecules, a is the radius of the bilayer, and s is the radius of the ENaC. This should be equal to 86.3 h. That is to say that in a two-dimensional random walk, it would take on average over 80 (!) hours for a CFTR molecule to encounter ENaC. Obviously, the corresponding time for a three-dimensional diffusion would be much larger.

The reason for this is that the space size/target-size ratio is almost 10^9. The diffusion factor could be reduced with the increase in the number of CFTR molecules and/or the increase of the target size. In any case these calculations show that the probability of a random collision of two proteins in a bilayer membrane is very low. Because the open probability of ENaC was lowered in *all* successful co-incorporations of ENaC and CFTR when these proteins were co-oriented in the membrane, the interaction between CFTR and ENaCs is not random. Furthermore, these considerations based on the theory of Brownian motion in membranes assume that the bilayer is flat. However, artificial bilayer lipid membranes, like those used in the above described experiments, in fact are not planar (i.e., flat), but rather concave structures, and the lateral mobility of molecules in curved membranes is anisotropic (Aizenbud and Gershon, 1982; Aizenbud and Gershon, 1985). It is possible that energetically favorable protein–protein interactions occur, especially between molecules that are predisposed for such interaction. Additional evidence supporting the hypothesis of direct interaction between CFTR and $\alpha\beta\gamma$-ENaC has been provided by the yeast two hybrid analyses of Kunzelmann *et al.* (1997), which actually leads into the discussion of the structural basis of ENaC-CFTR interaction.

V. ALPHA, BETA, OR GAMMA?

As it was mentioned above, analyses of the epithelial tissue obtained from the airways of both the CF patients and nonaffected individuals revealed the presence of all three α-, β-, and γ- ENaC subunits, although with a lower relative amount in CF (Burch *et al.,* 1995). These findings, together with the fact that ENaCs are expressed in different proportions along the respiratory tract (Farman *et al.,* 1997), raise the question: If CFTR interacts with ENaCs directly, what are the roles of each ENaC subunit in such an interaction?

This question was addressed in two independent studies, one involving the interaction of different domains of CFTR and rENaC subunits in a yeast two hybrid assay (Kunzelmann *et al.,* 1997), and another, examining the effects of CFTR on α-, $\alpha\beta$-, $\alpha\gamma$-, and $\alpha\beta\gamma$-rENaC in planar lipid bilayers (Ismailov *et al.,* 1997). In the first study, the intracellular domain of human CFTR (amino acids 394–830) was shown to interact with α-, but not with β-, or γ-ENaC. This finding was in contrast to the results of our planar lipid bilayer studies (Ismailov *et al.,* 1997). Some specifics of the experimental design of these latter experiments should be explained. Successfully employed in the initial studies of CFTR-induced inhibition of ENaC, the oocyte-to-bilayer technique has limited utility for detailed studies of regulatory interaction of ENaCs with CFTR because of aforementioned complexity of cell surface expression of these channels (Awayda *et al.,* 1997; Firsov *et al.,* 1998; Berdiev *et al.,* 1998). These circumstances, in essence, dictated the choice of the *in vitro* translated ENaC subunits co-reconstituted into proteoliposomes in different combinations prior to incorporation into the bilayer. Similar to that observed for ENaCs reconstituted into bilayers from oocyte vesicles, in the absence of the α-subunit, β- and γ-rENaCs separately or together did not display any channel activity whatsoever. As it was described for $\alpha\beta\gamma$-rENaC, biophysical properties of these *in vitro* translated α-. $\alpha\beta$-, $\alpha\gamma$-channels were essentially identical to those of α-, $\alpha\beta$-, $\alpha\gamma$-rENaC reconstituted into bilayers from the oocyte membranes. In these co-reconstitution experiments, we found that the immunopurified bovine tracheal CFTR effectively down-regulated $\alpha\beta$- and $\alpha\gamma$-, but not α-channels (Fig. 3).

While seemingly different from observations of Kunzelmann *et al.* (1997), taken together with them, our results can help to understand the reasons for an unsuccessful attempt of Adams *et al.* (1997) to coimmunoprecipitate human α-ENaC with CFTR coexpressed in Cos 7 cells. In these immunoprecipitation studies, the C-terminus of α-ENaC was labeled with the M2 (FLAG) epitope (last 8 C-terminal amino acids replaced). This replacement by itself could have dramatic consequences for interaction of this domain of the α-subunit (shown by Kunzelmann et al. (1997) as thc CFTR interacting site) with CFTR. On the other hand, although this replacement did not affect the assembly of the $\alpha\beta\gamma$-ENaC complex (which was the primary subject of this study), it is not entirely clear from the description of this experiment (which is listed as a control in the paper), whether these coimmunoprecipitations of α-ENaC were performed in the presence of β- and/or γ-ENaC subunits, or not. These questions, in fact, may relate the negative result of Adams *et al.* (1997) to our finding that β- and/or γ-rENaC subunits were required for the down-regulatory effect of CFTR on ENaC (Ismailov *et al.,* 1997).

Two other findings in molecular physiology of ENaCs are noteworthy with regard to these observations. First, as it was mentioned above, the α-ENaC(−/−) "knockout" mice neonates failed to clear their lungs of liquid, developed respiratory distress, and died within 40 h after the birth (Hummler *et al.,* 1996). Second, it has been demonstrated that premature truncation of the C-terminally located cytoplasmic tails of β- and γ- subunits, found in another human disease, namely, Liddle's syndrome, produces a constitutively active phenotype, most likely due to direct participation of these structural domains in a gating mechanism of the heterooligomeric channel (Hansson *et al.,* 1995a,b; Snyder *et al.,* 1995; Lifton, 1996; Firsov *et al.,* 1996). Based on these observations, one could imagine a scenario in which β- and γ- subunits do not physically interact

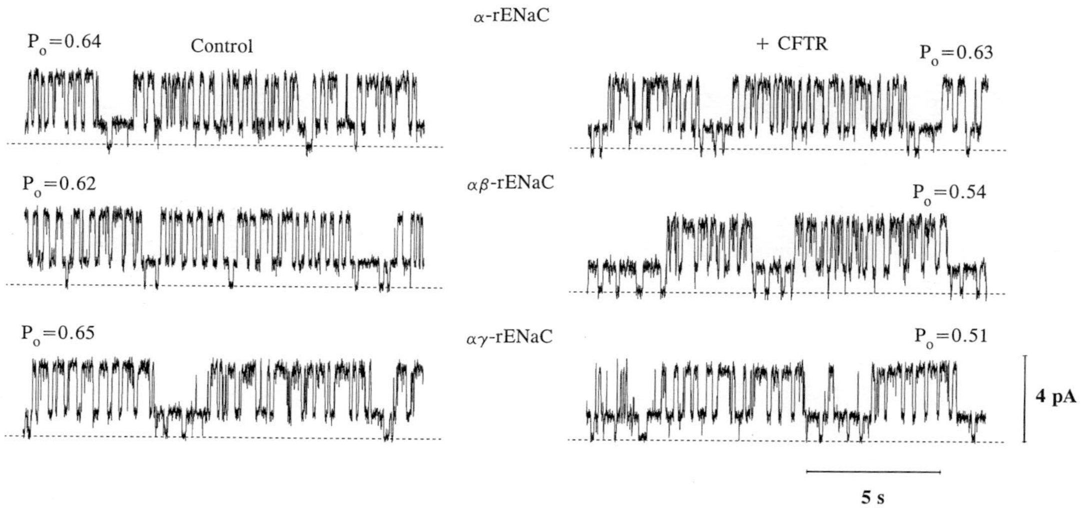

FIGURE 3 CFTR and single α-, $\alpha\beta$-, and $\alpha\gamma$-rENaC in lipid bilayers. Experimental conditions, see legend for Fig. 1. Modified from Ismailov *et al.* (1997).

with CFTR, yet they participate in down-regulation of the $\alpha\beta\gamma$-channel by modulating its gating via the C-termini. Alternatively, the C-terminus of the α-ENaC subunit may have an affinity to CFTR but is not able to interact with it unless β- and/or γ-ENaCs are associated. Studies to test all of these possibilities are yet to be performed.

VI. MUTATIONS OF CFTR FOUND IN CF AND INHIBITION OF $\alpha\beta\gamma$-ENaC

There are more than 500 naturally occurring mutations of CFTR known to date, causing disfunction of wet epithelial surfaces due to abnormalities of synthesis, processing, and both regulatory and the ion conduction functions of this molecule (Tsui *et al.,* 1993; Hamosh and Cutting, 1993; Welsh and Smith, 1993; Dean and Santis, 1994; Ferrari and Cremonesi, 1996). The question of how these mutations affect the rate of Na^+ transport in airway epithelia has become one of the most important problems in the CF field, and, in essence, triggered all forementioned studies of interaction between ENaC and CFTR.

The effects of the three naturally occuring mutants, namely ΔF508-, G551D-, and R117H-CFTR, on $\alpha\beta\gamma$-rENaC expressed in *Xenopus* oocytes have been examined to date (Mall *et al.,* 1996; Kunzelmann, *et al.,* 1997; Briel *et al.,* 1998). As it appears from these studies, amiloride-sensitive currents in the oocytes coexpressing ENaC and either one of two most common, and most lethal, disease causing mutants, ΔF508-CFTR or G551D-CFTR, are essentially in the same range of magnitude, independent of whether the eggs were stimulated with IBMX, or not (Mall *et al.,* 1996; Kunzelmann *et al.,* 1997; see summary in Table V). These observations are in obvious contrast with those when ENaCs were coexpressed with the WT-CFTR. Therefore, the authors concluded that the mechanism by which these mutations in CFTR molecule lead to Na^+ hyperabsorption in CF involves the abolishment cAMP-regulated difference in Na^+ conductance function. Consistent with this hypothesis, the oocytes coexpressing ENaCs together with the third naturally occurring mutation, R117H-CFTR, which has been associated with a relatively mild form of CF (Dean *et al.,* 1990; Al-Jader *et al.,* 1992; Kiesewetter *et al.,* 1993; Miller *et al.,* 1996), exhibited ~20% inhibition of amiloride-sensitive current when the eggs were stimulated with IBMX versus that in nonstimulated oocytes (Briel *et al.,* 1998).

To explain these findings, Briel *et al.* (1998) tested the hypothesis that the clinical characteristics of the disease (phenotype) and mutations in the CFTR gene (genotype) correlate with the ability of CFTR to act as a cAMP regulated Cl^- channel (reviewed by Welsh and Smith, 1993). The line of their considerations was as follows. The relative sequence of the severity of the disease (based on the degree of lung disease and sweat Cl^- concentration) is WT(no disease) < R117H << G551D ≈ ΔF508, so is the relative sequence of the ability of CFTR to induce Cl^- current (Drumm *et al.,* 1991; Sheppard *et al.,* 1993), and so is the extent of IBMX-induced inhibition of ENaC currents in oocytes (Mall *et al.,* 1996; Kunzelmann, *et al.,* 1997; Briel *et al.,* 1998). The outcome of the experiments testing whether this correlation bears physiological relevance or not is rather difficult to interpret because of the choice of two artificial mutants of CFTR, namely R347E and K335E, employed in these studies. The amplitude of cAMP-stimulated Cl^- current induced by either of these mutants is lower than that induced by the WT CFTR, but the relative sequence appears to be reversed in different types of cells: WT>R347E>K335E in HeLa cells (Anderson *et al.,* 1991), and WT>K335E>R347E in *Xenopus* oocytes (Hipper *et al.,* 1995). Furthermore, this hypothesis does not explain the fact that the basal amiloride-sensitive currents in oocytes coexpressing ENaC and either G551D- or ΔF508- CFTR were already approximately twice smaller than those in the presence of the WT-CFTR. In fact, these amplitudes are approximately four times smaller than that in ENaC alone expressing oocytes and are comparable with the amplitude of amiloride sensitive current in the WT-CFTR/ENaC oocytes stimulated with IBMX.

There is no apparent explanation for these data, nor is it clear if they have any relevance to the relative decrease in the extent of IBMX-induced inhibition of amiloride-sensitive current in the presence of CFTR, when CFTR was inhibited with diphenylamine-carboxylate or when the bath Cl^- was substituted with poorly permeant anions, like SCN^- or gluconate. The hypothesis that ENaC activity is regulated by CFTR via influx of Cl^- into oocytes, proposed to account for the latter observations, is indeed supported by the finding that the amplitude of amiloride-sensitive current in ENaC/CFTR oocytes stimulated with IBMX is higher when $[Cl^-]_{bath}$ is 5 mM than when $[Cl^-]_{bath}$ is 100 mM. However, this hypothesis does not withstand the fact that initial observations of decreased ENaC activity in the presence of CFTR in MDCK and 3T3 cells were made in Cl^- free solutions (Stutts *et al.,* 1995). The findings that changing $[Cl^-]_{bath}$ in the absence of CFTR, or stimulating Cl^- influx through endogenous Ca^{2+}-activated Cl^- currents did not have any effect on ENaC current suggest that even if Cl^- ions have any role in regulation of ENaC activity, at least in oocytes this effect appears to depend on the physical presence of CFTR.

Further, G551D- and ΔF508-mutations of CFTR involve amino acids located within the cytosolic domain identified in the aforementioned yeast two hybrid analyses as binding to α-ENaC (amino acids 394–830, Kunzelmann *et al.,* 1997). Moreover, the amiloride-sensitive currents in oocytes coexpressing $\alpha\beta\gamma$-ENaC together with the construct comprising this putative ENaC-binding site of the WT-CFTR decreased almost twofold following IBMX-stimulation of these oocytes (Kunzelmann *et al.,* 1997). On the other hand, unlike the full-length WT-CFTR, or ΔF508-CFTR and G551D-CFTR, this domain was not sufficient to lower the amplitude of basal amiloride-sensitive current, as compared to the ENaC-only oocytes. Furthermore, the same fragment but originated from the human G551D-CFTR, was not only unable to inhibit ENaC in *Xenopus* oocytes treated with IBMX, but, in fact, induced large basal amiloride-sensitive currents which were ~ tenfold higher than those in the presence of the full length WT- or G551D-CFTR, and ~ threefold higher than in the ENaC-only oocytes (Kunzelmann *et al.,* 1997). The mechanisms responsible for these phenomena are yet to be determined.

VI. CONCLUDING REMARKS

The understanding of the molecular basis of how the human body functions has been advanced rapidly in the past few years by the identification of disease related genes and by the integrative studies of the effects produced by mutations of these genes on molecular, cellular and higher levels of organization. Cystic fibrosis is not just one reference to this process, but an example of how the understanding of one illness can benefit from the knowledge gained in studies of other diseases. Although many questions about the pathogenesis of CF (some of which we tried to discuss in this article) are yet to be answered, the new knowledge of the mechanism of Na^+ transport in general and in the pathophysiology of human genetic hypertension and pseudohypoaldostronism, perceived since cloning of ENaCs by Canessa *et al.* (1993, 1994), will help to understand the mechanism of Na^+ hyperabsorption in CF airways, and vice versa.

Acknowledgments

The authors thank Dr. Dale J. Benos for his critical comments to the manuscript. This work was supported by the Cystic Fibrosis Foundation Grant Ismail97IO and NIH Grant DK 37206.

References

Adam, G., and Delbrück, M. (1967). Reduction of dimensionality in biological diffusion processes. *In* "Structural Chemistry and Molecular Biology" (A. Rich and N. Davidson, eds), pp. 198–215. Freeman, San Francisco.

Adams, C. M., Snyder, P. M., and Welsh, M. J. (1997). Interactions between subunits of the human epithelial sodium channel. *J. Biol. Chem.* **272,** 27295–27300.

Aizenbud, B. M., and Gershon, N. D. (1982). Diffusion of molecules on biological membranes of nonplanar form. A theoretical study. *Biophys. J.* **38,** 287–293.

Aizenbud, B. M., and Gershon, N. D. (1985). Diffusion of molecules on biological membranes of nonplanar form. II. Diffusion anisotropy. *Biophys. J.* **48,** 543–546.

Al-Jader, L. N., Meredith, A. L., Ryley, H. C., Cheadle, J. P., Maguire, S., Owen, G., Goodchild, M. C., and Harper, P. S. (1992). Severity of chest disease in cystic fibrosis patients in relation to their genotypes. *J. Med. Genet.* **29,** 883–887.

Alton, E. W., Hay, J. G., Munro, C., and Geddes, D. M. (1987). Measurement of nasal potential difference in adult cystic fibrosis, Young's syndrome, and bronchiectasis. *Thorax* **42,** 815–817.

Alton, E. W., Currie, D., Logan-Sinclair, R., Warner, J. O., Hodson, M. E., and Geddes, D. M. (1990). Nasal potential difference: A clinical diagnostic test for cystic fibrosis. *Eur. Respir. J.* **3,** 922–926.

Anderson, M. P., Gregory, R. J., Thompson, S., Souza, D. W., Paul, S., Mulligan, R. C., Smith, A. E., and Welsh, M. J. (1991). Demonstration that CFTR is a chloride channel by alteration of its anion selectivity. *Science* **253,** 202–205.

App, E. M., King, M., Helfesrieder, R., Kohler, D., and Matthys, H. (1990). Acute and long-term amiloride inhalation in cystic fibrosis lung disease. A rational approach to cystic fibrosis therapy. *Am. Rev. Respir. Dis.* **141,** 605–612.

Awayda, M. S., Ismailov, I. I., Berdiev, B. K., Fuller, C. M., and Benos, D. J. (1996). Protein kinase regulation of a cloned epithelial Na^+ channel. *J. Gen. Physiol.* **108,** 49–65.

Awayda, M. S., Tousson, A., and Benos, D. J. (1997). Regulation of a cloned epithelial Na^+ channel by its β and γ-subunits. *Am. J. Physiol.* **42,** C1889–C1899.

Ballard, S. T., Schepens, S. M., Falcone, J. C., Meininger, G. A., and Taylor, A. E. (1992). Regional bioelectric properties of porcine airway epithelium. *J. Appl. Physiol.* **73,** 2021–2027.

Barker, P. M., Gowen, C. W., Lawson, E. E., and Knowles, M. R. (1997). Decreased sodium ion absorption across nasal epithelium of very premature infants with respiratory distress syndrome. *J. Pediatr.* **130,** 373–377.

Berdiev, B. K., Prat, A. G., Cantiello, H. F., Ausiello, D. A., Fuller, C. M., Jovov, B., Benos, D. J., and Ismailov, I. I. (1996). Regulation of epithelial sodium channels by short actin filaments. *J. Biol. Chem.* **271,** 17704–17710.

Berdiev, B. K., Karlson, K. H., Jovov, B., Ripoll, P. J., Morris, R., Loffing-Cueni, D., Halpin, P., Stanton, B. A., Kleyman, T. R., and Ismailov, I. I. (1998). Subunit stoichiometry of a core conduction element in a cloned epithelial amiloride-sensitive Na^+ channel. *Biophys. J.* **75,** 2292–2301.

Berg, H. C., and Purcell, E. M. (1977). Physics of chemoreception. *Biophys. J.* **20,** 193–219.

Biwersi, J., and Verkman, A. S. (1994). Functions of CFTR other than as a plasma membrane chloride channel. *In* "Cystic Fibrosis—Current Topics" (J. A. Dodge, D. J. H. Brock, and J. H. Widdicombe, eds.), Vol. 2, pp. 155–171. Wiley, Chichester.

Biwersi, J., Emans, N., and Verkman, A.S. (1996). Cystic fibrosis transmembrane conductance regulator activation stimulates endosome fusion in vivo. *Proc. Natl. Acad. Sci. U.S.A.* **93,** 12484–12489.

Blank, U., Rückes, C., Clauss W., Hofmann T., Lindeman, H., Munker, G., and Weber, W.M. (1997). Cystic fibrosis and non-cystic fibrosis human nasal epithelium show analogous Na absorbtion and reversible block by phenamil. *Pfluegers Arch.* **434,** 19–24.

Boucher, R. C. (1994a). Human airway ion transport. Part 1. *Am. J. Respir. Crit. Care Med.* **150,** 271–281.

Boucher, R. C. (1994b). Human airway ion transport. Part 2. *Am. J. Respir. Crit. Care Med.* **150,** 581–593.

Boucher, R. C., Bromberg, P. A., and Gatzy, J. T. (1980). Airway transepithelial electric potential in vivo: Species and regional differences. *J. Appl. Physiol.* **48,** 169–176.

Boucher, R. C., Stutts, M. J., and Gatzy, J. T. (1981). Regional differences in bioelectric properties and ion flow in excised canine airways. *J. Appl. Physiol.* **51,** 706–714.

Boucher, R. C., Stutts, M. J., Knowles, M. R., Cantley, L., and Gatzy, J. T. (1986). Na^+ transport in cystic fibrosis respiratory epithelia. Abnormal basal rate and response to adenylate cyclase activation. *J. Clin. Invest.* **78,** 1245–1252.

Boucher, R. C., Cotton, C. U., Stutts, M. J., Knowles, M. R., and Gatzy, J. T. (1987a). Is cystic fibrosis respiratory epithelium characterized by a single defect (Cl^-) or multiple ion transport defects? *Prog. Clin. Biol. Res.* **254,** 89–100.

Boucher, R. C., Yankaskas, J. R., Cotton C. U., Knowles, M. R., and Stutts, M. J. (1987b). Cell culture approaches to the investigation of human airway ion transport. *Eur. J. Respir. Dis. Suppl.* **153,** 59–67.

Boucher, R. C., Cotton, C. U., Gatzy, J. T., Knowles, M. R., and Yankaskas, J. R. (1988). Evidence for reduced Cl^- and increased Na^+ permeability in cystic fibrosis human primary cell cultures. *J. Physiol.* (*London*) **405,** 77–103.

Bradbury, N. A., Jilling, T., Berta, G., Sorscher, E. J., Bridges, R. J., and Kirk, K. L. (1992). Regulation of plasma membrane recycling by CFTR. *Science* **256,** 530–532.

Briel, M., Greger, R., and Kunzelmann, K. (1998). Cl^- transport by cystic fibrosis transmembrane conductance regulator (CFTR) contributes to the inhibition of epithelial Na^+ channel (ENaCs) in *Xenopus* oocytes co-expressing CFTR and ENaC. *J. Physiol.* (*London*) **508,** 825–36.

Burch, L. H., Talbot, C., Knowles, M. R., Canessa, C., Rossier, B., and Boucher, R. (1995). Relative expression of the human epithelial Na^+ channel (ENaC) subunits in normal and cystic fibrosis airways. *Am. J. Physiol.* **269,** C511–C518.

Canessa, C. M., Horisberger, J.-D., and Rossier, B. C. (1993). Epithelial sodium channel related to proteins involved in neurodegeneration. *Nature* (*London*) **361,** 467–470.

Canessa, C. M., Schild, L., Buell, G., Thoreus, B., Gautschi, I., Horisberger, J.-D., and Rossier, B. C. (1994). Amiloride-sensitive epithelial Na^+ channel is made of three homologous subunits. *Nature* (*London*) **367,** 463–467.

Cantiello, H. F. (1995). Role of the actin cytoskeleton on epithelial Na^+ channel regulation. *Kidney Int.* **48,** 970–984.

Cantiello, H. F., Stow, J. L., Prat, A. G., and Ausiello, D. A. (1991). Actin filaments regulate epithelial Na^+ channel activity. *Am. J. Physiol.* **261,** C882–C888.

Cherksey, B. D. (1988). Functional reconstitution of an isolated sodium channel from bovine trachea. *Comp. Biochem. Physiol. A* **90,** 771–773.

Cherry, R. J. (1979). Rotational and lateral diffusion of membrane proteins. *Biochim. Biophys. Acta* **559,** 289–327.

Chinet, T. C., Fullton, J. M., Yankaskas, J. R., Boucher, R. C., and Stutts, M. J. (1993). Sodium-permeable channels in the apical membrane of human nasal epithelial cells. *Am. J. Physiol.* **265,** C1050–C1060.

Chinet, T. C., Fullton, J. M., Yankaskas, J. R., Boucher R. C., and Stutts, M. J. (1994). Mechanism of sodium hyperabsorption in cultured cystic fibrosis nasal epithelium: A patch-clamp study. *Am. J. Physiol.* **266,** C1061–C1068.

Collins, F. S. (1992). Cystic fibrosis: Molecular biology and therapeutic implications. *Science* **256,** 774–779.

Cotton, C. U., Lawson, E. E., Boucher, R. C., and Gatzy, J. T. (1983). Bioelectric properties and ion transport of airways excised from adult and fetal sheep. *J. Appl. Physiol.* **55,** 1542–1549.

Cotton, C. U., Stutts, M. J., Knowles, M. R., Gatzy, J. T., and Boucher, R. C. (1987). Abnormal apical cell membrane in cystic fibrosis respiratory epithelium. An *in vitro* electrophysiologic analysis. *J. Clin. Invest.* **79,** 80–85.

Cotton, C. U., Boucher, R. C., and Gatzy, J. T. (1988). Bioelectric properties and ion transport across excised canine fetal and neonatal airways. *J. Appl. Physiol.* **65,** 2367–2375.

Dean, M., and Santis, G. (1994). Heterogeneity in the severity of cystic fibrosis and the role of CFTR gene mutations. *Hum. Genet.* **93,** 364–368,.

Dean, M., White, M. B., Amos, J., Gerrard, B., Stewart, C., Khaw, K. T., and Leppert, M. (1990). Multiple mutations in highly conserved residues are found in mildly affected cystic fibrosis patients. *Cell (Cambridge, Mass.)* **61,** 863–870.

Delmarco, A., Pradal, U., Cabrini, G., Bonizzato, A., and Mastella, G. (1997). Nasal potential difference in cystic fibrosis patients presenting borderline sweat test. *Eur. Respir. J.* **10,** 1145–1149.

Drumm, M. L., Wilkinson, D. J., Smit, L. S., Worrell, R. T., Strong, T. V., Frizzell, R. A., Dawson, D. C., and Collins, F. S. (1991). Chloride conductance expressed by Δ F508 and other mutant CFTRs in Xenopus oocytes. *Science* **254,** 1797–1799.

Duszyk, M., French, A. S., and Man, S. F. (1989). Cystic fibrosis affects chloride and sodium channels in human airway epithelia. *Can. J. Physiol. Pharmacol.* **67,** 1362–1365.

Edidin, M. (1987). Rotational and lateral diffusion of membrane proteins and lipids: Phenomena and function. *Curr. Top. Membr. Transp.* **29,** 91–127.

Fahey, P. F., and Webb, W. W. (1978). Lateral diffusion in phospholipid bilayer membranes and multilamellar liquid crystals. *Biochemistry* **17,** 3046–3053.

Farman, N., Talbot, C. R., Boucher, R., Fay, M., Canessa, C., Rossier, B., and Bonvalet, J. P. (1997). Noncoordinated expression of α-, β-, and γ-subunit mRNAs of epithelial Na^+ channel along rat respiratory tract. *Am. J. Physiol.* **272,** C131–C141.

Feng, Z. P., Clark, R. B., and Berthiaume, Y. (1993). Identification of nonselective cation channels in cultured adult rat alveolar type II cells. *Am. J. Respir. Cell. Mol. Biol.* **9,** 248–254.

Ferrari, M., and Cremonesi, L. (1996). Genotype-phenotype correlation in cystic fibrosis patients. *Ann. Biol. Clin. (Paris)* **54,** 235–241.

Firsov, D., Schild, L., Gautschi, I., Merillat, A. M., Schneeberger, E., and Rossier, B. C. (1996). Cell surface expression of the epithelial Na channel and a mutant causing Liddle syndrome: A quantitative approach. *Proc. Natl. Acad. Sci. U.S.A.* **93,** 15370–15375.

Firsov, D., Gautschi, I., Merillat, A. M., Rossier, B. C., and Schild, L. (1998). The heterotetrameric architecture of the epithelial sodium channel (ENaC). *EMBO J.* **17,** 344–352.

Frizzell, R. A. (1995). Functions of the cystic fibrosis transmembrane conductance regulator protein. *Am. J. Respir. Crit. Care Med.* **151,** S54–S58.

Fuller, C. M., and Benos, D. J. (1992). CFTR! *Am. J. Physiol.* **263,** C267–C286.

Gallo, R. L. (1990). Aerosolized amiloride for the treatment of lung disease in cystic fibrosis. *N. Engl. J. Med.* **323,** 996–997 (Letter to the Editor).

Garty, H., and Palmer, L. G. (1997). Epithelial sodium channels: Function, structure and regulation. *Physiol. Rev.* **77,** 359–396.

Gowen, C. W., Lawson, E. E., Gingras-Leatherman, J., Gatzy, J. T., Boucher, R. C., and Knowles, M. R. (1986). Increased nasal potential difference and amiloride sensitivity in neonates with cystic fibrosis. *J. Pediatr.* **108,** 517–521.

Grubb, B. R., Vick, R. N., and Boucher, R. C. (1994). Hyperabsorption of Na^+ and raised Ca^{2+}-mediated Cl^- secretion in nasal epithelia of CF mice. *Am. J. Physiol.* **266,** C1478–C1483.

Hamosh, A., and Cutting, G. R. (1993). Genotype/phenotype relationships in cystic fibrosis. *In* "Cystic Fibrosis—Current Topics" (J. A. Dodge, D. J. H. Brock, and J. H. Widdicombe, eds.), Vol. 1, pp. 69–89. Wiley, Chichester.

Hanrahan, J. W., Tabcharani, J. A., Becq, F., Mathews, C. J., Augustinas, O., Jensen, T. J., Chang, X. B., and Riordan, J. R. (1995). Function and dysfunction of the CFTR chloride channel. *Soc. Gen. Physiol. Ser.* **50,** 125–137.

Hansson, J. H., Nelson-Williams, C., Suzuki, H., Schild, L., Shimkets, R., Lu, Y., Canessa, C., Iwasaki, T., Rossier B., and Lifton, R. P. (1995a). Hypertension caused by a truncated epithelial sodium channel γ subunit: Genetic heterogeneity of Liddle syndrome. *Nat. Genet.* **11,** 76–82.

Hansson, J. H. Schild, L., Lu Y., Wilson, T., Gautschi, I., Shimkets, R., Nelson-Williams, C., Rossier, B., and Lifton, R. (1995b). A *de novo* missense mutation of the β subunit of the epithelial sodium channel causes hypertension and Liddle syndrome, identifying a proline-rich segment critical for regulation of channel activity. *Proc. Natl. Acad. Sci. U.S.A.* **92,** 11495–11499.

Hay, J. G., and Geddes, D. M. (1985). Transepithelial potential difference in cystic fibrosis. *Thorax* **40,** 493–496.

Hay, J. G., McElvaney, N. G., Herena, J., and Crystal, R. G. (1995). Modification of nasal epithelial potential differences of individuals with cystic fibrosis consequent to local administration of a normal CFTR cDNA adenovirus gene transfer vector. *Hum. Gene Ther.* **6,** 1487–1496.

Higgins, C. F. (1992). Cystic fibrosis transmembrane conductance regulator (CFTR). *Br. Med. Bull.* **48,** 754–765.

Hipper, A., Mall, M., Greger, R., and Kunzelmann, K. (1995). Mutations in the putative pore-forming domain of CFTR do not change anion selectivity of the cAMP activated Cl^- conductance. *FEBS Lett.* **374,** 312–316.

Ho, L. P., Samways, J. M., Porteous, D. J., Dorin, J. R., Carothers, A., Greening, A. P. and Innes, J. A. (1997). Correlation between nasal potential difference measurements, genotype and clinical condition in patients with cystic fibrosis. *Eur. Respir. J.* **10,** 2018–2022.

Hummler, E., Barker, P., Gatzy, J., Beermann, F., Verdumo, C., Schmidt, A., Boucher, R., and Rossier, B. C. (1996). Early death due to defective neonatal lung liquid clearance in alpha-ENaC-deficient mice. *Nat. Genet.* **12,** 325–328.

Ismailov, I. I., Awayda, M. S., Berdiev, B. K., Bubien, J. K., Lucas, J. E., Fuller, C. M., and Benos, D. J. (1996a). Triple-barrel organization of ENaC, a cloned epithelial Na^+ channel. *J. Biol. Chem.* **271,** 807–816.

Ismailov, I. I., Awayda, M. S., Jovov, B., Berdiev, B. K., Fuller, C. M., Dedman, J. R., Kaetzel, M., and Benos, D. J. (1996b). Regulation of epithelial sodium channels by the cystic fibrosis transmembrane conductance regulator. *J. Biol. Chem.* **271,** 4725–4732.

Ismailov, I. I., Berdiev, B. K., Shlyonsky, V. G., Fuller, C. M., Prat, A. G., Jovov, B., Cantiello, H. F., Ausiello, D. A., and Benos, D. J. (1997). Role of actin in regulation of epithelial sodium channels by CFTR. *Am. J. Physiol.* **272,** C1077–1086.

Jefferson, D. M., Valentich, J. D., Marini, F. C., Grubman, S. A., Iannuzzi, M. C., Dorkin, H. L., Li, M., Klinger, K. W., and Welsh, M. J. (1990). Expression of normal and cystic fibrosis phenotypes by continuous airway epithelial cell lines. *Am. J. Physiol.* **259,** L496–L505.

Jilling, T., and Kirk, K. L. (1997). The biogenesis, traffic, and function of the cystic fibrosis transmembrane conductance regulator. *Int. Rev. Cytol.* **172,** 193–241.

Jorissen, M., Vereecke, J., Carmeliet, E., Van den Berghe, H., and Cassiman, J. J. (1990). Identification of a voltage- and calcium-dependent non-selective cation channel in cultured adult and fetal human nasal epithelial cells. *Pfluegers Arch.* **415,** 617–623.

Jovov, B., Ismailov, I. I., and Benos, D. J. (1995). Cystic fibrosis transmembrane conductance regulator is required for protein kinase A activation of an outwardly rectified anion channel purified from bovine tracheal epithelia. *J. Biol. Chem.* **270,** 1521–1528.

Kiesewetter, S., Macek, M., Jr., Davis, C., Curristin, S. M., Chu, C. S., Graham, C., Shrimpton, A. E., Cashman, S. M., Tsui, L. C., Mickle, J., Amos, J., Highsmith, W. E., Shuber, A.,

Witt, D. R., Crystal, R. G., and Cutting, G. R. (1993). A mutation in CFTR produces different phenotypes depending on chromosomal background. *Nat. Genet.* **5,** 274–278.

Knowles, M. R., Carson, J. L., Collier, A. M., Gatzy, J. T., and Boucher, R. C. (1981a). Measurements of nasal transepithelial electric potential differences in normal human subjects in vivo. *Am. Rev. Respir. Dis.* **124,** 484–490.

Knowles, M., Gatzy, J., and Boucher, R. (1981b). Increased bioelectric potential difference across respiratory epithelia in cystic fibrosis. *N. Engl. J. Med.* 305, 1489–1495.

Knowles, M. R., Buntin, W. H., Bromberg, P. A., Gatzy, J. T., and Boucher, R. C. (1982). Measurements of transepithelial electric potential differences in the trachea and bronchi of human subjects in vivo. *Am. Rev. Respir. Dis.* **126,** 108–112.

Knowles, M., Gatzy, J., and Boucher, R. (1983a). Relative ion permeability of normal and cystic fibrosis nasal epithelium. *J. Clin. Invest.* **71,** 1410–1417.

Knowles, M. R., Stutts, M. J., Spock, A., Fischer, N., Gatzy, J. T., and Boucher, R. C. (1983b). Abnormal ion permeation through cystic fibrosis respiratory epithelium. *Science* 221, 1067–1070.

Knowles, M., Murray, G., Shallal, J., Askin, F., Ranga, V., Gatzy, J., and Boucher, R. (1984). Bioelectric properties and ion flow across excised human bronchi. *J. Appl. Physiol.* **56,** 868–877.

Knowles, M. R., Church, N. L., Waltner, W. E., Yankaskas, J. R., Gilligan, P., King, M., Edwards, L. J., Helms, R. W., and Boucher, R. C. (1990a). A pilot study of aerosolized amiloride for the treatment of lung disease in cystic fibrosis. *N. Engl. J. Med.* 322, 1189–1194.

Knowles, M. R., Church, N. L., Waltner, W. E., Yankaskas, J. R., Gilligan, P., King, M., Edwards, L. J., Helms, R. W., and Boucher, R. C. (1990b). Aerosolized amiloride for the treatment of lung disease in cystic fibrosis. *N. Engl. J. Med.* **323,** 997–998.

Knowles, M. R., Church, N. L., Waltner, W. E., Gatzy, J. T., and Boucher, R. C. (1992). Amiloride in cystic fibrosis: Safety, pharmacokinetics, and efficacy in the treatment of pulmonary disease. *In* "Amiloride and Its Analogs: Unique Cation Transport Inhibitors" (E. J. Cragoe, Jr., T. R. Kleyman, and L. Simchowitz, eds.), pp. 301–316. VCH Publishers, New York.

Knowles, M. R., Paradiso, A. M., and Boucher, R. C. (1995a). *In vivo* nasal potential difference: Techniques and protocols for assessing efficacy of gene transfer in cystic fibrosis. *Hum. Gene Ther.* **6,** 445–455.

Knowles, M. R., Olivier, K., Noone, P., and Boucher, R. C. (1995b). Pharmacologic modulation of salt and water in the airway epithelium in cystic fibrosis. *Am. J. Respir. Crit. Care Med.* **151,** S65–S69.

Koefoed-Johnson, V., and Ussing, H. H. (1958). The nature of the frog skin potential. *Acta Physiol. Scand.* **42,** 298–308.

Kohler, D., App, E., Schmitz-Schumann, M., Wurtemberger, G., and Matthys, H. (1986). Inhalation of amiloride improves the mucociliary and the cough clearance in patients with cystic fibroses. *Eur. J. Respir. Dis., Suppl.* **146,** 319–326.

Kosari, F., Sheng, S., Li, J., Mak, D. O., Foskett, J. K., and Kleyman, T. R. (1998). Subunit stoichiometry of the epithelial sodium channel. *J. Biol. Chem.* **273,** 13469–13474.

Kunzelmann, K., Kathofer, S., and Greger, R. (1995). Na^+ and Cl^- conductances in airway epithelial cells: Increased Na^+ conductance in cystic fibrosis. *Pfleugers Arch.* **431,** 1–9.

Kunzelmann, K., Kathofer, S., Hipper, A., Gruenert, D. C., and Greger, R. (1996). Culture-dependent expression of Na^+ conductances in airway epithelial cells. *Pfluegers Arch.* **431,** 578–586.

Kunzelmann, K., Kiser, G. L., Schreiber, R., and Riordan, J. R. (1997). Inhibition of epithelial Na^+ currents by intracellular domains of the cystic fibrosis transmembrane conductance regulator. *FEBS Lett.* **400,** 341–344.

Letz, B., and Korbmacher, C. (1997). cAMP stimulates CFTR-like Cl^- channels and inhibits amiloride-sensitive Na^+ channels in mouse CCD cells. *Am. J. Physiol.* **272,** C657–C666.

Lifton, R. P. (1996). Molecular genetics of human blood pressure variation. *Science* **272,** 676–680.

Ling, B. N., Zuckerman, J. B., Lin, C., Harte, B. J., McNulty, K. A., Smith, P. R., Gomez, L. M., Worrell, R. T., Eaton, D. C., and Kleyman, T. R. (1997). Expression of the cystic fibrosis phenotype in a renal amphibian epithelial cell line. *J. Biol. Chem.* **272,** 594–600.

Lingueglia, E., Voilley, N., Waldmann, R., Lazdunski, M., and Barbry, P. (1993). Expression cloning of an epithelial amiloride-sensitive Na^+ channel. A new channel type with homologies to *Caenorhabditis elegans* degenerins. *FEBS Lett.* **318,** 95–99.

MacGregor, G. G., Olver, R. E., and Kemp, P. J. (1994). Amiloride-sensitive Na^+ channels in fetal type II pneumocytes are regulated by G proteins. *Am. J. Physiol.* **267,** L1–L8.

Mall, M., Hipper, A., Greger, R., and Kunzelmann, K. (1996). Wild type but not ΔF508 CFTR inhibits Na^+ conductance when coexpressed in Xenopus oocytes. *FEBS Lett.* **381,** 47–52.

Mall, M., Briel, M., Greger, R., Schreiber, R., and Kunzelmann, K. (1998). The amiloride-inhibitable Na^+ conductance is reduced by the cystic fibrosis transmembrane conductance regulator in normal but not in cystic fibrosis airways. *J. Clin. Invest.* **102,** 15–21.

Matsushita, K., McCray, P. B., Jr., Sigmund, R. D., Welsh, M. J., and Stokes, J. B. (1996). Localization of epithelial sodium channel subunit mRNAs in adult rat lung by in situ hybridization. *Am. J. Physiol.* **271,** L332–L339.

Mearns, M.B. (1993). Cystic fibrosis: The first 50 years. A review of the clinical problems and their management. *In* "Cystic Fibrosis—Current Topics" (J. A. Dodge, D. J. H. Brock, and J. H. Widdicombe, eds.), Vol. 1, pp. 217–250. Wiley, Chichester.

Middleton, P. G., Geddes, D. M., and Alton, E. W. (1993). Effect of amiloride and saline on nasal mucociliary clearance and potential difference in cystic fibrosis and normal subjects. *Thorax* **48,** 812–816.

Middleton, P. G., Geddes, D. M., and Alton, E.W . (1994). Protocols for *in vivo* measurement of the ion transport defects in cystic fibrosis nasal epithelium. *Eur. Respir. J.* **7,** 2050–2056.

Miller, P. W., Hamosh, A., Macek, M., Jr., Greenberger, P. A., MacLean, J., Walden, S. M., Slavin, R.G., and Cutting, G.R. (1996). Cystic fibrosis transmembrane conductance regulator (CFTR) gene mutations in allergic bronchopulmonary aspergillosis. *Am. J. Hum. Genet.* **59,** 45–51.

Orser, B. A., Bertlik, M., Fedorko, L., and O'Brodovich, H. (1991). Cation selective channel in fetal alveolar type II epithelium. *Biochim. Biophys. Acta* **1094,** 19–26.

Peters, R., and Cherry, R. J. (1982). Lateral and rotational diffusion of bacteriorodopsin in lipid bilayers: Experimental test of the Saffman-Delbrück equations. *Proc. Nat. Acad. Sci. U.S.A.* **79,** 4317–4321.

Prat, A. G., Bertorello, A. M., Ausiello, D. A., and Cantiello, H. F. (1993). Activation of epithelial Na^+ channels by protein kinase A requires actin filaments. *Am. J. Physiol.* **265,** C224–C233.

Quinton, P. M. (1990). Cystic fibrosis: A disease in electrolyte transport. *FASEB J.* **4,** 2709–2717.

Quinton, P. M., and Bijman, J. (1983). Higher bioelectric potentials due to decreased chloride absorption in the sweat glands of patients with cystic fibrosis. *N. Engl. J. Med.* **308,** 1185–1189.

Renard, S., Voilley, N., Bassilana, F., Lazdunski, M., and Barbry, P. (1995). Localization and regulation by steroids of the α, β, and γ subunits of the amiloride-sensitive Na^+ channel in colon, lung and kidney. *Pfluegers Arch.* **430,** 299–307.

Riordan, J. R. (1993). The cystic fibrosis transmembrane conductance regulator. *Annu. Rev. Physiol.* **55,** 609–630.

Rückes, C., Blank, U., Möller, K., Rieboldt, J., Lindemann, H., Münker, G., Clauss, W., and Weber, W.M. (1997). Amiloride-sensitive Na^+ channels in human nasal epithelium are different from classical epithelial Na^+ channels. *Biochem. Biophys. Res. Commun.* **237,** 488–491.

Saffman, P. G., and Delbrück, M. (1975). Brownian motion in biological membranes. *Proc. Natl. Acad. Sci. U.S.A.* **72,** 3111–3113.

Sauder, R. A., Chesrown, S. E., and Loughlin, G. M. (1987). Clinical application of transepithelial potential difference measurements in cystic fibrosis. *J. Pediatr.* **111,** 353–358.

Senyk, O., Ismailov, I., Bradford, A. L., Baker, R. R., Matalon, S., and Benos, D. J. (1995). Reconstitution of immunopurified alveolar type II cell Na^+ channel protein into planar lipid bilayers. *Am. J. Physiol.* **268,** C1148–C1156.

Sheppard, D. N., Rich, D. P., Ostedgaard, L. S., Gregory, R. J., Smith, A. E., and Welsh, M. J. (1993). Mutations in CFTR associated with mild-disease-form Cl^- channels with altered pore properties. *Nature* (*London*) **362,** 160–164.

Shimkets, R. A., Lifton, R., and Canessa, C. M. (1998). *In vivo* phosphorylation of the epithelial sodium channel. *Proc. Natl. Acad. Sci. U.S.A.* **95,** 3301–3305.

Smith, P. R., Stoner, L. C., Viggiano, S. C., Angelides, K. J., and Benos, D. J. (1995). Effect of vasopressin and aldosterone on the lateral mobility of epithelial Na channels in A6 renal epithelial cells. *J. Membr. Biol.* **147,** 195–205.

Snyder, P. M., Price, M. P., McDonald, F. J., Adams, C. M., Volk, K. A., Zeiher, B. G., Stokes, J. B., and Welsh, M. J. (1995). Mechanism by which Liddle's syndrome mutations increase activity of a human epithelial Na channel. *Cell* (*Cambridge, Mass.*) **83,** 969–978.

Snyder, P. M., Cheng, C., Prince, L. S., Rogers, J. C., and Welsh, M. J. (1998). Electrophysiological and biochemical evidence that DEG/ENaC cation channels are composed of nine subunits. *J. Biol. Chem.* **273,** 681–684.

Stutts, M. J., Canessa, C. M., Olsen, J. C., Hamrick, M., Cohn, J. A., Rossier, B. C., and Boucher, R. C. (1995). CFTR as a cAMP-dependent regulator of sodium channels. *Science* **269,** 847–850.

Stutts, M. J., Rossier, B. C., and Boucher, R .C. (1997). Cystic fibrosis transmembrane conductance regulator inverts protein kinase A- mediated regulation of epithelial sodium channel single channel kinetics. *J. Biol. Chem.* **272,** 14037–14040.

Tohda, H., and Marunaka, Y. (1995). Insulin-activated amiloride-blockable nonselective cation and Na^+ channels in the fetal distal lung epithelium. *Gen. Pharmacol.* **26,** 755–763.

Tomkiewicz, R. P., App, E. M., Zayas, J. G., Ramirez, O., Church, N., Boucher, R. C., Knowles, M. R., and King, M. (1993). Amiloride inhalation therapy in cystic fibrosis. Influence on ion content, hydration, and rheology of sputum. *Am. Rev. Respir. Dis.* **148,** 1002–1007.

Tsui, L.-C., and Buchwald, M. (1991). Biochemical and molecular genetics of cystic fibrosis. *Adv. Hum. Genet.* **20,** 153–166.

Tsui, L.-C., Markiewics D., Zielenski, J., Corey M., and Durie, P. (1993). Mutation analysis in cystic fibrosis. *In* "Cystic Fibrosis—Current Topics" (J. A. Dodge, D. J. H. Brock, and J. H. Widdicombe, eds.), Vol. 1, pp. 27–44. Wiley, Chichester.

Voilley, N., Lingueglia, E., Champigny, G., Mattei, M. G., Waldmann, R., Lazdunski, M., and Barbry, P. (1994). The lung amiloride-sensitive Na^+ channel: Biophysical properties, pharmacology, ontogenesis, and molecular cloning. *Proc. Natl. Acad. Sci. U.S.A.* **91,** 247–251.

Welsh, M. J., and Smith, A. E. (1993). Molecular mechanisms of CFTR chloride channel dysfunction in cystic fibrosis. *Cell* (*Cambridge, Mass.*) **73,** 1251–1254.

Widdicombe, J. H., and Welsh, M. J. (1980). Anion selectivity of the chloride-transport process in dog tracheal epithelium. *Am. J. Physiol.* 239, C112–117.

Willumsen, N. J., and Boucher, R. C. (1991a). Sodium transport and intracellular sodium activity in cultured human nasal epithelium. *Am. J. Physiol.* **261,** C319–C331.

Willumsen, N. J., and Boucher, R. C. (1991b). Transcellular sodium transport in cultured cystic fibrosis human nasal epithelium. *Am. J. Physiol.* **261,** C332–C341.

Yamaya, M., Finkbeiner, W. E., Chun, S. Y., and Widdicombe, J. H. (1992). Differentiated structure and function of cultures from human tracheal epithelium. *Am. J. Physiol.* **262,** L713–724.

Yankaskas, J. R., Cotton, C. U., Knowles, M. R., Gatzy, J. T., and Boucher, R. C. (1985a). Culture of human nasal epithelial cells on collagen matrix supports. A comparison of bioelectric properties of normal and cystic fibrosis epithelia. *Am. Rev. Respir. Dis.* **132,** 1281–1287.

Yankaskas, J. R., Knowles, M. R., Gatzy, J. T., and Boucher, R. C. (1985b). Persistence of abnormal chloride ion permeability in cystic fibrosis nasal epithelial cells in heterologous culture. *Lancet* **1,** 954–956.

Yankaskas, J. R., Stutts, M. J., Cotton, C. U., Knowles, M. R., Gatzy, J. T., and Boucher, R. C. (1987). Cystic fibrosis airway epithelial cells in primary culture: Disease-specific ion transport abnormalities. *Prog. Clin. Biol. Res.* **254,** 139–149.

Yue, G., Hu, P., Oh, Y., Jilling, T., Shoemaker, R. L., Benos, D. J., Cragoe, E. J., Jr., and Matalon, S. (1993). Culture-induced alterations in alveolar type II cell Na+ conductance. *Am. J. Physiol.* **265,** C630–640.

Yue, G., Shoemaker, R. L., and Matalon S. (1994). Regulation of low-amiloride-affinity sodium channels in alveolar type II cells. *Am. J. Physiol.* **267,** L94–L100.

Index

A

B

C

D

V

W

X

Y

ISBN 0-12-153347-6